Innate Immunity in Health and Disease

Edited by Shailendra K. Saxena and Hridayesh Prakash

Published in London, United Kingdom

IntechOpen

Supporting open minds since 2005

Innate Immunity in Health and Disease
http://dx.doi.org/10.5772/intechopen.82921
Edited by Shailendra K. Saxena and Hridayesh Prakash

Contributors
Atif Iqbal Ahmed Shaikh, Appasamy Thirumal Prabhakar, Rizwan Ahmad, Haseeb Ahsan, Daishu Han, Fei Wang, Ran Chen, Kavitha Sankaranarayanan, Bhuvaneshwari Sampath, Priyadarshan Kathirvelu, Irina V. Nesterova, Svetlana V. Kovaleva, Valentina V. Malynovskaya, Galina A. Chudilova, Tatyana V. Rusinova, Shalini Malhotra, Renu Kumari Yadav, Dhruv Kumar, Vaishali Chandel, Marcela Catalina Fandiño Vargas, Manju Jain, Nadeem Akhtar, Lovlesh Thakur, Aklank Jain, Hridayesh Prakash, Adam W. Bartlett, Mirvat Said, Linda Popella, Alexander Steinkasserer, Majid Mahmood, Marwa Mohammed Ali Jassim, Murtada Hafedh Hussein, Yutaka Tsutsumi, Ayodeji Folorunsh Ajayi, Oluwadunsin, Adebayo, Emmanuel Adebayo, Neslihan Edeer Edeer Karaca, Alaa Fadhel Fadhel Hassan, Shailendra K. Saxena, Swatantra Kumar, Rajni Nyodu, Vimal K Maurya, Nandini Duggal

First published in London, United Kingdom, 2021 by IntechOpen
IntechOpen is the global imprint of INTECHOPEN LIMITED, registered in England and Wales, registration number: 11086078, 5 Princes Gate Court, London, SW7 2QJ, United Kingdom
Printed in Croatia

British Library Cataloguing-in-Publication Data
A catalogue record for this book is available from the British Library

Additional hard and PDF copies can be obtained from orders@intechopen.com

Innate Immunity in Health and Disease
Edited by Shailendra K. Saxena and Hridayesh Prakash
p. cm.
Print ISBN 978-1-83880-765-8
Online ISBN 978-1-83880-766-5
eBook (PDF) ISBN 978-1-83880-767-2

For EU product safety concerns:
IN TECH d.o.o., Prolaz Marije Krucifikse Kozulić 3, 51000 Rijeka, Croatia, info@intechopen.com or visit our website at intechopen.com.

Meet the editors

Prof. Dr. Shailendra K. Saxena is a vice dean and professor at King George's Medical University, Lucknow, India. His research interests involve understanding the molecular mechanisms of host defense during human viral infections and developing new predictive, preventive, and therapeutic strategies for them using Japanese encephalitis virus (JEV), HIV, and emerging viruses as a model via stem cell and cell culture technologies. His research work has been published in various high-impact factor journals (*Science*, *PNAS*, *Nature Medicine*) with a high number of citations. He has received many awards and honors in India and abroad including various Young Scientist Awards, BBSRC India Partnering Award, and Dr. JC Bose National Award of Department of Biotechnology, Min. of Science and Technology, Govt. of India. Dr. Saxena is a fellow of various prestigious international societies/academies including the Royal College of Pathologists, United Kingdom; Royal Societies of Biology and Chemistry, London, United Kingdom; and Academy of Translational Medicine Professionals, Austria. He was named a Global Leader in Science by *The Scientist*. He is also an international opinion leader/expert in the vaccination for *Japanese encephalitis* by IPIC (*International Primary Immunodeficiencies Congress*).

Dr. Hridayesh Prakash is a fellow of the Royal Society of Biology, London. Currently, he is an associate professor at the Institute of Virology and Immunology, Amity University, NOIDA. He has expertise in innate immunity with a special interest in macrophage immunobiology, tumor immunology/immunotherapy, cell-based immunotherapies, pulmonary infection biology, and radiation biology. Dr. Prakash conducts research to exploit various immunotherapeutics for managing persistent bacterial and viral Infections and gastric cancer. He is unraveling the therapeutic potential of M1 effector macrophages against solid tumors. He is also studying various mechanisms that certain pathogens like *Helicobacter pylori*, *Chlamydia*, and *Mycobacteria* are exploiting for polarizing M1 effector macrophages towards the M2 phenotype during chronic and persistent infections. Under this major objective, he is now validating the therapeutic impact of M1 effector macrophages for the control of persistent infection-driven cancer (adenocarcinoma) progression. Dr. Prakash is also exploring the palliative potential of macrophages against autoimmunity and chronic inflammatory disorders like IBD, radio-pneumonitis, pulmonary fibrosis, and radiation syndrome.

Contents

Preface

This book provides readers with a comprehensive overview of recent trends in innate immunity as well as general concepts of immuno-biology of infections, immune-pathology, immuno-microbiology, immuno-genomics, and immuno-pharmacology. It also examines current clinical recommendations in the management of various diseases, highlighting ongoing issues, recent advances, and future directions of diagnostic approaches and therapeutic strategies.

The book focuses on various aspects and properties of innate immunity, whose deep understanding is very important for safeguarding the human race from further loss of resources and economies due to innate immune response-mediated diseases. Throughout this book we will examine the individual mechanisms by which the innate immune response acts to protect the host from pathogenic infectious agents and other non-communicable diseases.

Macrophages are ubiquitous and integrated parts of both innate and adaptive immunity. These cells have been researched extensively in different contexts. Macrophages display a range of plasticity in their phenotype in different pathological conditions. Together, peripheral and tissue macrophages constitute the reticuloendothelial system where they play a major role in sensing pathogens and tumor antigens for their effective eradication. Macrophages display a range of plasticity, which qualifies them as potential target cells for managing various human diseases clinically. Due to their plastic nature, these cells are potentially involved in most immunological and physiological responses.

Several groups including ours have demonstrated that several multidrug resistant (MDR)/extensively drug resistant (XTR) bacteria polarize M1 effector alveolar macrophages towards their M2 phenotype during persistent infection. This seems to be a potential link to the sensitization for infection and possibly the development of cancer.

Current research in the field focuses on managing the M1/M2 imbalance to minimize the risk of cancer arising from chronic and persistent lung infection with intracellular pathogens like *Chlamydia* or *Mycobacteria*. This may be achieved by targeting major signaling pathways that drive the M2 phenotype and are involved in cancer development (e.g., sphingolipids, Th2/Th17 responses).

In view of the preceding, this book discusses research methodologies, resources, and technologies for identifying the molecular signature involved in the polarization of M1 effector macrophages to M2 macrophages during disease. It also explores how selective phenotypes of macrophages can improve existing therapies, with special emphasis on infection and cancer, particularly lung cancers, and various gastric inflammatory diseases like inflammatory bowel disease.

Written by experts in the field, this volume is a self-contained collection of scholarly papers targeting an audience of practicing researchers, academics, Ph.D. students, and other scientists.

Dr. Shailendra K. Saxena
Professor,
Center for Advanced Research (CFAR),
King George's Medical University (KGMU),
Lucknow, India

Dr. Hridayesh Prakash
Amity Institute of Virology and Immunology,
Amity University,
Noida, India

Section 1

Innate Immunity in Infection Management

Chapter 1

Pathogenesis and Host Immune Response during Japanese Encephalitis Virus Infection

Swatantra Kumar, Rajni Nyodu, Vimal K. Maurya and Shailendra K. Saxena

Abstract

Japanese Encephalitis Virus (JEV) is a mosquito borne flavivirus infection. Transmission of JEV starts with the infected mosquito bite where human dermis layer act as the primary site of infection. Once JEV makes its entry into blood, it infects monocytes wherein the viral replication peaks up without any cell death and results in production of TNF-α. One of the most characteristics pathogenesis of JEV is the breaching of blood brain barrier (BBB). JEV propagation occurs in neurons that results in neuronal cell death as well as dissemination of virus into astrocytes and microglia leading to overexpression of proinflammatory cytokines. JEV infection results in host cells mediated secretion of various types of cytokines including type-1 IFN along with TNF-α and IFN-γ. Molecule like nitrous oxide (NO) exhibits antiviral activities against JEV infection and helps in inhibiting the viral replication by blocking protein synthesis and viral RNA and also in virus infected cells clearance. In addition, the antibody can also acts an opsonizing agent in order to facilitate the phagocytosis of viral particles, which is mediated by Fc or C3 receptor. This chapter focuses on the crucial mechanism of JEV induced pathogenesis including neuropathogenesis viral clearance mechanisms and immune escape strategies.

Keywords: Japanese encephalitis virus, Neuropathogenesis, Dendritic cells, Macrophages, Dendritic cells

1. Introduction

Japanese Encephalitis Virus (JEV) infection is a mosquito-borne zoonotic infection in human which is the most common cause of viral encephalitis in Southeast Asia [1]. The first case of JEV was reported in Japan in the year of 1871. The virus was first isolated in the year 1935 from human brain, which was a fatal case. JEV is responsible for causing a high morbidity and high mortality specifically in the pediatrics age group [2]. Transmission cycle of JEV includes pigs which act as the reservoir/amplifying-host, water bird as carriers and mosquitoes as vector and humans are considered as the dead-end host. JEV is transmitted into human via infected Culex mosquitoes bite and thereby infected individuals develop viremia [3]. Transmission cycle starts mostly post-monsoon where the chance of mosquito breeding increases in paddy fields. JEV is single stranded positive- sense envelope RNA virus and is a member of family *Flaviviridae*. The genome of JEV encodes a one single open reading frame (ORF) encoding single polyprotein that gets cleaved into three structural proteins namely as Capsid (C), Envelope (E), precursor to membrane protein (prM) and seven non-structural proteins including NS1, NS2A, NS2B, NS3, NS4A, NS4B, NS5 [4]. Based on the genomic variability, JEV has been categorized into five genotypes. The most prevalent genotypes seen in the northern region are I and III and that in southern region are II and IV and also, a putative genotype V [5]. The incubation period of JEV infection is 5–15 days and symptoms includes from febrile illness to a severe disease with patients showing meningoencephalitis, aseptic meningitis or a polio-like acute flaccid paralysis [6].

2. Immune cell targets employed by JEV in peripheral and central nervous system

Transmission of JEV starts with the infected mosquito bite where human dermis layer act as the primary site of infection. JEV replication occurs in peripheral system including PBMCs wherein the macrophages, dendritic cells (DCs) and monocytes become infected [7]. Such infection in peripheral system gets cleared off due activation of immune system, and that is the reason for low level of viremia in the blood [8]. During any viral infection, antigen presenting cells (APCs) such as dendritic cells (DCs) and macrophages are the first cell types that trigger the cellular immune responses. They produce various cytokines, which includes IL-6 and TNF-α, and many other pro-inflammatory cytokines. Once JEV makes its entry into blood, it infects monocytes wherein the viral replication peaks up without any cell death and results in production of TNF-α [9] that results in the activation and differentiation of monocytes into monocyte-derived dendritic cells (MDDCs) and monocyte-derived macrophages (MDMs). JEV has developed various immune escape strategies. JEV impairs with the process of DC maturation and where immature human monocyte-derived DCs (im-MDDC) helps in viral replication which takes place by surface expression of co-stimulatory cytokines/chemokine surface receptors [10]. JEV replication has been shown to take place in DCs via reducing the expression of co-stimulatory cytokines, hindering the T-cell activation and by escalating the Treg cells differentiation [11].

During JEV infection, interaction of host-pathogen in the monocyte cell lineage such as monocyte-derived macrophages (MDMs) increases the severity of the disease [12]. Macrophages acts as a hub of viral replication but in case of JEV infection, the productive replication of virus is limited followed by the increase sensitivity to the IFN response [13]. JEV modulates macrophages and DCs in distinguishing pattern. Macrophages get modulated through classical pathway by up regulating co-stimulatory molecules. DCs infected by JEV produce one of the anti-inflammatory cytokine, IL-10 and some of the pro-inflammatory cytokines such as TNF-α, IL-12 and IL-6, whereas macrophages infected by JEV does not produces IL-10 [14]. Such modulation of DCs and macrophages induces an inflammatory environment which then helps in permeability of BBB (blood brain barrier) and hence, the virus tends to spread into central nervous system (CNS). Infection of CNS causes functional damage to DCs including splenic DCs. Since dendritic cells helps in activating naïve T cells, their damage leads to an increase in viral circulation in CNS and hence, reduces the $CD4^+$ and $CD8^+$ T-cells response [15]. Although the mechanism of viral entry into the brain is not well understood but once it enters the brain cells, JEV is detected in cerebrospinal fluid (CSF) and in the nervous tissue [16]. One of the most characteristics pathogenesis of JEV is the breaching of BBB [17]. Neuron being the most important target cell during JEV however, when the infection gets into CNS, along with the neuronal cells, astrocytes also gets infected,

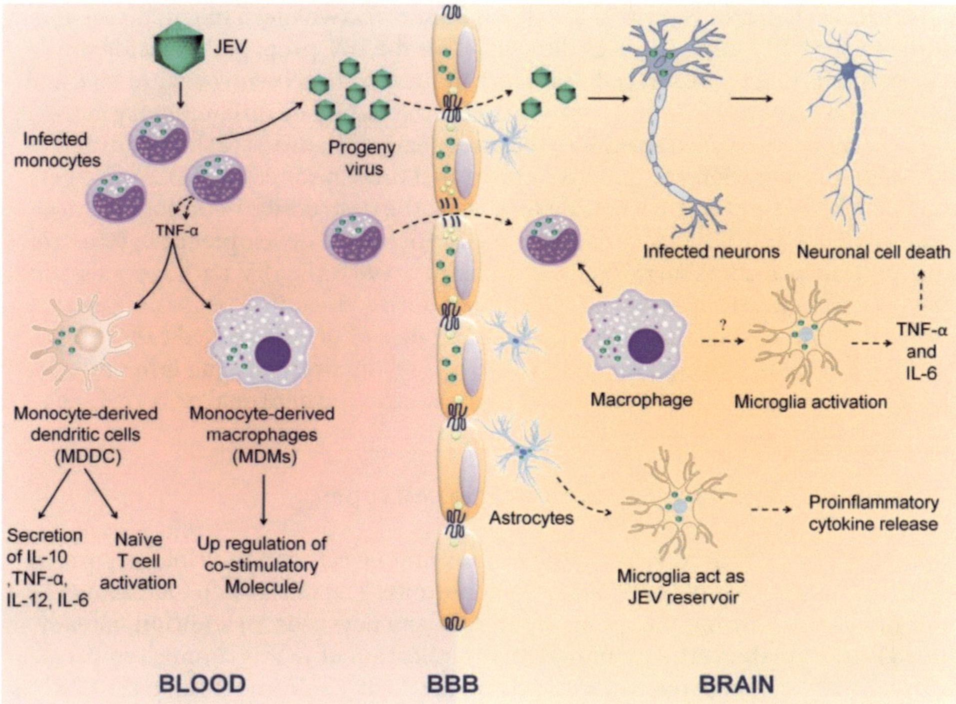

Figure 1.
Mechanism of Japanese encephalitis virus infection and involvement of immune cells. In blood JEV primarily infects monocytes and results in TNF-α production that results in the activation and differentiation of monocytes into monocyte-derived dendritic cells (MDDCs) and monocyte-derived macrophages (MDMs). These cells are involved in naïve T cell activation, secretion of TNF-α and IL-6. JEV can cross the blood brain barrier (BBB) via direct or transmigration of virus harboring monocytes. After entry into the brain, JEV can infect neuronal cells that results in cell death. The transmigrated monocytes differentiate into macrophages which may disseminate the virus to microglia which act as the viral reservoir. Astrocytes are known to disseminate the virus to microglia which results in the activation of microglia leading to over expression of proinflammatory cytokine release.

which is a constituent of BBB and an important part of CNS. Astrocytes are also considered to be helping in the transmission of JEV to the cerebrospinal fluid from peripheral tissues. The microglial cell that is considered to be the resident immune cells/macrophage of CNS is also infected by JEV. Microglial cells play a very significant role in CNS during the JEV infection via acting as a virus reservoir [18]. Upon activation microglia produces proinflammatory cytokines like TNF-alpha and IL-6, which induce death of neuronal cell (**Figure 1**) [19].

3. Neuropathogenesis during JEV infection

The pathogenesis of JEV needs to be explored at dual phases in human which initiates at the peripheral tissues and then, involvement of central nervous system (CNS). Before entering into CNS, JEV replicates in the langerhans cells (skin dendritic cells), which gets transported into the lymphatic and peripheral tissues which results in increased viremia. During the initial infection in periphery tissue, the $CD8^+$ T cell response prevents the dissemination of the JEV into the CNS. Lymphocytes harboring JEV can cross the BBB and via endocytosis process to penetrate the endothelial surface of CNS [20]. However, inability of host to produce antibodies against the infection and the immune evasion strategies of the virus makes this infection lethal. JEV propagation occurs in neurons that results in neuronal cell death. Neuronal cell death occurs via two mechanism; direct and indirect neuronal killing. Direct killing involves the JEV propagation inside the neuronal cells that results in cell death and indirect killing involves aggressive and intense inflammatory responses leading to up-regulation of inflammatory cytokines and reactive oxygen species that causes death of neurons [21]. In addition to cell death, proliferation and growth of neuronal progenitor cells (NPCs) also gets affected which could be the possible reason for the destructive neurological cases in JE survivors [22]. JEV can also cause abnormal neuronal development in fetus via crossing transplacental barrier [23]. In order to prevent the JEV pathogenesis, virus clearance from the peripheral nervous tissues during the initial phase of infection is crucial for designing effective therapy. Clearance of virus-infected cells and recovery during JEV infection relies on the several factors including IgM antibodies, T-lymphocytes and CXCL10 mediated viral clearance by neuronal cells [24].

4. Clearance of JEV by diverse immune cell types

During JEV infection, viral clearance via immune cells is a multiple step process which involves both innate and adaptive immunity. The initial step focuses on the inhibition or on limiting the spread of virus to any new cells. In addition, already infected cells are then either eliminated or replication of JEV is suppressed permanently. However, mechanism of virus clearance during JEV infection in the CNS tissue requires immense understanding of the level of JEV infection in the CNS tissue. One of the most reliable methods is cytolysis, either immune cytolysis or virus-induced. This method involves complete removal or elimination of virus infected cells or cells where the virus is propagating. The immunological processes that are required for clearance of virus are cell-type specific. However, in case JEV infection, the virus invade the host immune cells by cytolytic mechanism and hence, inhibiting the progression of NCP (neural progenitor cells pool). In order to combat such invasion, the activation of brain macrophages is crucial which gets initiated with the help of nerve cells. These macrophages then mediate non-cytolytic viral clearance by producing IFN-β and by supporting production of T-cells that eventually produces IFN-γ [25]. Further, virus secretes proteins/factors and makes cytokine imbalance and suppresses MHC-I present on the membrane surface.

In response to the JEV infection, several mechanisms of innate immune response get activated. After getting infection, host cells starts producing various types of cytokines including type-1 IFN along with TNF-α and IFN-γ. These cytokines induces inflammatory responses and hence, inhibits the viral replication. Furthermore, the IFN-α and IFN-β binds to the NK cells and initiates the lytic activity and hence, kills the JEV infected cells. This antiviral activity gets initiated by one of the cytokine IL-12 which produced at an early phase of infection. IFN-γ then activates the brain macrophages that expresses MHC-II molecules and subsequently, helps in more cytokine production that result in inhibition of viral replication [26]. Other than the cytokines, molecule like nitrous oxide (NO) also evidently exhibits antiviral activities against JEV infection and helps in inhibiting the viral replication by blocking protein synthesis and viral RNA and also in virus infected cells clearance [27]. Adaptive immune response is highly specific involving antigenic specificity display, self/non-self recognition, and immunologic memory. Generally, during flavivirus infection, the antibodies produced by the host cells along with the complement proteins helps in the destruction of the viral particles [28]. However, in case of JEV infection, the virus tends to evade and slips through the complement mediated mechanism of host cells and by inhibition of classical pathway. Additionally, the receptor present on the macrophages interacts with the components of the viral antigens and helps in generating soluble proteins, which then triggers adaptive immune responses promoting clearance of virus infected cells.

5. Cell-mediated immune mechanisms for JEV Clearance

In the process of clearance or elimination of JEV infected cells, cell mediated immunity plays a vital role. Cytokines playing the lead in this mechanism is IFN-γ and IL-2 secreted by T-helper (Th) cells or T-cytotoxic (TC) cells. IL-2 helps in the alteration of naïve T cells into virus-specific cytotoxic T lymphocytes (CTL) generation, which then eventually causes killing of virus infected cells [29]. However, in case of JEV or any flaviviral infection, post exposure to the infected cells, the virus controls the release of CTL and other cytokines. The stimulated Th cells generates cytokines which includes IFN-γ, IL-2, IL-6 and TNF-α which tends to disturbs the cellular activities of JEV and hence, protecting the host from viral infection. These effector molecules are produced by TH1 cells, $CD4^+$ and $CD8^+$ Tc cells, which mediate anti-viral response in order to initiate cell-mediated immune responses. NO amongst these cytokines, both IFN-γ and TNF-α could possibly help in peripheral virus clearance but not from the CNS. IFN-γ helps in maintaining anti viral properties in host cells and IL-2 then, converts naive T cell (CTL) into effector T cell and hence, activates NK cells which in return eliminates virus infected cells or virions.

6. Humoral immune mechanism for JEV clearance

During JEV infection, the host cell recruits humoral immune mechanism, which is a very significant mechanism in the process of protection against the infection in human. Once the host cells get infected by the virus, host humoral responses initiates the process by identification of virus with the recruitment of Th cells that responds to the viral antigens. These Th cells then present these viral antigens or proteins to the B cells along with the help of macrophages. Subsequently, the B cells loaded with viral antigen then get converted to plasma cells and post expansion, starts secreting Abs after few days post JE infection. Thus, this is the initiation of the humoral immune response mediated by antibody production [30]. The antibody binds to the epitopes necessary for the fusion of viral envelope with the plasma membrane and thus, blocking the penetration of virus molecules into the host cells. Furthermore, the antibody can also acts an opsonizing agent in order to facilitate the phagocytosis of viral particles, which is mediated by Fc or C3 receptor. Thus, the mechanism of inhibition of virus propagation and reduction of virus generated cytopathic effects is shown by the JE infected neutralizing antibodies. Host immune responses are triggered 4–7 days post infection resulted after structural and non-structural proteins of virus and host cells interaction.

7. Complement system against JEV infection

Viruses are a kind of pathogen that depends completely upon the host for its survival and its replication. Hence, in order to survive the virus has developed immune escape mechanism from complement system by secreting many inhibitory proteins/cytokines. However, the complement system has an adversary part to play in cases of any *Flavivirus* infection including JE due to it's of multi-component system. This adverse action takes place by either restricting the viral propagation and hence, protecting the host or by triggering an intense inflammatory response, which increases the disease severity. The complement system helps in destruction of the viral particles by complement-dependent lysis of virus, by opsonization, by modulating functions of B-cell and T-cell and also, by phagocytosis [31]. The other pathways in which the viral particles or antigens gets destructed by the complement system is the antibody-dependent and antibody-independent pathways. This action of viral recognition and then its clearance by complement system is carried out by a group of serum proteins along with molecules presented on cell surface. Most of the proteins are presented in an inactive form which then gets activated by three main patterns through which complement system gets activated. These are the classical pathway, alternative pathway and lectin pathway. All the three pathways get started when C3, a proteolytic fragment promotes the cell uptake, which initiates the complement system. The lectin and the alternative pathway get activated by the binding of the mannose-binding lectin (MBL) present on the surface of cell and by the hydrolysis of C3, respectively. These complement system is comprised of four kinds of serum proteins; C3, factor D, factor b and properdin. A more stabilized complex C3b Bb is formed when C3b binds to B (serum protein) with the help of Mg + ions. Also, the unhydrolyzed C3 can possibly produce more of C3b, which gets deposited on the cell surface. Further, C3b Bb 3b complex known to show C5 convertase activity, and eventually, C3b-C5 and 5b component is formed commencing the first step of viral lysis [32]. This complement-mediated pathway to inhibit the contact target chosen by JEV or any flavivirus may help to understand the possible way of combating the infection [33]. The alternative pathway amplifies the activation triggered by the former two pathways and hence, heavily destroys the virus particles. The classical pathway is an antibody dependent process wherein the antibody stimulates the phagocytic cells. The initiation step is the binding of C1q (a polyvalent molecule) with the antigen–antibody complexes, which is present on the surface of a pathogen or by binding directly to the viral protein. Further, C1 complex binds to the cleaved molecules of C4 (C4a and C4b) and furthermore, C2 can complex with C4b and form a new product C4b2b that shows efficient C3 convertase activity, which is essentially required to activate C3 protein. Finally, along with C3, taking into account, C3b and C4b2b with C5 enzyme, a very efficient complex is formed for the destruction of viral particles. Thus, all three pathways converse in a common sequence of events and cause heavy cell lysis of pathogen.

8. Conclusions

Flaviviruses like JEV has taken the infection strategy into an advanced level by evading the detection machinery of the host-immune mechanism which is in responses to any viral infection to kill the virus- infected cells. This is the point where it has become a necessity of the moment to carry extensive studies on this subject. JEV modulates the host machinery in dual ways that is, by virus-mediated damage and by host- immune responses. JEV alters or inhibit both the innate and adaptive immune responses of the host. Since, no viral antigen is presented by the macrophages which are infected during the early infection phase, no adaptive immune responses is triggered during this phase. Escaping all the immune responses, JEV manages to disseminate into the CNS and causing damage to the CNS is what makes JEV infection more lethal. Any renewal and replenishment of tissues in CNS after infection is challenging. Host responses like CTL activation has been efficient in combating the virus induced MHC molecules. However, JEV comes with a counter viral strategy by activating the non-classical MHC molecules, since, these non -classical MHC molecules inhibits NK cells by binding to its receptors.

NK cells are the crucial cells so as to say, which are efficient to kill the virus infected cells. To get a clear understanding of how to combat the viral immune escape strategy, many viral antigens/proteins are profiled and characterized. Such studies are a new approach towards generating effective vaccines against JEV along with other flaviviruses. Because of all these challenging viral strategies, it has become a necessary research step to reach to a point where viral eradication shall be feasible. In order to reach to this point, basic understanding of the JEV strategy on how JEV manages to trigger imbalance between host's immunopathological and shielding mechanisms is very important. To combat spread of such lethal infection, vaccination against it should be made mandatory in the entire endemic region. Effective surveillance in the endemic region has to be considered especially in pediatric age, since; this age group is the mostly effected and has proved to be lethal.

Acknowledgements

The authors are grateful to the Vice Chancellor, King George's Medical University (KGMU), Lucknow, India for the encouragement for this work. The authors have no other relevant affiliations or financial involvement with any organization or entity with a financial interest in or financial conflict with the subject matter or materials discussed in the manuscript apart from those disclosed.

Conflict of interest

The authors declare no conflict of interest.

Author details

Swatantra Kumar†, Rajni Nyodu†, Vimal K. Maurya and Shailendra K. Saxena*†
Centre for Advanced Research (CFAR), Faculty of Medicine, King George's Medical University (KGMU), Lucknow, India

*Address all correspondence to: shailen@kgmcindia.edu

† These authors contributed equally to this work as first author.

References

[1] Japanese encephalitis. World Health Organization. (https://www.who.int/news-room/fact sheets/detail/japaneseencephalitis#:~:text=Permanent%20neurologic%20or%20psychiatric%20sequelae,no%20cure%20for%20the%20disease) [Accessed on 04 Aril 2021]

[2] Solomon T, Ni H, Beasley DW, Ekkelenkamp M, Cardosa MJ, Barrett AD. Origin and evolution of Japanese encephalitis virus in southeast Asia. J Virol. 2003;77(5):3091-3098. DOI:10.1128/jvi.77.5.3091-3098.2003

[3] Tu T, Xu K, Xu L, Gao Y, Zhou Y, He Y, Liu Y, Liu Q, Ji H, Tang W. Association between meteorological factors and the prevalence dynamics of Japanese encephalitis. PLoS One. 2021; 16(3):e0247980.

[4] Yasui K, Miyamoto M, Kimura-Kuroda J, et al. Analysis of Japanese encephalitis (JE) virus genome and implications for recombinant JE vaccine. Southeast Asian J Trop Med Public Health. 1990;21(4):663-669.

[5] Gao X, Liu H, Li X, et al. Changing Geographic Distribution of Japanese Encephalitis Virus Genotypes, 1935-2017. Vector Borne Zoonotic Dis. 2019;19(1):35-44. DOI:10.1089/vbz.2018.2291

[6] Saxena SK, Kumar S, Maurya VK. Pathogen-associated acute encephalitis syndrome: therapeutics and management. Future Microbiol. 2019;14:259-262. DOI:10.2217/fmb-2018-0334

[7] Nikitina E, Larionova I, Choinzonov E, Kzhyshkowska J. Monocytes and Macrophages as Viral Targets and Reservoirs. Int J Mol Sci. 2018;19(9):2821. Published 2018 Sep 18. DOI:10.3390/ijms19092821

[8] Solomon, Tom. Flavivirus Encephalitis. New England Journal of Medicine. 2004 351(4):370.DOI:10.1056/NEJMra030476

[9] Sooryanarain H, Ayachit V, Gore M. Activated CD56(+) lymphocytes (NK+NKT) mediate immunomodulatory and anti-viral effects during Japanese encephalitis virus infection of dendritic cells in-vitro. Virology. 2012;432(2):250-260. DOI:10.1016/j.virol.2012.05.013.

[10] Aleyas AG, George JA, Han YW, et al. Functional modulation of dendritic cells and macrophages by Japanese encephalitis virus through MyD88 adaptor molecule-dependent and -independent pathways. J Immunol. 2009;183(4):2462-2474. DOI:10.4049/jimmunol.0801952

[11] Gupta N, Hegde P, Lecerf M, et al. Japanese encephalitis virus expands regulatory T cells by increasing the expression of PD-L1 on dendritic cells. Eur J Immunol. 2014;44(5):1363-1374. DOI:10.1002/eji.201343701

[12] Sooryanarain H, Sapkal GN, Gore MM. Pathogenic and vaccine strains of Japanese encephalitis virus elicit different levels of human macrophage effector functions. Arch Virol. 2012;157(10):1905-1918. DOI:10.1007/s00705-012-1386-8

[13] Kundu K, Dutta K, Nazmi A, Basu A. Japanese encephalitis virus infection modulates the expression of suppressors of cytokine signaling (SOCS) in macrophages: implications for the hosts' innate immune response. Cell Immunol. 2013;285(1-2):100-110. DOI:10.1016/j.cellimm.2013.09.005

[14] Lannes N, Summerfield A, Filgueira L. Regulation of inflammation in Japanese encephalitis. J Neuroinflammation. 2017;14(1):158. Published 2017 Aug 14. DOI:10.1186/s12974-017-0931-5

[15] Cao S, Li Y, Ye J, et al. Japanese encephalitis Virus wild strain infection suppresses dendritic cells maturation and function, and causes the expansion of regulatory T cells. Virol J. 2011;8:39. DOI:10.1186/1743-422X-8-39

[16] Hsieh JT, St John AL. Japanese encephalitis virus and its mechanisms of neuroinvasion. PLoSPathog. 2020; 16(4):e1008260. DOI:10.1371/journal.ppat.1008260

[17] Al-Obaidi MMJ, Bahadoran A, Har LS, et al. Japanese encephalitis virus disrupts blood-brain barrier and modulates apoptosis proteins in THBMEC cells. Virus Res. 2017;233:17-28. DOI:10.1016/j.virusres.2017.02.012

[18] Chen CJ, Ou YC, Lin SY, et al. Glial activation involvement in neuronal death by Japanese encephalitis virus infection. J Gen Virol. 2010;91(Pt 4):1028-1037. DOI:10.1099/vir.0.013565-0

[19] Lannes N, Neuhaus V, Scolari B, et al. Interactions of human microglia cells with Japanese encephalitis virus. Virol J. 2017;14(1):8. Published 2017 Jan 14. DOI:10.1186/s12985-016-0675-3

[20] Li F, Wang Y, Yu L, et al. Viral Infection of the Central Nervous System and Neuroinflammation Precede Blood-Brain Barrier Disruption during Japanese Encephalitis Virus Infection. J Virol. 2015;89(10):5602-5614. DOI:10.1128/JVI.00143-15

[21] Chen SO, Chang TJ, Stone G, Chen CH, Liu JJ. Programmed cell death induced by Japanese encephalitis virus YL vaccine strain or its recombinant envelope protein in varied cultured cells. Intervirology. 2006;49(6):346-351. DOI:10.1159/000095154

[22] Das S, Basu A. Japanese encephalitis virus infects neural progenitor cells and decreases their proliferation. J Neurochem. 2008;106(4):1624-1636. DOI:10.1111/j.1471-4159.2008.05511.x

[23] Mathur A, Arora KL, Chaturvedi UC. Transplacental Japanese encephalitis virus (JEV) infection in mice during consecutive pregnancies. J Gen Virol. 1982;59(Pt 1):213-217. DOI:10.1099/0022-1317-59-1-213

[24] Pan CH, Chen HW, Huang HW, Tao MH. Protective mechanisms induced by a Japanese encephalitis virus DNA vaccine: requirement for antibody but not CD8(+) cytotoxic T-cell responses. J Virol. 2001;75(23):11457-11463. DOI:10.1128/JVI.75.23.11457-11463.2001

[25] Larena M, Regner M, Lobigs M. Cytolytic effector pathways and IFN-γ help protect against Japanese encephalitis. Eur J Immunol. 2013;43(7): 1789-1798. DOI:10.1002/eji.201243152

[26] Chhatbar C, Detje CN, Grabski E, et al. Type I Interferon Receptor Signaling of Neurons and Astrocytes Regulates Microglia Activation during Viral Encephalitis. Cell Rep. 2018;25(1):118-129.e4. DOI:10.1016/j.celrep.2018.09.003

[27] Saxena SK, Singh A, Mathur A. Antiviral effect of nitric oxide during Japanese encephalitis virus infection. Int J Exp Pathol. 2000;81(2):165-172. DOI:10.1046/j.1365-2613.2000.00148.x

[28] Li Y, Counor D, Lu P, Duong V, Yu Y, Deubel V. Protective immunity to Japanese encephalitis virus associated with anti-NS1 antibodies in a mouse model. Virol J. 2012;9:135. Published 2012 Jul 24. DOI:10.1186/1743-422X-9-135

[29] Konishi E, Kurane I, Mason PW, et al. Induction of Japanese encephalitis virus-specific cytotoxic T lymphocytes in humans by poxvirus-based JE vaccine candidates [published correction appears in Vaccine 1999 Feb 5;17(5):I].

Vaccine. 1998;16(8):842-849. DOI:10.1016/s0264-410x(97)00265-x

[30] Ochsenbein AF, Pinschewer DD, Odermatt B, Carroll MC, Hengartner H, Zinkernagel RM. Protective T cell-independent antiviral antibody responses are dependent on complement. J Exp Med. 1999;190(8): 1165-1174. DOI:10.1084/jem.190.8.1165

[31] Thielens NM, Tacnet-Delorme P, Arlaud GJ. Interaction of C1q and mannan-binding lectin with viruses. Immunobiology. 2002;205(4-5):563-574. DOI:10.1078/0171-2985-00155

[32] Agrawal P, Nawadkar R, Ojha H, Kumar J, Sahu A. Complement Evasion Strategies of Viruses: An Overview. Front Microbiol. 2017;8:1117. DOI:10.3389/fmicb.2017.01117

[33] Avirutnan P, Mehlhop E, Diamond MS. Complement and its role in protection and pathogenesis of flavivirus infections. Vaccine. 2008;26(Suppl 8):I100-I107. DOI:10.1016/j.vaccine.2008.11.061

Chapter 2

How Human Herpesviruses Subvert Dendritic Cell Biology and Function

Linda Popella and Alexander Steinkasserer

Abstract

In the last decades, a multitude of distinct herpesvirus-mediated immune evasion mechanisms targeting dendritic cell (DC) biology were uncovered. Within this chapter, we summarize the current knowledge how herpesviruses, especially the α-herpesviruses HSV-1, HSV-2, varicella-zoster virus (VZV), and the β-herpesvirus HCMV, shape and exploit the function of myeloid DCs in order to hamper the induction of potent antiviral immune responses. In particular, the main topics covering herpesvirus-mediated immune evasion will involve: (i) the modulation of immature DC (iDC) phenotype, (ii) modulation of iDC apoptosis, (iii) the inhibition of DC maturation, (iv) degradation of the immune-modulatory molecule CD83 in mature DCs (mDCs), (v) interference with the negative regulator of β2 integrin activity, cytohesin-1 interaction partner (CYTIP), (vi) resulting in modulation of adhesion and migration of mDCs, (vii) autophagic degradation of lamins to support productive HSV-1 replication in iDCs, (viii) the release of uninfectious L-particles with immune-modulatory potential from HSV-1-infected mDCs, and (ix) the implications of DC subversion regarding T lymphocyte activation.

Keywords: dendritic cells, HSV-1, HSV-2, VZV, HCMV, CD83, CYTIP, adhesion, migration, lamins, autophagy, H-particles, L-particles, T lymphocyte activation

1. Introduction

Herpesviridae constitute an extremely successful virus family, evident from the considerable prevalence among the world's population [1]. During co-evolution of herpesviruses with its human host, not only the host's immune system was compelled to mount efficient antiviral defense mechanisms but also the virus has evolved a multitude of sophisticated strategies to dampen those immune responses [2–6]. Thus, herpesviral infections of men represent a tug of war, in which the host's antiviral responses are faced with the virus-mediated immune evasion mechanisms. The probably most intriguing strategy of herpesviral immune subversion is the establishment of latency in immune-privileged niches in the host, leading to lifelong persistent infections accompanied by episodes of viral reactivation [7, 8]. An additional cornerstone contributing to the success of human herpesviruses is the potent infection of a plethora of distinct cell types *in vitro* and *in vivo*, including the manipulation of vital functions of nonimmune as well as immune cells, many of them targeting dendritic cell (DCs) as described below [3, 5, 9–12].

DCs are specialized leukocytes that are highly efficient to antigen specifically activate T lymphocytes, and thus link the innate with the adaptive arm of our immune system [13–16]. In the past four decades, several groups identified DCs as being a rather heterogeneous cell population comprising distinct subsets [17–21]. Those subsets greatly differ in their expression of distinct surface markers, function, anatomical localization as well as migratory capability [22–24]. In general, two distinct DC classes can be defined: myeloid conventional/classical (cDC1—CD141$^+$ and cDC2—CD1c$^+$) and plasmacytoid DCs (pDCs) [25–27]. pDCs play a crucial role during viral infections, since they secrete high amounts of type I interferons upon toll-like receptor (TLR) activation [28–30]. The conventional/classical DCs are specialized in antigen presentation and comprise distinct DC subsets with spatial differences, i.e., blood or lymphoid as well as nonlymphoid tissues. Noteworthy, a third main group among the DC lineage, which arises from monocytes, is called monocyte-derived DCs and reflects inflammatory DCs [31]. Within this chapter, we will mainly focus on monocyte-derived DCs or conventional DCs and their interplay with distinct human herpesviruses.

Another important feature among DCs is that these cells exist in two distinct activation states. In essence, immature DCs (iDCs) reside and patrol in the vast majority of tissues under steady-state conditions, seeking for (nonhost) antigens [32]. Upon antigen uptake and antigen recognition via, e.g., engagement of pathogen recognition receptors, or the perception of "danger" signals, including inflammatory cytokines released from adjacent infected cells, DCs undergo maturation [33]. While sessile iDCs possess a strong phagocytic but low antigen-presenting capacity, mature DCs (mDCs) turn into efficient migrating and antigen-presenting cells (APCs) [32]. DC activation is characterized by an elevated production of type I and III interferons (IFNs) as well as pro-inflammatory cytokines, such as IL-6, TNF-α, or IL-12 [34–37]. Moreover, mDCs are equipped with high surface levels of MHC class I and II molecules [38, 39], and abundantly expose the co-stimulatory molecules CD80, CD86, and CD40, which are important for proper T lymphocyte activation [32]. In this regard, the interaction of DC-expressed CD40 with CD40 ligand (CD40L) expressed on T lymphocytes will result in DC-derived IL-12 production, which is an important Th1 cytokine [28].

In addition, the glycoprotein CD83 is massively expressed on the surface of mDCs, thus serving as reliable marker of mature DCs. More importantly, CD83 is crucial for T lymphocyte development as well as activation, based on its inherent potent immune-modulatory properties [40–44]. Beyond that, migration of mDCs toward T lymphocyte-rich areas in lymphoid organs is facilitated by a switch in the chemokine receptor repertoire during DC activation. In particular, the C–C chemokine receptor 7 (CCR7) is one of the driving forces that chemotactically guides mDCs toward lymphoid-expressed C–C motif chemokine 19 (CCL19) and CCL21. Once arrived in the lymph node, mDCs present their peripheral-acquired antigens to T lymphocytes, which harbor the cognate antigen receptor, to subsequently prime an adaptive immune response [45–50].

Given the pivotal role of DCs during the induction of an adaptive immune response, it is not surprising that herpesviruses efficiently infect DCs and hijack vital DC-inherent functions, such as migration and antigen presentation, to hamper the antiviral host defense. Many of the *in vitro* HSV infection studies, that have been performed, involve the analysis of murine bone marrow-derived DCs (BMDCs). Additionally, human blood monocyte-derived DCs serve as a second widely used *in vitro* model system for the elucidation of herpesviral-mediated modulations of DC biology and function. This is due to the development of appropriate settings for the generation of monocyte-derived DCs in large numbers, for their subsequent highly efficient infection with specific human herpesviruses [51–56]. Concerning *in vitro*

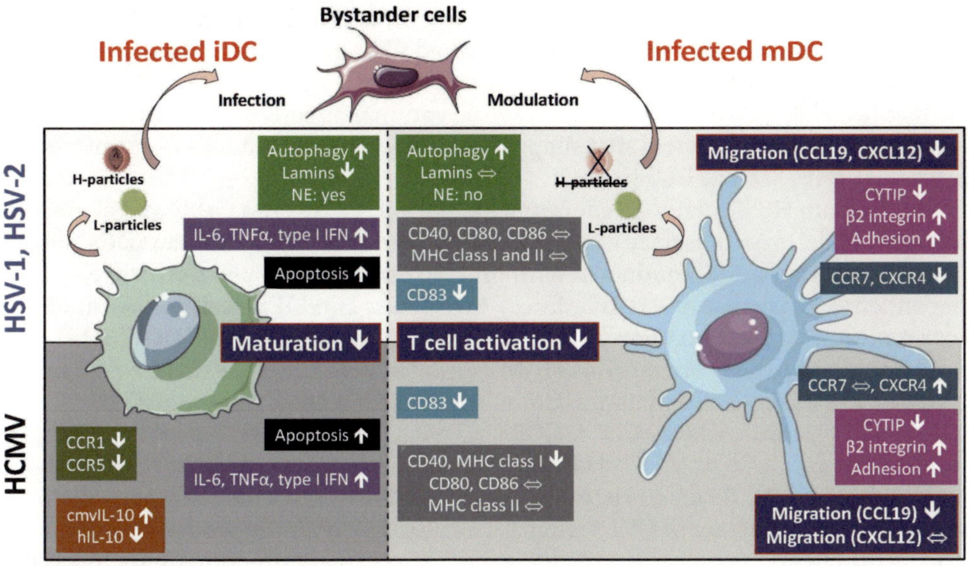

Figure 1.
Schematic summary of HSV- and HCMV-mediated immune evasion strategies targeting DCs. The upper panel depicts immune evasion mechanisms observed for HSV-1- as well as HSV-2-infected iDCs (left panel) and mDCs (right panel). The lower panel illustrates those observed for HCMV-infected iDCs (left panel) and mDCs (right panel). Abbreviations: iDC: immature dendritic cell, mDC: mature dendritic cell, H-particles: heavy (infectious) particles, L-particles: light (noninfectious) particles, NE: nuclear egress, IL-6: interleukin-6; TNFα: tumor necrosis factor α, IFN: interferon, CCR1/CCR5/CCR7: C–C chemokine receptor type 1/type 5/type 7, cmvIL-10: cytomegalovirus-encoded interleukin-10, hIL-10: human cellular IL-10, CCL19: CC-chemokine ligand 19, CXCL12: C-X-C motif chemokine ligand 12, CXCR4: C-X-C motif chemokine receptor 4, CYTIP: cytohesin-1 interacting protein. (Graphics modified from SMART: Servier Medical Art).

HCMV infection studies of DCs, the infection efficiency varies depending on the viral strain used. This is due to the absence versus presence of the genetic locus comprising UL128-UL131, which is crucial for endothelial cell- and leukotropism [57–59].

Within this chapter, we summarize how herpesviruses, especially the α-herpesviruses HSV-1, HSV-2, VZV, and the β-herpesvirus HCMV, shape and exploit the function of classical DCs (summarized in **Figure 1**).

2. Interaction of herpesviruses and immature DCs

2.1 Impediment of iDC biology by herpesviruses

Well-studied examples of functional paralysis of DCs are the herpesviral interference with the expression levels of important surface molecules on iDCs, viability of iDCs, and DC maturation when infecting myeloid iDCs. Regarding the first, complete infection with a clinical isolate of the α-herpesvirus HSV-1 (MC1) asynchronously inhibits the surface expression of CD1a, CD40, CD54 (intercellular adhesion molecule 1, ICAM-1), CD80, and CD86 on iDCs. Apart from this, CD11c, MHC class I and class II surface exposure is unaltered upon HSV-1 MC1 infection of iDCs, indicative of a selective targeting of distinct surface molecules on iDCs [60]. Inconsistent with this, a recombinant disabled infectious single-cycle (DISC)-HSV-1 strain, deleted for glycoprotein H (gH) and thus rendering viral progeny noninfectious, does, however, reduce MHC class I but induce CD86, MHC class II as well as CD1a surface expression on directly infected iDCs, when using a low MOI [61, 62]. Importantly, CD80 and CD83 are unaffected on directly infected iDCs, whereas these molecules are strongly induced on uninfected bystander cells that additionally show an increase in MHC class I and class II, CD54, and CD86 levels [62]. Moreover, supernatants of HSV-1-infected iDCs are sufficient to trigger partial phenotypic maturation of iDCs, mirrored by an increase in MHC class II and CD86 surface levels [63]. Given the latter two observations, directly HSV-1-infected iDCs might secrete soluble factors that shape the phenotype (and function) of uninfected bystander cells, such as DCs [63].

Apart from HSV-1, HSV-2 reduces CD40, CD80, CD86, and CD83 and slightly hampers MHC class II surface expression on macaque as well as human iDCs [64]. Noteworthy, one study, conducted with murine BMDCs, revealed a serotype-dependent and age-specific regulation of MHC class I and II as well as co-stimulatory molecule expression upon an HSV-1 versus HSV-2 infection of iDCs [65].

Additionally, the β-herpesvirus HCMV significantly alters the surface protein repertoire on iDCs. In particular, HCMV-infected iDCs show reduced CD1a, CD11c, CD13, CD33, CD40, CD54, CD58, CD80, CD83, and MHC class I expression levels, whereas CD86 or MHC class II surface expression is only slightly decreased by HCMV [66, 67]. Beyond this, an HCMV infection of iDCs also dampens their migratory capacity toward CCL3 and CCL5 due to the UL18-dependent internalization of the chemokine receptors CCR1 and CCR5, without affecting CCR7 [68, 69].

Interestingly, and in sharp contrast to the abovementioned herpesviruses, the α-herpesvirus VZV does not disturb the surface expression of important immune molecules on iDCs, such as CD1a, CD40, CD86, MHC class I, and MHC class II [70].

Beyond targeting surface molecule expression on iDCs, a herpesviral infection additionally impacts the expression and release of cytokines and interferons via replication-dependent vs. -independent pathways [71, 72]. Regarding HSV-1, directly infected murine iDCs produce increased amounts of IL-12, which is dependent on viral replication [72–74]. By contrast, IL-12 levels are barely detectable or decrease upon an HSV-1 infection of human iDCs [62, 63, 72]. Moreover, IL-12 production is negatively influenced in the presence of an additional inflammatory stimulus, e.g., LPS or CD40, upon an HSV-1 infection of both murine and human origin [62, 63, 65, 73]. This reduction in IL-12 secretion during DC activation might partially explain the reduced T lymphocyte stimulatory capacity upon infection, since IL-12 is an important cytokine for Th1 responses and highly induced upon DC activation [63, 75]. Moreover, CD40 downregulation on HSV-1-infected iDCs might be involved in the impairment in IL-12 secretion during DC activation [60, 62, 76]. Furthermore, TNFα production by HSV-1-infected iDCs increases strain specifically, which is further dependent on viral replication [77], while IL-6 production is also elevated [72, 73].

Apart from directly infected iDCs, uninfected bystander iDCs, as well as iDCs treated with the supernatant derived from infected cultures, exhibit elevated levels of IL-12, but not IL-6, with or without the presence of the additional inflammatory stimulus LPS, and upregulate CD86 as well as MHC class II expression [62, 63, 74]. Regarding this bystander effect, type I IFNs, which are secreted by infected cells for paracrine perception, are one of the most important mediators [71, 78]. Notably, recombinant IFN-α is sufficient to induce partial maturation and IL-12 secretion by iDCs [78]. Type I interferons (IFN α/β) are cytokines with potent antiviral properties, which are predominantly secreted by pDCs early upon infection, but also by cDCs, and other cell types, during a second wave of innate antiviral response [79–82]. Concerning the latter, type I IFN production upon an HSV-1 infection of iDCs is independent from TLR9 signaling and viral replication, but dependent on viral entry, involving recognition of distinct viral glycoproteins, and very likely includes the sensing of virion-associated DNA [78, 81, 83]. Regarding the aforementioned findings, another study shows that HSV-1 strain KOS isolate-dependently triggers TLR2 activation, while only a minority of these substrains as well as clinical isolates are capable of doing so [74]. In particular, this TLR2 induction upon an HSV-1 infection results in highly elevated levels of IL-6 as well as IL-12, independent of viral replication. Moreover, UV-inactivated HSV-1 isolates, possessing TLR2-activating property, induce IL-6 and IL-12 production via single and sequential TLR2- as well as TLR9-dependent mechanisms in iDCs. By contrast, replication-competent HSV-1, capable of inducing TLR2 signaling, induces IL-6 and IL-12 expression via MyD88 signaling either through TLR2- or TLR9-dependent mechanisms in iDCs. Moreover, HSV-1 subspecies that do not activate the TLR2 pathway are mostly recognized via TLR9-dependent mechanisms in iDCs. Among others, HSV-1 is recognized via specific mechanisms in iDCs involving two distinct membrane-bound TLRs, i.e., TLR2 and TLR9, whose involvement vary among different HSV-1 isolates and the presence versus absence of viral replication [74]. Also, other pathogen recognition receptors (PRRs), such as DNA sensors, contribute to the recognition of HSV-1, not only in iDCs, and corporately act to induce a potent antiviral response [4, 81, 84, 85].

Notably, the cytosolic DNA sensor DDX41 and its downstream mediator stimulator of interferon genes (STING) play an important role in mediating a type I IFN response upon HSV-1 infection of iDCs [85]. By contrast, elevated levels of pro-inflammatory cytokines and type I IFNs upon HSV-1 infection of iDCs are not triggered via RIG-I-like receptor (RLR)/mitochondrial antiviral signaling protein (MAVS)-dependent mechanisms, which sense pathogen-derived RNA species [71]. These combined observations reveal that iDCs undergo IFN signaling-dependent as well as -independent changes upon an HSV-1 infection [71].

To counteract the antiviral response of iDCs upon viral recognition, HSV-1-encoded virion host shutoff (vhs, UL41) dampens the production of type I IFN as well as TLR-independent release of pro-inflammatory cytokines (TNFα, IL-6, and IL-12) immediately upon infection of human as well as murine iDCs [71, 72, 86]. In particular, HSV-1 vhs inhibits the early replication-independent activation of NFκB, which is an essential transcription factor for IFN expression and consequent IFN signaling, leading to the induction of interferon-stimulated genes (ISGs) [71, 72]. Moreover, ICP27 additionally counteracts NFκB as well as interferon regulatory factor 3 (IRF3) activation to hamper early antiviral immune responses, as at least shown in macrophages [77]. Apart from this, a multitude of distinct viral proteins target specific steps during HSV-1 recognition, such as ICP0; however, most of these data are based on studies using cell types others than DCs [5, 87].

Also, HSV-2 shapes the production of cytokines and IFNs upon infection of iDCs. In particular, HSV-2 strongly and replication-dependently induces the production of IL-6 and TNFα by murine and human iDCs, while the latter also increases independent of viral replication [88–90]. The elevated secretion of TNFα has been connected to support a co-infection with HIV-1 and might act in trans, very likely via promoting the expression of HIV-1 co-receptor CCR5 on bystander cells [88]. Apart from this, IFNβ and IFNγ are specifically upregulated during the incubation of iDCs with UV-inactivated HSV-2, but not with replication-competent HSV-2, showing higher cytokine levels in the human system [88, 90]. Contrasting elevated amounts of IL-12 upon HSV-1 exposure of murine iDCs, HSV-2 does not influence the secretion of this cytokine in the absence of any additional stimulus [90]. However, LPS-induced IL-12 production by murine iDCs is also hampered upon exposure to HSV-2 or UV-inactivated HSV-2 virions, reminiscent of HSV-1 [65].

Among the β-*Herpesvirinae*, also HCMV modulates cytokine expression of iDCs upon infection. During early responses, HCMV triggers the production of the pro-inflammatory cytokines IL-6 and TNFα, chemokines, such as CCL5, CXCL10, and CXCL11, as well as the TLR3-independent production of type I IFN by infected iDCs [67, 91–93]. Consistent with the HSV-mediated suppression of

IL-12 production in the presence of LPS or CD40L by iDCs, also HCMV hampers the induction of IL-12 during DC activation [67, 92]. In addition, HCMV encodes a multitude of chemokines/cytokines and chemokine/cytokine receptors itself, thereby hijacking the host's immune responses [94, 95]. One of the best-characterized viral encoded chemokines, which has been implicated in the modulation of DC biology, is cmvIL-10 that shares functional analogy with its human IL-10 homolog [96]. Notably, HCMV-infected iDCs show a decreased production of cellular IL-10, which is an important anti-inflammatory immune dampening cytokine [91, 97]. Noteworthy, the expression of cmvIL-10 increases during the course of infection and, apart from influencing DC maturation (see also Section 2.3), it inhibits the expression of IL-6, IL-12, and TNFα, when iDCs are exposed to cmvIL-10-containing supernatants, derived from HCMV AD169-infected fibroblasts [98].

2.2 Herpesviruses modulate cell survival of iDCs

Another functional impairment of iDC biology is the enhanced apoptosis observed in HSV-1-, HSV-2-, and HCMV-infected iDCs [60, 63, 65, 90, 99, 100]. Concerning HSV-1, upon an initial anti-apoptotic phase, which is most likely important for viral replication, infected iDCs show a subsequent early increase in apoptosis concomitant with higher caspase-3 activity, which is dependent on viral gene expression [60, 63, 99, 101]. Mechanistically, HSV-1 triggers a strong decline in the cellular FLICE-inhibitory protein (c-FLIP) expression, a pro-survival protein, in a cell type-independent manner, which strongly correlates with reduced cell survival of iDCs [102]. However, differential regulation of apoptosis in iDCs versus epithelial cells, whereas in the latter, viral ICP27 plays a dominant role to prevent premature cell death [103, 104], is further associated with lower levels of anti-apoptotic latency-associated transcript (LAT) sequences in iDCs [102, 105]. Since LATs are able to block caspase-8-triggered apoptosis and can partially compensate for c-FLIP downmodulation, lower LAT abundancies in iDCs are insufficient to counterbalance HSV-1-induced apoptosis in iDCs [106, 107]. Given the temporal and cell type-dependent regulation of apoptosis, HSV-1 adopts its anti-apoptotic factors to ensure efficient viral replication in, e.g., epithelial cells, by inhibiting apoptosis, and simultaneously avoid DC-mediated antigen presentation, by promoting premature cell death of these cells.

Consistent with HSV-1, also HSV-2 strongly induces apoptosis of infected and bystander iDCs, while it is not fully clear whether viral gene expression plays an essential role during this process [64, 65, 90, 99]. In particular, HSV-2 mediates an increase in caspase-3 activity, transient induction in caspase-8 protein levels, and decrease in c-FLIP expression, which is accelerated in comparison to HSV-1 [90, 99, 102]. While HSV-1 and HSV-2 seem to dampen the presentation of viral antigens by infected iDCs via induction of premature cell death, uninfected bystander DCs are capable of cross-presenting engulfed antigens derived from apoptotic cells and to stimulate $CD8^+$ T lymphocytes (further discussed in Section 4).

Apart from HSVs, also the β-herpesvirus HCMV induces iDC apoptosis and/or necrosis early upon infection, which triggers maturation of uninfected bystander iDCs [100]. Regarding HCMV-associated apoptosis in iDCs, a viral-encoded IL-10 homolog (cmvIL-10) mediates the downregulation of c-FLIP expression upon LPS stimulation of iDCs [108]. However, cmvIL-10-triggered apoptosis is absent in unstimulated iDC cultures [109].

In sharp contrast, VZV-infected iDCs do not undergo apoptosis, suggesting that VZV benefits from viable directly infected iDCs for virus dissemination and persistence [70, 110]. In this regard, VZV selectively downregulates Fas on the surface of infected iDCs and mDCs, very likely to inhibit apoptosis of these cells [111].

2.3 Perturbation of DC maturation during herpesviral infections

Apart from directly influencing the phenotype and viability of iDCs, herpesviruses additionally evolved mechanisms to suppress DC maturation upon recognition of the virus, since this step is associated with the switch into an antigen-presenting phenotype, for efficient priming of adaptive (antiviral) immune responses. Regarding this, HSV-1 blocks the expression of important molecules, such as CD80, CD83, CD86, MHC class II, CCR7, and CXCR4, in directly infected DCs in the absence or presence of additional stimuli, i.e., LPS or pro-inflammatory cytokines [62, 73, 86, 112, 113]. Thus, HSV-1-infected iDCs are hampered in their maturation capacity and therefore unable to efficiently stimulate T lymphocytes [62, 113]. Two HSV-1-encoded proteins are known so far to be involved in the inhibition of DC maturation, i.e., the viral encoded virulence factor ICP34.5 and vhs [72, 73, 86, 113, 114].

Regarding the first, ICP34.5 is essential and sufficient to partially perturb LPS-induced DC maturation via blocking IFN-α/β secretion *in vitro* as well as *in vivo* [113]. Noteworthy, the N-terminal domain of ICP34.5 interacts with and suppresses TANK-binding kinase 1 (TBK1) to block IRF3 phosphorylation and ultimately IFN and IFN-stimulated gene induction [114, 115]. Apart from this, ICP34.5 additionally targets the IκB kinase complex to potently abrogate NFκB activation in DCs upon TLR4 stimulation. Particularly, ICP34.5 recruits protein phosphatase 1 (PP1) to dephosphorylate IκB kinase in order to tightly control NFκB activation during an HSV-1 infection of DCs [73]. Notably, inhibiting DC activation by HSV-1 ICP34.5 sufficiently promotes viral replication in a murine corneal infection model [113]. Furthermore, due to its attenuated replication efficiency and its incapability to inhibit DC maturation, an engineered HSV-1 ICP34.5 mutant induces protective immunity in a DC-dependent way upon lethal challenge in mice and thus constitutes a promising vaccination candidate [114].

In contrast to the ICP34.5-mediated inhibition of TLR-dependent DC activation, the tegument-associated viral protein vhs suppresses TLR-independent pathways that induce DC maturation upon viral recognition [72, 86]. Concerning its involvement in suppressing DC maturation, vhs exerts its inhibitory function by targeting replication-dependent and -independent cellular responses, the latter involving the blockade of NFκB activation [71, 72]. Vhs is a ribonuclease that degrades viral as well as cellular mRNAs upon infection and is thus implicated in interfering with a variety of distinct pathways [116, 117]. Since tegument proteins are directly released into the infected cell, vhs might immediately suppress DC activation prior to ICP34.5. Also, HSV-1 vhs mutant strains might possess a promising potential for vaccine development, as these strains are highly attenuated *in vivo* [118–121].

HSV-1 is only one example among *Herpesviridae* that potently blocks activation of DCs upon infection, since also its family member HSV-2 aims to inhibit DC maturation [90, 112]. Similar to HSV-1, HSV-2 suppresses the activation of DCs in directly infected cells, but not in their uninfected counterparts [90]. However, the precise underlying mechanisms are yet undefined.

Also, VZV has evolved strategies to avoid the activation-driven upregulation of functionally important surface molecules on DCs upon infection [70]. In essence, VZV interferes with the NFκB signaling pathway that strongly regulates the expression levels of maturation-associated proteins in DCs. While the upstream receptors for NFκB signal perception remain unaffected, both NFκB subunits p50 and p65 are trapped in the cytoplasm of VZV-infected DCs to avoid signaling via this pathway. Moreover, the E3 ubiquitin ligase domain of ORF61 seems to inhibit IκBα degradation in DCs, as demonstrated in a TNFα-stimulated NFκB reporter assay in HEK293FT cells [122].

Notably, also the β-herpesvirus HCMV potently blocks maturation of DCs upon an inflammatory stimulus [66, 67]. Particularly, the viral encoded IL-10 homolog (cmvIL-10; UL111a) does not only induce apoptosis of iDCs or the surface exposure of DC-SIGN to promote HCMV infection but also negatively affects DC maturation via IL-10 receptor perception [98, 108, 123]. Upon cmvIL-10 stimulation of iDCs, the IL-10 signaling pathway is induced, reflected by significant activation of STAT3, an intrinsic key factor implicated in the control of DC maturation [108, 124]. Thus, cmvIL-10 functionally resembles the human IL-10 homolog and thus dampens DC-induced antiviral immune responses.

Given the distinct regulation of surface proteins, implicated in immune activation, as well as the inhibition of DC activation, it seems reasonable to assume that different herpesviral species evolved specific and independent strategies to hijack DC biology and function to support efficient replication and favor the establishment of latency.

2.4 HSV-1 manipulates autophagy in a cell type-dependent manner

Macroautophagy (henceforth autophagy) is a conserved cellular machinery that delivers intracellular constituents, such as proteins or whole cellular organelles, to lysosomal digestion, both under homeostatic or stress-related conditions. In essence, autophagy induction, upon, e.g., starvation or stress-related stimuli, provides a source of amino acids from degraded proteins for de novo protein biosynthesis. Furthermore, autophagy is also important for antigen presentation, since it represents an additional route to process cytoplasmic and nuclear antigens, e.g., during viral infections, for MHC class II-mediated presentation. Moreover, autophagy is also involved in cross-presentation of exogenous antigens via MHC class I molecules [125–127]. Thus, autophagic degradation plays an important role during antiviral defense mechanisms in infected cells. Apart from classical autophagy, a process called xenophagy is characterized by the specific autophagic sequestration of foreign pathogen-derived contents, such as whole viral particles, to limit viral replication [128–130]. However, this chapter focuses on classical autophagy as well as its modulation during herpesviral infections of DCs.

Mechanistically, mammalian autophagy involves the coordinated interplay of different autophagy-related proteins (ATG) [131, 132]. During initiation of the phagophore, i.e., the initial autophagosome membrane, a complex containing UNC-51-like kinase 1 (ULK1), focal adhesion kinase family interacting protein of 200 kDa (FIP200), ATG13, and ATG101 is formed. Subsequently, phagophore nucleation involves the activation of the PI3KC3 complex I, which among others includes Beclin 1 or class III phosphatidylinositol 3-kinase (PI3K), and a ubiquitin-like conjugation system consisting of different ATG proteins. The expansion of the autophagophore is among others characterized by lipidation of microtubule-associated protein light chain 3 (LC3)-I. In particular, the attachment of phosphatidyl-ethanolamine (PE) to LC3B-I generates LC3B-II, which is inserted into the nascent autophagosomal membrane. Thus, the LC3B-I to LC3B-II conversion indicates autophagy induction, based on the increased presence of mature autophagosomes. The final steps are the fusion of mature autophagosomes with lysosomes, and the subsequent degradation of the resulting autophagolysosomes including their cargo, by, e.g., hydrolysis [125, 133].

It is well established that autophagy is triggered in various cell types upon a herpesviral infection [134]. As a viral countermeasure, HSV-1 evolved strategies to manipulate autophagy, however, in a cell type- and infection stage-dependent way. As such, HSV-1 induces autophagy very early upon infection [135], whereas the viral encoded protein ICP34.5, classified as leaky late gene product, subsequently suppresses autophagy via targeting Beclin 1 or dephosphorylating eIF2α, while US11, a late gene, inhibits protein kinase R (PKR) to block eIF2α phosphorylation, in, e.g., fibroblasts or neurons [136–140]. In this regard, PKR and eIF2α are two key factors, in, e.g., fibroblasts, that participate during the induction of autophagy upon an HSV-1 infection [141, 142].

Apart from this, the interplay between HSV-1 and autophagy in myeloid antigen-presenting cells, such as DCs, underlies a different regulation. Very interestingly, induction of autophagy in HSV-1-infected murine DCs follows a PKR/eIF2α-independent mechanism, which is not counteracted by ICP34.5. Particularly, in infected murine BMDCs HSV-1, genomic DNA is sensed via a STING-dependent pathway, but independent of viral replication and leads to the transient induction of autophagy [143]. Beyond this, in the context of an HSV-1 infection of murine BMDCs or a related cell line, i.e., DC2.4, ICP34.5 does not block autophagy induction, but rather blocks the maturation of autophagosomes, and in turn autophagic flux. This ICP34.5-dependent mechanism suppresses the autophagy-dependent processing of viral antigens for presentation via the MHC class I as well as class II pathway [144, 145]. More precisely, HSV-1 ICP34.5-encoded Beclin 1-binding domain is responsible for the aberrant autophagosome maturation and thus subversion of $CD4^+$ T cell stimulation in murine DCs [145].

2.5 HSV-1 exploits cellular autophagy in infected human monocyte-derived iDCs

Noteworthy, one interesting example of how herpesviruses hijack cellular autophagy to promote viral replication, i.e., nuclear egress, comes from the interplay of HSV-1 with human monocyte-derived DCs [146]. In general, after generation of nuclear progeny capsids, these viral structures have to cross the nuclear membrane [147]. However, during nuclear egress, the nuclear lamina, which is a dense meshwork inside the nucleus, represents the main barrier for nucleocapsids to get access to the inner nuclear membrane. Lamins and other membrane-associated proteins are the main constituents of the nuclear lamina. Lamins are a group of type V intermediate filament proteins and are grouped into types A, B, and C. While lamin B connects the nuclear lamina with the inner nuclear membrane, lamin A/C—products of alternative splicing—supports the stiffness of the nuclear envelope [148]. In proliferating cells, such as fibroblasts, the nuclear lamina undergoes reversible disassembly during mitosis or during the nuclear export of large messenger ribonucleoprotein (mRNP) complexes. Mechanistically, lamina disassembly is initiated by site-specific phosphorylation of lamin A/C [149–151]. During co-evolution, HSV-1 has evolved a nuclear egress complex (NEC), including viral protein kinases as well as cellular effectors, such as Pin1, to mediate a similar phosphorylation-triggered destabilization of the nuclear lamina and budding of the capsid at the nuclear envelope, a process reminiscent of the nuclear export of large mRNPs [150, 152]. This process is triggered in permissive proliferating cells, such as HFF or Vero cells, in which an HSV-1 infection results in the release of considerable amounts of HSV-1 virions [153].

In sharp contrast, human monocyte-derived mDCs only barely release infectious progeny virus into the supernatant [154], despite the efficient release of significant amounts of noninfectious light (L-) particles void of the capsid [155], further discussed in Section 3.3. By contrast, iDCs promote complete replication of HSV-1, with the final release of infectious heavy- (H-) particles into the supernatant. Notably, the nuclear egress of HSV-1 capsids in iDCs is facilitated by autophagy-dependent lamin degradation [146] and is thus fundamentally different from other cell types, such as fibroblasts. Furthermore, the loss of lamin protein expression is dependent on viral replication, but independent from viral-encoded vhs in infected iDCs. By contrast, autophagic degradation in mDCs is hampered based on an intrinsic inhibition. In essence, elevated kinesin family member 1 B (KIF1B) and KIF2A expression levels block the fusion of autophagosomes with lysosomes, which is an essential step during autophagic degradation, and thus inhibit the nuclear egress of viral capsids due to stable lamin expression. Apart from this, HSV-1 ICP34.5 is not involved in the differential regulation of autophagic turnover in human monocyte-derived iDCs versus mDCs and does not interfere with lamin degradation [146].

3. Herpesviruses and mature DCs

Upon DC activation, these vital immune cells undergo a phenotypic and functional switch, and thus become equipped with several functionally important molecules. Among others, molecules for (i) antigen presentation, i.e., MHC class I and II molecules, (ii) co-stimulation as well as modulation of T cell stimulation, e.g., CD40, CD80, CD83, CD86, or (iii) adhesion and migration, e.g., CCR7, CXCR4, and cytohesin-1 interaction partner (CYTIP), are highly expressed by mDCs [14, 44, 156]. However, herpesviruses aim to avoid potent induction of adaptive antiviral immune responses and thus aim to alter the expression of several of these proteins.

In essence, HSV interfere with CD83, CCR7, CXCR4, and CYTIP protein expression in mDCs [112, 157–159]. By contrast, CD40, CD80, CD86, and MHC class I as well as class II expression is mostly unaltered upon an HSV infection of mDCs [112, 154, 157]. Noteworthy, HSV-1-mediated MHC class I as well as class II evasion, however, occurs in distinct cell types others than APCs; however, this will not be further discussed within this chapter [160–163].

Regarding VZV, infected mDCs show decreased levels of CD80, CD83, CD86, and MHC class I surface expression, whereas the abundance of MHC class II surface molecules is not affected [164]. Among the β-*Herpesvirinae*, HCMV strongly hampers CD40, CD83, and MHC class I surface expression on mDCs, while only slightly affecting, if at all, CD80, CD86, or MHC class II expression, or leaving other important molecules, such as CCR7, unaffected [12, 66, 165, 166].

In the following sections, we will highlight the functional consequences of a herpesviral infection of mDC regarding the CD83 protein expression, mDC adhesion as well as migration.

3.1 Modulation of CD83 expression in mDCs

Intensive research has proven the vital role of the glycoprotein CD83 during the development of the mammalian immune system as well as during the priming and controlling of immune responses. In this regard, several *in vitro* and *in vivo* studies revealed the immune-modulatory potential of the two known CD83 isoforms, i.e., the membrane-bound and soluble CD83 (sCD83). Particularly, the membrane-bound form of CD83 is pivotal for the thymic $CD4^+$ T lymphocyte selection, via stabilizing MHC class II surface expression on thymic epithelial cells, and essential to suppress overshooting immune responses during the development or resolution of autoimmune disorders [42, 167–169]. Apart from this, sCD83 possesses an interesting therapeutic potential in order to prevent/resolve autoimmune disorders and to inhibit transplant rejection, which is mediated via the induction of regulatory mechanisms including indoleamine 2,3-dioxygenase (IDO)-induced regulatory T lymphocytes [170–173]. Considering this, it is not surprising that herpesviruses target CD83 to combat the induction of an antiviral immune response.

One well-known example regarding the modulation of CD83 protein expression in infected mDCs is mediated by the α-herpesvirus HSV-1. Upon infection of mDCs, HSV-1 inhibits both the expression of cell membrane-bound and intracellular CD83 protein [157, 174]. In particular, the HSV-1 encoded immediate-early expressed infected cell protein 0 (ICP0) triggers the proteasome-dependent, but ubiquitin-independent degradation of CD83 in mDCs. The same is true using a HEK293T co-transfection model. Thus, ICP0 cell type-independently mediates CD83 degradation without the need of any additional viral factor [174].

Notably, an HSV-1 infection does not only hamper CD83 expression on directly infected mDCs but also on their uninfected bystander counterparts. This bystander effect is due to the release of uninfectious light (L-) particles which are void of the capsid and thus the viral genome (discussed further in Section 3.3), but contain viral proteins, including ICP0, to modulate the function of adjacent uninfected cells [155, 175]. Noteworthy, apart from the degradation of CD83 in HSV-1-infected mDCs, the infection does not provoke the release of sCD83 molecules, excluding the involvement of CD83 shedding [157]. However, the precise molecular mechanism, how HSV-1 ICP0 triggers CD83 degradation, is still under investigation.

The obvious importance of CD83 during the induction of an antiviral immune response becomes even more evident from the fact that also other α-herpesviruses, i.e., HSV-2 and varicella-zoster virus (VZV) mediate a strong reduction of CD83 protein expression in infected mDCs [112, 164]. Regarding VZV, CD83 is trapped inside discrete cytoplasmic compartments and fails to get transported to the cell surface in infected mDCs [164]. Thus, VZV shapes the surface molecule repertoire of infected mDCs to efficiently spread inside the host and to avoid proper T cell activation [164, 176]. Also, HSV-2 strongly inhibits CD83 surface expression upon infection of mDCs via a proteasome-dependent degradation of CD83, reminiscent of its family member HSV-1 [112]. The nature of VZV- and HSV-2-triggered CD83 modulation is currently unclear and requires further investigations.

Strikingly, apart from CD83 degradation upon human α-herpesvirus infections of mDCs, also the β-herpesvirus human cytomegalovirus (HCMV) significantly hampers CD83 protein expression by mDCs [165, 166]. In this respect, the HCMV-mediated reduction of CD83 expression by mDCs is dependent on the major immediate early protein 2 (IE2)-triggered proteasomal degradation and closely resembles the HSV-1 ICP0-dependent degradation of CD83 [157, 165, 174]. Contrasting findings were reported regarding the sCD83 levels upon an HCMV infection of mDCs. While there is evidence for increased levels of sCD83 in the supernatants of HCMV Bob-U/Bob-B-infected mDC cultures [166], concomitant with an impaired T lymphocyte-stimulatory capacity of these mDCs [166], mDC infection with HCMV TB40E does not increase the release of sCD83 into the supernatant [165].

Since different herpesviral members have independently evolved mechanisms to suppress CD83 expression by mDCs, it is reasonable to assume that CD83 possess a vital role in controlling (persistent) viral infections based on its inherent modulatory role during T lymphocyte activation.

3.2 Herpesviruses differentially modulate the migratory capacity of mDCs

During the initiation of an adaptive antiviral immune response, APCs, such as DCs, must present their acquired antigens to cognate T lymphocytes. To do so, DCs undergo a maturation process and are chemotactically guided toward T lymphocyte-rich areas inside lymph nodes. As a prerequisite for directed migration, mDCs loosen their adhesive forces and express specific chemokine receptors, i.e., CCR7 and CXCR4, which perceive the lymphoid-expressed chemokines

CCL19/CCL21 and CXCL12, respectively [32, 45, 46, 177]. These chemotactic cues, among migration-promoting signals, trigger intracellular signal transduction pathways for cell polarization [178].

Importantly, leukocytes, and especially mDCs, possess a fundamentally different regulation of their three-dimensional migration mode compared to other cell types. In general, while their two-dimensional migration is dependent on adhesive contacts mediated by integrins, their three-dimensional migration follows the "amoeboid" adhesion-independent paradigm [179–183]. Thus, mDCs can switch between their dependence on integrin-mediated adhesive contacts with specific integrin ligands versus the rapid migration along chemokine gradients without the need of preformed integrin ligand tracks. Considering this, integrin-mediated adhesive contacts require a very tight regulation to avoid aberrant adhesion, and thus immobilization, of mDCs in two- as well as three-dimensional environments [182, 184, 185].

Noteworthy, DCs are vital for efficient priming of adaptive anti-HSV-1 immune responses [186, 187]. Thus, HSV-1 evolved sophisticated strategies to hamper the migration of directly infected DCs toward draining lymph nodes, in order to hamper antigen presentation by these cells and delay the immune response. In particular, HSV-1 can efficiently infect DCs not only *in vitro* but also *in vivo* which is mirrored by the presence of HSV-1-infected DCs in primary skin lesions. However, these infected DCs do not migrate to the draining lymph nodes. By contrast, uninfected bystander skin-derived DCs acquire viral antigens and transport them to lymph node resident cells [188–192].

Based on several *in vitro* studies using human monocyte-derived mDCs, HSV-mediated mechanisms have been discovered that aim to suppress mDC migration. Concerning this, HSV-1- or HSV-2-infected mDCs reveal a rapid and very strong inhibition of their migratory capacity toward CCL19 and CXCL12 chemokine gradients, in transwell assays as well as in three-dimensional collagen matrices [112, 158, 159]. One important countermeasure of HSV-1 and also HSV-2 to hamper mDC migration is the downregulation of CCR7 surface expression [112, 158]. Considering that CCR7 orchestrates cell migration along CCL19 chemokine gradients, essential for directed migration toward T lymphocyte-rich zones in draining lymph nodes, downmodulation of this receptor constitutes an important strategy to subvert chemokine-mediated DC migration to draining lymph nodes [47, 48]. Apart from this, HSV-1 additionally hampers CXCR4 protein expression levels on mDCs, to inhibit the perception of CXCL12, a chemokine also expressed in lymphoid organs or the bone marrow [193].

However, since the inhibition of chemokine receptor expression is timely delayed in respect to the inhibited migration of HSV-infected mDCs, an additional mechanism has been suggested. Indeed, HSV-1 and HSV-2 additionally induce the adhesion of infected mDCs via amplifying the activity of β2 integrins, especially lymphocyte function-associated antigen 1 (LFA-1), despite unaffected expression levels of the respective integrin subunits [112, 159]. Compared to other integrin families, β2 integrins are the predominant integrin family expressed on leukocytes and thus possess an exceptional role in regulating mDC adhesion [194–196]. Furthermore, integrin activity and thus the ligand binding status rely on bi-directional regulatory mechanisms, i.e., inside-out and outside-in signaling events [197, 198].

In this section, we will focus on the regulation of β2 integrin activity via inside-out signaling in mDCs. This includes the direct intracellular binding of either of two specific proteins, i.e., talin or cytohesin-1, to the CD18 chain, which is common to all β2 integrins [196, 199–201]. In contrast to talin, cytohesin-1 specifically regulates β2 integrin activity and, upon CD18 binding, promotes the conformational switch into its ligand-binding high-affinity state, thus mediating cell adhesion [196, 202, 203]. To avoid an overshoot in β2 integrin activity, the CYTIP abrogates the cytohesin-1-mediated β2 integrin activation by regulating its intracellular localization [204]. This becomes evident from an siRNA-mediated approach in which CYTIP expression was ablated in mDCs, which causes the induction of adhesion and inhibition of migration [156, 159]. Considering this inverse regulation of β2 integrin activity by cytohesin-1 and CYTIP, HSV-1 and HSV-2 have evolved an elaborate strategy to potently enhance β2 integrin activity via mediating the rapid proteasome- and ubiquitin-dependent degradation of CYTIP. Functionally, this leads to increased adhesion with subsequently inhibited DC migration and thus very likely to an impaired antiviral T lymphocyte stimulation [112, 159].

Interestingly, and in contrast to the observations for HSV-infected mDCs, VZV, another member among α-*Herpesvirinae*, does not interfere with mDC migration, but hijacks mDCs to successfully disseminate inside the host and to hide from immune recognition. By using mDCs as trojan horses, VZV facilitates its access into lymphoid organs for subsequent infection of T lymphocytes, which are strongly modulated and finally used as ferries for further viral spread, which ultimately facilitates the establishment of latency [70, 176, 205–207].

Beyond this, also the β-herpesvirus HCMV differentially shapes the migration capacity of mDCs. In this respect, HCMV-infected mDCs are inhibited in their CCL19- but not CXCL12-dependent migratory capacity, despite unaltered surface expression of CCR7 but transient induction of CXCR4 surface expression levels [12]. Hence, apart from solely inhibiting the upregulation of CCR7 expression during maturation, as observed for infected iDCs [67], HCMV reduces mDC migration via a distinct mechanism, beyond CCR7-targeting. Accordingly, also HCMV triggers an increased mDC adhesion via the induction of β2 integrin activity, which is mechanistically mediated by the proteasomal degradation of CYTIP, reminiscent of the scenario observed for HSV-infected mDCs [12, 112].

Regarding the upregulation of CXCR4 surface expression, it is known that HCMV encodes a variety of chemokines and chemokine receptor homologs [9, 95], while four of them differentially modulate CXCR4 signaling, i.e., pUS27, pUS28, pUL33, and pUL78. It is likely that HCMV-encoded CXCR4 chemokine receptor homologs regulate CXCR4 expression as well as its signaling axis in a cell type-dependent manner [12, 208, 209]. Since the cognate chemokine CXCL12 is not only expressed in the lymph node but also abundantly produced by osteoblasts in the bone marrow [193], HCMV appears to shape migration of infected mDCs to an ecological niche, the bone marrow, which is highly populated with potential target cells for the establishment of latency [210]. Furthermore, since HCMV can reactivate during differentiation/maturation into mDCs [211], HCMV-positive monocytes as well as mDCs might constitute important vehicles for viral dissemination *in vivo* [66, 212].

Given the differential regulation of mDC migration by distinct herpesviruses, it highlights the importance to hijack this vital function during infection in order to establish latency or to delay the antiviral immune response.

3.3 Abortive replication of HSV-1 in mDCs accompanied by the generation of noninfectious light (L-) particles

As mentioned earlier, HSV-1-infected iDCs release considerable amounts of infectious progeny virus into the culture supernatants [60, 146], whereas HSV-1-infected mDCs do not promote productive replication of HSV-1 [146, 154]. Apart from the successful initiation of the viral tripartite gene expression cascade both in iDCs and mDCs [112, 213], these apparent contrary findings—concerning the replication outcome—can be explained by the inhibition of HSV-1 nuclear egress in mDCs only [146]. This is based on the scarce autophagic degradation of nuclear lamins in mDCs, which facilitates the nuclear egress of HSV-1 and thus the generation of infectious virions in iDCs.

However, and noteworthy, HSV-1-infected mDCs release significant amounts of light (L-) particles which are void of the capsid and thus noninfectious [155]. Beyond this, L-particles contain virion-associated tegument constituents and the glycoprotein-scattered envelope. Furthermore, these virion-like structures are suggested to share similar maturation steps and might hijack the same cellular entry receptors for attachment and fusion as their infectious counterparts do [175, 214–216]. It has indeed been shown that L-particles can efficiently deliver their viral content toward bystander target cells. Thus, L-particles might foster HSV-1 infectivity, via, e.g., shaping surrounding cells to increase their permissiveness, complementing functional defective virions, or modulating the cellular micro-environment for immune evasion [175, 214, 217, 218]. Regarding the latter, L-particles derived from HSV-1-infected BHK21 cells as well as mDCs are able to decrease CD83 surface expression on mDCs and therefore modulate uninfected adjacent cells during an infection in benefit of the virus. Based on the fact that a whole variety of viral proteins is incorporated into L-particles, such as ICP0 which is sufficient to mediate CD83 downregulation in mDCs, these viral structures transmit important viral components to uninfected bystander cells and modulate their functions [155].

Beyond this, the generation of noninfectious particles is not unique to HSV-1 but also observed for other herpesviruses, such as HSV-2, HCMV ("dense bodies"), or VZV, and also for other distinct viruses, such as hepatitis B virus ("subviral particles"), when infecting different host cell types [214, 219–223]. Thus, herpesvirus-derived noninfectious particles possess one or more important functions to modulate antiviral immune responses and thereby foster viral replication and spread.

4. Conclusions and implications of herpesviral-mediated modulations of DC biology for T lymphocyte activation

The entire spectrum of herpesviral-mediated modulations of DC biology and function aims to delay/hamper the proper activation of T lymphocytes, which would otherwise generate a potent antiviral immune response to eliminate the virus and avoid the establishment of latency. In general, the virally induced inhibition of iDC maturation, suppression of surface expression of co-stimulatory or antigen-presentation molecules on mDCs, interference with cytokine release, or the inhibition of DC migration constitute efficient immune evasion strategies [224, 225]. On the other hand, the host counteracts these strategies and mounts an adaptive antiviral immune response, mirrored by the generation of antigen-specific $CD4^+$ and $CD8^+$ T lymphocytes as well as the production of antibodies [3, 225–227]. Regarding HSV and HCMV infections, DC-dependent cross-presentation of viral antigens represents, among others, a crucial way to induce cytotoxic T lymphocyte (CTL) responses [228–231]. Moreover, there is strong evidence that bystander migratory submucosal and lymph node-resident DCs sequentially (cross-)present HSV-derived antigens in the lymph node. By contrast, directly infected DCs are most likely not involved in the activation of $CD4^+$ and $CD8^+$ T lymphocytes [188, 189, 191, 231–234].

In summary, it is obvious that herpesviruses, including HSV-1, HSV-2, HCMV, and VZV manipulate the function of infected DCs, which are the most potent APCs, for immune evasion and subversion of antiviral immune responses [3, 176, 186].

Acknowledgements

This work was supported by the German Research Foundation (Deutsche Forschungsgemeinschaft, DFG), grant number STE 432/11-1, awarded to AS.

Conflict of interest

The authors declare no conflict of interest.

Author details

Linda Popella* and Alexander Steinkasserer*
Department of Immune Modulation, Universitätsklinikum Erlangen, Erlangen, Germany

*Address all correspondence to: linda.popella@uk-erlangen.de
and alexander.steinkasserer@uk-erlangen.de

References

[1] Schleiss MR. Persistent and recurring viral infections: The human herpesviruses. Current Problems in Pediatric and Adolescent Health Care. 2009;**39**(1):7-23

[2] Griffin BD, Verweij MC, Wiertz EJ. Herpesviruses and immunity: The art of evasion. Veterinary Microbiology. 2010;**143**(1):89-100

[3] Smith C, Khanna R. Immune regulation of human herpesviruses and its implications for human transplantation. American Journal of Transplantation: Official Journal of the American Society of Transplantation and the American Society of Transplant Surgeons. 2013;**13**(Suppl 3):9-23; quiz

[4] Su C, Zhan G, Zheng C. Evasion of host antiviral innate immunity by HSV-1, an update. Virology Journal. 2016;**13**(1):38

[5] Tognarelli EI, Palomino TF, Corrales N, Bueno SM, Kalergis AM, Gonzalez PA. Herpes simplex virus evasion of early host antiviral responses. Frontiers in Cellular and Infection Microbiology. 2019;**9**:127

[6] Gewurz BE, Vyas JM, Ploegh HL. Herpesvirus evasion of T-cell immunity. In: Arvin A, Campadelli-Fiume G, Mocarski E, editors. Human Herpesviruses: Biology, Therapy, and Immunoprophylaxis. Cambridge: Cambridge University Press; 2007

[7] Whitley RJ, Roizman B. Herpes simplex virus infections. Lancet (London, England). 2001;**357**(9267):1513-1518

[8] Sinclair J, Sissons P. Latency and reactivation of human cytomegalovirus. The Journal of General Virology. 2006;**87**(Pt 7):1763-1779

[9] Alcami A, Koszinowski UH. Viral mechanisms of immune evasion. Trends in Microbiology. 2000;**8**(9):410-418

[10] Tortorella D, Gewurz BE, Furman MH, Schust DJ, Ploegh HL. Viral subversion of the immune system. Annual Review of Immunology. 2000;**18**(1):861-926

[11] Kobelt D, Lechmann M, Steinkasserer A. The interaction between dendritic cells and herpes simplex virus-1. In: Steinkasserer A, editor. Dendritic Cells and Virus Infection. Berlin, Heidelberg: Springer Berlin Heidelberg; 2003. pp. 145-161

[12] Grosche L, Kummer M, Steinkasserer A. What goes around, comes around - HSV-1 replication in monocyte-derived dendritic cells. Frontiers in Microbiology. 2017;**8**:2149

[13] Banchereau J, Briere F, Caux C, Davoust J, Lebecque S, Liu YJ, et al. Immunobiology of dendritic cells. Annual Review of Immunology. 2000;**18**:767-811

[14] Banchereau J, Steinman RM. Dendritic cells and the control of immunity. Nature. 1998;**392**(6673):245-252

[15] Steinman RM. The dendritic cell system and its role in immunogenicity. Annual Review of Immunology. 1991;**9**:271-296

[16] Lanzavecchia A, Sallusto F. The instructive role of dendritic cells on T cell responses: Lineages, plasticity and kinetics. Current Opinion in Immunology. 2001;**13**(3):291-298

[17] Haniffa M, Collin M, Ginhoux F. Ontogeny and functional specialization of dendritic cells in human and mouse. Advances in Immunology. 2013;**120**:1-49

[18] Shortman K, Liu YJ. Mouse and human dendritic cell subtypes. Nature Reviews. Immunology. 2002;**2**(3):151-161

[19] Ueno H, Schmitt N, Klechevsky E, Pedroza-Gonzalez A, Matsui T, Zurawski G, et al. Harnessing human dendritic cell subsets for medicine. Immunological Reviews. 2010;**234**(1):199-212

[20] Villadangos JA, Schnorrer P. Intrinsic and cooperative antigen-presenting functions of dendritic-cell subsets in vivo. Nature Reviews. Immunology. 2007;7(7):543-555

[21] Ziegler-Heitbrock L, Ancuta P, Crowe S, Dalod M, Grau V, Hart DN, et al. Nomenclature of monocytes and dendritic cells in blood. Blood. 2010;**116**(16):e74-e80

[22] Hart DN. Dendritic cells: Unique leukocyte populations which control the primary immune response. Blood. 1997;**90**(9):3245-3287

[23] Lipscomb MF, Masten BJ. Dendritic cells: Immune regulators in health and disease. Physiological Reviews. 2002;**82**(1):97-130

[24] Worbs T, Hammerschmidt SI, Förster R. Dendritic cell migration in health and disease. Nature Reviews. Immunology. 2017;**17**(1):30-48

[25] Collin M, Bigley V. Human dendritic cell subsets: An update. Immunology. 2018;**154**(1):3-20

[26] McKenna K, Beignon AS, Bhardwaj N. Plasmacytoid dendritic cells: Linking innate and adaptive immunity. Journal of Virology. 2005;**79**(1):17-27

[27] Merad M, Sathe P, Helft J, Miller J, Mortha A. The dendritic cell lineage: Ontogeny and function of dendritic cells and their subsets in the steady state and the inflamed setting. Annual Review of Immunology. 2013;**31**:563-604

[28] Cella M, Jarrossay D, Facchetti F, Alebardi O, Nakajima H, Lanzavecchia A, et al. Plasmacytoid monocytes migrate to inflamed lymph nodes and produce large amounts of type I interferon. Nature Medicine. 1999;**5**(8):919-923

[29] Colonna M, Trinchieri G, Liu YJ. Plasmacytoid dendritic cells in immunity. Nature Immunology. 2004;**5**(12):1219-1226

[30] Palucka K, Banchereau J. Cancer immunotherapy via dendritic cells. Nature Reviews. Cancer. 2012;**12**(4):265-277

[31] Sichien D, Lambrecht BN, Guilliams M, Scott CL. Development of conventional dendritic cells: From common bone marrow progenitors to multiple subsets in peripheral tissues. Mucosal Immunology. 2017;**10**(4):831-844

[32] Villadangos JA, Heath WR. Life cycle, migration and antigen presenting functions of spleen and lymph node dendritic cells: Limitations of the Langerhans cells paradigm. Seminars in Immunology. 2005;**17**(4):262-272

[33] Steinman RM. Dendritic cells: Understanding immunogenicity. European Journal of Immunology. 2007;**37**(S1):S53-S60

[34] Blanco P, Palucka AK, Pascual V, Banchereau J. Dendritic cells and cytokines in human inflammatory and autoimmune diseases. Cytokine & Growth Factor Reviews. 2008;**19**(1):41-52

[35] Dudek AM, Martin S, Garg AD, Agostinis P. Immature, semi-mature, and fully mature dendritic cells:

Toward a DC-cancer cells Interface that augments anticancer immunity. Frontiers in Immunology. 2013;**4**:438

[36] Kaka AS, Foster AE, Weiss HL, Rooney CM, Leen AM. Using dendritic cell maturation and IL-12 producing capacity as markers of function: A cautionary tale. Journal of Immunotherapy (Hagerstown, Md.: 1997). 2008;**31**(4):359-369

[37] Kany S, Vollrath JT, Relja B. Cytokines in inflammatory disease. International Journal of Molecular Sciences. 2019;**20**(23):6008

[38] Inaba K, Turley S, Iyoda T, Yamaide F, Shimoyama S, e Sousa CR, et al. The formation of immunogenic major histocompatibility complex class II–peptide ligands in lysosomal compartments of dendritic cells is regulated by inflammatory stimuli. The Journal of Experimental Medicine. 2000;**191**(6):927-936

[39] Ackerman AL, Cresswell P. Regulation of MHC class I transport in human dendritic cells and the dendritic-like cell line KG-1. Journal of Immunology (Baltimore, Md.: 1950). 2003;**170**(8):4178-4188

[40] Berchtold S, Jones T, Mühl-Zürbes P, Sheer D, Schuler G, Steinkasserer A. The human dendritic cell marker CD83 maps to chromosome 6p23. Annals of Human Genetics. 1999;**63**(2):181-183

[41] Lechmann M, Berchtold S, Steinkasserer A, Hauber J. CD83 on dendritic cells: More than just a marker for maturation. Trends in Immunology. 2002;**23**(6):273-275

[42] Prechtel AT, Steinkasserer A. CD83: An update on functions and prospects of the maturation marker of dendritic cells. Archives of Dermatological Research. 2007;**299**(2):59-69

[43] Zhou LJ, Schwarting R, Smith HM, Tedder TF. A novel cell-surface molecule expressed by human interdigitating reticulum cells, Langerhans cells, and activated lymphocytes is a new member of the Ig superfamily. Journal of Immunology (Baltimore, Md.: 1950). 1992;**149**(2):735-742

[44] Zhou LJ, Tedder TF. Human blood dendritic cells selectively express CD83, a member of the immunoglobulin superfamily. Journal of Immunology (Baltimore, Md.: 1950). 1995;**154**(8):3821-3835

[45] Christensen JE, Thomsen AR. Co-ordinating innate and adaptive immunity to viral infection: Mobility is the key. APMIS (Acta Pathologica, Microbiologica et Immunologica Scandinavica). 2009;**117**(5-6):338-355

[46] Sozzani S, Allavena P, D'Amico G, Luini W, Bianchi G, Kataura M, et al. Differential regulation of chemokine receptors during dendritic cell maturation: A model for their trafficking properties. Journal of Immunology (Baltimore, Md.: 1950). 1998;**161**(3):1083-1086

[47] Förster R, Schubel A, Breitfeld D, Kremmer E, Renner-Müller I, Wolf E, et al. CCR7 coordinates the primary immune response by establishing functional microenvironments in secondary lymphoid organs. Cell. 1999;**99**(1):23-33

[48] Ohl L, Mohaupt M, Czeloth N, Hintzen G, Kiafard Z, Zwirner J, et al. CCR7 governs skin dendritic cell migration under inflammatory and steady-state conditions. Immunity. 2004;**21**(2):279-288

[49] Randolph GJ, Angeli V, Swartz MA. Dendritic-cell trafficking to lymph nodes through lymphatic vessels. Nature Reviews. Immunology. 2005;**5**(8):617-628

[50] Weber M, Hauschild R, Schwarz J, Moussion C, de Vries I, Legler DF, et al. Interstitial dendritic cell guidance by haptotactic chemokine gradients. Science. 2013;**339**(6117):328-332

[51] Bender A, Sapp M, Schuler G, Steinman RM, Bhardwaj N. Improved methods for the generation of dendritic cells from nonproliferating progenitors in human blood. Journal of Immunological Methods. 1996;**196**(2):121-135

[52] Caux C, Vanbervliet B, Massacrier C, Dezutter-Dambuyant C, de Saint-Vis B, Jacquet C, et al. CD34+ hematopoietic progenitors from human cord blood differentiate along two independent dendritic cell pathways in response to GM-CSF+TNF alpha. The Journal of Experimental Medicine. 1996;**184**(2):695-706

[53] Chapuis F, Rosenzwajg M, Yagello M, Ekman M, Biberfeld P, Gluckman JC. Differentiation of human dendritic cells from monocytes in vitro. European Journal of Immunology. 1997;**27**(2):431-441

[54] Freudenthal PS, Steinman RM. The distinct surface of human blood dendritic cells, as observed after an improved isolation method. Proceedings of the National Academy of Sciences of the United States of America. 1990;**87**(19):7698-7702

[55] Lutz MB, Strobl H, Schuler G, Romani N. GM-CSF monocyte-derived cells and langerhans cells as part of the dendritic cell family. Frontiers in Immunology. 2017;**8**(1388)

[56] Sallusto F, Lanzavecchia A. Efficient presentation of soluble antigen by cultured human dendritic cells is maintained by granulocyte/macrophage colony-stimulating factor plus interleukin 4 and downregulated by tumor necrosis factor alpha. The Journal of Experimental Medicine. 1994;**179**(4):1109-1118

[57] Gerna G, Percivalle E, Lilleri D, Lozza L, Fornara C, Hahn G, et al. Dendritic-cell infection by human cytomegalovirus is restricted to strains carrying functional UL131-128 genes and mediates efficient viral antigen presentation to CD8+ T cells. The Journal of General Virology. 2005;**86** (Pt 2):275-284

[58] Gerna G, Percivalle E, Sarasini A, Baldanti F, Campanini G, Revello MG. Rescue of human cytomegalovirus strain AD169 tropism for both leukocytes and human endothelial cells. The Journal of General Virology. 2003;**84**(Pt 6):1431-1436

[59] Gerna G, Revello MG, Baldanti F, Percivalle E, Lilleri D. The pentameric complex of human cytomegalovirus: Cell tropism, virus dissemination, immune response and vaccine development. The Journal of General Virology. 2017;**98**(9):2215-2234

[60] Mikloska Z, Bosnjak L, Cunningham AL. Immature monocyte-derived dendritic cells are productively infected with herpes simplex virus type 1. Journal of Virology. 2001;**75**(13):5958-5964

[61] Forrester A, Farrell H, Wilkinson G, Kaye J, Davis-Poynter N, Minson T. Construction and properties of a mutant of herpes simplex virus type 1 with glycoprotein H coding sequences deleted. Journal of Virology. 1992;**66**(1):341-348

[62] Salio M, Cella M, Suter M, Lanzavecchia A. Inhibition of dendritic cell maturation by herpes simplex virus. European Journal of Immunology. 1999;**29**(10):3245-3253

[63] Pollara G, Speidel K, Samady L, Rajpopat M, McGrath Y, Ledermann J,

et al. Herpes simplex virus infection of dendritic cells: Balance among activation, inhibition, and immunity. The Journal of Infectious Diseases. 2003;**187**(2):165-178

[64] Peretti S, Shaw A, Blanchard J, Bohm R, Morrow G, Lifson JD, et al. Immunomodulatory effects of HSV-2 infection on immature macaque dendritic cells modify innate and adaptive responses. Blood. 2005;**106**(4):1305-1313

[65] Jones CA, Fernandez M, Herc K, Bosnjak L, Miranda-Saksena M, Boadle RA, et al. Herpes simplex virus type 2 induces rapid cell death and functional impairment of murine dendritic cells in vitro. Journal of Virology. 2003;**77**(20):11139-11149

[66] Beck K, Meyer-Konig U, Weidmann M, Nern C, Hufert FT. Human cytomegalovirus impairs dendritic cell function: A novel mechanism of human cytomegalovirus immune escape. European Journal of Immunology. 2003;**33**(6):1528-1538

[67] Moutaftsi M, Mehl AM, Borysiewicz LK, Tabi Z. Human cytomegalovirus inhibits maturation and impairs function of monocyte-derived dendritic cells. Blood. 2002;**99**(8):2913-2921

[68] Varani S, Frascaroli G, Homman-Loudiyi M, Feld S, Landini MP, Soderberg-Naucler C. Human cytomegalovirus inhibits the migration of immature dendritic cells by down-regulating cell-surface CCR1 and CCR5. Journal of Leukocyte Biology. 2005;**77**(2):219-228

[69] Wagner CS, Walther-Jallow L, Buentke E, Ljunggren H-G, Achour A, Chambers BJ. Human cytomegalovirus-derived protein UL18 alters the phenotype and function of monocyte-derived dendritic cells. Journal of Leukocyte Biology. 2008;**83**(1):56-63

[70] Abendroth A, Morrow G, Cunningham AL, Slobedman B. Varicella-zoster virus infection of human dendritic cells and transmission to T cells: Implications for virus dissemination in the host. Journal of Virology. 2001;**75**(13):6183-6192

[71] Cotter CR, Kim WK, Nguyen ML, Yount JS, López CB, Blaho JA, et al. The virion host shutoff protein of herpes simplex virus 1 blocks the replication-independent activation of NF-κB in dendritic cells in the absence of type I interferon signaling. Journal of Virology. 2011;**85**(23):12662-12672

[72] Cotter CR, Nguyen ML, Yount JS, López CB, Blaho JA, Moran TM. The virion host shut-off (vhs) protein blocks a TLR-independent pathway of herpes simplex virus type 1 recognition in human and mouse dendritic cells. PLoS One. 2010;**5**(2):e8684

[73] Jin H, Yan Z, Ma Y, Cao Y, He B. A herpesvirus virulence factor inhibits dendritic cell maturation through protein phosphatase 1 and IκB kinase. Journal of Virology. 2011;**85**(7):3397-3407

[74] Sato A, Linehan MM, Iwasaki A. Dual recognition of herpes simplex viruses by TLR2 and TLR9 in dendritic cells. Proceedings of the National Academy of Sciences of the United States of America. 2006;**103**(46):17343-17348

[75] Cella M, Scheidegger D, Palmer-Lehmann K, Lane P, Lanzavecchia A, Alber G. Ligation of CD40 on dendritic cells triggers production of high levels of interleukin-12 and enhances T cell stimulatory capacity: T-T help via APC activation. The Journal of Experimental Medicine. 1996;**184**(2):747-752

[76] Snijders A, Kalinski P, Hilkens CM, Kapsenberg ML. High-level IL-12 production by human dendritic cells requires two signals. International Immunology. 1998;**10**(11):1593-1598

[77] Melchjorsen J, Siren J, Julkunen I, Paludan SR, Matikainen S. Induction of cytokine expression by herpes simplex virus in human monocyte-derived macrophages and dendritic cells is dependent on virus replication and is counteracted by ICP27 targeting NF-kappaB and IRF-3. The Journal of General Virology. 2006;**87**(Pt 5):1099-1108

[78] Pollara G, Jones M, Handley ME, Rajpopat M, Kwan A, Coffin RS, et al. Herpes simplex virus type-1-induced activation of myeloid dendritic cells: The roles of virus cell interaction and paracrine type I IFN secretion. The Journal of Immunology. 2004;**173**(6):4108-4119

[79] Garcia-Sastre A, Biron CA. Type 1 interferons and the virus-host relationship: A lesson in detente. Science. 2006;**312**(5775):879-882

[80] Taniguchi T, Takaoka A. The interferon-alpha/beta system in antiviral responses: A multimodal machinery of gene regulation by the IRF family of transcription factors. Current Opinion in Immunology. 2002;**14**(1):111-116

[81] Rasmussen SB, Sørensen LN, Malmgaard L, Ank N, Baines JD, Chen ZJ, et al. Type I interferon production during herpes simplex virus infection is controlled by cell-type-specific viral recognition through toll-like receptor 9, the mitochondrial antiviral signaling protein pathway, and novel recognition systems. Journal of Virology. 2007;**81**(24):13315-13324

[82] Tailor P, Tamura T, Ozato K. IRF family proteins and type I interferon induction in dendritic cells. Cell Research. 2006;**16**(2):134-140

[83] Reske A, Pollara G, Krummenacher C, Katz DR, Chain BM. Glycoprotein-dependent and TLR2-independent innate immune recognition of herpes simplex virus-1 by dendritic cells. Journal of Immunology (Baltimore, Md.: 1950). 2008;**180**(11):7525-7536

[84] Ishikawa H, Ma Z, Barber GN. STING regulates intracellular DNA-mediated, type I interferon-dependent innate immunity. Nature. 2009;**461**(7265):788-792

[85] Zhang Z, Yuan B, Bao M, Lu N, Kim T, Liu Y-J. The helicase DDX41 senses intracellular DNA mediated by the adaptor STING in dendritic cells. Nature Immunology. 2011;**12**(10):959-965

[86] Samady L, Costigliola E, MacCormac L, McGrath Y, Cleverley S, Lilley CE, et al. Deletion of the virion host shutoff protein (vhs) from herpes simplex virus (HSV) relieves the viral block to dendritic cell activation: Potential of vhs- HSV vectors for dendritic cell-mediated immunotherapy. Journal of Virology. 2003;**77**(6):3768-3776

[87] Suazo PA, Tognarelli EI, Kalergis AM, Gonzalez PA. Herpes simplex virus 2 infection: Molecular association with HIV and novel microbicides to prevent disease. Medical Microbiology and Immunology. 2015;**204**(2):161-176

[88] Marsden V, Donaghy H, Bertram KM, Harman AN, Nasr N, Keoshkerian E, et al. Herpes simplex virus type 2-infected dendritic cells produce TNF-α, which enhances CCR5 expression and stimulates HIV production from adjacent infected cells. The Journal of Immunology. 2015;**194**(9):4438-4445

[89] Martinelli E, Tharinger H, Frank I, Arthos J, Piatak M Jr, Lifson JD, et al. HSV-2 infection of dendritic cells amplifies a highly susceptible HIV-1 cell target. PLoS Pathogens. 2011;**7**(6):e1002109

[90] Stefanidou M, Ramos I, Mas Casullo V, Trépanier JB, Rosenbaum S, Fernandez-Sesma A, et al. Herpes simplex virus 2 (HSV-2) prevents dendritic cell maturation, induces apoptosis, and triggers release of proinflammatory cytokines: Potential links to HSV-HIV synergy. Journal of Virology. 2013;**87**(3):1443-1453

[91] Grigoleit U, Riegler S, Einsele H, Laib Sampaio K, Jahn G, Hebart H, et al. Human cytomegalovirus induces a direct inhibitory effect on antigen presentation by monocyte-derived immature dendritic cells. British Journal of Haematology. 2002;**119**(1):189-198

[92] Karimi MH, Shariat A, Yaghobi R, Mokhtariazad T, Moazzeni SM. Role of cytomegalovirus on the maturation and function of monocyte derived dendritic cells of liver transplant patients. World Journal of Transplantation. 2016;**6**(2):336-346

[93] Mezger M, Bonin M, Kessler T, Gebhardt F, Einsele H, Loeffler J. Toll-like receptor 3 has no critical role during early immune response of human monocyte-derived dendritic cells after infection with the human cytomegalovirus strain TB40E. Viral Immunology. 2009;**22**(6):343-351

[94] McSharry BP, Avdic S, Slobedman B. Human cytomegalovirus encoded homologs of cytokines, chemokines and their receptors: Roles in immunomodulation. Viruses. 2012;**4**(11):2448-2470

[95] Murphy PM. Viral exploitation and subversion of the immune system through chemokine mimicry. Nature Immunology. 2001;**2**(2):116-122

[96] Kotenko SV, Saccani S, Izotova LS, Mirochnitchenko OV, Pestka S. Human cytomegalovirus harbors its own unique IL-10 homolog (cmvIL-10). Proceedings of the National Academy of Sciences. 2000;**97**(4):1695-1700

[97] Moore KW, de Waal MR, Coffman RL, O'Garra A. Interleukin-10 and the interleukin-10 receptor. Annual Review of Immunology. 2001;**19**(1):683-765

[98] Chang WL, Baumgarth N, Yu D, Barry PA. Human cytomegalovirus-encoded interleukin-10 homolog inhibits maturation of dendritic cells and alters their functionality. Journal of Virology. 2004;**78**(16):8720-8731

[99] Bosnjak L, Miranda-Saksena M, Koelle DM, Boadle RA, Jones CA, Cunningham AL. Herpes simplex virus infection of human dendritic cells induces apoptosis and allows cross-presentation via uninfected dendritic cells. The Journal of Immunology. 2005;**174**(4):2220-2227

[100] Raftery MJ, Schwab M, Eibert SM, Samstag Y, Walczak H, Schönrich G. Targeting the function of mature dendritic cells by human cytomegalovirus: A multilayered viral defense strategy. Immunity. 2001;**15**(6):997-1009

[101] Novak N, Peng W. Dancing with the enemy: The interplay of herpes simplex virus with dendritic cells. Clinical and Experimental Immunology. 2005;**142**(3):405-410

[102] Kather A, Raftery MJ, Devi-Rao G, Lippmann J, Giese T, Sandri-Goldin RM, et al. Herpes simplex virus type 1 (HSV-1)-induced apoptosis in human dendritic cells as a result of downregulation of cellular FLICE-inhibitory protein and reduced expression of HSV-1 antiapoptotic latency-associated transcript sequences. Journal of Virology. 2010;**84**(2):1034-1046

[103] Aubert M, O'Toole J, Blaho JA. Induction and prevention of apoptosis in human HEp-2 cells by herpes simplex virus type 1. Journal of Virology. 1999;**73**(12):10359-10370

[104] Sanfilippo CM, Chirimuuta FN, Blaho JA. Herpes simplex virus type 1 immediate-early gene expression is required for the induction of apoptosis in human epithelial HEp-2 cells. Journal of Virology. 2004;**78**(1):224-239

[105] Müller DB, Raftery MJ, Kather A, Giese T, Schönrich G. Frontline: Induction of apoptosis and modulation of c-FLIPL and p53 in immature dendritic cells infected with herpes simplex virus. European Journal of Immunology. 2004;**34**(4):941-951

[106] Henderson G, Peng W, Jin L, Perng GC, Nesburn AB, Wechsler SL, et al. Regulation of caspase 8- and caspase 9-induced apoptosis by the herpes simplex virus type 1 latency-associated transcript. Journal of Neurovirology. 2002;**8**(Suppl 2):103-111

[107] Jin L, Peng W, Perng GC, Brick DJ, Nesburn AB, Jones C, et al. Identification of herpes simplex virus type 1 latency-associated transcript sequences that both inhibit apoptosis and enhance the spontaneous reactivation phenotype. Journal of Virology. 2003;**77**(11):6556-6561

[108] Raftery MJ, Wieland D, Gronewald S, Kraus AA, Giese T, Schönrich G. Shaping phenotype, function, and survival of dendritic cells by cytomegalovirus-encoded IL-10. The Journal of Immunology. 2004;**173**(5):3383-3391

[109] Chang WLW, Baumgarth N, Eberhardt MK, Lee CYD, Baron CA, Gregg JP, et al. Exposure of myeloid dendritic cells to exogenous or endogenous IL-10 during maturation determines their longevity. The Journal of Immunology. 2007;**178**(12):7794-7804

[110] Gerada C, Campbell TM, Kennedy JJ, McSharry BP, Steain M, Slobedman B, et al. Manipulation of the innate immune response by varicella zoster virus. Frontiers in Immunology. 2020;**11**:1

[111] Hu H, Cohen JI. Varicella-zoster virus open reading frame 47 (ORF47) protein is critical for virus replication in dendritic cells and for spread to other cells. Virology. 2005;**337**(2):304-311

[112] Grosche L, Mühl-Zürbes P, Ciblis B, Krawczyk A, Kuhnt C, Kamm L, et al. Herpes simplex virus type-2 paralyzes the function of monocyte-derived dendritic cells. Viruses. 2020;**12**(1):112

[113] Jin H, Ma Y, Prabhakar BS, Feng Z, Valyi-Nagy T, Yan Z, et al. The γ134.5 protein of herpes simplex virus 1 is required to interfere with dendritic cell maturation during productive infection. Journal of Virology. 2009;**83**(10):4984-4994

[114] Ma Y, Chen M, Jin H, Prabhakar BS, Valyi-Nagy T, He B. An engineered herpesvirus activates dendritic cells and induces protective immunity. Scientific Reports. 2017;**7**(1):41461

[115] Verpooten D, Ma Y, Hou S, Yan Z, He B. Control of TANK-binding kinase 1-mediated signaling by the γ134.5 protein of herpes simplex virus 1. The Journal of Biological Chemistry. 2009;**284**(2):1097-1105

[116] Smiley JR. Herpes simplex virus virion host shutoff protein: Immune evasion mediated by a viral RNase? Journal of Virology. 2004;**78**(3):1063-1068

[117] Suazo PA, Ibanez FJ, Retamal-Diaz AR, Paz-Fiblas MV, Bueno SM, Kalergis AM, et al. Evasion of early antiviral responses by herpes simplex viruses. Mediators of Inflammation. 2015;**2015**:593757

[118] Geiss BJ, Smith TJ, Leib DA, Morrison LA. Disruption of virion

host shutoff activity improves the immunogenicity and protective capacity of a replication-incompetent herpes simplex virus type 1 vaccine strain. Journal of Virology. 2000;**74**(23):11137-11144

[119] Keadle TL, Morrison LA, Morris JL, Pepose JS, Stuart PM. Therapeutic immunization with a virion host shutoff-defective, replication-incompetent herpes simplex virus type 1 strain limits recurrent herpetic ocular infection. Journal of Virology. 2002;**76**(8):3615-3625

[120] Strelow LI, Leib DA. Role of the virion host shutoff (vhs) of herpes simplex virus type 1 in latency and pathogenesis. Journal of Virology. 1995;**69**(11):6779-6786

[121] Walker J, Leib DA. Protection from primary infection and establishment of latency by vaccination with a herpes simplex virus type 1 recombinant deficient in the virion host shutoff (vhs) function. Vaccine. 1998;**16**(1):1-5

[122] Sloan E, Henriquez R, Kinchington PR, Slobedman B, Abendroth A. Varicella-zoster virus inhibition of the NF-κB pathway during infection of human dendritic cells: Role for open reading frame 61 as a modulator of NF-κB activity. Journal of Virology. 2012;**86**(2):1193-1202

[123] Avdic S, McSharry BP, Slobedman B. Modulation of dendritic cell functions by viral IL-10 encoded by human cytomegalovirus. Frontiers in Microbiology. 2014;**5**:337

[124] Melillo JA, Song L, Bhagat G, Blazquez AB, Plumlee CR, Lee C, et al. Dendritic cell (DC)-specific targeting reveals Stat3 as a negative regulator of DC function. Journal of Immunology (Baltimore, Md.: 1950). 2010;**184**(5):2638-2645

[125] Dikic I, Elazar Z. Mechanism and medical implications of mammalian autophagy. Nature Reviews. Molecular Cell Biology. 2018;**19**(6):349-364

[126] English L, Chemali M, Duron J, Rondeau C, Laplante A, Gingras D, et al. Autophagy enhances the presentation of endogenous viral antigens on MHC class I molecules during HSV-1 infection. Nature Immunology. 2009;**10**(5):480-487

[127] Dengjel J, Schoor O, Fischer R, Reich M, Kraus M, Muller M, et al. Autophagy promotes MHC class II presentation of peptides from intracellular source proteins. Proceedings of the National Academy of Sciences of the United States of America. 2005;**102**(22):7922-7927

[128] Gatica D, Lahiri V, Klionsky DJ. Cargo recognition and degradation by selective autophagy. Nature Cell Biology. 2018;**20**(3):233-242

[129] Sharma V, Verma S, Seranova E, Sarkar S, Kumar D. Selective autophagy and xenophagy in infection and disease. Frontiers in Cell and Development Biology. 2018;**6**:147

[130] Klionsky DJ, Cuervo AM, Dunn WA Jr, Levine B, van der Klei I, Seglen PO. How shall I eat thee? Autophagy. 2007;**3**(5):413-416

[131] Bento CF, Renna M, Ghislat G, Puri C, Ashkenazi A, Vicinanza M, et al. Mammalian autophagy: How does it work? Annual Review of Biochemistry. 2016;**85**:685-713

[132] Wesselborg S, Stork B. Autophagy signal transduction by ATG proteins: From hierarchies to networks. Cellular and Molecular Life Sciences: CMLS. 2015;**72**(24):4721-4757

[133] Ghislat G, Lawrence T. Autophagy in dendritic cells. Cellular & Molecular Immunology. 2018;**15**(11):944-952

[134] Miszczak D, Cymerys J. A game of survival: Herpesvirus strategies

of autophagy manipulation. Postępy Higieny i Medycyny Doświadczalnej (Online). 2014;**68**:1406-1414

[135] McFarlane S, Aitken J, Sutherland JS, Nicholl MJ, Preston VG, Preston CM. Early induction of autophagy in human fibroblasts after infection with human cytomegalovirus or herpes simplex virus 1. Journal of Virology. 2011;**85**(9):4212-4221

[136] Orvedahl A, Alexander D, Talloczy Z, Sun Q, Wei Y, Zhang W, et al. HSV-1 ICP34.5 confers neurovirulence by targeting the Beclin 1 autophagy protein. Cell Host & Microbe. 2007;**1**(1):23-35

[137] Lussignol M, Queval C, Bernet-Camard MF, Cotte-Laffitte J, Beau I, Codogno P, et al. The herpes simplex virus 1 Us11 protein inhibits autophagy through its interaction with the protein kinase PKR. Journal of Virology. 2013;**87**(2):859-871

[138] Cavignac Y, Esclatine A. Herpesviruses and autophagy: Catch me if you can! Viruses. 2010;**2**(1):314-333

[139] Dong X, Levine B. Autophagy and viruses: Adversaries or allies? Journal of Innate Immunity. 2013;**5**(5):480-493

[140] O'Connell D, Liang C. Autophagy interaction with herpes simplex virus type-1 infection. Autophagy. 2016;**12**(3):451-459

[141] Siddiqui MA, Malathi K. RNase L induces autophagy via c-Jun N-terminal kinase and double-stranded RNA-dependent protein kinase signaling pathways. The Journal of Biological Chemistry. 2012;**287**(52):43651-43664

[142] Talloczy Z, Jiang W, Virgin HW, Leib DA, Scheuner D, Kaufman RJ, et al. Regulation of starvation- and virus-induced autophagy by the eIF2alpha kinase signaling pathway. Proceedings of the National Academy of Sciences of the United States of America. 2002;**99**(1):190-195

[143] Rasmussen SB, Horan KA, Holm CK, Stranks AJ, Mettenleiter TC, Simon AK, et al. Activation of autophagy by alpha-herpesviruses in myeloid cells is mediated by cytoplasmic viral DNA through a mechanism dependent on stimulator of IFN genes. Journal of Immunology (Baltimore, Md.: 1950). 2011;**187**(10):5268-5276

[144] Budida R, Stankov MV, Döhner K, Buch A, Panayotova-Dimitrova D, Tappe KA, et al. Herpes simplex virus 1 interferes with autophagy of murine dendritic cells and impairs their ability to stimulate CD8+ T lymphocytes. European Journal of Immunology. 2017;**47**(10):1819-1834

[145] Gobeil PAM, Leib DA. Herpes simplex virus γ34.5 interferes with autophagosome maturation and antigen presentation in dendritic cells. MBio. 2012;**3**(5):e00267-e00212

[146] Turan A, Grosche L, Krawczyk A, Mühl-Zürbes P, Drassner C, Düthorn A, et al. Autophagic degradation of lamins facilitates the nuclear egress of herpes simplex virus type 1. The Journal of Cell Biology. 2019;**218**(2):508-523

[147] Rechenchoski DZ, Faccin-Galhardi LC, Linhares REC, Nozawa C. Herpesvirus: An underestimated virus. Folia Microbiologica. 2017;**62**(2):151-156

[148] Bigalke JM, Heldwein EE. Nuclear exodus: Herpesviruses lead the way. Annual Review of Virology. 2016;**3**(1):387-409

[149] Heald R, McKeon F. Mutations of phosphorylation sites in lamin A that prevent nuclear lamina disassembly in mitosis. Cell. 1990;**61**(4):579-589

[150] Montpetit B, Weis K. Cell biology. An alternative route for nuclear mRNP

export by membrane budding. Science. 2012;**336**(6083):809-810

[151] Kochin V, Shimi T, Torvaldson E, Adam SA, Goldman A, Pack C-G, et al. Interphase phosphorylation of lamin A. Journal of Cell Science. 2014;**127**(12):2683-2696

[152] Milbradt J, Hutterer C, Bahsi H, Wagner S, Sonntag E, Horn AHC, et al. The prolyl isomerase Pin1 promotes the herpesvirus-induced phosphorylation-dependent disassembly of the nuclear lamina required for nucleocytoplasmic egress. PLoS Pathogens. 2016;**12**(8):e1005825

[153] Everett RD, Boutell C, Orr A. Phenotype of a herpes simplex virus type 1 mutant that fails to express immediate-early regulatory protein ICP0. Journal of Virology. 2004;**78**(4):1763-1774

[154] Kruse M, Rosorius O, Krätzer F, Stelz G, Kuhnt C, Schuler G, et al. Mature dendritic cells infected with herpes simplex virus type 1 exhibit inhibited T-cell stimulatory capacity. Journal of Virology. 2000;**74**(15):7127-7136

[155] Heilingloh CS, Kummer M, Mühl-Zürbes P, Drassner C, Daniel C, Klewer M, et al. L particles transmit viral proteins from herpes simplex virus 1-infected mature dendritic cells to uninfected bystander cells, inducing CD83 downmodulation. Journal of Virology. 2015;**89**(21):11046-11055

[156] Hofer S, Pfeil K, Niederegger H, Ebner S, Nguyen VA, Kremmer E, et al. Dendritic cells regulate T-cell deattachment through the integrin-interacting protein CYTIP. Blood. 2006;**107**(3):1003-1009

[157] Kummer M, Turza NM, Mühl-Zürbes P, Lechmann M, Boutell C, Coffin RS, et al. Herpes simplex virus type 1 induces CD83 degradation in mature dendritic cells with immediate-early kinetics via the cellular proteasome. Journal of Virology. 2007;**81**(12):6326-6338

[158] Prechtel AT, Turza NM, Kobelt DJ, Eisemann JI, Coffin RS, McGrath Y, et al. Infection of mature dendritic cells with herpes simplex virus type 1 dramatically reduces lymphoid chemokine-mediated migration. The Journal of General Virology. 2005;**86** (Pt 6):1645-1657

[159] Theodoridis AA, Eich C, Figdor CG, Steinkasserer A. Infection of dendritic cells with herpes simplex virus type 1 induces rapid degradation of CYTIP, thereby modulating adhesion and migration. Blood. 2011;**118**(1):107-115

[160] Früh K, Ahn K, Djaballah H, Sempe P, van Endert PM, Tampe R, et al. A viral inhibitor of peptide transporters for antigen presentation. Nature. 1995;**375**(6530):415-418

[161] Hill A, Jugovic P, York I, Russ G, Bennink J, Yewdell J, et al. Herpes simplex virus turns off the TAP to evade host immunity. Nature. 1995;**375**(6530):411-415

[162] Trgovcich J, Johnson D, Roizman B. Cell surface major histocompatibility complex class II proteins are regulated by the products of the gamma(1)34.5 and U(L)41 genes of herpes simplex virus 1. Journal of Virology. 2002;**76**(14):6974-6986

[163] Wiertz EJ, Devlin R, Collins HL, Ressing ME. Herpesvirus interference with major histocompatibility complex class II-restricted T-cell activation. Journal of Virology. 2007;**81**(9):4389-4396

[164] Morrow G, Slobedman B, Cunningham AL, Abendroth A. Varicella-zoster virus productively

infects mature dendritic cells and alters their immune function. Journal of Virology. 2003;**77**(8):4950-4959

[165] Heilingloh CS, Grosche L, Kummer M, Mühl-Zürbes P, Kamm L, Scherer M, et al. The major immediate-early protein IE2 of human cytomegalovirus is sufficient to induce proteasomal degradation of CD83 on mature dendritic cells. Frontiers in Microbiology. 2017;**8**:119

[166] Senechal B, Boruchov AM, Reagan JL, Hart DN, Young JW. Infection of mature monocyte-derived dendritic cells with human cytomegalovirus inhibits stimulation of T-cell proliferation via the release of soluble CD83. Blood. 2004;**103**(11):4207-4215

[167] von Rohrscheidt J, Petrozziello E, Nedjic J, Federle C, Krzyzak L, Ploegh HL, et al. Thymic CD4 T cell selection requires attenuation of March8-mediated MHCII turnover in cortical epithelial cells through CD83. The Journal of Experimental Medicine. 2016;**213**(9):1685-1694

[168] Wild AB, Krzyzak L, Peckert K, Stich L, Kuhnt C, Butterhof A, et al. CD83 orchestrates immunity toward self and non-self in dendritic cells. JCI Insight. 2019;**4**(20):e126246

[169] Zinser E, Naumann R, Wild AB, Michalski J, Deinzer A, Stich L, et al. Endogenous expression of the human CD83 attenuates EAE symptoms in humanized transgenic mice and increases the activity of regulatory T cells. Frontiers in Immunology. 2019;**10**:1442

[170] Bock F, Rossner S, Onderka J, Lechmann M, Pallotta MT, Fallarino F, et al. Topical application of soluble CD83 induces IDO-mediated immune modulation, increases Foxp3+ T cells, and prolongs allogeneic corneal graft survival. Journal of Immunology (Baltimore, Md.: 1950). 2013;**191**(4):1965-1975

[171] Ge W, Arp J, Lian D, Liu W, Baroja ML, Jiang J, et al. Immunosuppression involving soluble CD83 induces tolerogenic dendritic cells that prevent cardiac allograft rejection. Transplantation. 2010;**90**(11):1145-1156

[172] Li Z, Ju X, Silveira PA, Abadir E, Hsu W-H, Hart DNJ, et al. CD83: Activation marker for antigen presenting cells and its therapeutic potential. Frontiers in Immunology. 2019;(10):1312

[173] Royzman D, Andreev D, Stich L, Rauh M, Bäuerle T, Ellmann S, et al. Soluble CD83 triggers resolution of arthritis and sustained inflammation control in IDO dependent manner. Frontiers in Immunology. 2019;**10**:633

[174] Heilingloh CS, Mühl-Zürbes P, Steinkasserer A, Kummer M. Herpes simplex virus type 1 ICP0 induces CD83 degradation in mature dendritic cells independent of its E3 ubiquitin ligase function. The Journal of General Virology. 2014;**95**(Pt 6):1366-1375

[175] Heilingloh CS, Krawczyk A. Role of L-particles during herpes simplex virus infection. Frontiers in Microbiology. 2017;**8**:2565

[176] Schönrich G, Raftery MJ. Dendritic cells as Achilles' heel and Trojan horse during varicella zoster virus infection. Frontiers in Microbiology. 2015;**6**:417

[177] Randolph GJ, Ochando J, Partida-Sanchez S. Migration of dendritic cell subsets and their precursors. Annual Review of Immunology. 2008;**26**:293-316

[178] Devreotes P, Horwitz AR. Signaling networks that regulate cell migration. Cold Spring Harbor Perspectives in Biology. 2015;7(8):a005959

[179] Lämmermann T, Bader BL, Monkley SJ, Worbs T, Wedlich-Soldner R, Hirsch K, et al. Rapid leukocyte migration by integrin-independent flowing and squeezing. Nature. 2008;**453**(7191):51-55

[180] Lämmermann T, Sixt M. Mechanical modes of 'amoeboid' cell migration. Current Opinion in Cell Biology. 2009;**21**(5):636-644

[181] Schmidt S, Friedl P. Interstitial cell migration: Integrin-dependent and alternative adhesion mechanisms. Cell and Tissue Research. 2010;**339**(1):83-92

[182] Cougoule C, Lastrucci C, Guiet R, Mascarau R, Meunier E, Lugo-Villarino G, et al. Podosomes, but not the maturation status, determine the protease-dependent 3D migration in human dendritic cells. Frontiers in Immunology. 2018;**9**:846

[183] Roediger B, Ng LG, Smith AL, de St Groth BF, Weninger W. Visualizing dendritic cell migration within the skin. Histochemistry and Cell Biology. 2008;**130**(6):1131-1146

[184] Friedl P, Weigelin B. Interstitial leukocyte migration and immune function. Nature Immunology. 2008;**9**(9):960-969

[185] Renkawitz J, Sixt M. Mechanisms of force generation and force transmission during interstitial leukocyte migration. EMBO Reports. 2010;**11**(10):744-750

[186] Bedoui S, Greyer M. The role of dendritic cells in immunity against primary herpes simplex virus infections. Frontiers in Microbiology. 2014;**5**:533

[187] Kassim SH, Rajasagi NK, Zhao X, Chervenak R, Jennings SR. In vivo ablation of CD11c-positive dendritic cells increases susceptibility to herpes simplex virus type 1 infection and diminishes NK and T-cell responses. Journal of Virology. 2006;**80**(8):3985-3993

[188] Allan RS, Waithman J, Bedoui S, Jones CM, Villadangos JA, Zhan Y, et al. Migratory dendritic cells transfer antigen to a lymph node-resident dendritic cell population for efficient CTL priming. Immunity. 2006;**25**(1):153-162

[189] Bedoui S, Whitney PG, Waithman J, Eidsmo L, Wakim L, Caminschi I, et al. Cross-presentation of viral and self antigens by skin-derived CD103+ dendritic cells. Nature Immunology. 2009;**10**(5):488-495

[190] Eidsmo L, Allan R, Caminschi I, van Rooijen N, Heath WR, Carbone FR. Differential migration of epidermal and dermal dendritic cells during skin infection. The Journal of Immunology. 2009;**182**(5):3165-3172

[191] Hor Jyh L, Whitney Paul G, Zaid A, Brooks Andrew G, Heath William R, Mueller SN. Spatiotemporally distinct interactions with dendritic cell subsets facilitates CD4+ and CD8+ T cell activation to localized viral infection. Immunity. 2015;**43**(3):554-565

[192] Lee HK, Zamora M, Linehan MM, Iijima N, Gonzalez D, Haberman A, et al. Differential roles of migratory and resident DCs in T cell priming after mucosal or skin HSV-1 infection. The Journal of Experimental Medicine. 2009;**206**(2):359-370

[193] Ponten F, Jirstrom K, Uhlen M. The Human Protein Atlas - A tool for pathology. The Journal of Pathology. 2008;**216**(4):387-393

[194] Hyun YM, Lefort CT, Kim M. Leukocyte integrins and their ligand interactions. Immunologic Research. 2009;**45**(2-3):195-208

[195] Harris ES, McIntyre TM, Prescott SM, Zimmerman GA.

The leukocyte integrins. The Journal of Biological Chemistry. 2000;**275**(31):23409-23412

[196] Quast T, Tappertzhofen B, Schild C, Grell J, Czeloth N, Förster R, et al. Cytohesin-1 controls the activation of RhoA and modulates integrin-dependent adhesion and migration of dendritic cells. Blood. 2009;**113**(23):5801-5810

[197] van Kooyk Y, Figdor CG. Avidity regulation of integrins: The driving force in leukocyte adhesion. Current Opinion in Cell Biology. 2000;**12**(5):542-547

[198] Hynes RO. Integrins: Bidirectional, allosteric signaling machines. Cell. 2002;**110**(6):673-687

[199] Shattil SJ, Kim C, Ginsberg MH. The final steps of integrin activation: The end game. Nature Reviews. Molecular Cell Biology. 2010;**11**(4):288-300

[200] Kolanus W. Guanine nucleotide exchange factors of the cytohesin family and their roles in signal transduction. Immunological Reviews. 2007;**218**:102-113

[201] Wegener KL, Partridge AW, Han J, Pickford AR, Liddington RC, Ginsberg MH, et al. Structural basis of integrin activation by Talin. Cell. 2007;**128**(1):171-182

[202] Kolanus W, Nagel W, Schiller B, Zeitlmann L, Godar S, Stockinger H, et al. αLβ2 integrin/LFA-1 binding to ICAM-1 induced by cytohesin-1, a cytoplasmic regulatory molecule. Cell. 1996;**86**(2):233-242

[203] Weber KSC, Weber C, Ostermann G, Dierks H, Nagel W, Kolanus W. Cytohesin-1 is a dynamic regulator of distinct LFA-1 functions in leukocyte arrest and transmigration triggered by chemokines. Current Biology. 2001;**11**(24):1969-1974

[204] Boehm T, Hofer S, Winklehner P, Kellersch B, Geiger C, Trockenbacher A, et al. Attenuation of cell adhesion in lymphocytes is regulated by CYTIP, a protein which mediates signal complex sequestration. The EMBO Journal. 2003;**22**(5):1014-1024

[205] González-Motos V, Jürgens C, Ritter B, Kropp KA, Durán V, Larsen O, et al. Varicella zoster virus glycoprotein C increases chemokine-mediated leukocyte migration. PLoS Pathogens. 2017;**13**(5):e1006346

[206] Ku C-C, Padilla JA, Grose C, Butcher EC, Arvin AM. Tropism of varicella-zoster virus for human tonsillar CD4+ T lymphocytes that express activation, memory, and skin homing markers. Journal of Virology. 2002;**76**(22):11425-11433

[207] Ku C-C, Zerboni L, Ito H, Graham BS, Wallace M, Arvin AM. Varicella-zoster virus transfer to skin by T cells and modulation of viral replication by epidermal cell interferon-α. The Journal of Experimental Medicine. 2004;**200**(7):917-925

[208] Lecointe D, Dugas N, Leclerc P, Hery C, Delfraissy JF, Tardieu M. Human cytomegalovirus infection reduces surface CCR5 expression in human microglial cells, astrocytes and monocyte-derived macrophages. Microbes and Infection. 2002;**4**(14):1401-1408

[209] Warner JA, Zwezdaryk KJ, Day B, Sullivan DE, Pridjian G, Morris CA. Human cytomegalovirus infection inhibits CXCL12-mediated migration and invasion of human extravillous cytotrophoblasts. Virology Journal. 2012;**9**:255

[210] Sinclair J, Poole E. Human cytomegalovirus latency and reactivation in and beyond the myeloid lineage. Future Virology. 2014;**9**(6):557-563

[211] Reeves MB, MacAry PA, Lehner PJ, Sissons JG, Sinclair JH. Latency, chromatin remodeling, and reactivation of human cytomegalovirus in the dendritic cells of healthy carriers. Proceedings of the National Academy of Sciences of the United States of America. 2005;**102**(11):4140-4145

[212] Reeves MB, Sinclair JH. Circulating dendritic cells isolated from healthy seropositive donors are sites of human cytomegalovirus reactivation in vivo. Journal of Virology. 2013;**87**(19):10660-10667

[213] Goldwich A, Prechtel AT, Mühl-Zürbes P, Pangratz NM, Stössel H, Romani N, et al. Herpes simplex virus type I (HSV-1) replicates in mature dendritic cells but can only be transferred in a cell–cell contact-dependent manner. Journal of Leukocyte Biology. 2011;**89**(6):973-979

[214] McLauchlan J, Addison C, Craigie MC, Rixon FJ. Noninfectious L-particles supply functions which can facilitate infection by HSV-1. Virology. 1992;**190**(2):682-688

[215] Szilagyi JF, Cunningham C. Identification and characterization of a novel non-infectious herpes simplex virus-related particle. The Journal of General Virology. 1991;**72**(Pt 3):661-668

[216] Russell T, Bleasdale B, Hollinshead M, Elliott G. Qualitative differences in capsidless L-particles released as a by-product of bovine herpesvirus 1 and herpes simplex virus 1 infections. Journal of Virology. 2018;**92**(22):e01259-18

[217] Dargan DJ, Subak-Sharpe JH. The effect of herpes simplex virus type 1 L-particles on virus entry, replication, and the infectivity of naked herpesvirus DNA. Virology. 1997;**239**(2):378-388

[218] Subak-Sharpe JH, Dargan DJ. Hsv molecular biology: General aspects of herpes simplex virus molecular biology. Virus Genes. 1998;**16**(3):239-251

[219] Buckingham EM, Jarosinski KW, Jackson W, Carpenter JE, Grose C. Exocytosis of varicella-zoster virus virions involves a convergence of endosomal and autophagy pathways. Journal of Virology. 2016;**90**(19):8673-8685

[220] Jiang B, Himmelsbach K, Ren H, Boller K, Hildt E. Subviral hepatitis B virus filaments, like infectious viral particles, are released via multivesicular bodies. Journal of Virology. 2015;**90**(7):3330-3341

[221] Meckes DG Jr, Raab-Traub N. Microvesicles and viral infection. Journal of Virology. 2011;**85**(24): 12844-12854

[222] Mohammad AA, Costa H, Landazuri N, Lui WO, Hultenby K, Rahbar A, et al. Human cytomegalovirus microRNAs are carried by virions and dense bodies and are delivered to target cells. The Journal of General Virology. 2017;**98**(5):1058-1072

[223] Sauer C, Klobuch S, Herr W, Thomas S, Plachter B. Subviral dense bodies of human cytomegalovirus stimulate maturation and activation of monocyte-derived immature dendritic cells. Journal of Virology. 2013;**87**(20):11287-11291

[224] Huch JH, Cunningham AL, Arvin AM, Nasr N, Santegoets SJAM, Slobedman E, et al. Impact of varicella-zoster virus on dendritic cell subsets in human skin during natural infection. Journal of Virology. 2010;**84**(8):4060-4072

[225] Kinchington PR, Leger AJS, Guedon J-MG, Hendricks RL. Herpes simplex virus and varicella zoster virus, the house guests who never leave. Herpesviridae. 2012;**3**(1):5

[226] Clementi N, Cappelletti F, Criscuolo E, Castelli M, Mancini N, Burioni R, et al. Role and potential therapeutic use of antibodies against herpetic infections. Clinical Microbiology and Infection. 2017;**23**(6): 381-386

[227] Rajasagi NK, Kassim SH, Kollias CM, Zhao X, Chervenak R, Jennings SR. CD4+ T cells are required for the priming of CD8+ T cells following infection with herpes simplex virus type 1. Journal of Virology. 2009;**83**(10):5256-5268

[228] Belz GT, Carbone FR, Heath WR. Cross-presentation of antigens by dendritic cells. Critical Reviews in Immunology. 2002;**22**(5-6):439-448

[229] Martin H, Mandron M, Davrinche C. Interplay between human cytomegalovirus and dendritic cells in T cell activation. Medical Microbiology and Immunology. 2008;**197**(2):179-184

[230] Nopora K, Bernhard C, Ried C, Castello A, Murphy K, Marconi P, et al. MHC class I cross-presentation by dendritic cells counteracts viral immune evasion. Frontiers in Immunology. 2012;**3**:348

[231] Theisen D, Murphy K. The role of cDC1s in vivo: CD8 T cell priming through cross-presentation. F1000Research. 2017;**6**:98

[232] Silvin A, Yu CI, Lahaye X, Imperatore F, Brault J-B, Cardinaud S, et al. Constitutive resistance to viral infection in human CD141+ dendritic cells. Science Immunology. 2017;**2**(13):eaai8071

[233] Whitney PG, Makhlouf C, MacLeod B, Ma JZ, Gressier E, Greyer M, et al. Effective priming of herpes simplex virus-specific CD8+ T cells in vivo does not require infected dendritic cells. Journal of Virology. 2018;**92**(3):e01508-e01517

[234] Kitano M, Yamazaki C, Takumi A, Ikeno T, Hemmi H, Takahashi N, et al. Imaging of the cross-presenting dendritic cell subsets in the skin-draining lymph node. Proceedings of the National Academy of Sciences. 2016;**113**(4):1044-1049

Chapter 3

Cytological Diagnosis of Infectious Diseases: Identification of Pathogens and Recognition of Cellular Reactions

Yutaka Tsutsumi

Abstract

Cytological diagnosis of infectious diseases is as important as the cytodiagnosis of malignancies, because the detection of pathogens in cytological specimens is crucially valuable for prompt and appropriate patients' treatment. When compared with histological diagnosis, cytology is strong at detecting microbes under Papanicolaou and Giemsa stains. Host response against the infectious agent can be estimated by the type of background inflammatory cells. Patterns of the inflammatory cellular responses against extracellular and intracellular pathogens should be recognized. Immunocytochemical and molecular approaches can be applied, even when we have only one cytology specimen in hand. The cell transfer technique is useful to create plural material from one glass slide for immunocytochemistry and other techniques. In case of transmissible disorders including sexually transmitted diseases, the prompt and appropriate diagnosis will avoid avoidable transmission of infectious agents among people, and eventually contribute to the safety of the human society.

Keywords: cell transfer technique, cytodiagnosis, defense mechanism, host response, immunocytochemistry, infectious diseases, inflammatory cells, pathogens

1. Introduction

In the daily practice of cytological diagnosis, cytopathologists tend to focus on the diagnosis of premalignant and malignant diseases. Generally speaking, the cytology practice functions as screening for malignancy. However, the cytodiagnosis of infectious diseases and the identification of pathogens in cytological preparations must not be undervalued. The correct cytodiagnosis of infectious diseases leads patients to prompt and appropriate treatment. Histopathological diagnosis is strong at recognizing host responses against pathogens, while pathogens are more easily identified in the cytology specimen than the histology specimen. When infectious diseases are clinically suspected, it is better for us to perform Giemsa staining in addition to routine Papanicolaou staining.

In the present review article, the author presents varied aspects of cytomorphology of infectious diseases, in addition to general remarks for the defense mechanisms against infectious microorganisms. Immunocytochemistry significantly contributes to the definite and final cytodiagnosis. Often times, only one cytology specimen is available in the daily practice, so that the special techniques "how we can detect pathogens in only one cytology preparation" are needed for evaluating with additional staining. Please refer to the previous articles, textbooks and web sites of the author, describing the cytological diagnosis of infectious diseases [1–7]. It is most regrettable that some of them were written in Japanese.

2. Defense mechanisms against infection

Defense mechanisms against infection are categorized into two types: nonspecific and specific. Both types cooperatively function as an effective anti-infection system. Varied inflammatory cells are involved in the processes [8–10].

2.1 Types of inflammatory cells

Types of inflammatory cells and their properties are briefly summarized in **Table 1**. Function of the cells and their proliferative and migratory activity are shown.

Representative light microscopic and electron microscopic features of the inflammatory cells are illustrated in **Figures 1** and **2**. Of note is that cytokines mediate intercellular communication with which the immune cells talk to each other [11]. Cytokines include interferons, interleukins, chemokines, lymphokines and tumor necrosis factors.

2.2 Nonspecific defense mechanisms against infection

2.2.1 Physical barriers

The epidermis of the skin and the surface mucosal layer on the mucosal membrane play an effective physical barrier against invasion of the pathogen. The cilia on the pseudostratified mucosa of the airway effectively excrete the pathogen.

Cell type	Function	Proliferative activity	Migratory potential
Granulocyte			
Neutrophil	Phagocytosis	None	Migratory
Eosinophil	Allergy, anti-helminth function	None	Migratory
Basophil	Histamine production	None	Migratory
Mast cell	Histamine production	Proliferative	Migratory
Monocyte	Phagocytosis	Proliferative	Migratory
Macrophage	Phagocytosis, granuloma reaction	Proliferative	Migratory
B-lymphocyte	Humoral immunity	Proliferative	Migratory
T-lymphocyte	Cellular immunity/helper activity	Proliferative	Migratory
NK cell	Innate immunity	Proliferative	Migratory
Plasma cell	Antibody production	None	None
Dendritic cell	Antigen presentation	Proliferative	None

Table 1.
Inflammatory cells and their properties.

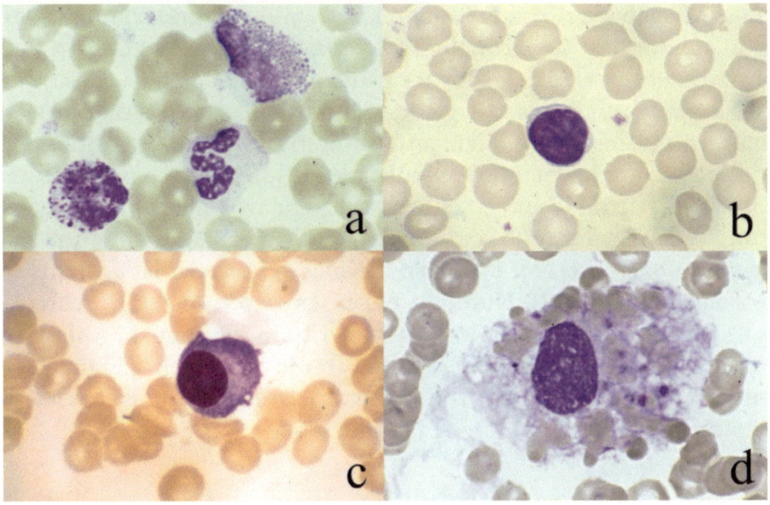

Figure 1.
Types of inflammatory cells (may-Giemsa). a: Three kinds of granulocytes (from left to right: Basophil, neutrophil and eosinophil) seen in the bone marrow smear. b: Small lymphocyte, c: Plasma cell, d: Hemophagocytic (activated) macrophage. Compare the cytoplasmic granules in the granulocytes. The cytoplasm of the small lymphocyte is scanty, and the plasma cell contains basophilic cytoplasm with a prominent Golgi area. The macrophage actively phagocytizes red cells and platelets.

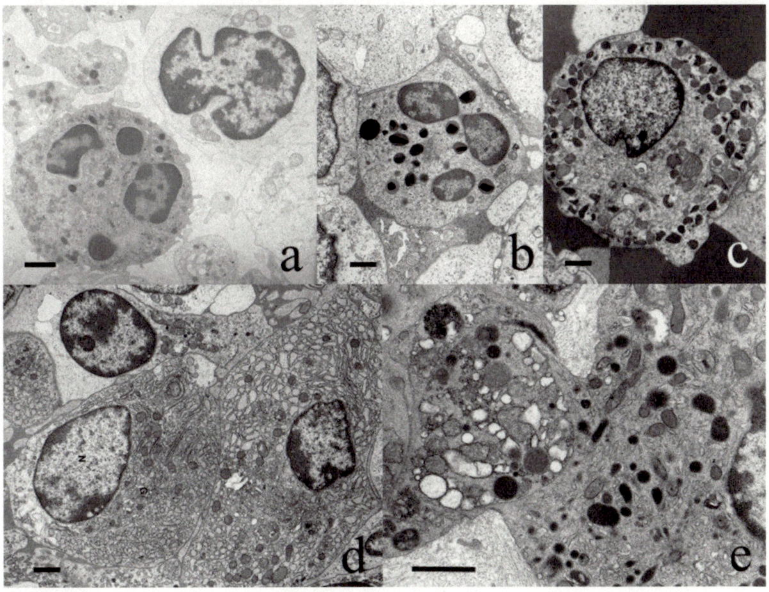

Figure 2.
Electron microscopic appearance of inflammatory cells. a: Neutrophil (left) and lymphocyte (right), b: Eosinophil, c: Basophil, d: Two plasma cells in the small bowel mucosa, e: Activated macrophage in soft tissue. The lymphocyte has an indented nucleus, while the granulocytes possess segmented nuclei. The cytoplasmic granules feature the respective granulocytes: Small-sized granules in the neutrophil, large crystalline granules in the eosinophil, and large rounded granules often with a fingerprint image in the basophil. The plasma cells contain a round nucleus with peripherally condensed heterochromatin and the cytoplasm rich in rough endoplasmic reticulum. The large-sized macrophage possesses an ameboid cytoplasmic process and numbers of electron-dense lysosomal granules. Bars indicate 1 μm.

2.2.2 Antibacterial secretory proteins

The secretory juice secreted from secretory glands contains varied antibacterial proteins such as lactoferrin, lysozyme (muramidase) and defensins [12].

Lactoferrin shows a bacteriostatic function by combining and competing trivalent ferric ions mandatory for the growth of bacteria and fungi. Lactoferrin is secreted from the lactating breast, serous salivary glands, lacrimal glands, eccrine sweat glands, gastric glands and prostatic glands. Of particular note is that protease digestion of lactoferrin yields lactoferricin and lactoferrampin, potent antimicrobial peptides derived from the lactoferrin molecule [13]. Lysozyme operates as a bactericidal or bacteriolytic molecule by cutting the joint sequence of *N*-acetylglucosamine and *N*-acetylmuramic acid in the peptide glycan network on the cell wall of Gram-positive bacteria [14]. Defensins belong to bactericidal proteins strongly binding to phospholipids [12]. Numbers of serous secretory glands secrete both lysozyme and defensins, together with lactoferrin. In the small bowel mucosa, lysozyme and defensins are actively secreted from Paneth cells (**Figure 3**). Representative microscopic features of production of lactoferrin and lysozyme in varied secretory epithelial cells are displayed in **Figure 4**.

2.2.3 Phagocytes and natural killer (NK) cells

Bacteria are nonspecifically phagocytized by phagocytes such as neutrophils and macrophages [15]. **Figure 5** exhibits bacteria phagocytized by neutrophils. Because of the lack of proliferative activity, neutrophils are predominantly seen in acute inflammation. Macrophages are proliferative, so that they mainly appear in chronic inflammation. Myeloperoxidase, lysozyme and defensins show bactericidal activities in the phagocytic vacuole (primary granule) of the neutrophil [16]. In the secondary (specific) granule of neutrophils, lactoferrin is contained. The main bactericidal enzyme functioning in the macrophage is lysozyme (see **Figure 3b**). NK cells correspond to CD56-positive large granular lymphocytes [17]. The cytoplasmic granules of the NK cell contain bactericidal, antiviral and apoptosis-inducing proteins common with the CD8-positive killer (cytotoxic) T-lymphocyte, such as perforin, granzymes (A and B) and T-cell intracellular antigen-1 (TIA-1).

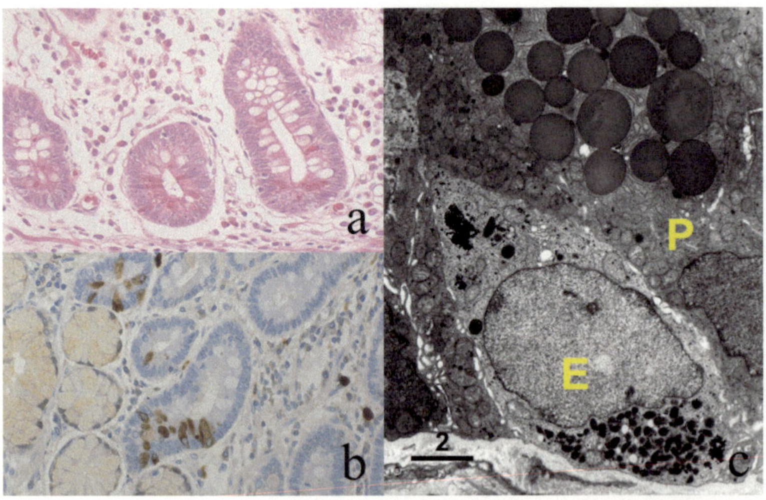

Figure 3.
Paneth cells in the duodenal mucosa. a: H&E, b: Lysozyme immunostaining, c: Electron microscopy. Paneth cells are distributed at the bottom of the intestinal crypt. Coarse eosinophilic granules are accumulated in the luminal side, and strongly immunoreactive for lysozyme. Macrophages scattered in the lamina propria mucosae express lysozyme immunoreactivity. Ultrastructurally, large-sized (around 2 μm in diameter) and round-shaped exocrine granules in the supranuclear cytoplasm of a Paneth cell (P) are homogeneously electron-dense. Compare them with infranuclear small-sized, electron-dense secretory granules of an endocrine cell (E). Bar = 2 μm.

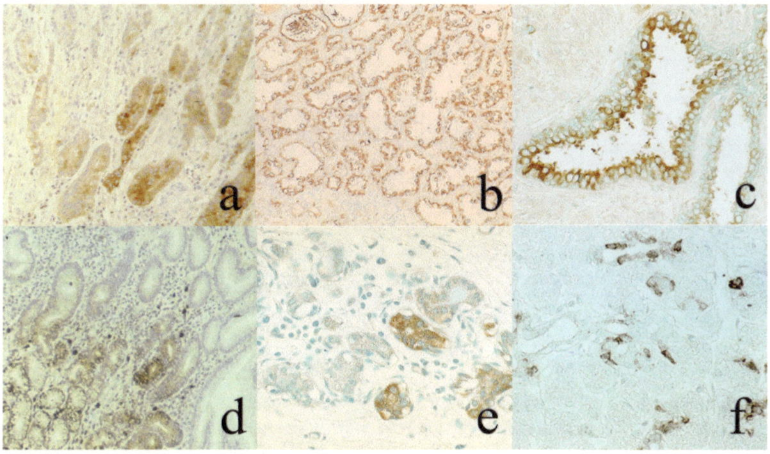

Figure 4.
Distribution of lactoferrin (upper panels: a–c) and lysozyme (lower panels: d–f) in secretory glands (immunostaining). a: Gastric fundic gland, b: Lactating breast, c: Prostatic gland, d: Gastric pyloric gland, e: Non-lactating breast, f: Parotid gland. Lactoferrin is actively expressed in the fundic gland, lactating breast and prostatic acini. The pyloric gland in the gastric antrum, normal mammary ductules and serous acinar cells of the salivary gland are immunoreactive for lysozyme.

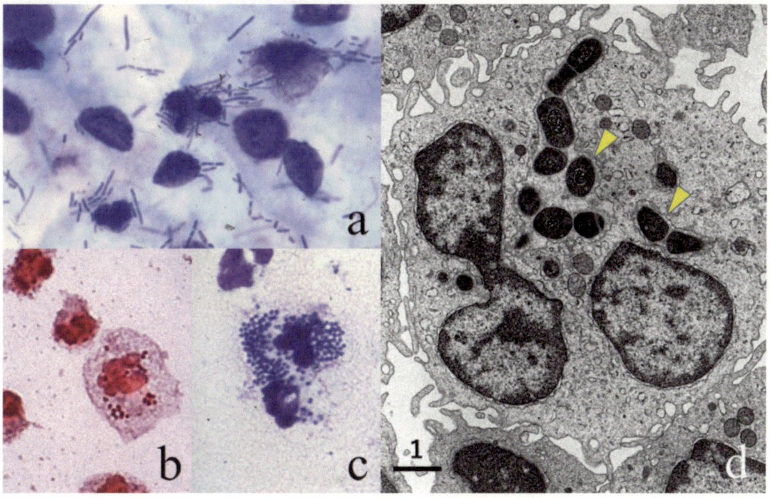

Figure 5.
Bacteria phagocytized by neutrophils. a: Escherichia coli *in the urine (acute cystitis; Giemsa),* Neisseria meningitidis *in the cerebrospinal fluid (gram), c:* Neisseria gonorrhoeae *in the urethral discharge (Giemsa), d: Electron microscopy of a neutrophil phagocytizing bacilli in the pleural effusion. Neutrophils actively phagocytize the bacilli (a and d) and cocci (b and c). Note that the bacilli are localized in lysosomal granules (d, arrowheads). Bar = 1 μm.*

These cells play significant roles in the host defense against pathogens for the initial two weeks after infection, until the establishment of the "specific" (humoral and/or cellular) immune reaction.

It should be noted that neutrophils form neutrophil extracellular traps (NETs), a filamentous spiderweb-like network entrapping bacteria, after cell death called NETtosis [18]. NETs are composed of DNA stretches and anti-bacterial proteins, including lactoferrin and myeloperoxidase [19]. NETs are richly formed in the abscess lesion, as shown in **Figure 6**.

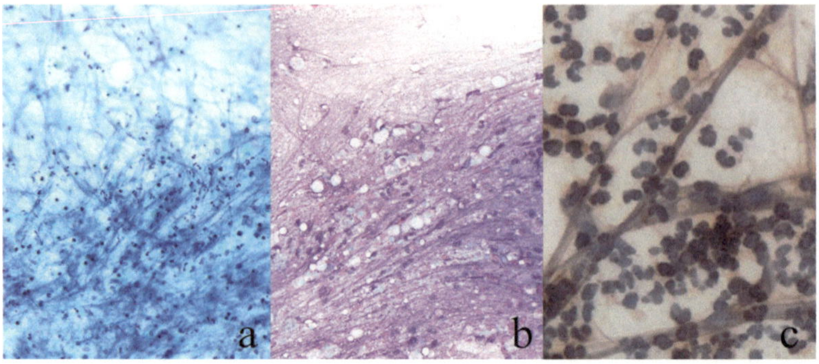

Figure 6.
Neutrophil extracellular traps (NETs) seen among aspirated neutrophilic exudates aspirated from a mammary abscess lesion. a: Papanicolaou, b: Giemsa, c: Lactoferrin immunostaining. Long and basophilic filamentous structures are formed among the neutrophils. NETs, composed of DNA and anti-bacterial proteins such as lactoferrin, entrap the causative bacteria with a spiderweb-like network to avoid them from spreading.

2.2.4 Innate immunity and Toll-like receptors

Acute viral infection usually calms down in one week. The strong anti-viral mediators are type I interferons (IFN-alpha and IFN-beta). The IFNs are produced by the keratinocyte of the epidermis and squamous mucosa, the columnar cells of the intestinal and airway mucosa and Langerhans (dendritic) cells distributed among the epithelial cells [20, 21]. Toll-like receptors (TLRs) expressed on these cells specifically recognize microbe-derived components such as lipoproteins, lipopolysaccharide, viral double-stranded RNA, non-methylated CpG islands of DNA and flagellin to induce IFN secretion. Toll means great and curious in German. In the human being, there are 10 kinds of TLRs. The TLR-mediated innate immunity, as well as phagocytosis by neutrophils and macrophages and the NK cell-mediated defense, comprise major functions of the vertebrate intrinsic system for the exclusion of the pathogen.

2.3 Specific defense mechanisms against infection

The specific acquired immunity consists of humoral immunity and cell-mediated (cellular) immunity [8, 9]. Production of specific antibodies by B-lymphocytes is the key mechanism of the humoral immunity. Serum complements secreted from the liver activate neutralizing activity of specific antibodies. The key players of the cell-mediated immunity are cytotoxic (killer) T-lymphocytes and activated macrophages. It takes a certain period (usually two weeks to one month) until establishing the specific acquired immunity.

The specific defense mechanisms against infection should be divided into two categories: the systemic immunity versus local (mucosal) immunity. The pathogen invading the inside of the body are specifically protected by IgG-mediated humoral immunity and also by CD8-positive cytotoxic T-lymphocyte-mediated cellular immunity.

The mucosal immunity provides a defense mechanism protecting invasion of the pathogen across the mucosa [22]. Dimeric secretory IgA (sIgA) functions as a mediator of the mucosal humoral immunity, but it hardly shows a neutralizing (killing) activity. The lamina propria mucosae contains numbers of IgA-producing plasma cells. sIgA is secreted onto the mucosal surface after coupling with "secretory component (SC)" produced by the columnar epithelial cells. **Figure 7** schematically displays the process of formation of sIgA. Microscopic features of IgA secretion in

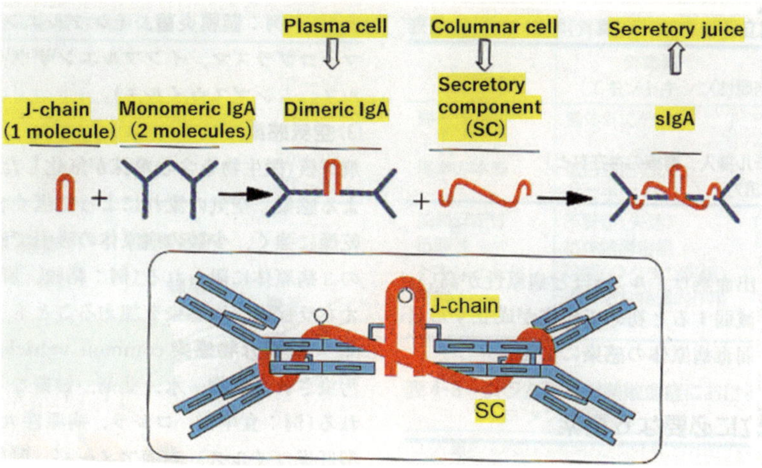

Figure 7.
Schematic presentation of the process of formation of secretory IgA (sIgA). IgA is secreted onto the mucosa after binding with secretory component (SC), a product of mucosa-lining columnar epithelial cells. Dimeric IgA consists of two molecules of monomeric IgA and J-chain. IgA-producing plasma cells are richly distributed in the normal bowel mucosa.

intestinal metaplasia of the stomach are demonstrated in **Figure 8**. sIgA is uniquely resistant to protease digestion. It is of note that IgG can also be secreted onto the mucosal surface after coupling with IgG Fc-binding protein, a unique IgG Fc receptor of secretory type, produced by mucin-secreting cells in the mucosa [23, 24].

Extrathymic T-lymphocytes are distributed among the mucosal columnar cells as "intraepithelial lymphocytes" predominantly expressing CD8 and T-cell receptor gamma/delta on the cell surface [25] (**Figure 9**). Intraepithelial lymphocytes significantly increase in certain mucosal infections such as *Giardia lamblia* infection (giardiasis). The extrathymic T-lymphocytes locally recruit in the "crypt patch" located in the lamina propria mucosae. Because of the lack of education in the thymus, the extrathymic T-lymphocytes, self-reactive to provoke apoptosis of the

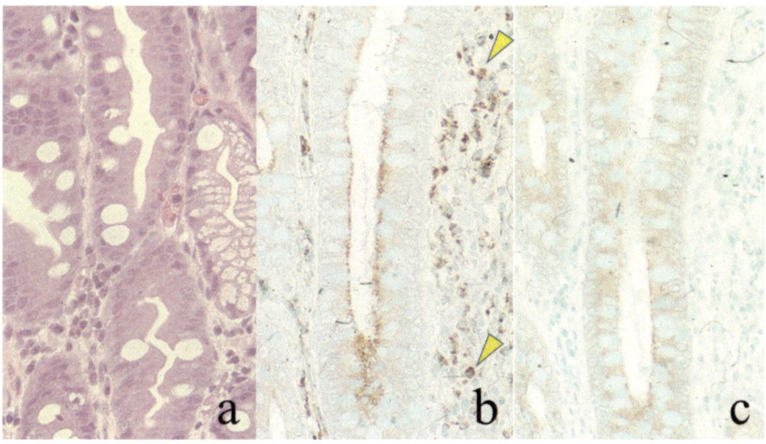

Figure 8.
Intestinal metaplasia of the gastric mucosa showing secretory IgA transportation. a: H&E, b: IgA, c: Secretory component (SC). Goblet cells and absorptive-type cells with brush borders are observed in the metaplastic gland. The columnar cells of the intestinal type produce SC, which traps dimeric IgA secreted from IgA plasma cells (arrowheads) in the lamina propria mucosae. The apical cytoplasm of the columnar cells is immunoreactive for IgA, representing the intracellular transportation of sIgA.

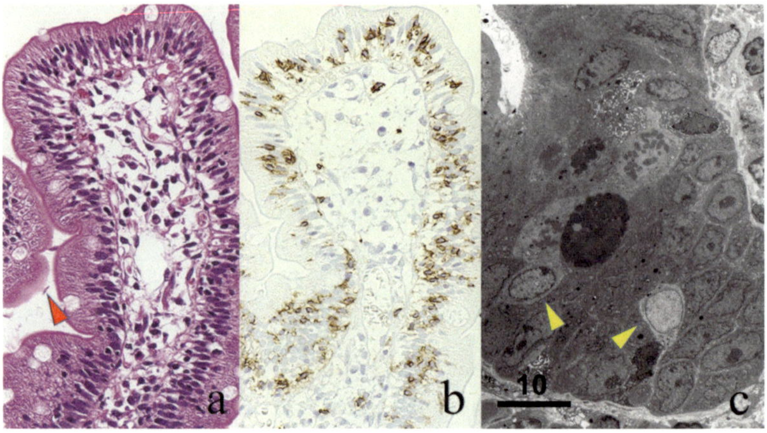

Figure 9.
Intraepithelial lymphocytes (IELs) in the duodenal mucosa, accentuated by Giardia lamblia *infection. a: H&E, b: CD8 immunostaining, c: Electron microscopy. Marked increase of IELs is noted in the intestinal villi. The red arrowhead indicates a giardian protozoa. CD8 immunostaining clearly demonstrates the dense distribution of the killer-type T-lymphocytes among the columnar epithelial cells. Ultrastructurally, small lymphocytes are seen in the intercellular space among the columnar cells (yellow arrowheads). Small cytoplasmic granules are scattered in the cytoplasm. Bar = 10 μm.*

columnar epithelial cells, may control the number and function of indigenous bacterial flora living in the lumen.

The mucosa-associated lymphoid tissue (MALT) is distributed in the intestinal and airway mucosa. The largest MALT is called as Peyer's patch in the ileal mucosa (**Figure 10**). The B-lymphocyte-rich lymphoid follicles with activated germinal centers are covered with dome-shaped columnar epithelial cells without villous structures. In contrast to the other part of the gut mucosa, B-lymphocytes and microfold (M) cells are distributed among the dome columnar cells. The M cells are special epithelial cells suited for efficient endocytosis and transcytosis, and function as gateways to the mucosal immune system [26]. The MALT is known to play a central role in the mucosal homing of B-lymphocytes destined to secrete IgA.

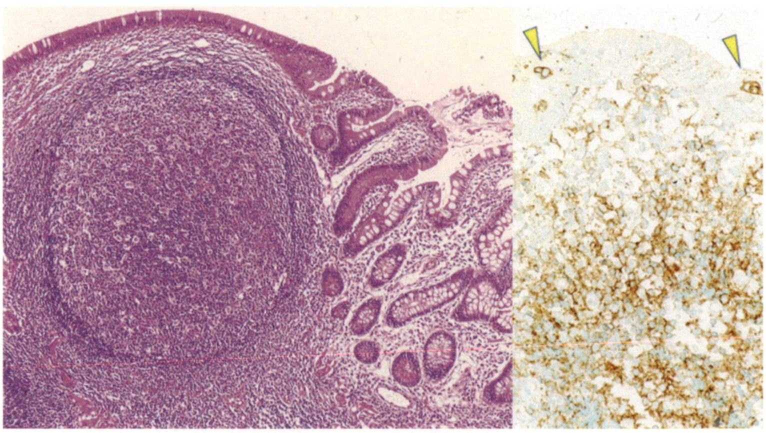

Figure 10.
Peyer patch in the ileal mucosa. Left: H&E, right: CD20 immunostaining. The lymphoid follicle with an activated germinal center is covered with dome-shaped columnar epithelial cells without villous structures. CD20-positive B-lymphocytes are distributed not only in the lymphoid follicle but also among the dome columnar cells (arrowheads).

2.4 Three major patterns of host responses against infection

From the pathological point of view, there are three major mechanisms of host responses against infection, depending on the type of pathogens and the mode of infection (either extracellular or intracellular infection).

1. Neutrophilic reaction against extracellular pathogens

2. Cellular immune reaction against intracellular pathogens

3. Humoral immunity via neutralizing antibody reaction

Table 2 summarizes the features of the defense mechanisms and host responses against pathogens. Patterns of the host response against pathogens are listed up in **Table 3.**

2.4.1 Neutrophilic reaction against extracellular pathogens

The extracellular pathogens growing outside the host cell, such as suppurative bacteria (*Staphylococcus*, *Streptococcus*, *Escherichia coli*, *Pseudomonas aeruginosa*, *Actinomyces*, etc.) and hypha-forming fungi (*Candida*, *Aspergillus* and *Mucorales*),

Defensing cell type	Pathogen	Host response	Compromised condition
Neutrophils	Extracellular pathogens	Abscess/phlegmone	Neutropenia
T-cells/Macrophages	Intracellular pathogens	Granuloma/lymphocytic infiltration	Cellular immunodeficiency*
B-cells/Neutralizing antibodies	Bacterial capsule/exotoxin viremic viruses	Not specified	Complement deficiency

**Steroid administration and acquired immunodeficiency (AIDS).*

Table 2.
Defense mechanism, pathogens and host responses.

I. **Neutrophilic accumulation** (abscess/phlegmone) against extracellular pathogens: Infection by suppurative bacteria, Enterobacteria, anaerobic bacteria, Actinomyces, Candida, Aspergillus and Mucorales provokes neutrophilic reactions.

II. **Cellular immunity** reactive against intracellular (cytozoic) pathogens:
   1. **Granuloma formation**: tuberculosis, Hansen's disease (tuberculoid leprosy), syphilis (stage III), typhoid fever, cryptococcosis, histoplasmosis, coccidioidomycosis, toxoplasmosis and leishmaniasis
   2. **Lymphocytic infiltration**: viral infection and rickettsiosis
   3. **Plasmacytic infiltration**: syphilis (stages I and II)
   4. **Macrophage infiltration**: xanthogranuloma/malakoplakia, *Legionella* pneumonia, lepromatous leprosy

III. **Host responses of other types**
   1. **Suppurative granuloma**: cat scratch disease (bartonellosis), lymphogranuloma venereum (chlamydiosis), listeriosis, yersiniosis, melioidosis, tularemia, brucellosis, sporotrichosis and chromomycosis
   2. **Eosinophilic infiltration**: infestation of nematodes and trematodes
   3. **Eosinophilic granuloma**: allergic bronchopulmonary aspergillosis
   4. **Foreign body granuloma**: parasitic ova (worm egg tubercle), primary *Anisakis* infestation

Table 3.
Host response against pathogens.

are principally phagocytized by neutrophils [27]. Infection of the extracellular pathogen thus results in abscess and phlegmonous inflammation, solely composed of neutrophils. In case of infection by non-invasive (extracellular) protozoa like *Trichomonas vaginalis*, neutrophilic exudation is activated to characteristically form so-called "cannon balls" (clusters of neutrophils). When the specific antibodies against the pathogen and complements are present in the body fluid, the phagocytic activity of neutrophils is significantly enhanced through an opsonin effect. Capsule-forming microbes frequently escape phagocytosis by neutrophils. In case of infection by anaerobic bacteria, massive ischemic necrosis is commonly associated.

Representative cytological features of neutrophil-mediated inflammation are illustrated in **Figure 11**. Accumulated neutrophils grossly correspond to pus (pyogenic exudates).

2.4.2 Cellular immune reaction against intracellular pathogens

Neutrophils and antibodies are ineffective against intracellular (cytozoic) pathogens. Instead, cell-mediated immunity functions as the major defense mechanism [28]: The infected host cells themselves are eliminated. The intracellular pathogen represents virus, chlamydia, rickettsia, protozoa, yeast-form fungi (*Cryptococcus*, *Histoplasma*, *Coccidioides*, etc.) and certain types of bacteria such as *Mycobacterium*, *Legionella* and *Salmonella*. Microscopically, lymphocytic infiltration (CD8-positive T-cell reaction) or granulomatous reaction is seen.

When macrophages are activated by the T-lymphocytes, epithelioid granulomas are formed. The epithelioid cells derive from activated macrophages. Multinucleated giant cells of Langhans type are often formed by plasma membrane fusion of the macrophages. Typical examples include tuberculosis, cryptococcosis and stage III syphilis. Necrosis is often associated with the granulomatous reaction, and in case of tuberculosis, caseous necrosis is characteristic. Typically, epithelioid granuloma with central necrosis is surrounded by small lymphocytes. Examples of cytology appearance of the granulomatous reaction in tuberculosis are shown in **Figure 12**.

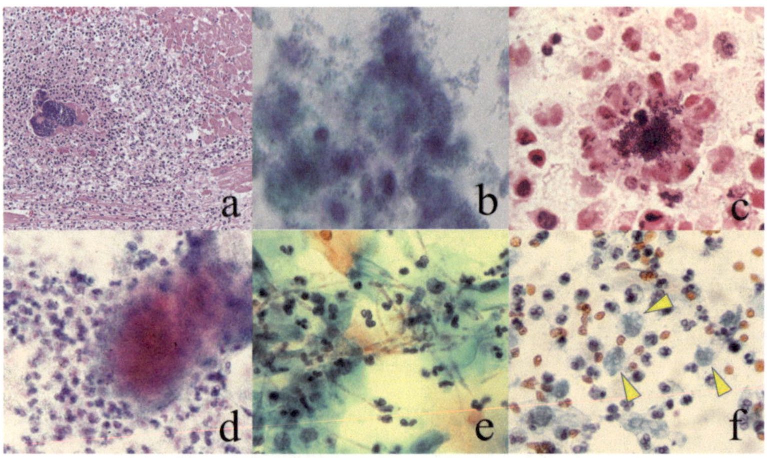

Figure 11.
Neutrophilic response against extracellular pathogens. a: MRSA-induced microabscess in the heart muscle in septicemia (H&E), b: Enterococcus faecium*-induced acute cystitis (urine sediment, Papanicolaou), c:* Streptococcus milleri*-induced pyothorax (Gram), d:* Actinomyces israelii*-induced endometritis (Papanicolaou), e: Hypha-forming* Candida albicans*-induced vaginitis (Papanicolaou), f:* Trichomonas vaginalis*-induced vaginitis (Papanicolaou). Extracellular pathogens, including* T. vaginalis *(arrowheads), are attacked by neutrophils. Neutrophils can fight against microbes larger than themselves (e and f).*

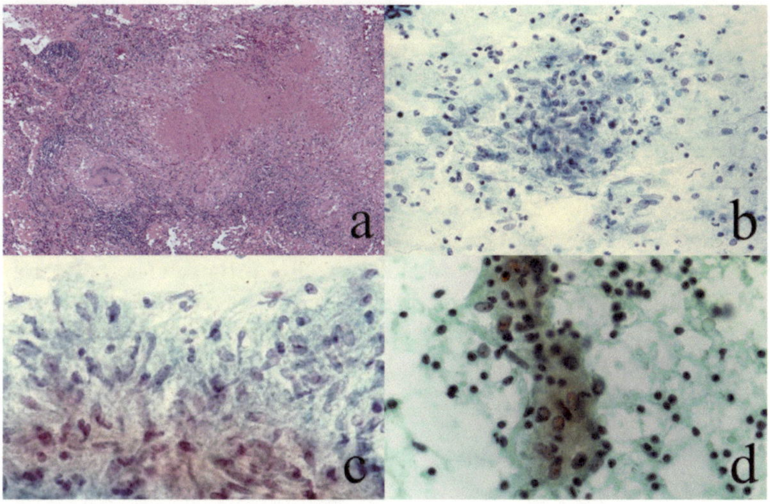

Figure 12.
Granulomatous inflammation against Mycobacterium tuberculosis. *a: Caseous granuloma in the lung (H&E), b and c: Epithelioid granuloma (Papanicolaou), d: A Langhans-type giant cell with lymphocytic infiltration. Three-layered structure is seen in caseous granuloma: Central caseous necrosis, epithelioid granuloma and lymphocytic cuffing (a). The epithelioid cells are rounded-spindle-shaped (c). Necrotic background is observed in b, and lymphocytes surround the multinucleated giant cell in d.*

In contrast, viral and rickettsial infection provokes infiltration of small T-lymphocytes without granulomatous reaction (**Figure 13**). In case of viral meningitis, lymphocytes are the major component in the cerebrospinal fluid. When lymphocytes are predominantly seen in the urine, the possibility of follicular cystitis caused by persistent infection of beta-hemolytic streptococci should be considered [29]. Similarly, the cervical smear preparation may contain lymphocytic reactions to be diagnosed as follicular cervicitis, and the possibility of *Chlamydia trachomatis* infection should be considered [30].

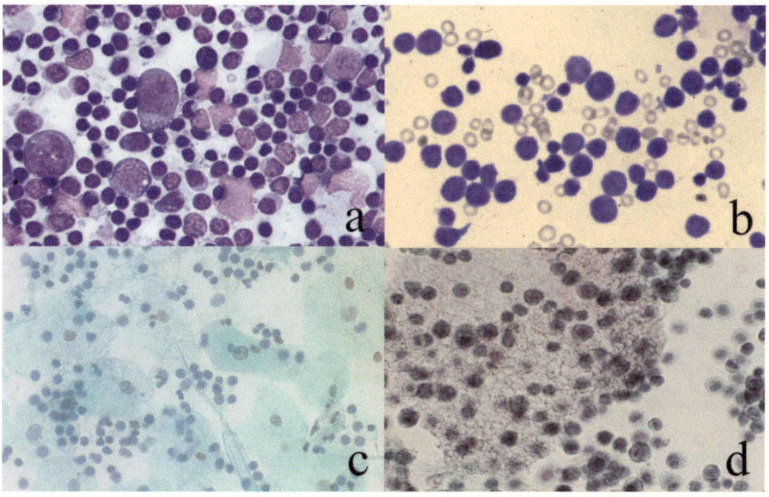

Figure 13.
Lymphocytic reaction in cytology specimens. a: Stamp smear of the reactive lymph node (Giemsa), b: Infiltration of activated lymphocytes in cerebrospinal fluid in a case of varicella-zoster virus-induced meningitis (Giemsa), c: Small lymphocytes seen in urine (Papanicolaou), d: Pneumocystis jirovecii *Pneumonia (Papanicolaou). In the lymph node, activated large-sized blastic cells are intermingled with small lymphocytes (a). Lymphocytes in viral meningitis reveal activated features with enlarged nuclei (b). Small lymphocytes are rich in the urine in a case of follicular cystitis (c). Small lymphocytes are seen around a cluster of* P. jirovecii, *resembling clustered hemolytic red cells (d).*

In case of *Legionella* pneumonia, macrophages comprise the main cellular reactant scarcely with lymphocytic infiltration. In cutaneous and visceral leishmaniasis, activated macrophages actively phagocytize the protozoan bodies. In visceral leishmaniasis (kala azar), the microbes are seen in activated Kupffer cells. In stages I and II syphilis, dense infiltration of plasma cells is characteristic. The gastrointestinal mucosa with chronic active gastritis and inflammatory bowel disease, as well as the nasal mucosa with chronic rhinitis and the gingival tissue with periodontitis or radicular dentigerous cyst, are also densely distributed by plasma cells. Plasma cells are often clustered in inflammatory foci of subacute inflammation or a subacute phase of inflammation [31]. Cytological features are represented in **Figure 14**. The cellular immunity-mediated removal of the infected parenchymal cells may cause functional insufficiency of the organs and tissues: Examples include hepatitis and encephalitis.

2.4.3 Humoral immunity via neutralizing antibody reaction

Neutralizing antibodies in the serum effectively eliminate pathogens that are distributed extracellularly. Typically, the neutralizing antibodies are produced against bacterial exotoxins, bacterial capsules and viral virions. Anti-viral antibodies are effective against the viral particles in the blood (during viremia) and body fluid. These features are applied to vaccination practice, and permanent immunity can be expected [32]. Vaccines injected subcutaneously induce IgG-type neutralizing antibodies in the serum. Oral vaccines such as Sabin vaccine against poliovirus and Rotavirus vaccine induce secretory IgA in the gut lumen. It is of note that IgG-type neutralizing antibodies in the serum can be transported with mucin by the IgG Fc-binding protein secreting from mucin-producing cells [23, 24]. Individuals with inherited complement deficiency, particularly the deficiency of C3 (the major opsonin), are vulnerable to recurrent pyogenic infections especially with

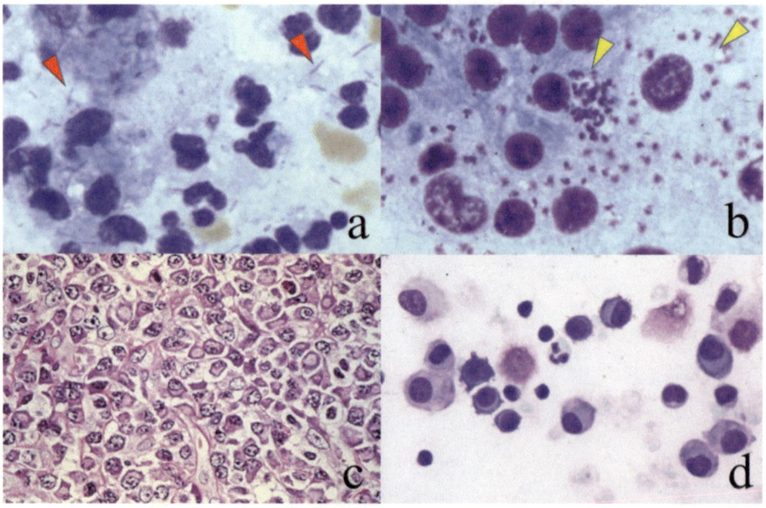

Figure 14.
Macrophage activation and plasma cell infiltration. a: Stamp preparation of Legionella pneumophila *pneumonia (Giemsa), b: Stamp preparation of the spleen with visceral leishmaniasis, c: Stage II syphilis (skin biopsy, H&E), d: Plasma cell infiltration in pleural effusion (Giemsa). Macrophages are activated but with minimal lymphocytic response in legionnaire's pneumonia and in leishmaniasis. The pathogens, rods (red arrowheads) in a and protozoan bodies (yellow arrowheads) in b, are phagocytized by the macrophages. Dense infiltration of plasma cells is a microscopic hallmark of stages I and II syphilis (c). The appearance of plasma cells in pleural effusion is infrequently encountered, since the plasma cells are poorly migratory. A subacute phase of infective pleuritis is suggested.*

encapsulated bacteria, including *Streptococcus pneumoniae* and *Neisseria meningitidis*. It is of note that sporadic meningococcal meningitis in adults may accompany inherited complement deficiency [33].

The activated humoral immunity is microscopically represented by follicular hyperplasia with enlarged germinal center formation in the lymph node and a variety of organs and tissues [34]. In fact, follicular hyperplasia is a microscopic feature of autoimmune disorders [35]. It has been clarified that a variety of cytokines are secreted from immunocytes to communicate each other and to secrete immunoglobulins. In particular, tumor necrosis factor (TNF) receptor-1 signaling is required for the differentiation of follicular dendritic cells, germinal center formation and antibody responses [36]. The germinal center also contains stimulated lymphocytes secreting interleukin-2 (IL-2) and interferon-gamma (IFN-γ) [37]. Representative examples of activated humoral immunity are demonstrated in **Figure 15**.

2.5 Host reactions against pathogens of other types

Host reactions against pathogens of other types are commented below.

1. Suppurative granuloma: An intermediate form of abscess and granuloma, that is called "suppurative granuloma" (abscess surrounded by granuloma), is seen in cat scratch disease (Bartonella infection), tularemia, listeriosis, yersiniosis, melioidosis and cutaneous mycosis (sporotrichosis and chromomycosis) [38]. Microscopic features of cat scratch disease involving the spleen are demonstrated in **Figure 16**.

2. Xanthogranuloma and malakoplakia: In certain situations, low-virulent extracellular pathogens, particularly *E. coli*, may grow intracellularly in the cytoplasm of macrophages, resulting in formation of xanthogranuloma (yellow-colored granuloma) [39]. Malakoplakia, a special form of the

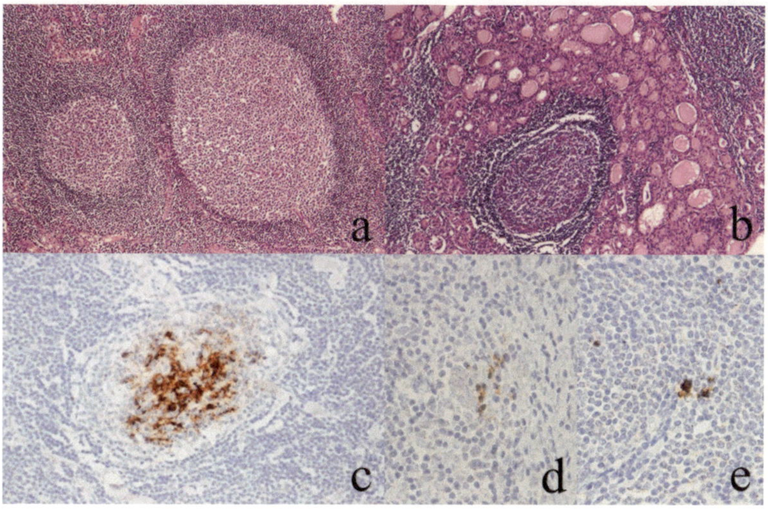

Figure 15.
Follicular (germinal center) hyperplasia, representing enhanced humoral immunity. a: Enlarged cervical lymph node (H&E), b: Hashimoto thyroiditis (H&E), c: TNF-alpha, d: IFN-gamma, e: IL-2. Lymphoid follicle formation with enlarged germinal centers microscopically indicates enhanced B-lymphocyte activation. Autoimmune disorders like Hashimoto thyroiditis often show follicular (germinal center) hyperplasia. TNF-alpha is densely deposited on the follicular dendritic cells (c). Lymphocytes immunoreactive for IFN-gamma and IL-2 are observed in the activated germinal center.

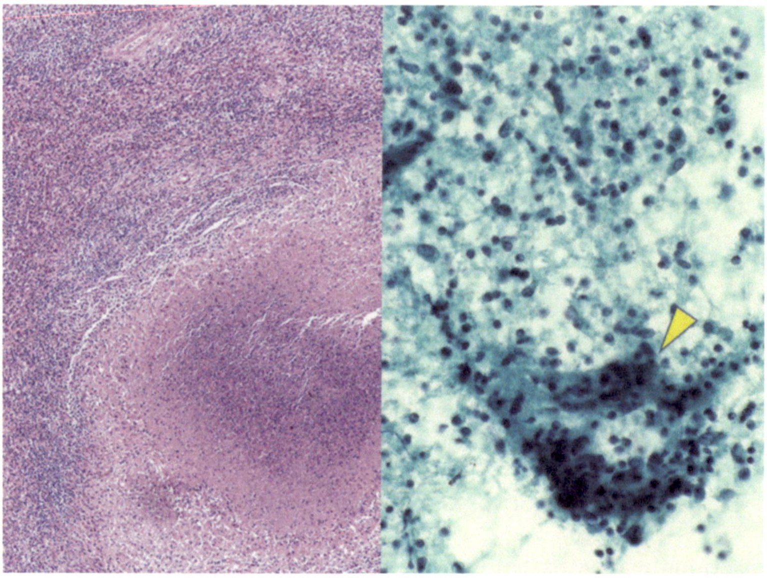

Figure 16.
*Suppurative granuloma seen in cat scratch disease (*Bartonella henselae *infection). Left: H&E (spleen), right: Stamp preparation (Papanicolaou). The surgically resected spleen grossly contains plural splenic abscesses. Histologically, suppurative granulomas, consisting of central abscess and surrounding epithelioid granuloma, are noted. Stamp cytological preparation demonstrates a cluster of epithelioid cells (arrowhead) in mildly necrotic background admixed with neutrophils and lymphocytes.*

xanthogranuloma, is microscopically featured by Michaelis-Gutmann bodies (round-shaped calcified and basophilic cytoplasmic inclusions). These lesions may be seen in the kidney, urinary bladder, epididymis, colon and gallbladder. Examples of malakoplakia and xanthogranulomatous inflammation are illustrated in **Figure 17**.

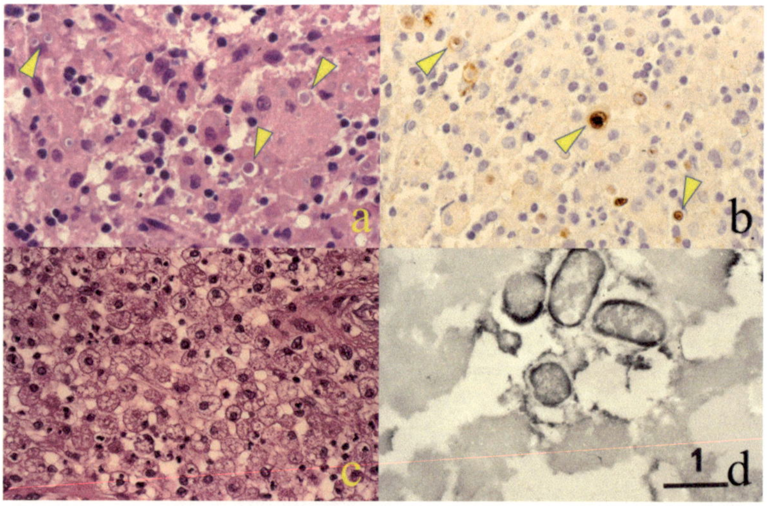

Figure 17.
Malakoplakia of the rectal mucosa (a, b) and xanthogranulomatous epididymitis (c, d). a and c: H&E, b and d: Immunostaining for E. coli *antigen (pre-embedding immunoelectron microscopy using a paraffin section in d). Malakoplakia is microscopically featured by Michaelis-Gutmann bodies, rounded basophilic cytoplasmic inclusions immunoreactive for* E. coli *antigens (arrowheads). Xanthogranuloma consists of accumulated foamy macrophages. Both lesions are caused by* E. coli *infection under an immunocompromised condition. Rod-shaped bacteria with cell wall labeling are proven in the cytoplasm of the foamy macrophage (d). Bar = 1 μm.*

3. Eosinophilic infiltration: Infestation of parasites, particularly round worm (nematode) and fluke (trematode), provokes infiltration of eosinophils and IgE-type immune response, common with the type 1 allergic reaction [40] (**Figure 18**). Cestode (tapeworm) infestation usually lacks the eosinophilic reaction. In case of allergic lung reaction against *Aspergillus* (allergic bronchopulmonary aspergillosis), eosinophilic granuloma (granuloma with eosinophilic infiltration) is observed [41]. Occasionally, foreign body reactions against worm bodies and ova are observed. Formation of multinucleated foreign body giant cells is characteristic (**Figure 19**). Parasitic ova induce foreign body granulomas to form so-called "worm egg tubercles". When eosinophilic infiltration is associated, immune-mediated eosinophilic granulomas are formed [42]. In case of anisakiasis, foreign body reactions without eosinophilic infiltration are seen against the Anisakis larva if the infestation occurs for the first time in a non-immunized patient after eating raw sea fish [43]. The mechanisms may be similar to the nonspecific phagocytic action of macrophages against genuine foreign bodies such as asbestos bodies and injected paraffin by augmentation mammoplasty.

4. Acellular hemorrhagic necrosis: In opportunistic infection associated with neutropenia and cellular immunodeficiency, the inflammatory cellular reactions are poorly provoked, resulting in hemorrhagic necrosis of the tissue [44].

3. Use of immunocytochemistry for cytology specimens

You must not give up additional staining even when you have one and only cytology specimen in hand. The resources for applying immunocytochemistry to the one and only cytology specimen are presented below. In case of liquid-based cytology (LBC), additional plural specimens are easily available. In Japan, however, the LBC procedure is still underdeveloped because of the low cost-performance. Therefore, these techniques are practically valid and helpful [45].

3.1 Cell block preparation

The sediments of liquid specimens such as pleural and pericardial effusions, ascites, the content of cystic lesions and urine can be kept for a long period of time as cell blocks after formalin fixation and paraffin embedding [46]. Immunostaining and *in situ* hybridization (ISH) method can be performed by preparing multiple paraffin sections from the cell block. So far, several technical inventions have been reported how to prepare cell blocks [47].

In **Figure 20**, chronic active Epstein–Barr virus (EBV) infection seen in a male case aged at his 20's is presented [48]. The patient manifested collagen disease-like signs and symptoms, such as fever, skin rash, muscle weakness, liver dysfunction and eosinophilia. In the ascites cytology specimen, a number of large granular lymphocytes (a form of atypical lymphocytes) were detected in the background with red cells, eosinophils and hemophagocytic macrophages. A cell block was prepared to know the nature of the lymphoid cells. The large-sized lymphocytes expressed natural killer cell markers such as CD45 and CD56, and EBV-encoded small nuclear RNA (EBER) was demonstrated in the nuclei by the ISH technique. The final diagnosis was chronic active EBV infection with virus-associated hemophagocytic syndrome. The prognosis of this disease is poor. In fact, the patient died of duodenal ulcer perforation seven months later. Of note is that EBV does not

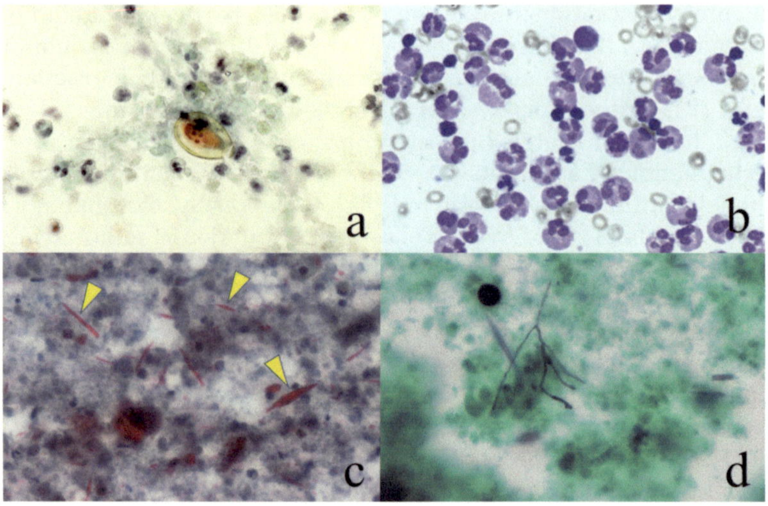

Figure 18.
Eosinophilic infiltration. a: Bile cytology in clonorchiasis (Papanicolaou), b: Eosinophilia in pleural effusion in tuberculosis (Giemsa), c and d: Allergic bronchopulmonary aspergillosis (sputum cytology, c: Papanicolaou and d: Grocott). A small-sized ovum of Clonorchis sinensis *with miracidium formation is seen in the bile and surrounded by eosinophils with bilobed nuclei (a). Eosinophils densely seen in the hemorrhagic pleural effusion in a case of tuberculosis may represent an allergic reaction against acid-fast bacilli (b). In allergic bronchopulmonary aspergillosis, rhomboid-shaped and red-colored Charcot-Leiden's crystals (arrowheads) deriving from eosinophilic granules are seen. Degraded eosinophils are observed in the background (c). Grocott stain identifies a few distorted* Aspergillus *hyphae phagocytized by a multinucleated giant cell (d).*

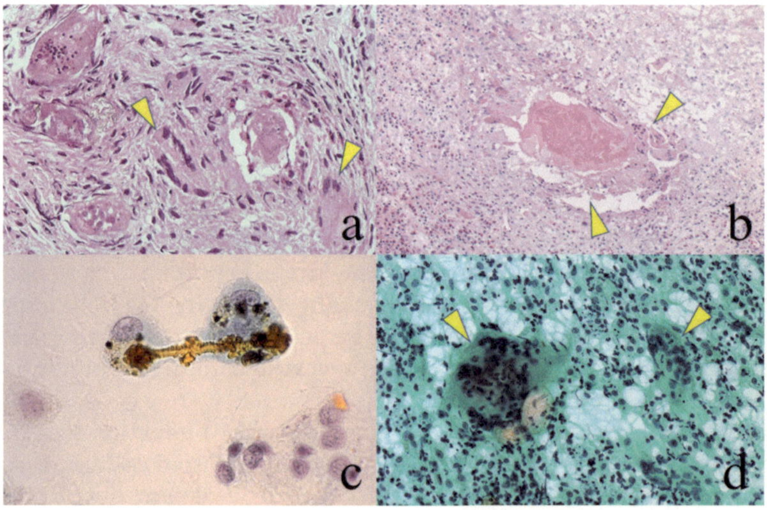

Figure 19.
Foreign body granulomatous reactions. a: A worm egg tubercle formed in the colonic submucosa in Schistosoma haematobium *infestation (H&E), b: An omental nodule by* Anisakis *larva migration (H&E), c: Asbestos body in the sputum (Papanicolaou), d: Fine needle aspiration from a nodular lesion post augmentation mammoplasty (Papanicolaou). Foreign body reactions with multinucleated giant cells are noted around schistosoma eggs with miracidium formation (a) and a dead nematode larva (b). Eosinophilic reactions are scarcely seen. For the comparison, two examples of genuine foreign bodies are shown. Asbestos bodies (c) and paraffin oil droplets (d) injected by augmentation mammoplasty are phagocytized by macrophages. A long, brown-colored asbestos fibril is engulfed by two macrophages in c. vacuolated cytoplasm filled with lipid-soluble substances is characteristic, and multinucleated giant cells are dispersed in d. Arrowheads indicate multinucleated giant cells.*

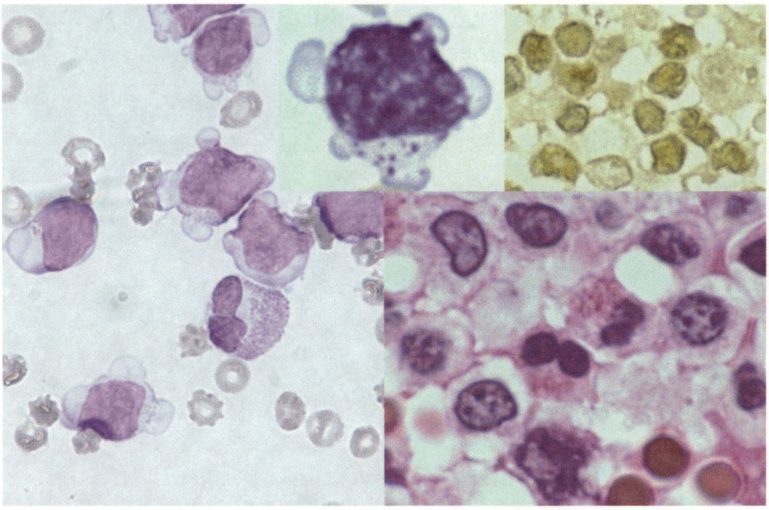

Figure 20.
Chronic active EBV infection (ascites cytology, left, Giemsa, left inset: High-powered view, right: Cell block H&E, right inset: EBER). Hemorrhagic ascites contains large granular lymphocytes and eosinophils. Azurophilic granules are noted in the cytoplasm of the large granular (atypical) lymphocytes. In cell clock preparation, the nuclei of atypical lymphocytes are positive with EBER-ISH method. No intranuclear inclusions are visible.

produce viral particles in the infected cell, so that no intranuclear inclusions are formed and thus the EBER technique is needed.

3.2 Re-staining method

A re-staining method is applicable to the single (one and only) cytology specimen [49, 50]. At first, the cells or areas of target should be marked on the back side of the glass slide with a diamond-tip pen and then photomicrographed. After removal of the coverslip in xylene, stained dyes can be removed by dipping in acid alcohol solution (50% ethanol containing 0.5% hydrochloric acid) for hours or simply dipping specimens in tap water overnight. The immunostained cells or areas of target are re-photomicrographed for comparison.

In case of Giemsa-stained glass slides or immunostained preparations on trimethoxy[3-(phenylamino)propyl]silane-coated glass slides, the re-staining method is especially valuable, since the cell transfer technique described below is not applicable due to tight attachment of the cells.

In **Figure 21** showing scraping cytology of herpes simplex virus (HSV) infection on the vulva, the re-staining method visualizes intranuclear and intracytoplasmic viral antigens. A commercially available polyclonal antibody was used for immunostaining. Vulvar HSV infection belongs to sexually transmitted disease (STD).

Figure 22 demonstrates chlamydial cytoplasmic inclusions, so-called "nebular inclusion bodies", in the scraping cytology from the uterine cervix. Chlamydiosis also represents STD. The inclusions are clearly re-stained with a monoclonal antibody B104.1 against *Chlamydia trachomatis*. Tiny cytoplasmic inclusions, visualized with immunostaining, are scarcely recognizable in the Papanicolaou-stained preparation.

3.3 Cell transfer technique

If you have one and only glass slide of Papanicolaou-stained cytology specimen or hematoxylin and eosin (H&E)-stained histology specimen and you want to

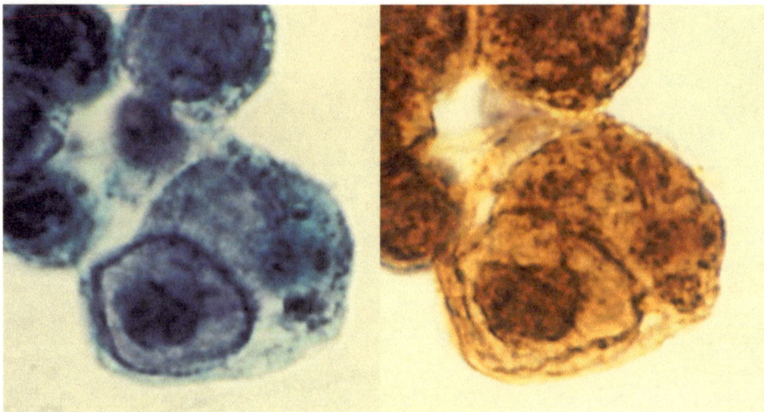

Figure 21.
Herpes simplex virus infection (scraped from vulva, left: Papanicolaou, right: HSV immunostaining). With re-staining method, viral antigen immunostained with a polyclonal antibody are localized both in the haloed nuclei and in the cytoplasm.

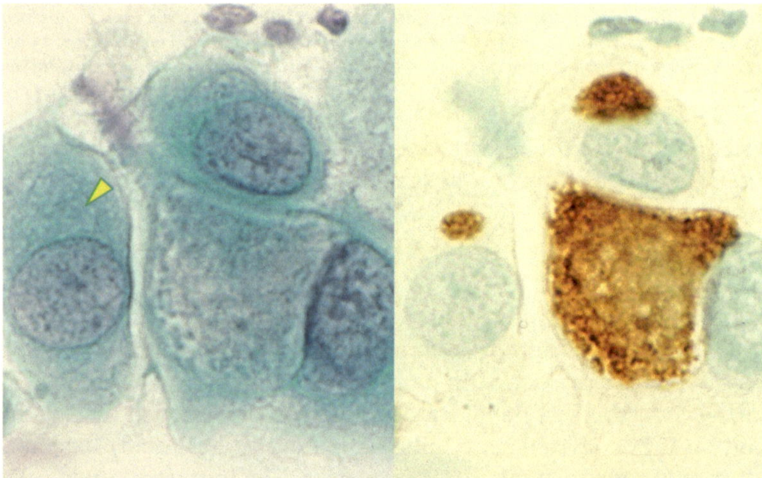

Figure 22.
Chlamydial infection (cervical smear, left: Papanicolaou, right: Immunostaining with monoclonal antibody). With re-staining method, the nevular inclusion bodies in the cytoplasm are clearly labeled for the chlamydial antigen. A tiny inclusion in the left-sided cell (arrowhead) is scarcely recognizable in the pap smear.

evaluate the expression of immunocyto/histochemical markers, the cell transfer technique [51] is quite useful and valuable (**Figure 23**). Firstly, cover slips must be removed by dipping in warmed xylene. Secondly, the specimen is covered with a mounting medium/resin at 2–3 mm thickness, in order to form a coating film of the solidified mounting medium in a warm incubator overnight. Then, the film of the solidified resin should be peeled off the glass slide by dipping in warm water for one hour to get the cells or tissues transferred onto the film side. The solidified resin film is placed in water on the silane-coated glass slide to be dried in a warm incubator. Finally, the resin component can be removed by dipping in xylene to get the cells and tissues transferred to a new glass slide. You can obtain plural glass slides if the solidified resin is cut by scissors into several pieces.

Cells smeared outside the cover slip can be transferred to another glass slide without removing the cover slip (**Figure 23**). This is particularly useful in case of gynecological cytology specimens.

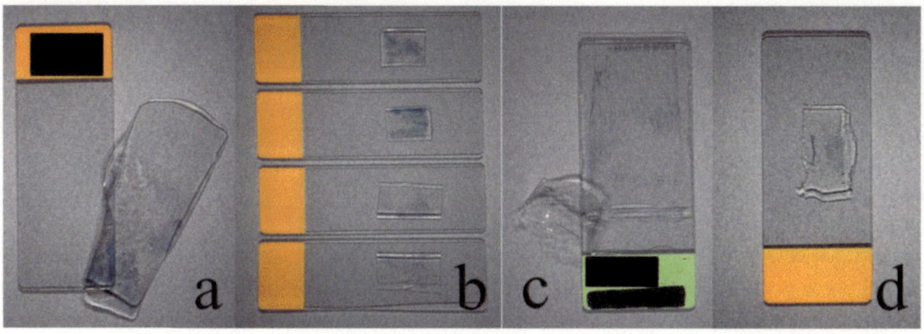

Figure 23.
Cell transfer technique. a&b: Papanicolaou-stained cells in the cervical smear are transferred to solidified resin film, and the film was cut into pieces to get plural specimens placed on silane-coated glass slides. c&d: Cells smeared outside the cover slip can be transferred to another glass slide without removing the cover slip.

By a conventional technique of cell transfer, it takes time to have the cells transferred. Itoh et al. [52] invented a time-saving method to get the cells transferred in one hour (**Table 4**). In brief, the mounting medium should be diluted two-fold by xylene, and the rein film should be solidified on a hot plate at 70–80C.

When the target cells in the specimen are few in number, it is recommended to have the cells marked with a diamond-tip pen on the back side of the glass slide before removal of the cover slip. When an archival long-kept and fully dried specimen is used, the cover slip removal is not easily achieved. In order to accelerate the removal, xylene solution should be warmed up to 70–80C and/or the cover slip should be cracked with a diamond-tip pen.

Harada et al. [53] reported that detachment of the cover slip is accelerated by using a packaging duct tape, as summarized in **Table 5**. **Figure 24** illustrates the step of Harada's method for rapid removal of the cover slip. The method is applicable to archival glass slides long kept at room temperature for 20 years. It takes only one hour to remove the cover slip. By combining Harada's method with the above-mentioned Itoh's quick method for the cell transfer, old cytology glass slides become ready for immunocytochemical analysis within a few hours.

1. *Removal of cover slip*
   The cover slip can be removed by dipping in warm xylene
2. *Application of mounting medium*
   The mounting medium (e.g. Malinol®, Muto Chemicals, Tokyo) should be two-fold diluted by xylene and one mL of the diluted resin is placed onto the smeared cells on the glass slide.
3. *Solidification*
   The resin should be solidified for 30 minutes on a hot plate at 70–80C.
4. *Softening*
   The glass slide with an adherent resin film should be soaked for 15 minutes in warm water at 50–60C.
5. *Detachment*
   The softened resin film can be peeled off the glass slide with forceps. The detached film can be divided into several pieces with scissors.
6. *Pasting*
   The resin film is soaked in water and pasted onto a new silane-coated glass slide. Caution is needed for the recognition of the right side of the film. After removal of excessive water, the cell-transferred glass slide should be dried on a hot plate at 70–80C. You can prepare plural glass slides when the resin film is cut into several pieces by scissors.
7. *Removal of solidified resin*
   The solidified resin can be removed by dipping in xylene. Hydration is then achieved through placing the glass slide in ethanol. The specimen is ready for immunostaining.

Table 4.
Itoh's time-saving cell transfer technique.

1. *Preparing warm xylene* A glass container is kept warm at 62–64C in an incubator.	
2. *Preparing a packaging duct tape cut to the cover slip size* The packaging duct tape should be cut to fit the cover slip, a little bit larger than the size of the cover slip.	
3. *Soaking of the glass slide in warm xylene* The glass slide is soaked in warm xylene for 40 minutes.	
4. *Wiping out xylene* Warm xylene is wiped completely from the cover slip with KimWipes® (Kimberly Clark Corp, Irving, Texas, USA).	
5. *Sticking the packaging duct tape* A piece of the packaging duct tape is evenly stuck onto the cover slip. Confirm uniform sticking of the packaging duct tape (without bubbling) by observing from the back side.	
6. *Peeling off the cover slip* The packaging duct tape should be removed together with the cover slip at a breath without hesitation.	
7. *Soaking the glass slide in xylene* The glass slide is then soaked in xylene at room temperature, being ready for the next step.	
Notice: The steps 4–7 should be performed as quick as possible.	

Table 5.
Harada's method for rapid removal of cover slip.

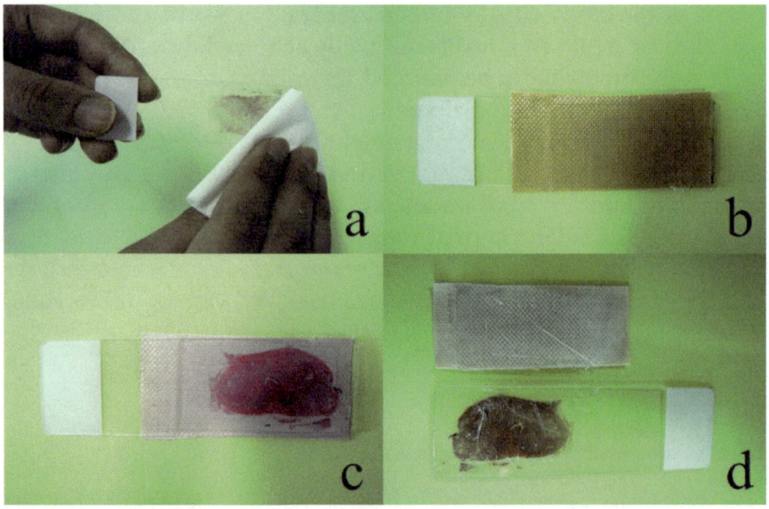

Figure 24.
Harada's rapid removal of cover slip using packaging duct tape (H&E-stained prostate section kept at room temperature for 20 years). a: Wipe off warm xylene with KimWipes®, b: Stick a piece of the packaging duct tape cut a little bit larger than the cover slip evenly onto cover slip, c: Confirm from the back side that the tape was uniformly stuck without bubbles, d: Peel off the packaging duct tape together with the cover slip at a breath.

Figure 25 displays identification of human papillomavirus (HPV) type 16 genome in the nuclei of severely dysplastic cervical squamous cells. The cells in the routine cervical smear were transferred onto the silane-coated glass slide, and the ISH technique was applied to localizing the viral genome in the nuclei of the dysplastic cells. The cell transfer step is essentially requested to avoid detaching the cells during the staining step, because heating pretreatment of the specimen is inevitable for the ISH technique. In this way, the Papanicolaou-stained cytomorphology and the state of HPV infection can directly be compared in the same cervical cells.

The cervical smear was prepared from a postmenopausal lady, and we had one and only glass slide in hand. The cells on the glass slide were transferred to another silane-coated glass slide. The clustered atypical parabasal keratinocytes in the background of senile colpitis were positively stained for p16-INK4a, indicating the

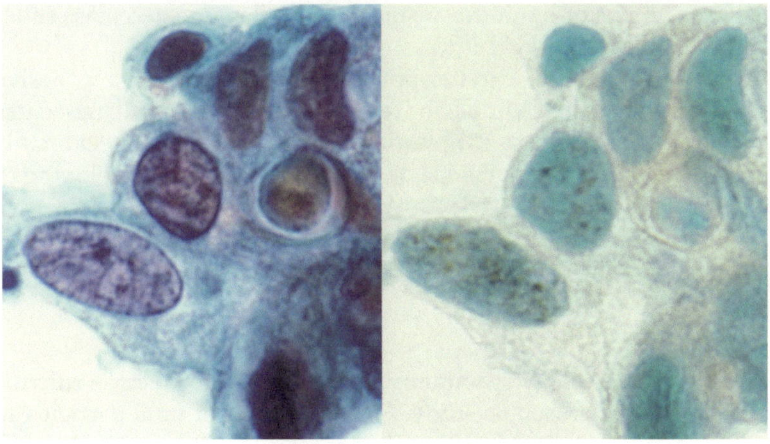

Figure 25.
Severe dysplasia of uterine cervix (left: Papanicolaou, right: ISH for HPV, type 16 genome). The cells were transferred onto a silane-coated glass slide to localize HPV, type 16 genome by ISH technique, requiring heating pretreatment. Dotted signals are seen in the dysplastic nuclei. The microscopic features of HPV infected cells can directly be compared with those of pap staining.

carcinogenic HPV infection (HPV-infected genuine dysplasia or high-squamous intraepithelial lesion) [45], as illustrated in **Figure 26**. In this way, genuine HPV-related dysplasia was distinguished from reactive (benign) atypia of the parabasal keratinocytes secondary to senile atrophy. Heating pretreatment is an essential step for immunolocalizing p16-INK4a that is a tumor suppressor gene product for modulating the cell cycle. Carcinogenic HPV infection inactivates retinoblastoma (RB) gene leading to the overexpression of p16-INK4a. In other words, the p16-INK4a is a marker of HPV-infected cells in the uterine cervix [54].

It should be noted that the cell transfer technique is applicable to paraffin sections, as well as Papanicolaou-stained cytology specimens in the gynecologic and respiratory fields, but Giemsa-stained cytology preparations employing dry fixation

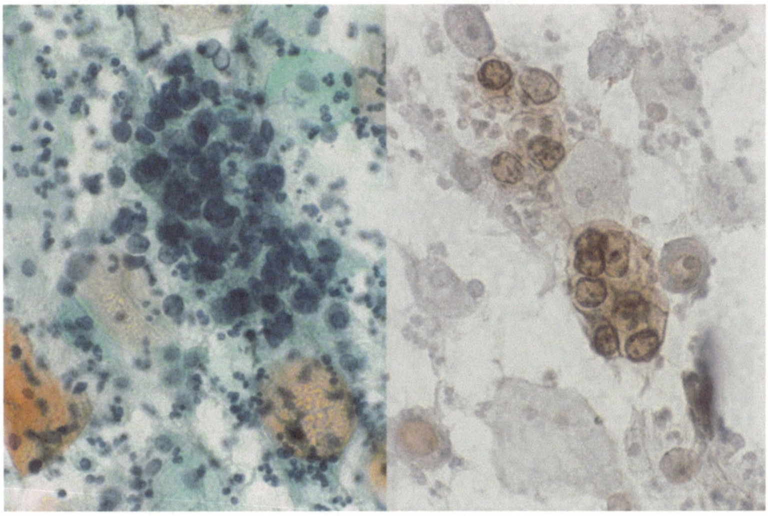

Figure 26.
Moderate dysplasia of uterine cervix after menopause (left: Papanicolaou, right: Immunostaining for p16-INK4a). Dysplastic change and reactive atypia secondary to senile colpitis should be distinguished. The expression of p16-INK4a confirmed the precancerous state of the cervix in this postmenopausal female patient. After the cell transfer onto the silane-coated glass slide, p16-INK4a was immunostained by employing heat-induced antigen retrieval.

and cytology specimens of liquid form smeared on silane-coated glass slides are not suitable for the cell transfer technique.

The cell transfer technique can be applied to the transfer of the cohesive cells cultured on the plastic slide (Nunc Lab-Tek® chamber) to the silane-coated glass slide. Xylene is not applicable to the plastic slide and the cover slip cannot be placed on the plastic slide. The broken glass slides can be repaired by employing the cell transfer technique (**Figure 27**) [55].

3.4 Application of cytology specimens to ultrastructural study and polymerase chain reaction analysis

The ethanol-fixed cytology specimens can be applied to electron microscopic and immunoelectron microscopic study. Fine structures of viral particles and chlamydial bodies are preserved even after ethanol fixation. **Figure 28** demonstrates immunoelectron microscopic observation of chlamydial antigen in a uterine cervical columnar cell. In the nebular inclusion body in the ethanol-fixed cell, both elementary bodies and reticulate bodies of the chlamydial microbe are clearly observed. The plasma membrane of the particles is positively labeled with the monoclonal antibody [56]. **Figure 29** schematically illustrates the cell cycle of *C. trachomatis* in the infected cell. Smaller-sized elementary bodies represent the infectious particles, while larger-sized reticulate bodies belong to the proliferative form. ISH technique can also be applied to cytology preparations [57].

Pathogens are often localized in a limited part in the specimen. It is practical and convenient for us to focus target on the infected cell for (immuno)electron microscopic study. At first, immunostaining with diaminobenzidine color reaction should be performed at the light microscopic level. After taking photomicrographs, the cover slip is removed, the specimen is re-fixed in 1% osmium tetroxide solution, and the cells are targeted for epoxy resin (Epon) block preparation by employing the inverted beam capsule method. Ethanol fixation accelerates penetration of antibody molecules into the cell, so that routine method for the light microscopy gives us an excellent result also at the ultrastructural level. Fine morphology of particulated microbes is well preserved even after ethanol fixation.

The cytology specimen can be analyzed with polymerase chain reaction (PCR) analysis by employing the cell transfer technique. The solidified resin film prepared

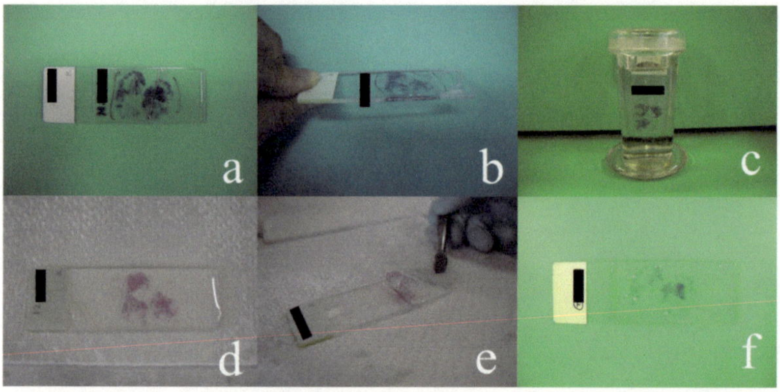

Figure 27.
Repair of a broken glass slide employing the cell transfer technique. a: The glass slide was broken, b: The broken slide is supported by another glass slide underneath adhered with epoxy glue, c: Cover slip is removed in warm xylene, d: Mounting medium (resin) is covered on the glass slide, e: Solidified resin film is peeled off after dipping in warm water, f: The resin film is pasted onto a new glass slide. After enough drying, the resin is removed in xylene and then a cover slip is set.

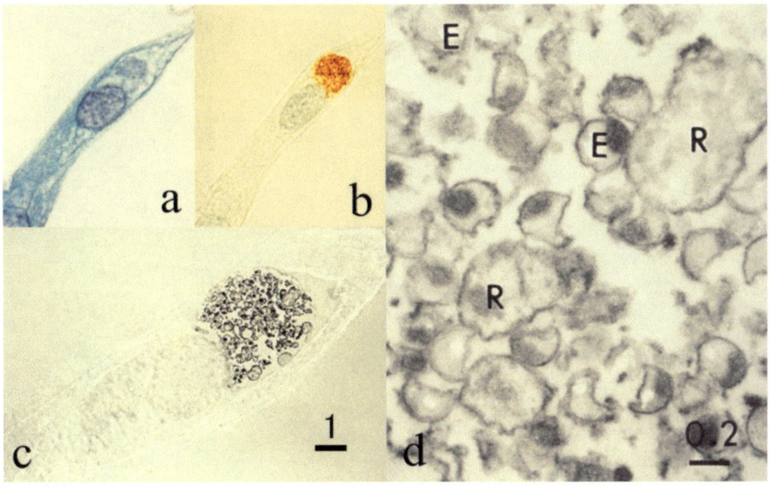

Figure 28.
Immunoelectron microscopic study for chlamydial antigen using an ethanol-fixed cervical smear. a: Papanicolaou, b: Immunostaining for Chlamydia trachomatis *antigen, c&d: Immunoelectron microscopy using the pre-embedding method, c: Low power, d: High power, E: Elementary body, R: Reticulate body, bars = 1 µm (c) and 0.2 µm (d). The chlamydial antigen is localized on the particle membrane. Reticulate bodies are larger than elementary bodies. Fine morphology is fairly well preserved even after ethanol fixation.*

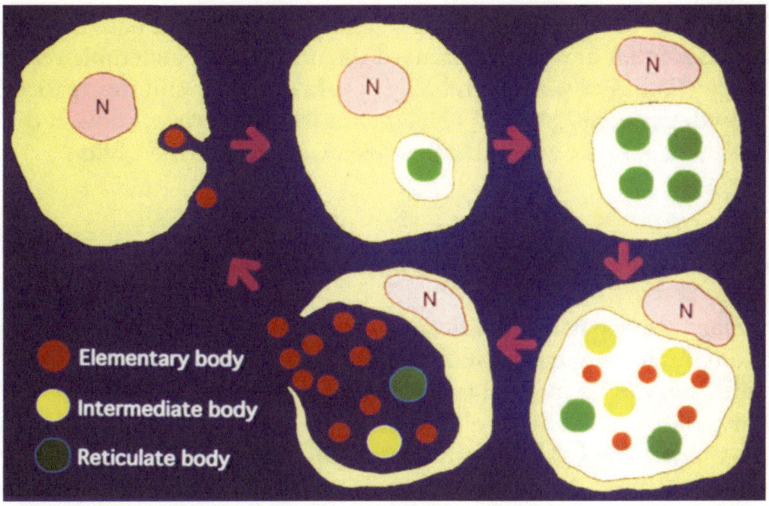

Figure 29.
Schematic presentation of the growth cycle of Chlamydia trachomatis. *Smaller-sized elementary bodies (red) infect the cell, and larger-sized reticulate bodies (green) proliferate to form intracytoplasmic nevular inclusions. Intermediate bodies (yellow) indicate a transitional form between the reticulate ad elementary bodies. N = nucleus.*

from a Papanicolaou-stained smear should be cut by scissors. Parts of the specimen are kept as Papanicolaou-stained slides, while DNA or RNA can be extracted from the other parts after xylene treatment [45].

Figure 30 illustrates *Entamoeba gingivalis* colonization in inflamed exudate seen around an intrauterine contraceptive device (IUD). Microscopically, amebic trophozoites are scattered around actinomycotic grains [58]. The patient aged in her 50's complained of white-colored fluor genitalis. After removal of the IUD, cytological specimens were sampled from the surface of the device. The postmenopausal lady totally forgot the artificial material inserted in her uterus. In order to confirm the diagnosis, PCR analysis was performed by utilizing the cell transfer technique.

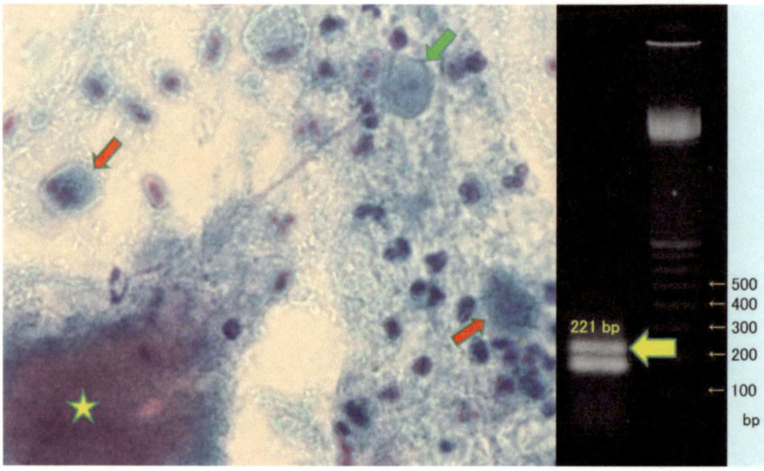

Figure 30.
Entamoeba gingivalis *colonized around an intrauterine contraceptive device, analyzed by PCR employing the cell transfer technique. Left: Papanicolaou, right: Electrophoresis banding of the amplified product. Amoeba-like cells phagocytizing neutrophils (red arrows) are seen around the actinomycotic grain (asterisk). The centrally located nucleus with a prominent karyosome is seen (green arrow). DNA was extracted from cell-transferred pieces. PCR was run for 35 cycles. DNA sequence of* E. gingivalis *was proven from the 221 base-paired band (yellow arrow). Sense primer: 5′-tcagataccgtcgtagtcct-3′, antisense primer: 5′-cctggtgtgcccttccgt-3′.*

The DNA sequence of *E. gingivalis* was identified from the 221 based band on the gel. It was evident that oral sex had caused the infection of anaerobic residents of the oral cavity (both *Actinomyces* and *E. gingivalis*) [59] around the artificial material. The growth of *Actinomyces*, an obligate anaerobic microbe, allowed the survival of *E. gingivalis*, an obligate anaerobic protozoan, in the uterine cavity.

4. Cytodiagnosis of bacterial infection

The cytology service has a significant role in the detection and presumptive identification of microorganisms [60]. Generally speaking, the cytodiagnosis of bacterial infection can be reached more easily for extracellular pathogens than for intracellular pathogens. It should also be noted that bacteria are more steadily observed in Giemsa-stained preparations than in Papanicolaou-stained preparations. For immunocytochemical confirmation, the techniques mentioned above (the usage of cell block and re-staining or cell transfer technique) are valuable. Representative examples of the cytodiagnosis of bacterial infection are described below.

4.1 Bacterial vaginosis

Large-sized Gram-positive rods, *Lactobacillus* or so-called Döderlein bacilli, are the normal flora (indigenous microbiota) of the vagina and maintain the local acidity at pH 3.8–4.5 by producing lactic acid. Their length ranges from 2 to 9 μm, with the width of 0.5 to 0.8 μm. The lactic flora produces hydrogen peroxide and antimicrobial peptides (bacteriocins) to inhibit growth of other microbes [61]. The number of the non-mobile bacilli is increased around the period of ovulation through the secretory phase of the menstrual cycle. Döderlein bacilli are seen in healthy mature (premenopausal) women, but after menopause without hormonal activity, they are no longer observed in the cytology preparation. In Papanicolaou-stained preparations, they look like homogeneously basophilic large rods without spore formation. The background generally shows paucity of inflammation, but they are occasionally phagocytized by neutrophils [62].

In case of bacterial vaginosis (vaginitis), abnormal bacteria grow and Döderlein bacilli are markedly decreased or totally disappear [63, 64]. Representative example is Gardnerella vaginitis caused by infection of *Gardnerella vaginalis*, a small (1 to 1.5 μm-sized) Gram-negative non-mobile coccobacillus. The small bacteria, *G. vaginalis*, often cluster on the squamous cells of the superficial type to form so-called "clue cells". *Gardnerella* infection is often evident in a proliferative phase of the menstrual cycle. The infection is commonly associated with neutrophilic reaction, but poor neutrophilic response may be seen in some cases, hence the term "bacterial vaginosis", instead of bacterial vaginitis.

Another microbe causing bacterial vaginosis is *Mobiluncus*, spp., a V-shaped or crescentic, mobile, obligate anaerobic bacillus with unstable Gram reactivity. The size is intermediate between Döderlein bacillus and *G. vaginalis*. *Atopobium (Fannyhessea) vaginae*, a small-sized (less than 1 μm), obligate anaerobic Gram-positive elliptical coccobacillus often forming a short chain (somewhat resembling streptococcus), is a recently reported member causing bacterial vaginosis [65]. The growth of filamentous long-shaped bacillus, *Leptothrix*, may be associated with bacterial vaginosis. After menopause, these bacteria may often be replaced by enterobacteriae such as *Escherichia coli* and *Klebsiella pneumoniae*. *Pseudomonas aeruginosa* may colonize the vaginal mucosa, accompanying biofilm formation (refer to **Figure 37**). All these microbes provoke neutrophilic exudation.

Figure 31 displays representative cytomorphology of vaginal bacteria in Papanicolaou-stained preparations.

4.2 Chlamydial infection

Chlamydiosis is a representative and common STD. Symptomatic non-gonococcal urethritis is seen in male patients, while symptoms are mild in females. Columnar epithelial cells infected with *Chlamydia trachomatis* contain round-shaped cytoplasmic inclusion bodies named as nevular inclusion bodies [66]. The

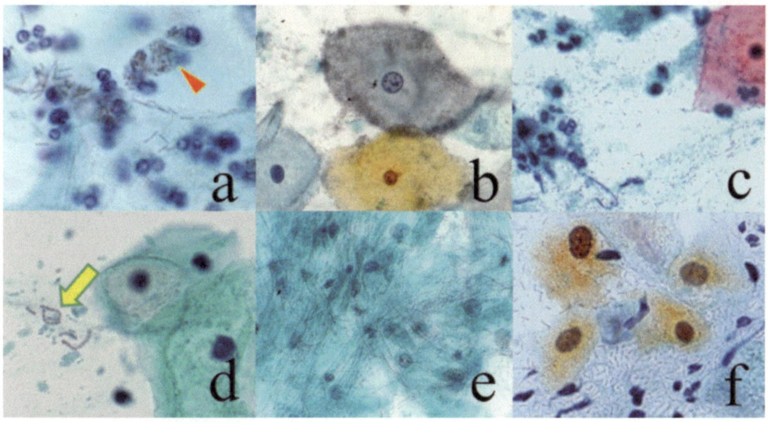

Figure 31.
*Döderlein bacillus (*Lactobacillus*) and microbes causing bacterial vaginosis (cervical smears, Papanicolaou). a: Döderlein bacillus, b:* Gardnerella vaginalis, *c:* Mobiluncus, *d:* Atopobium vaginae, *e:* Leptothrix, *f:* Klebsiella pneumoniae. *Döderlein bacilli, a normal vaginal resident, is large-sized and occasionally phagocytized by neutrophils (arrowhead).* G. vaginalis *is small-sized and often clustered on the superficial keratinocytes to form "clue cells".* Mobiluncus *is intermediate-sized and V- or crescent-shaped.* A. vaginae *appears as chained coccoid microbes, resembling* Streptococcus *(arrow).* Leptothrix *is non-pathogenic filamentous bacteria.* K. pneumoniae *Is a capsule-forming, large-sized bacillus mainly seen on the postmenopausal vaginal mucosa.*

life cycle of Chlamydia is shown in **Figure 29**. Immunocytochemistry using the restaining method or cell transfer technique is quite effective for making a diagnosis of chlamydiosis (refer to **Figures 22** and **28**). In most cases of chlamydiosis, bacterial vaginosis is associated, so that a variable number of neutrophils are seen in the background. Chlamydial infection commonly causes lymphoid follicle formation in the mucosa: Lymphocytic background may be seen in the cervical smear preparation, as *C. trachomatis*-associated follicular cervicitis [67].

Chlamydial inclusions are also seen in cytology specimens scraped from male urethra. *C. trachomatis* causes epididymitis and salpingitis. Extragenital chlamydiosis should be of notice [68]. Chlamydial pharyngitis and proctitis are mediated by oral sex and anal sex, respectively. Acute chlamydial conjunctivitis occurs in sexually active young men, and the cytoplasmic inclusions are demonstrated by quick Giemsa (Diff-Quik) staining. Representative features are shown in **Figure 32**.

4.3 Gonococcal infection

Gonorrhea is a classic example of STD. *Neisseria gonorrhoeae*, a Gram-negative paired coccus (diplococcus), causes acute urethritis in male, and it shows high affinity to urethral epithelial cells [69]. **Figure 33** demonstrates urine cytology from a 28 year-old single Japanese male patient. It should be of note that paired cocci are specifically attached onto the large-sized squamous cells of urethral origin. Typically, diplococci are phagocytized by neutrophils. Cytological diagnosis of gonorrhea can be made immediately.

4.4 Bacteria growing in liquid specimens and effusions

Acute cystitis is common in women, most often caused by *Escherichia coli* infection. Pyruria is an important sign of bacterial cystitis. In case of acute cystitis in

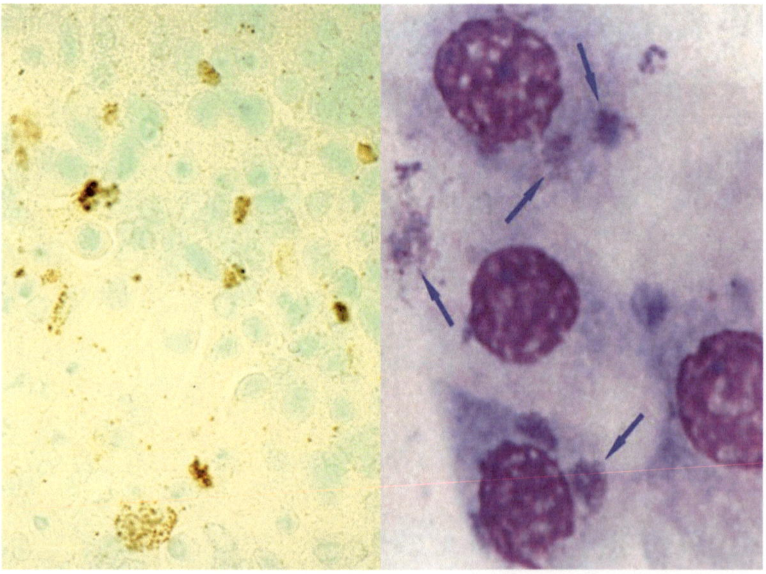

Figure 32.
Chlamydial infection (left: Immunostaining for chlamydial antigen in scraped male urethra with methylgreen counterstain, right: Giemsa-stained scraping cytology of conjunctiva). Numbers of urethral and conjunctival epithelial cells possess chlamydial cytoplasmic inclusion bodies. Note extragenital infection of C. trachomatis *on the eye (arrows).*

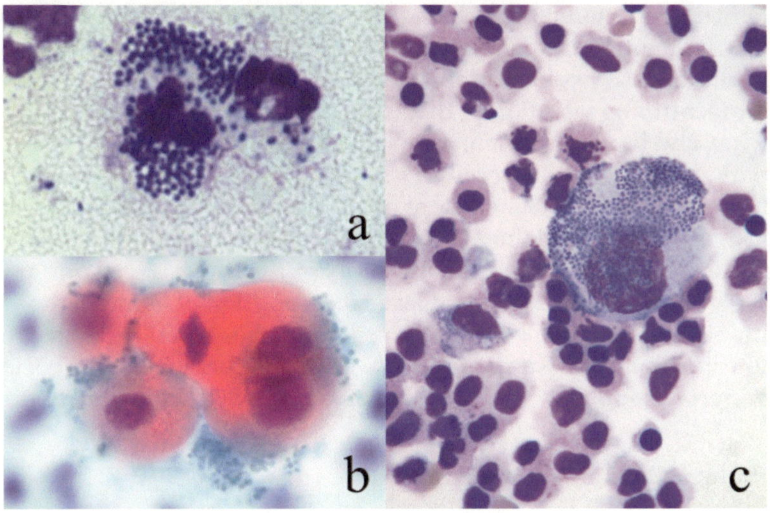

Figure 33.
Gonococcal infection (a: Urethral discharge, Giemsa, b&c: Urine cytology, Papanicolaou [b] and Giemsa [c]). Paired cocci are phagocytized by a neutrophil in pyogenic urethral discharge (a). Neisseria gonorrhoeae *reveals specific affinity to urethral squamous epithelial cells (b&c). The background urothelial cells of urinary bladder origin are devoid of colonization. By courtesy of Mr. Tomohiro Watanabe, Chuken Kumamoto, Japan.*

aged men, the association of prostatic gland hyperplasia causing urethral stenosis should be considered. Urinary bladder cancer may often associate bacterial growth in the urine. Rods mostly represent *E. coli*, while chained cocci usually belong to *Enterococci* (**Figure 34**). Refer also to **Figure 11b**, where cocci (*Enterococcus faecalis*) are actively phagocytized by neutrophils in urine. Particularly when neutrophilic reaction is evident, the diagnosis of bacterial cystitis should be added to that of urothelial carcinoma.

Similarly, bile cytology specimens may reveal bacillary growth around adenocarcinoma cells. The possibility of ascending purulent cholangitis due to malignant

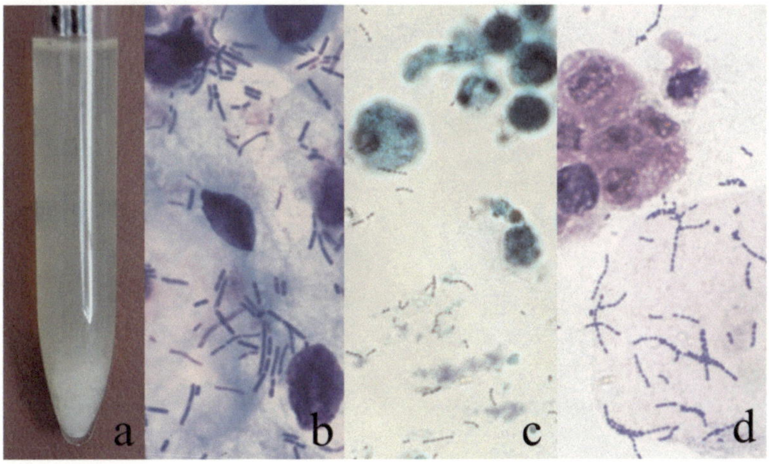

Figure 34.
Bacterial acute cystitis (a: Gross appearance of pyuria, b: Rods, Giemsa, c&d: Chained cocci with urothelial cancer cells, Papanicolaou [c] and Giemsa [d]). White-colored urine sediment (consisting of neutrophils) is formed at the bottom of the test tube. Rods causing cystitis commonly belong to Escherichia coli *(a&b).* Enterococcus faecalis *or* E. faecium *also causes acute cystitis (c&d). It is of note that urinary bladder cancer often accompanies secondary bacterial cystitis.*

bile duct obstruction should be excluded. It may indicate an emergency state requiring prompt antibiotics therapy. Therefore, the cytodiagnosis must be adenocarcinoma plus bacillary colonization. Giemsa staining is superior to Papanicolaou staining for identifying infection of the extracellular bacteria.

When you find bacilli in specimens of ascites or pleural effusion, you should check how the specimen was kept until the centrifugation procedure to get the sediment [70]. If the specimen was kept overnight at room temperature, bacterial grew after the specimen sampling. Neutrophilic response is absent. In the urine sample left for a prolonged time, urease activity of the bacteria, yielding ammonium irons, provokes urine alkalization that leads to deposition of ammonium-magnesium phosphate crystals and non-crystalline phosphate. Representative cytological features are shown in **Figure 35**. Compare them with the specimen of genuine bacterial pleuritis caused by *Streptococcus milleri* as shown in **Figure 11c**.

4.5 Morphological change of Gram-negative rods

Administration of wide-spectrum penicillin and cefem antibiotics may provoke considerable morphological changes of the Gram-negative *Enterobacteriae* in the bile and urine. These include filamentous deformation and spheroplastic change. The beta-lactam antibiotics bound to the penicillin-binding proteins on the bacterial cell membrane hamper the bacterial growth, leading to the shape changes [71, 72]. In the bile shown in **Figure 36**, *Klebsiella pneumoniae* accompanied marked elongation and spheroid change. Microbial culture of the bile was positive for *K. pneumoniae*. The morphologically altered bacteria somewhat resemble fungi. The filaments and spheroplasts are negative with Gram and Grocott stains. *Pseudomonas aeruginosa* in the urine may also show marked filamentous change. Because of the effect of antibiotics treatment, neutrophilic response may be suppressed.

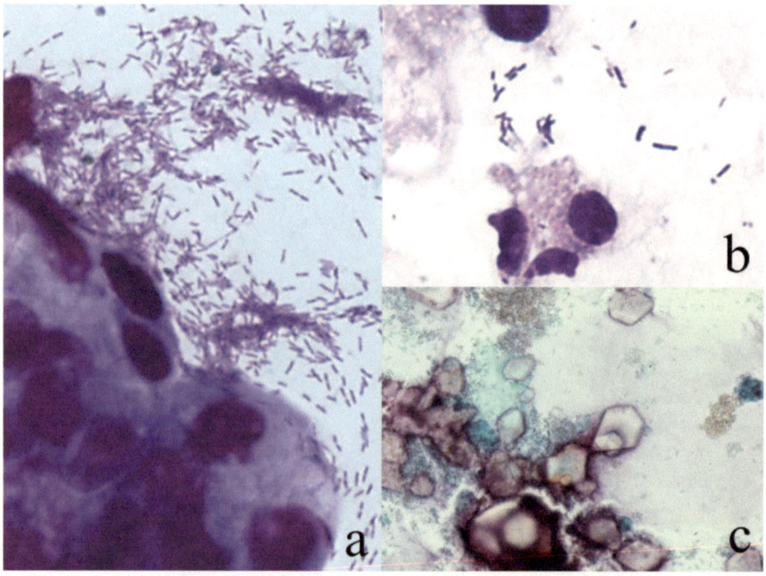

Figure 35.
Bacterial growth in liquid cytology material (a: Rods growing with adenocarcinoma of the bile duct, bile cytology, Giemsa, b: Rods growing in ascitic fluid, Giemsa, c: Ammonium-magnesium phosphate crystals in alkaline urine induced by bacillary growth). In the bile, marked growth of rods is seen around adenocarcinoma cells (a). The possibility of secondary ascending infection of Enterobacteriae *should be excluded. The growth of rods in the ascitic fluid may have occurred after specimen sampling (b). The specimen was kept overnight at room temperature, and neutrophils appear to be autolytic. Deposition of crystals occurs when the urine sample was left for a prolonged time (c). Bacterial urease activity accelerates alkalization of the urine.*

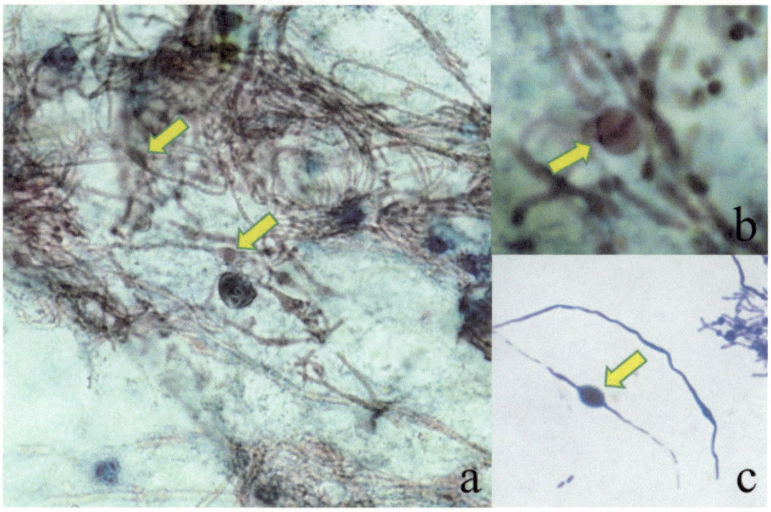

Figure 36.
Antibiotics-induced shape changes of Klebsiella pneumoniae *in the bile: Formation of filaments and spheroplasts (a&b: Papanicolaou, c: Giemsa). The rounded form is called as spheroplast (arrows). Beta-lactam antibiotics bound to penicillin-binding protein on the bacterial cell membrane provokes shape changes of the gram-negative rods. Microbial culture confirmed infection of* K. pneumoniae *in this case. Distinction from fungal colonization is requested.*

Under an immunocompromised condition, *Enterobacteria* such as *E. coli* and *K. pneumoniae* may proliferate within the cytoplasm of macrophages in the digestive and urinary organs to manifest xanthogranulomatous inflammation and malakoplakia, as mentioned above (**Figure 17**).

4.6 Biofilm infection

Pseudomonas aeruginosa of mucoid form commonly accompanies biofilm infection. Biofilm-forming bacteria stick to each other and also to the surface of material or injured mucosa. The adherent bacteria become embedded in a slimy (mucoid) extracellular matrix or secretory capsule. The biofilm protects the microbe from the attack by neutrophils, antibodies, complements and antibiotics: biofilm infection represents a state of resistance of the bacteria to antibiotics therapy [73, 74]. The biofilm-forming *P. aeruginosa* may thus cause persistent and intractable infection particularly in the airway. The neutrophilic host response is thus often minor in degree. Representative examples of biofilm infection are displayed in **Figure 37**. Refer also to **Figure 44f**. Gallbladder adenocarcinoma was cytologically associated with biofilm infection of rods, *P. aeruginosa*, embedded in mucoid matrix. In the vagina of the aged after hysterectomy, infection of *P. aeruginosa* of mucoid-type is proven cytologically.

Biofilm may also be formed by capsule-forming bacteria such as streptococci, staphylococci and enterococci.

4.7 Scraping/touch smear cytology of autopsied lung

We pathologists commonly encounter pneumonia in autopsy cases. Nosocomial (hospital-acquired) pneumonia is often caused by Methicillin-resistant *Staphylococcus aureus* (MRSA) or *Enterobacteriae*, while community-acquired pneumonia may result from infection of *Streptococcus pneumoniae*, *Haemophilus influenzae*,

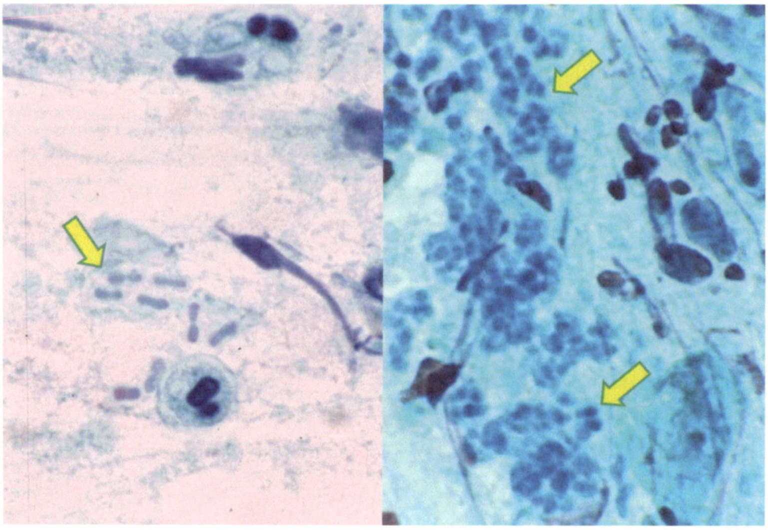

Figure 37.
Biofilm infection of Pseudomonas aeruginosa *(Papanicolaou, left: Bile, right: Vaginal smear in the aged). Mucoid-type colonies (arrows) are formed in both the bile and vagina. Adenocarcinoma was found in the left case, while a history of hysterectomy was recorded in the right case aged in her 70's. The rods are embedded in thick capsule. Microbial culture identified* P. aeruginosa *in both cases. The biofilm infection is resistant to chemotherapy.*

Moraxella (Branhamella) catarrhalis, etc. In case of lethal lobar pneumonia, candidate causative microbes include *S. pneumoniae* and *Legionella pneumophila* [75].

Giemsa-stained scraping or touch smear cytology sampled from the pneumonia lesion is practical in determining the causative microorganism during autopsy services. It is important for pathologists to avoid biohazard. *S. pneumoniae* is transmissible by droplet transmission, while *L. pneumophila* does not show human-to-human transmission. Giménez staining is also useful for demonstrating the microbe. **Figure 38** illustrates scraping cytology sampled from lethal lobar pneumonia in the aged patient. See also **Figure 14a**. Rods were phagocytized by macrophages, so that the causative microbe was identified as *L. pneumophila*, an intracellular microorganism. It is of note that the main cellular reactants against *L. pneumophila* are macrophages. Because of the paucity of lymphocytic response, granulomas are not formed. The importance of *L. pneumophila* as a cause of community-acquired lobar pneumonia of the aged should be emphasized [76].

4.8 Nocardiosis

Nocardiosis is usually encountered in immunocompromised patients [77, 78]. *Nocardia asteroides*, a Gram-positive filamentous and aerobic bacterium, can be demonstrated in bronchial brushing cytology specimens. A young male suffering from ulcerative colitis under steroid treatment complained of fever, coughing and phlegm. Cavitation was radiologically identified in his left upper lobe of the lung, and mycotic infection was clinically suspected. A quick-witted cytotechnologist performed Grocott staining in the cytology preparation. Grocott-stained filamentous bacteria were identified in the background of neutrophilic response, and the diagnosis of nocardiosis was made. The filamentous bacteria were not easily recognized in Papanicolaou-stained preparation, because they do not form aggregated grains. They were additionally positive with Gram and Ziehl-Neelsen's stains. Gram positivity and weak acid-fastness characterize *Nocardia*. **Figure 39** illustrates

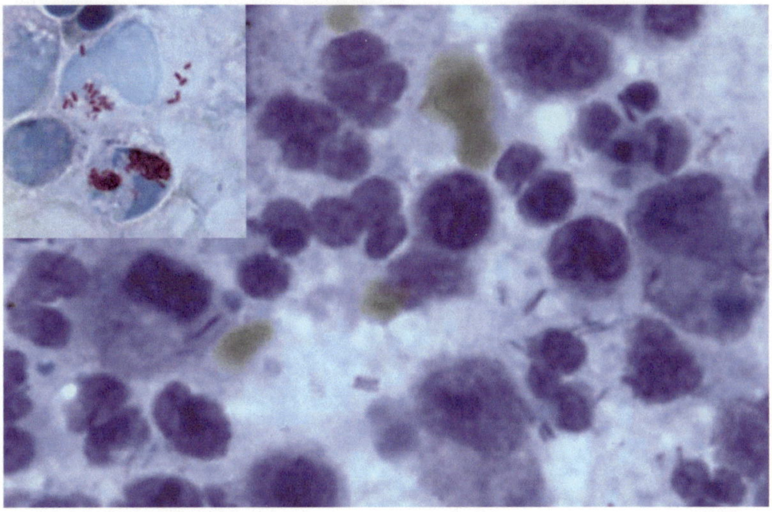

Figure 38.
Lobar pneumonia caused by Legionella pneumophila *(touch smear preparation of the autopsied lung, Giemsa, inset: Giménez). Rods are phagocytized by macrophages and neutrophils, indicating* L. pneumophila*-induced lobar pneumonia. Giménez stain is a simple method for visualizing the pathogen in red. Confirmation of the causative pathogen during autopsy assists at avoiding biohazard.* L. pneumophila *does not show human-to-human transmission, while* Streptococcus pneumoniae, *another causative candidate of lobar pneumonia, may infect the human by droplet transmission.*

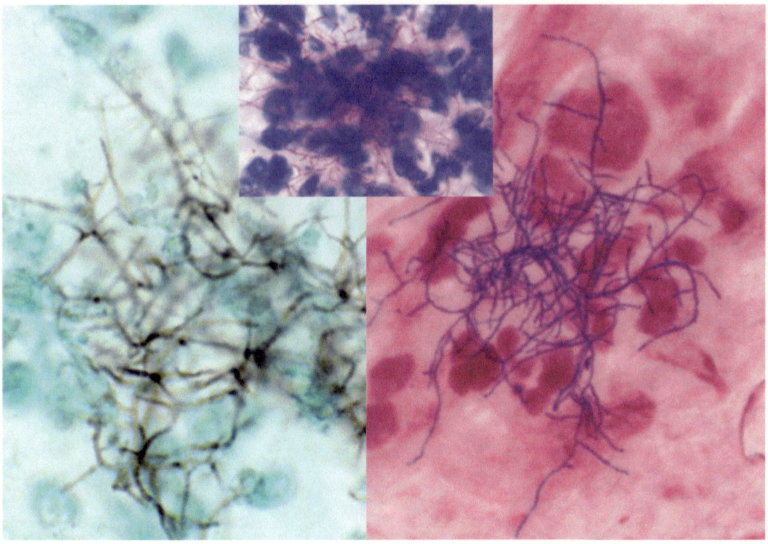

Figure 39.
Nocardiosis of lung (bronchial brushing cytology, left: Grocott, right: Gram, inset: Ziehl-Neelsen). In a young male patient with steroid treatment against ulcerative colitis, lung abscess was noticed. In the bronchial brushing cytology, clusters of filamentous bacteria are observed with Grocott and gram stains. Neutrophilic reaction against the pathogen is evident. The filaments are weakly acid-fast. It is often difficult to identify the filaments under Papanicolaou staining. Nocardia asteroids *was identified by microbial culture.*

cytopathologic appearance of lung nocardiosis. Microbial culture identified *N. asteroides*, and administration of sulfonamides was clinically quite effective.

4.9 Actinomycosis

Actinomycosis, infection of *Actinomyces israelii*, happens in immunocompetent individuals [79, 80], in sharp contrast to nocardiosis. Formation of sulfur granules, reaching 1–2 mm in size, is characteristic of actinomycosis (**Figure 40**). Refer also to **Figure 11d**. The sulfur granule is a dense cluster of obligately anaerobic filamentous bacteria embedded in the homogeneous, periodic acid-Schiff (PAS)-reactive matrix called Splendore-Hoeppli material. The filaments are visualized with Gram, PAS and Grocott stains. Active neutrophilic response against the grains can be observed. In the lung, the sulfur granules are commonly seen within the destroyed airway, and inflammatory pseudotumor may be formed as a result of severe inflammatory fibrosis. Actinomycosis is also encountered in the oral cavity, liver and pelvic organs, including the endometrium (see also **Figure 30**). Actinomycotic grains are often seen in the pit of the enlarged pharyngeal tonsil as a non-pathogenic resident microbe.

4.10 Tuberculosis and non-tuberculous mycobacteriosis

When epithelioid cell granuloma is seen in bronchial scraping cytology, the possibility of lung tuberculosis should be considered (**Figure 41**). See also **Figure 12**. Often times, necrotic background is associated [81, 82]. Infrequently, tuberculous pleuritis may induce eosinophilic exudation (**Figure 18b**).

It is an important mission of the cytopathologist to have the hospital staff noticed for the biohazard [83]. Under an immunosuppressed condition, numerous acid-fast bacilli are phagocytized by macrophages, and Giemsa staining discloses negatively stained long rods in their cytoplasm [84] (**Figure 42**). The outer membrane of the cell wall of mycobacteria contains large amounts of glycolipids, especially mycolic acids [85]. This unique cell wall structure not only gives acid-fastness but also inhibits the penetration of dyes in the Giemsa solution. *Mycobacterium tuberculosis*, a representative acid-fast bacillus, shows airborne transmission. Bronchial sampling is performed in the isolated room equipped for bronchofiberscopy, so that check-ups for the close contact persons are requested. Cytology laboratory may be contaminated with the transmissible dryness-resistant pathogen inside the droplet nucleus. The bacterial morphology is indistinguishable between tuberculosis

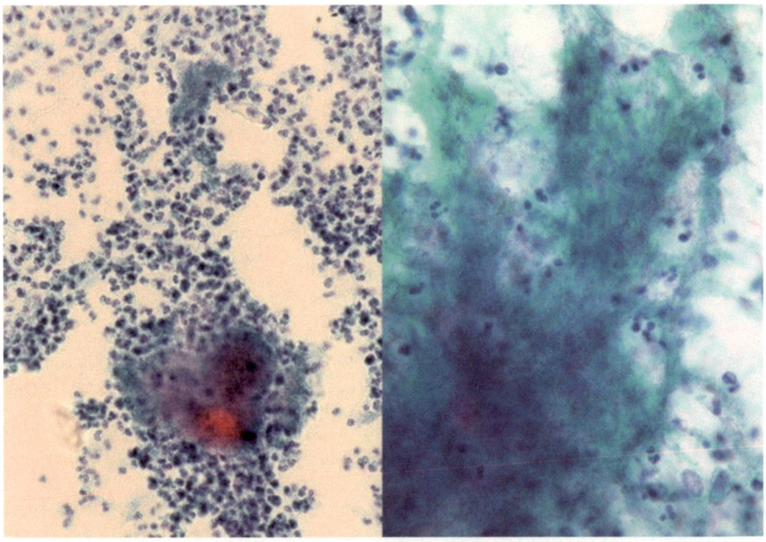

Figure 40.
Actinomycosis of the endometrium (scraping cytology, Papanicolaou, left: Low-power, right: High-power). Formation of sulfur granules is characteristic of Actinomyces israelii *infection. The granule is surrounded by neutrophils, and it consists of filamentous bacteria embedded in the hyaline matrix called Splendore-Hoeppli material. In contrast to nocardiosis, the diagnosis of actinomycosis can be reached with Papanicolaou staining. Endometrial actinomycosis may be provoked by the insertion of intrauterine contraceptive device.*

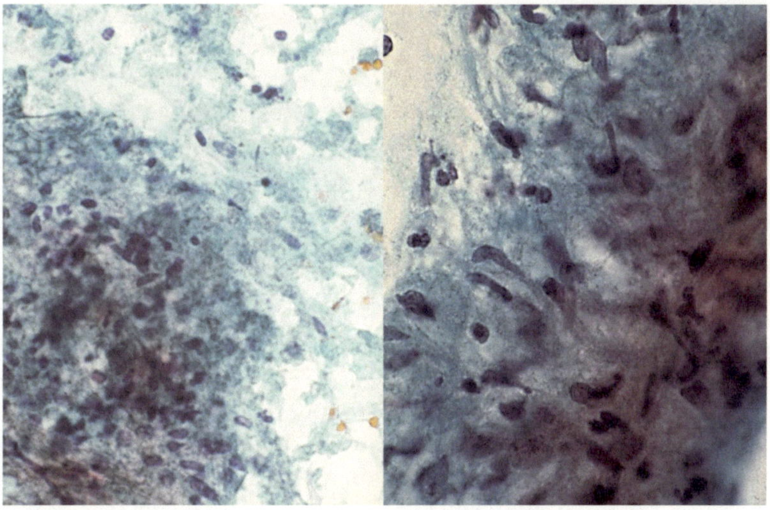

Figure 41.
*Pulmonary tuberculosis (bronchial brushing cytology, Papanicolaou, left: Low-power, right: High-power). Clusters of epithelioid cells represent a granulomatous reaction. The association of necrotic background (left) strongly suggests mycobacterial infection. It is difficult to distinguish tuberculosis (*Mycobacterium tuberculosis *infection) from non-tuberculous mycobacteriosis. Not only correct cytological diagnosis but also prompt warning against intrahospital biohazard are requested. Note also that non-tuberculous mycobacteria accompany no biohazard.*

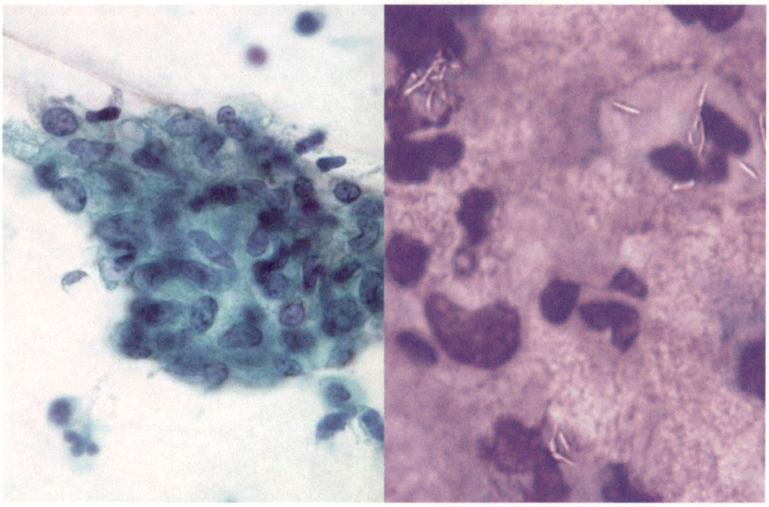

Figure 42.
Negative staining of mycobacteria phagocytized by macrophages (bronchial brushing cytology, left: Papanicolaou, right: Giemsa). Epithelioid granuloma is seen in Papanicolaou-stained preparation. Giemsa stain is useful to identify acid-fast bacilli, since the mycobacteria are resistant to be stained. Therefore, unstained bacillary images are clearly discernible in the cytoplasm of macrophages. No bacilli are visible in the pap smear. In this case, Mycobacterium avium *(a representative non-tuberculous mycobacterium) was cultured. By courtesy of Mr. Tomohiro Watanabe, Chuken Kumamoto, Japan.*

and non-tuberculous mycobacteriosis [82]. The distinction of the two is important since non-tuberculous mycobacteria do not show human-to-human transmission. Identification of *M. tuberculosis* by polymerase chain reaction, as well as the interferon gamma-releasing assay (QuantiFeron or T-Spot) [86] using the blood of the patient and close contact persons, are essentially important to avoid occupation-related infection. In case of tuberculosis, not only correct cytodiagnosis but also prompt warning against intrahospital biohazard are thus strongly requested.

When epithelioid granuloma and neutrophilic reaction are seen in the same specimen, the possibility of suppurative granuloma should be suspected. The typical example is cat scratch disease (bartonellosis) (**Figure 16**) caused by *Bartonella henselae* infection [87]. This tick-associated infection is commonly seen in the cervical or axillary lymph node and infrequently involving the spleen.

4.11 Bacteria seen in the blood

Some bacteria may be observed in the peripheral blood (**Figure 43**). Spiral microbes of *Borrelia recurrentis* are seen in the peripheral blood in an early stage of relapsing fever. The febrile disease is endemic in the African continent [88]. In case of bacterial septicemia, bacteria phagocytized by phagocytes (neutrophils and monocytes) are occasionally identified in peripheral blood preparations. In *Capnocytophaga canimorsus* septicemia caused by dog bite, a few bacilli are phagocytized by neutrophils [89]. *Streptococcus suis*, an important pathogen of pigs, may cause meningitis and lethal septicemia in the human who farms pigs or handles pork. The disease is endemic in southeastern Asia [90]. *In situ* hybridization (ISH) study of the buffy coat of the peripheral blood in septicemia infected with *Escherichia coli/Klebsiella pneumoniae*, *Staphylococcus aureus* or *Pseudomonas aeruginosa* exhibits bacterial DNA signals in the cytoplasm of neutrophils, even after chemotherapy [91].

4.12 Bacteria seen in sputum preparations: importance of Gram staining

Gram staining is cheap and quick technique to identify pathogens on smear preparations of the sputum, exudates, liquid materials and effusions. The importance of Gram staining in the diagnosis of pneumonia should be emphasized [92, 93]. It takes minutes to get results.

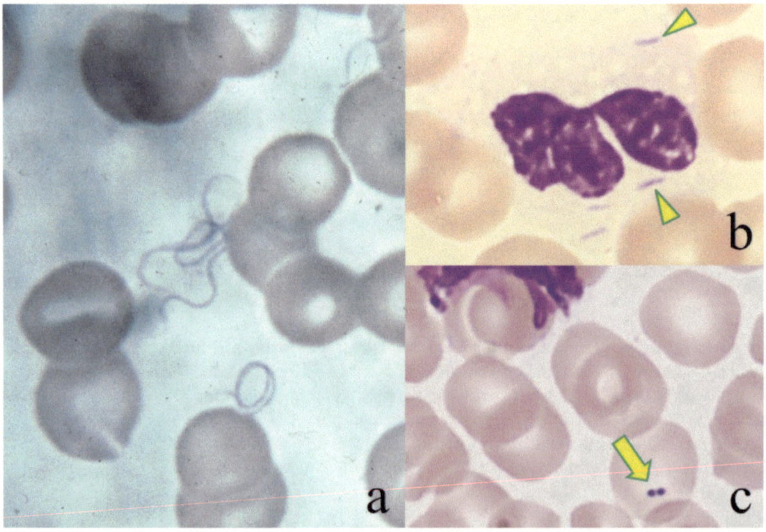

Figure 43.
Bacteria seen in the blood (Giemsa, a: Relapsing fever, b: Capnocytophaga canimorsus *septicemia, c:* Streptococcus suis *septicemia). Close observation of Giemsa-stained peripheral blood preparations occasionally identifies pathogens. In an early stage of relapsing fever, spiral pathogens,* Borrelia recurrentis, *appear in the blood. Fulminant and lethal septicemia of zoonotic origin is rare. Arrowheads indicate rods phagocytized by neutrophils in a case of dog bite-induced* C. canimorsus *infection. Diplococci are seen on a red cell (arrow) in case of* S. suis *septicemia seen in a pig breeder.*

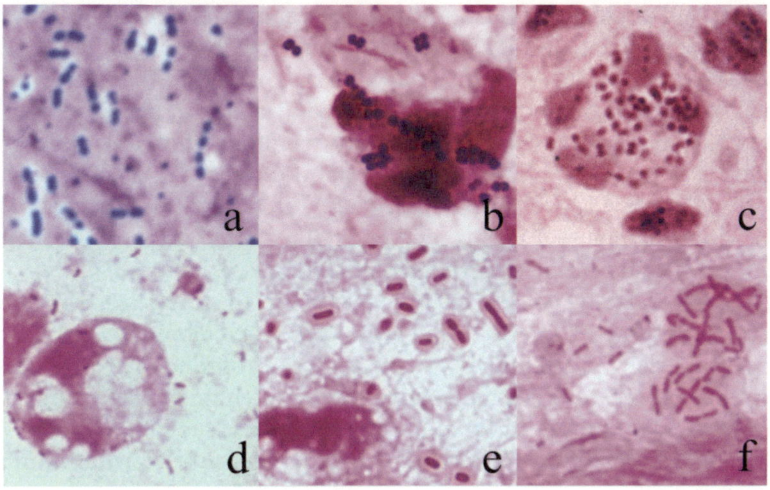

Figure 44.
Pathogens causing pneumonia in the sputum (Gram, a: Streptococcus pneumoniae, *b: Methicillin-resistant* Staphylococcus aureus *(MRSA), c:* Moraxella (Branhamella) catarrhalis, *d:* Haemophilus influenzae, *e:* Klebsiella pneumoniae, *f:* Pseudomonas aeruginosa*). Gram-stained preparations give us prompt identification of pathogens causing pneumonia. Cocci are seen in a-c, and rods in d-f. Gram stain is positive in a&b, but negative in c-f. the bacteria are phagocytized by neutrophils in b&c, while the capsule-forming pathogens are escaped from phagocytosis in a, d-f. mucoid form is observed in f. the rods in e&f are much larger than those in d.*

Typical microscopic appearance of Gram-stained sputum preparations is illustrated in **Figure 44**. These include *Streptococcus pneumoniae*, Methicillin-resistant *Staphylococcus aureus* (MRSA), *Moraxella (Branhamella) catarrhalis*, *Haemophilus influenzae*, *Klebsiella pneumoniae* and *Pseudomonas aeruginosa*. Gram positivity and the shape (cocci or bacilli), as well as the size, capsule formation and the pattern of appearance (paired, clustered or chained), provide us with valuable information for the microbes. The presentation of bacteria phagocytized by phagocytic cells (mainly neutrophils and occasionally macrophages) or those surrounded by neutrophils strongly suggests the etiologic (pathogenic) agent of pneumonia. Whether pneumonia is community-acquired or hospital-acquired is also quite important for determining the causative bacteria. The bacteria on and around squamous epithelia are regarded as the normal flora residing in the oral cavity. It is needed for us to evaluate the number of squamous epithelia and neutrophils in the specimens.

5. Cytodiagnosis of fungal infection

Invasive fungal infection is treatment-resistant and often lethal [94]. Fungi are commonly visualized with PAS reaction and Grocott (methenamine silver) staining. Gram staining may be positive. Fungi infectious to the human being are divided into two forms: yeast and hypha-forming types. Hypha-forming fungi belong to extracellular pathogens, and provoke neutrophilic reaction. Yeasts, round in shape and not forming hyphae, infect intracellularly and protected by cellular immunity provoking granulomatous cellular reaction. Candida accompanying both yeast and hypha-forming (myceliform) morphology is placed in an intermediate form [95].

5.1 Candidosis

Superficial candidosis (moniliasis) represents the most common mycosis. *Candida albicans*, the major species of Candida, is characterized by dimorphic appearance: ovoid yeast cells (germ spores) and filamentous pseudohyphae. *C. albicans* is a normal indigenous flora of the oral cavity, intestinal lumen and skin, residing as a form of yeasts.

Candida vaginitis is the most frequently encountered candidosis in cytology specimens [96]. Hypha-forming colonies are surrounded by neutrophils, and Döderlein bacilli, the normal flora, are no longer observed. Yeast form fungi are also intermingled (**Figure 45**). Refer also to **Figure 11e**. Vaginal candidosis is often associated with pregnancy, diabetes mellitus and acquired immunodeficiency syndrome (AIDS), and it is also experienced as a form of STD.

Candida (Torulopsis) glabrata can be identified in the Papanicolaou-stained cervical smear preparation as a form of paired and orange-colored yeasts. No hyphae are formed, Döderlein bacilli are preserved, and neutrophilic reaction is mild [97]. *C. glabrata* thus represents a normal vaginal flora and must not be misinterpreted as candida vaginitis. The simple and thoughtless comments such as "Candida-positive" may mislead the clinician to inappropriate and unnecessary treatment against candida vaginitis.

Candida is often seen in sputum and urine cytology (**Figure 46**). The appearance of yeast-form Candida without hypha formation in the sputum may represents the increased non-pathogenic flora. In fact, the neutrophilic response is minor in degree. In the pathogenic state, Candida consistently forms pseudohyphae. It should be recognized that mucosal candidosis is common in the oral cavity and upper airway, but candida scarcely provokes pneumonia. In case of candida cystitis, the urine cytology reveals both yeasts and pseudohyphae in the inflammatory (neutrophil-rich) background [95]. *Trichosporon cutaneum (beigelii)*, showing a microscopic resemblance with *Candida albicans*, may also cause mycotic cystitis [98]. Uneven PAS reactivity of *T. cutaneum* gives a hint for differentiation from *C. albicans*, but microbial culture is essential for confirming the causative fungus.

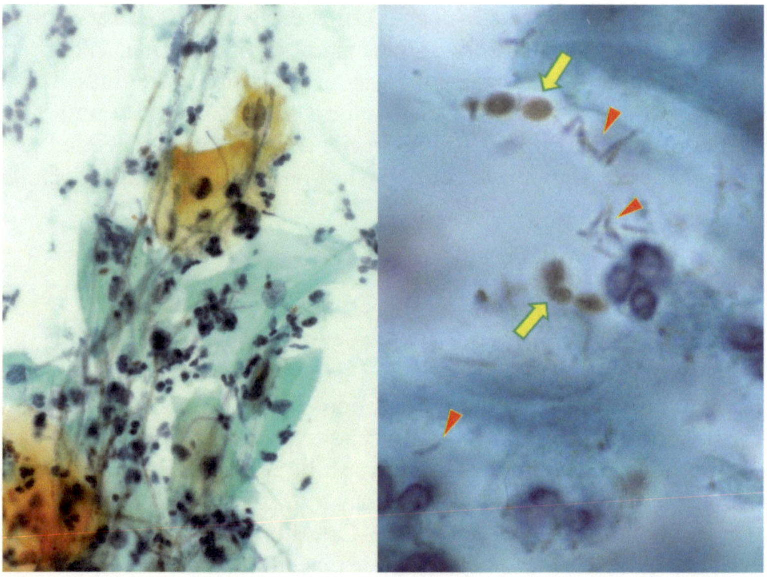

Figure 45.
Pathogenic and non-pathogenic Candida in cervical smear preparations (Papanicolaou, left: Candida vaginitis, right: Candida (Torulopsis) glabrata *as normal vaginal flora). Pathogenic* Candida albicans *forms pseudohyphae in the vagina and provokes neutrophilic reaction, causing candida vaginitis.* C. glabrata *forms paired and orange-colored yeasts without forming hyphae (arrows). The preservation of Döderlein bacilli (arrowheads) in the background is the proof for the lack of pathogenicity.*

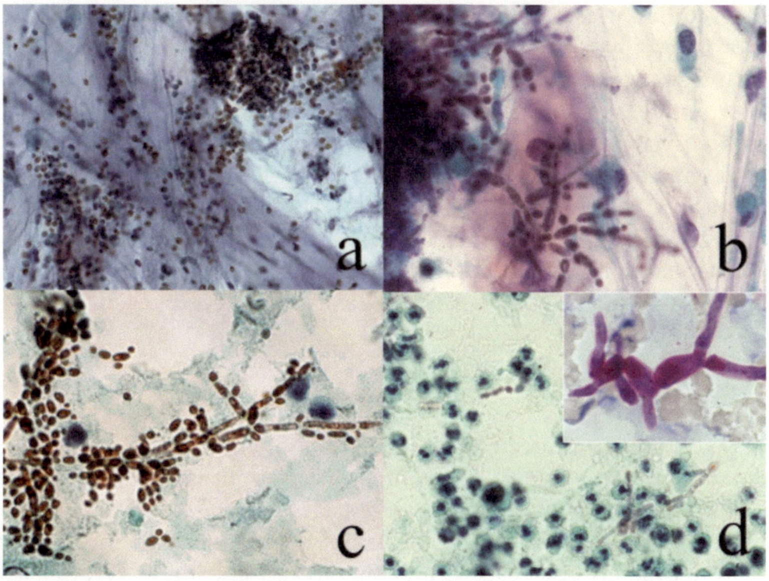

Figure 46.
Candidosis and trichosporonosis (Papanicolaou, a: Yeast form Candida in sputum, b: Candidosis in sputum, c: Candida cystitis, d: Trichosporon cystitis, inset; PAS). Candida yeasts often stain yellowish with Papanicolaou staining. When only yeast form is observed in the sputum, we can judge the microbe as non-pathogenic (a). Since Candida pneumonia is rare, hypha-forming Candida growth may occur in the oral cavity or pharynx (b). In the urine, typical orange-colored microscopic features of Candida infection, accompanying both yeasts and pseudohyphae, are noted (c). Trichosporon cutaneum, *microscopically resembling Candida, occasionally causes fungal cystitis (d). Uneven PAS reactivity is a feature of the Trichosporon species (inset).*

5.2 Cryptococcosis

Figure 47 demonstrates bronchial brushing cytology of pulmonary cryptococcal granuloma. Multinucleated giant cells of macrophage origin phagocytize small

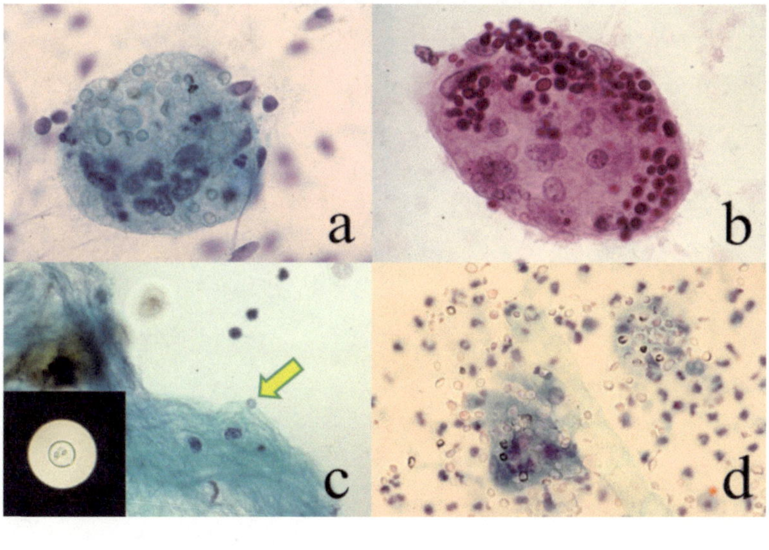

Figure 47.
Cryptococcosis (Papanicolaou [a,c&d], PAS [b], indian ink [inset], a&b: Sputum, c: Cerebrospinal fluid, d: Urine). In the sputum, multinucleated giant cells phagocytize numbers of translucent and PAS-positive small yeasts, representing cryptococcal granuloma of the lung. In the cerebrospinal fluid, only a few yeasts are seen (arrow). Indian ink method demonstrates an unstained yeast in the black background. In case of disseminated cryptococcosis, numerous yeasts are seen in the urine. Some are phagocytized by giant cells, but others are seen in a non-phagocytized free form.

transparent rounded yeasts in the cytoplasm. *Cryptococcus neoformans* grows in the cytoplasm of macrophages to provoke a granulomatous cellular reaction. It should be noted that cryptococcal infection is accelerated by impaired cellular immunity, e.g. after steroid therapy and in AIDS [99].

C. neoformans may infect the meningeal space [100]. Indian ink method clearly demonstrates capsule-forming yeasts in the cerebrospinal fluid (CSF). Usually, yeasts in the CSF are few in number (**Figure 47c**). In case of disseminated cryptococcosis seen in patients with suppressed cellular immunity, a large number of yeasts are observed, and they are often not phagocytized by macrophages. Urine cytology in disseminated cryptococcosis is displayed in **Figure 47d**.

5.3 Pneumocystosis

Pneumocystis jirovecii-induced acute interstitial pneumonia is seen in patients with severe suppression of cellular immunity after administration of steroid or in AIDS [101, 102]. *P. jirovecii* (previously called as *P. carinii*) is now categorized in the fungus, but unculturable *in vitro*. Grocott staining is essential for making the diagnosis of pneumocystosis in bronchial/alveolar lavage specimens (**Figure 48**). Pneumocystis pneumonia often manifests dry cough without formation of phlegm. Grocott-positive small cysts are clearly observed. PAS reactivity is negative. Lymphocytic reaction may be seen in the cytology specimen. In heavily infected specimens in AIDS patients, pathogens (cysts) look like clustered hemolytic red cells in Papanicolaou-stained preparations. With Giemsa staining, the nuclei of smaller-sized ameboid trophozoites are stained purple. Response of small lymphocytes may be provoked, as illustrated in **Figure 13d**.

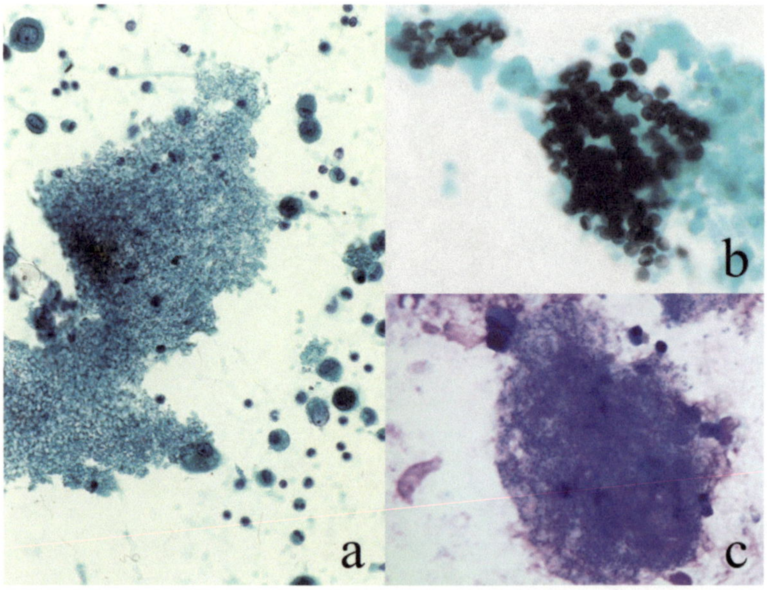

Figure 48.
Pneumocystis pneumonia (bronchial lavage, a: Papanicolaou, b: Grocott, c: Giemsa). Under suppressed cellular immunity, Pneumocystis jirovecii *appears in the airway as clustered translucent cysts in the pap smear, somewhat resembling hemolytic red cells (a). Lymphocytes and macrophages surround the pathogens. The cysts are clearly detected with Grocott stain (b). The cyst wall and dot-like structure within the cyst are stained black. Giemsa stains nuclei of small-sized trophozoites in purple (c).*

5.4 Aspergillosis

Aspergillus is a representative example of hypha-forming (myceliform) filamentous fungi, growing extracellularly [103]. Basophilic hyphae, typically accompanying Y-shaped bifurcation and septum formation, are arranged in the same direction (**Figure 49**). Non-viable hyphae show less basophilia. Neutrophils accumulate around the hyphae. The most common species is *Aspergillus fumigatus*. *A. flavus* occasionally secretes orange/red-colored pigment [104]. Melanin pigment is seen in *A. niger* that also produces calcium oxalate crystals [105].

In aspergilloma containing a fungus ball within the cavitated bronchus, conidial heads, globoid, radiate or broom-shaped, are formed in the aerobic (air-filled) cavity, and Grocott-positive conidia (conidiospores) may be seen in the bronchial lavage specimens. It should be noted that the round-shaped conidia closely resemble *Cryptococcus neoformans*. An important point of distinction is that the conidia floating in the air are not phagocytized by macrophages.

Aspergillus infrequently provokes an allergic reaction with eosinophilic granuloma formation. The status is called as allergic bronchopulmonary aspergillosis. A number of eosinophils and eosinophilic Charcot-Leiden crystals appear in the sputum, in association with a few injured fungal hyphae (see **Figure 18c** and **d**).

5.5 Mucormycosis

Mucormycosis (zygomycosis) is the infection by *Zygomycetes*, including *Mucor ramosissimus*, *Rhizomucor pusillus*, *Rhizopus oryzae*, etc. *Zygomycetes* is an opportunistic microbe mainly affecting premature babies and patients with neutropenia or severe diabetes mellitus. When compared with Aspergillus, the hyphae are less basophilic and thick and lack septum formation. The lamified angle of the hypha tends to be wide. The infection provokes neutrophilic responses. The main sites of

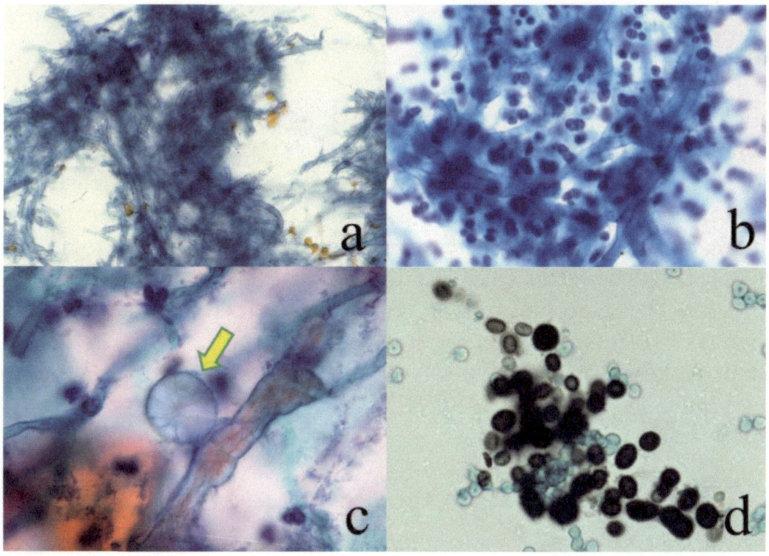

Figure 49.
Aspergillosis (a-c: Papanicolaou, sputum, d: Grocott, bronchial lavage). Basophilic branching hyphae are clustered in the sputum (a&b). Neutrophilic reaction is evident in b. calcium oxalate crystals (arrow) are closely associated with hyphae of Aspergillus niger *(by courtesy of Mr. Tomohiro Watanabe, Chuken Kumamoto, Japan). In the bronchial lavage specimen, Grocott-positive conidia (conidiospores) released from the conidial head formed in the cavity are seen. They resemble cryptococcal yeasts, but they are not phagocytized by macrophages.*

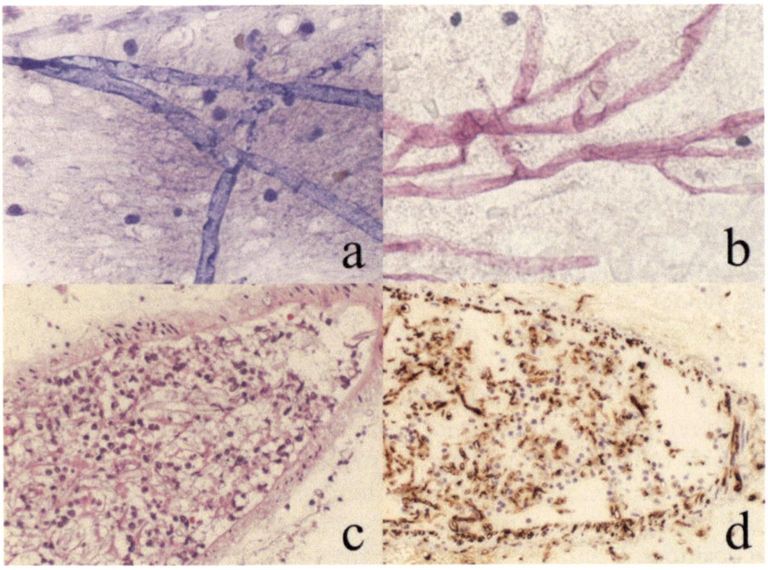

Figure 50.
Mucormycosis (invasive brain lesion, touch smear [a: Giemsa, b: PAS], angioinvasive histology [c: H&E, d: Immunostaining with a monoclonal antibody WSSA-RA-1]). The brain lesion of invasive mucormycosis is seen in a young boy. Thick branching hyphae without septum formation are noted. The brain biopsy specimen reveals vascular invasion of faintly basophilic hyphae that are clearly immunoreactive for Zygomycetes antigen.

involvement include the paranasal cavity and lung. Calcium oxalate crystal deposition and yellow pigment secretion may be associated with mucormycosis [106]. Lethal systemic dissemination may occur in neutropenic patients and in premature neonates [107]. **Figure 50** illustrates scraping cytology of the invasive brain lesion seen in a young boy. Thick branching hyphae without septum formation are noted.

6. Cytodiagnosis of viral infection

Intranuclear clusters of viral particles are recognized as intranuclear inclusion bodies. Intranuclear inclusion bodies characterize DNA virus infection, while most of the RNA viruses do not form inclusion bodies. Exceptionally, measles virus, an RNA virus, forms intranuclear inclusions. Some DNA viruses may cause cytoplasmic viral inclusions: hepatitis B virus to form ground-glassed hepatocytes and molluscum contagiosum virus (a family of pox viruses) to form molluscum bodies in keratinocytes. There are two types of intranuclear inclusion bodies, smudge (homogeneous or full) type and Cowdry A (haloed) type [108]. The viral infection principally provokes lymphocytic host response, when the patient is immunocompetent (see **Figure 13b**).

6.1 Viral inclusions

Representative examples of intranuclear viral inclusion bodies seen in cytology specimens are shown in **Figure 51**.

Human papillomavirus (HPV), a wart-forming DNA virus, provokes skin and mucosal lesions. Intranuclear inclusions are seen in the skin lesion (wart), but not observed in the mucosal lesions (refer to **Figures 25** and **26**). Both types associate koilocytosis, namely perinuclear haloe formation in the superficial keratinocytes. In gynecologic cytology specimens, koilocytosis is seen in condyloma acuminatum

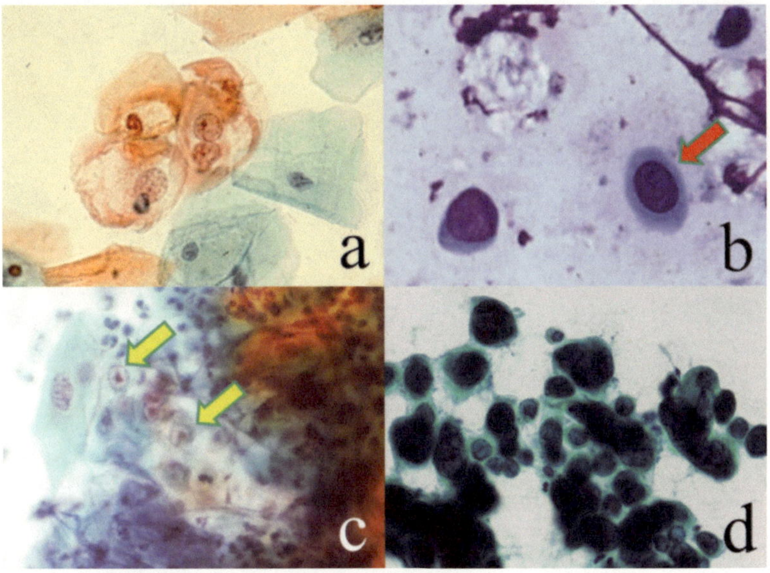

Figure 51.
Viral infections (Papanicolaou [a,c&d], Giemsa [b], a: Cervical smear, b: Conjunctival scraping, c: Corneal scraping, d: Touch smear from the nipple). Koilocytosis (perinuclear haloe formation) in the cervical superficial keratinocytes characterizes human papillomavirus infection (a). In condyloma acuminatum (a benign mucosal wart) caused by HPV 6/11, nuclear atypia is mild in degree. Intranuclear inclusions are seen in adenovirus infection in highly contagious epidemic keratoconjunctivitis (b) and in herpes simplex virus infection (c&d). Smudge (full)-type inclusion bodies (red arrow) are seen in b&d, while Cowdry type A inclusion bodies are noted in c (yellow arrows). Multinucleation of the infected keratinocytes is seen in d.

(a benign mucosal wart of STD type) and dysplastic (precancerous) cervical lesions. Cervical squamous cancer cells lack both intranuclear inclusions and koilocytosis [109].

Epidemic keratoconjunctivitis is a highly contagious eye disease caused by infection of adenovirus, types 8, 19 or 37. Quick Giemsa-stained epithelial cells scraped from the conjunctiva reveal intranuclear inclusion bodies of smudge type [110].

Infection of herpes simplex virus (HSV; human herpesvirus-1 or -2) typically accompanies intranuclear inclusion bodies of both smudge and Cowdry A types [111]. Scraping cytological features of herpetic keratitis and HSV infection of the nipple are illustrated. See also **Figure 21**, where vulvar keratinocytes are infected by HSV as a form of STD. Epstein-Barr virus (EBV, human herpesvirus 4) may cause neoplastic growth of lymphocytes and gastric epithelial cells, but intranuclear inclusions are not formed [112]. In case of chronic active EBV infection, the induction of large granular lymphocytes of NK cell lineage is characteristic, as demonstrated in **Figure 20**.

6.2 Opportunistic viral infections

Cytomegalovirus (CMV; human herpesvirus-5) is a representative example of the opportunistic virus activated in case of impaired cellular immunity. CMV provokes enlargement of the infected cells with formation of large basophilic (owl-eyed) intranuclear inclusion bodies. Granular intracytoplasmic inclusion bodies are also noted [113]. Hemorrhagic varicella (lethal infection of varicella-zoster virus [VZV; human herpesvirus-3) occurred in a child case of acute lymphoblastic leukemia after bone marrow transplantation. The cytology specimen aspirated from a hemorrhagic vesicle shows intranuclear inclusion bodies [114].

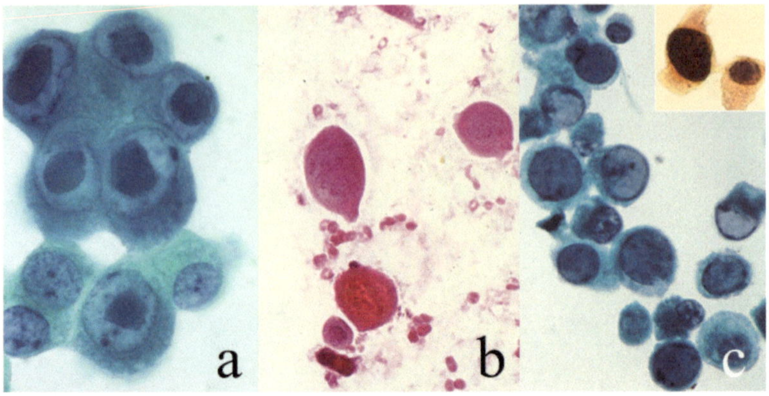

Figure 52.
Opportunistic viral infections (Papanicolaou [a&c], H&E [b], immunostaining for SV40 antigen [inset], a: Bronchial lavage, b: Aspiration from hemorrhagic vesicle, c: Urine cytology). In cytomegalovirus infection in AIDS, large basophilic intranuclear inclusion of Cowdry type a are diagnostic. Cytoplasmic granular inclusion bodies are also noted (a). In hemorrhagic varicella seen in a leukemic boy after bone marrow transplantation, intranuclear inclusion bodies of smudge type are seen in acantholytic keratinocytes in the aspirated vesicle fluid (b). BK virus infection provokes intranuclear inclusions of smudge type in urothelial cells, forming so-called "decoy cells" (c). Polyomavirus-specific SV40 antigen is proven in the nuclei (inset). The decoy cells showing nuclear enlargement should be distinguished from urothelial carcinoma.

BK virus-infected cells in the urine sediment are called "decoy cells", somewhat resembling urothelial cancer cells. Intranuclear inclusion bodies of smudge type are observed in the infected urothelial cells under suppressed cellular immunity. BK virus antigen or SV40 antigen can be demonstrated in the nuclei. Electron microscopy identified numerous round and small-sized viral particles in the nuclei [115].

Representative cytological features of opportunistic viral infection are displayed in **Figure 52**. The cellular (lymphocytic) response is scarcely seen in these immunocompromised cases.

7. Cytodiagnosis of protozoan infection

Protozoa, unicellular microbes usually measuring 5–20 μm, may accompany pseudopodia, flagellae or cilia. Most protozoa infect not only human but also animals, categorized in zoonotic infection [116]. Some protozoa such as *Entamoeba histolytica*, *Giardia lamblia* and *Trichomonas vaginalis* lack mitochondria [117].

7.1 Protozoa in cytology specimens

Representative features are demonstrated in **Figure 53**.

The most common protozoan experienced in routine cytology services is *Trichomonas vaginalis* in cervical smear preparations. This STD pathogen is seen adjacent to glycogen-rich superficial keratinocytes. Döderlein bacilli are no longer observed in the background. Neutrophils are often clustered to form so-called cannon (pus) balls [118]. *T. vaginalis* is a non-invasive protozoan and grows extracellularly, so that neutrophilic response is induced (**Figure 11f**). Cannon ball formation (clustering of neutrophils as cannon balls) is a microscopic hallmark of trichomoniasis.

Giardia lamblia commonly colonizes the duodenal and gallbladder mucosae. Bile cytology preparations may contain flagellated, symmetrical pear-shaped protozoan cells, characteristically having paired nuclei [119]. *G. lamblia*, non-invasive protozoan, commonly induces lymphoid follicle formation with marked increase intraepithelial lymphocytes in the duodenal mucosa [120], as displayed in **Figure 9**.

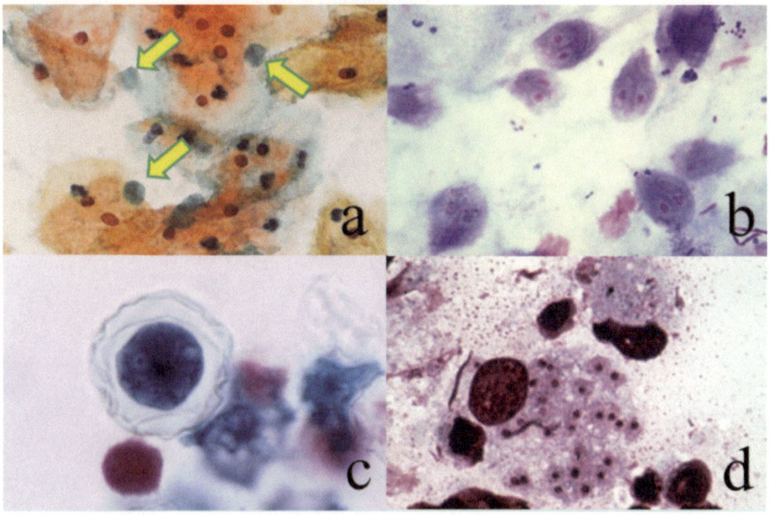

Figure 53.
Protozoan infections (Papanicolaou [a&c], Giemsa [b&d], a: Cervical smear, b: Bile, c: Corneal scraping, d: Touch smear of skin biopsy). Green-colored Trichomonas vaginalis *(arrows) commonly attaches to the superficial (orange-colored) keratinocytes (a). Bacterial vaginosis is associated. Flagellated binucleated trophozoites of* Giardia lamblia *are seen in the bile (b). Bacteria are co-infected. An acanthamebic cyst with thick wrinkled wall formation is seen in the painful cornea (c). Acanthamebic keratitis is caused by inappropriate usage of contact lens. Round-shaped protozoa phagocytized by macrophages represent amastigotes of* Leishmania tropica *(d). Small kinetoplasts are seen just adjacent to the centrally located nuclei.*

Cellular reaction in the bile is usually poor, but lymphocytic background may be associated. Regarding the enteric co-infection of *G. lamblia* and *Entamoeba histolytica* in a AIDS case, refer to **Figure 55**. *Acanthamoeba* is a free-living protozoan widely seen in environmental water. When one wears contaminated contact lenses, painful keratitis may happen. Scraping cytology from the turbid and eroded cornea identified cysts and trophozoites of *Acanthamoeba*, spp. [121]. Touch smear preparation or fine needle aspiration sampled from a biopsied skin tissue of cutaneous leishmaniasis demonstrates amastigotes of *Leishmania tropica* growing in the cytoplasm of macrophages [122]. Both round basophilic nuclei and small-sized kinetoplasts are observed in the non-flagellated protozoan cells. Indistinguishable cytological features are seen in visceral (systemic) leishmaniasis (kala azar), as shown in **Figure 14b**.

7.2 Protozoa in blood preparations

Blood smear preparations may contain protozoa. Malaria, Babesia and Trypanosoma should be of notice (**Figure 54**). Regarding the methods for detecting blood parasites (protozoa and nematode larvae), refer to review articles [123, 124].

Falciparum malaria and tertian malaria, mosquito-mediated febrile diseases, are caused by infection of *Plasmodium falciparum* and *P. vivax*, respectively. In falciparum malaria, ring forms are seen in normal-sized red cells, and often times two or more ring forms infect one red cell. Black-colored malaria pigment is associated. Neither ameboid forms nor schizonts appear in the peripheral blood. The red cells infected by the ameboid form of *P. falciparum* strongly express cell adhesion molecules on the surface, so that they adhere to the capillary endothelial cells expressing CD36 and intercellular cell adhesion molecule-1 (ICAM-1). This process of capillary obstruction is the direct cause of cerebral (malignant) malaria [125]. In tertian malaria, ring forms are fewer in number, and the ring forms are associated

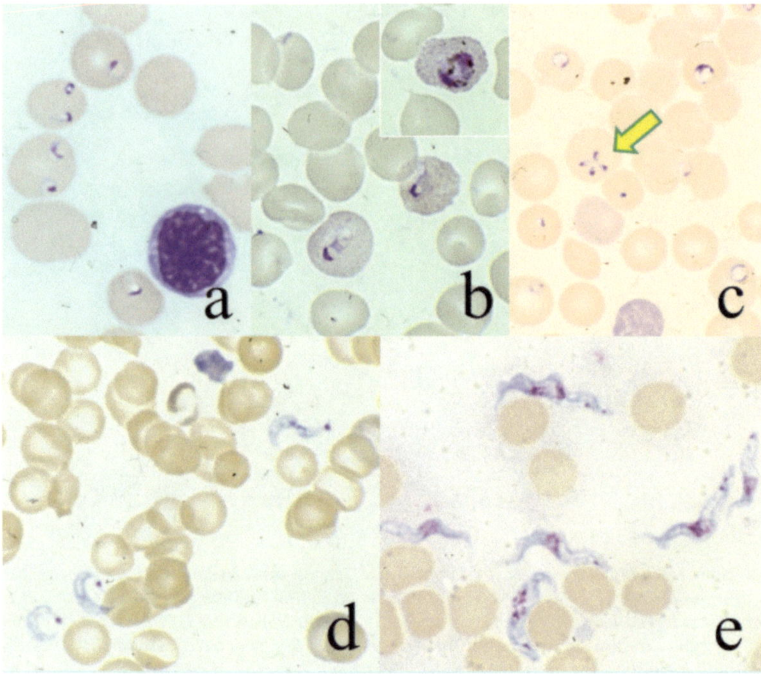

Figure 54.
Protozoa seen in the blood (Giemsa, a: Malaria falciparum, *b:* Malaria vivax, *c:* Babesia microti, *d:* Trypanocoma cruzi, *e:* Trypanosoma gambiense*). In falciparum malaria, two ring form parasites are often seen in a single red cell. More than half of red cells are infected (a). In tertian (vivax) malaria, the infected red cells are enlarged and contain Schüffner's spots (b). Inset indicates an ameboid form. In babesiosis, ring form parasites resemble those of malaria. Tetrads imitating a maltese cross (arrow) are pathognomonic of* Babesia *(c). Trypanosomiasis is featured by the appearance of flagellated trypomastigotes in the peripheral blood. Tropomastigotes of* T. cruzi, *C-shaped and having a kinetoplast at the end (d), are smaller than twisted ones of* T. gambiense *(e).*

with cytoplasmic granules (Schüffner spots) in enlarged red cells. Ameboid forms and schizonts are scattered [126].

Infection of *Babesia microti* is mediated by tick bite. Babesiosis is mainly seen in animal blood, and human cases are rarely encountered, particularly after splenectomy. Ring forms are seen in red cells, and formation of cruciform bodies (tetrads), resembling a maltese cross, is pathognomonic [127].

Trypanosoma cruzi, mediated by Triatoma bite, is identified in the peripheral blood in an acute stage of infection. C-shaped trypomastigotes with a distinct kinetoplast at the end, 20 μm in length, are seen outside the red cells. The clinical course is benign. In chronic stage, *T. cruzi* infects the cardiomyocytes causing chronic heart failure by Chagas disease. In African sleeping disease, lethal meningoencephalitis occurs. The flagellated trypomastigotes of *T. gambiense* are larger (20–30 μm in length) than those of *T. cruzi* [128].

7.3 Opportunistic protozoan infection

Opportunistic protozoan infection is commonly complicated by impaired cellular immunity (particularly in AIDS). These include amebic dysentery, giardiasis, cryptosporidiosis, toxoplasmosis and microsporidiosis. The inflammatory cellular response is poor in immunocompromised patients.

Trophozoites of *Entamoeba histolytica* are identified in cytology preparations aspirated from amebic liver abscess. The environment allowing the growth of obligate anaerobic trophozoites lacking mitochondria results in karyorrhexis of neutrophils and loss of PAS reactivity in the background fluid due to advanced anoxia. Karyosomes (aggregated chromatins centrally located in the nucleus) are pathognomonic of protozoan cells. The plump cytoplasm of the trophozoite consists of thick perinuclear endoplasm and thin peripheral ectoplasm [129]. They often phagocytize red cells. A monoclonal antibody EHK153 detects the ameba in cell block preparations. No cyst form is discerned in the lesion. Rectal lavage from an AIDS patient complaining of severe diarrhea demonstrates opportunistic co-infection of *G. lamblia* and *E. histolytica*. **Figure 55** illustrates cytological features of amebiasis.

As reference, poorly pathogenic *Entamoeba gingivalis*, a normal and anaerobic resident in the oral cavity, may appear in the sputum (see **Figure 62c**), and neutrophils are typically phagocytized by the trophozoites [59]. *E. gingivalis* may colonize the endometrium around an intrauterine contraceptive device (IUD) in healthy women, and co-infection with *Actinomyces israelii* is needed, as illustrated in **Figure 30**.

Cysts of *Cryptosporidium parvum* in diarrheal discharge in an AIDS patient show acid-fastness, red-colored with Ziehl-Neelsen's staining. The acid-fast cysts are small-sized, measuring 3–5 μm. Cryptosporidiosis in AIDS is lethal due to the lack of effective therapeutic drugs [130].

A patient with acute myeloid leukemia post bone marrow transplantation complained myalgia the leg. Fine needle aspiration was performed from the painful muscle. Clustered tachyzoites (pseudocysts) are seen in the cytoplasm of striated muscle cells. The diagnosis of *Toxoplasma gondii*-induced myositis was made [131].

Microsporidiosis caused by *Encephalitozoon cuniculi* is seen in ascites fluid of an immunosuppressed mouse. Giemsa staining clarifies the nuclei of small cysts clustered in the cytoplasm of macrophages. Microsporidiosis may be encountered in the

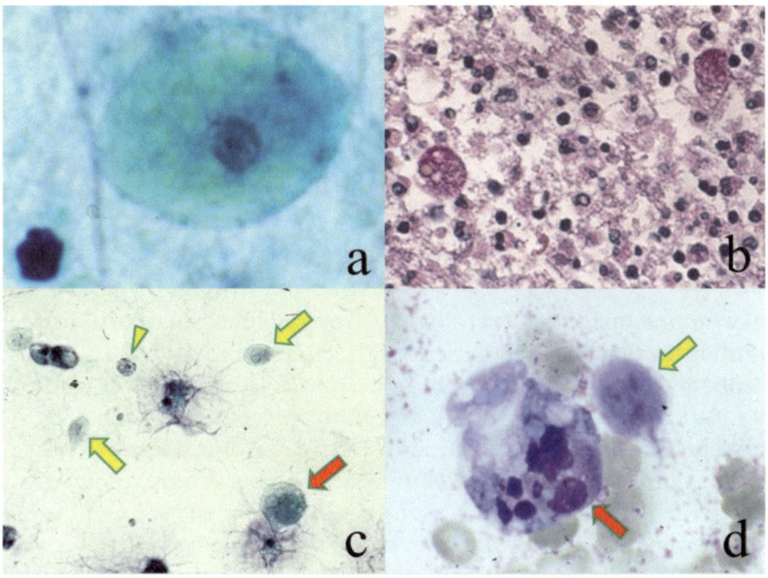

Figure 55.
Entamoeba histolytica *infection (Papanicolaou [a&c], Giemsa [d], PAS for cell block preparation [b]). The aspirate from liver abscess contains trophozoites of* E. histolytica. *They are characterized by a karyosome, (a centrally located chromatin aggregate) unique in protozoan nuclei, and plump cytoplasm consisting of thick perinuclear endoplasm and thin peripheral ectoplasm (a). Trophozoites in the cell block preparation reveal PAS reactivity. Neutrophils are devoid of glycogen because of anaerobic environment (b). Rectal lavage from an AIDS patient complaining of severe diarrhea demonstrates opportunistic co-infection of* E. histolytica *and* Giardia lamblia *(c&d). Yellow arrows indicate trophozoites of* G. lamblia, *and its cyst form is shown by the yellow arrowhead. The large-sized trophozoite of* E. histolytica *(red arrows) phagocytizes neutrophils.*

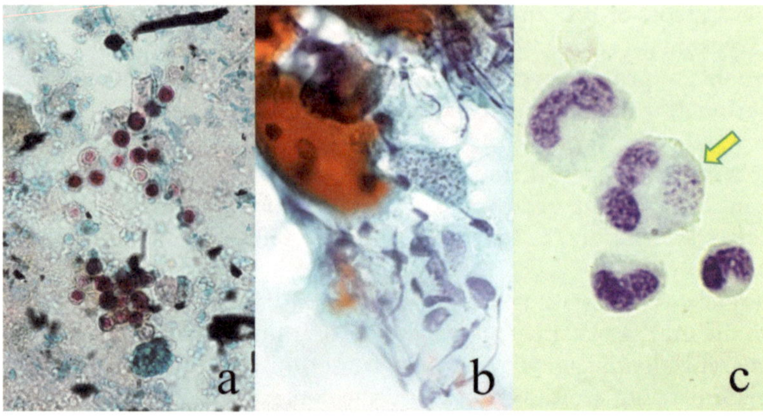

Figure 56.
Opportunistic protozoan infections (a: Ziehl-Neelsen, diarrheal feces in cryptosporidiosis, b: Papanicolaou, needle aspirate from toxoplasma myositis, c: Giemsa, microsporidiosis in ascites fluid of immunosuppressed mouse). Cryptosporidiosis provokes lethal watery diarrhea in AIDS patients. Small-sized (3–6 μm in siameter) cysts of Cryptosporidium parvum *in the fecal excretion are acid-fast (a). A patient with acute myeloid leukemia post bone marrow transplantation complained of myalgia in his leg, and the painful muscle was fine needle-aspirated. Tachyzoites of* Toxoplasma gondii *are clustered in the cytoplasm of the striated muscle cell (b).* Encephalitozoon cuniculi *infects a macrophage in ascites fluid of an immunosuppressed mouse (c). Tiny microsporidium bodies are clustered in a cytoplasmic inclusion (arrow).*

intestine, striated muscle and brain as an opportunistic complication in AIDS patients [132]. Recent studies indicate that the genus microsporidium belongs to the specialized fungus, instead of the protozoan.

Representative microscopic appearance of the latter three infections is demonstrated in **Figure 56**.

8. Cytodiagnosis of helminthic infection

Larval parasites (nematodes) and parasitic ova are occasionally experienced in cytology specimens. It should be noted that manifesting helminthic parasitosis is mostly caused by visceral larva migrans in the human body.

8.1 Larval nematodes in cytology specimens

Typical example includes disseminated strongyloidiasis, opportunistically happening in immunocompromised patients suffering from AIDS or adult T-cell leukemia/lymphoma [133, 134]. *Strongyloides stercoralis* shows percutaneous infestation of the larva via normal skin in tropical and subtropical areas. In Japan, the disease is endemic in southern Okinawa and Amami districts. Adult worms (nematodes), 2–3 mm in length, infest the small bowel mucosa, and persistent autoinfestation occurs through direct intraluminal hatching to infective larva, up to 600 μm in length. In disseminated strongyloidiasis, larval nematodes migrate to a variety of organs and tissues, and they may be seen in cytology specimens of the sputum, urine and body fluids (**Figure 57**). The cellular response against the worm is poor.

8.2 Larval nematodes in blood preparations

In human filariasis encompassing several types [135], microfilariae, 200–400 μm in length, are observed in the peripheral blood smears (**Figure 58**). Bancroftian (lymphatic) filariasis caused by *Wuchereria bancrofti* is seen worldwide, and scrotal

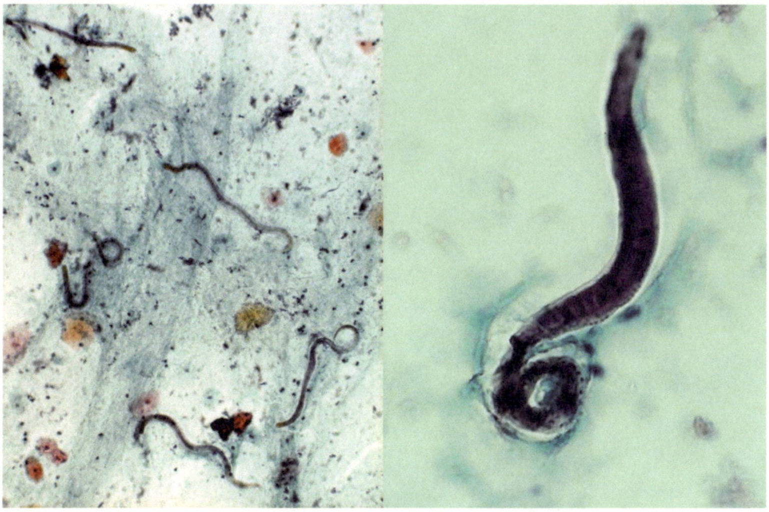

Figure 57.
Disseminated strongyloidiasis (Papanicolaou, sputum cytology, left: Low-power, right: High-power). In disseminated strongyloidiasis, larval nematodes of Strongyloides stercoralis *migrate to a variety of organs and tissues, and plural numbers of larvae are seen in sputum cytology specimen. Infective larva measures up to 600 μm in length.*

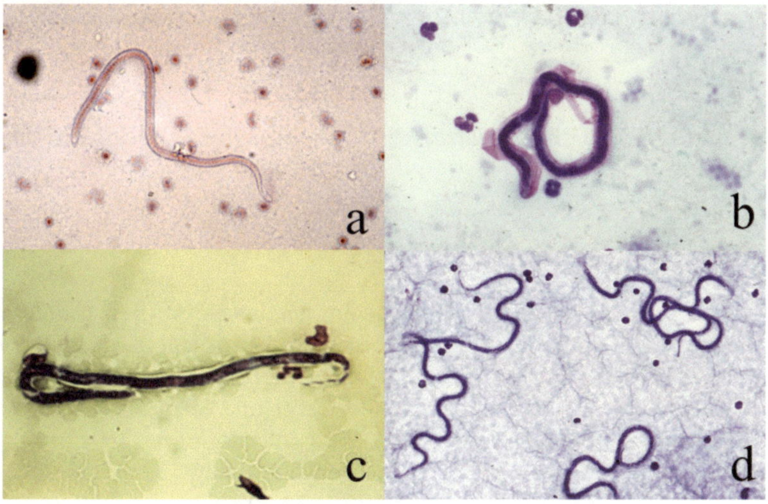

Figure 58.
Larval nematodes seen in the blood (Giemsa, a: Wuchereria bancrofti*, b:* Brugia malayi*, c:* Loa loa*, d:* Dirofilaria immitis *in dog). In human filariasis, microfilariae, 200–400 μm in length, are observed in the peripheral blood smears. The microfilariae are sheathed in bancroftian (lymphatic) filariasis (a) and in brugian filariasis endemic in subtropical Asia (b). Conjunctival African eye worm disease, Loa Loa filariasis, also accompanies sheathed microfilariae in the blood. In canine dirofilariasis, numerous unsheathed microfilariae appear in the peripheral blood.*

swelling and elephantiasis of the lower extremities are clinically featured. Brugian filariasis caused by *Brugia malayi* is endemic in subtropical Asia. The microfilariae are sheathed in both forms. Conjunctival infestation of *Loa loa*, an African eye worm, provokes sheathed microfilariae in the peripheral blood. A transparent, 2–7 mm-long adult worm is seen beneath the conjunctival mucosa. In onchocerciasis causing river blindness in the highland of central America and tropical Africa and mediated by blackfly bite, unsheathed microfilariae of *Onchocerca volvulus* appear in the peripheral blood and preferably invade the eye ball. In canine filariasis,

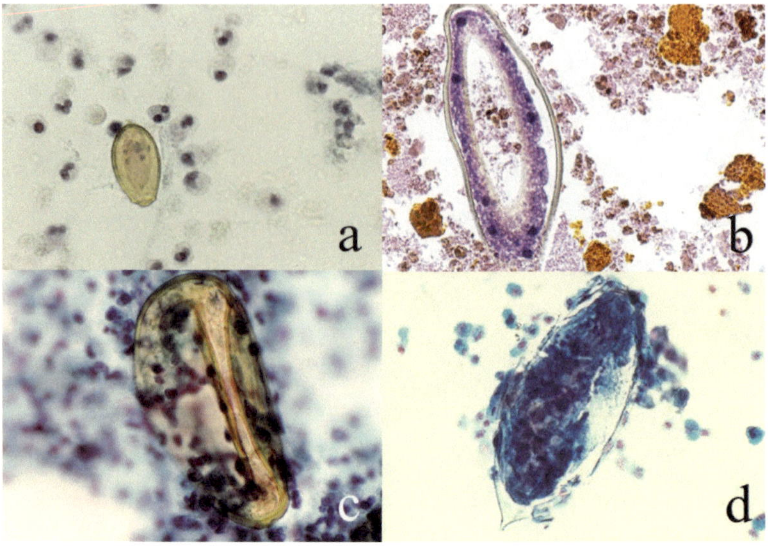

Figure 59.
Parasitic ova seen in cytology specimens (Papanicolaou [a,c&d], Giemsa [b], a: Clonorchis sinensis *Ovum in the bile, b:* Fasciola hepatica *Ovum in the bile, c:* Paragonimus westermani *ovum in the sputum, d:* Schistosoma haematobium *ovum in the urine). The ovum of* C. sinensis *is smallest, while the ovum of* F. hepatica *is largest. Large-sized asymmetric ovum of* P. westermani *is yellow/golden-colored. The large-sized ovum of* S. haematobium *is spiked at one end. The ova of* C. sinensis *and* S. haematobium *contain multinucleated and ciliated miracidium. The ova of* F. hepatica *and* P. westermani *contain unembryonated yolk cells without miracidium formation. Eosinophilic background is observed in a, c and d.*

unsheathed microfilariae of *Dirofilaria immitis* appear in the peripheral blood. See the review articles [123, 124].

8.3 Parasitic ova in cytology specimens

Parasitic ova may appear in cytology specimens. Based on their unique morphology, parasitosis of adult helminthic worms can be indicated (**Figure 59**). Small ova (30 μm in length) of *Clonorchis sinensis* [136] and large ova (130 μm in length) of *Fasciola hepatica* [137] may be seen in the bile. Large ova (around 100 μm in length) of *Paragonimus westermani* [138] and *Schistosoma haematobium* [139] may appear in the hemosputum and hemorrhagic urine, respectively. Regarding ova of *C. sinensis* in the bile, refer also to **Figure 18a**. Eosinophilic background is often associated. Foreign body granulomatous reaction (so-called egg tubercle) is provoked against ova of *S. haematobium* as illustrated in **Figure 19a**.

The ova of *C. sinensis* is the smallest one, while the ova of *F. hepatica* is the largest. Large-sized asymmetric ova of *P. westermani* are yellow/golden-colored. The large-sized ova of *S. haematobium* are spiked at one end. The ova of *C. sinensis* and *S. haematobium* contain multinucleated and ciliated miracidium. The ova of *F. hepatica* and *P. westermani* contain unembryonated yolk cells without miracidium formation.

9. Structures confusing with infectious agents

Certain microscopic structures seen in cytology specimens are occasionally confusing with infectious agents [140]. Representative examples are shown below.

9.1 Incidental airborne contamination during specimen sampling

Incidental contaminants during the process of specimen preparation should be noticeable (**Figure 60**). A variety of living bodies floating in the air may attach onto cytology specimens rich in sticky mucinous material. These include pollen [141], non-pathogenic fungi (conidia of *Alternaria alternata* [142] and hyphae of *Helicosporium* [143]) and mites [144] in house dust. Hairs of carpet beetle larvae may be contaminated from cotton swabs or wooden spatulas [145]. Star-shaped algae commonly found in fresh water marsh may be contaminated in cytology specimens via laboratory water supply [146]. They are positive with PAS and Grocott stains. Ointment matrix may be contaminated in gynecologic cytology sampled from patients suffering from vaginal candidosis. The important notice is the absence of cellular response against the substances.

9.2 Larval nematodes incidentally contaminated in cytology specimens

Sputum cytology preparations may contain a larval nematode [147]. The larva is microscopically indistinguishable from pathogenic *S. stercoralis*, but the patient is asymptomatic with negativity of human immunodeficiency virus antibody. Only one larva is observed in the specimen, and repeated examination fails to show the nematode any longer. In such a case, the patient inhaled an egg of the free-living nematode in the soil, and the ovum hatched to larva in the airway. Nematode larvae may be directly contaminated from the soil in pediatric urine preparations and scraping cytology specimens sampled from severe-degree eroded athlete foot. Representative pictures are displayed in **Figure 61**.

9.3 Structures confusing with pathogenic microbes

Certain microscopic structures may resemble pathogenic microbes [140], as shown in **Figure 62**. Sharp-margined vacuoles formed in the cytoplasm of uterine

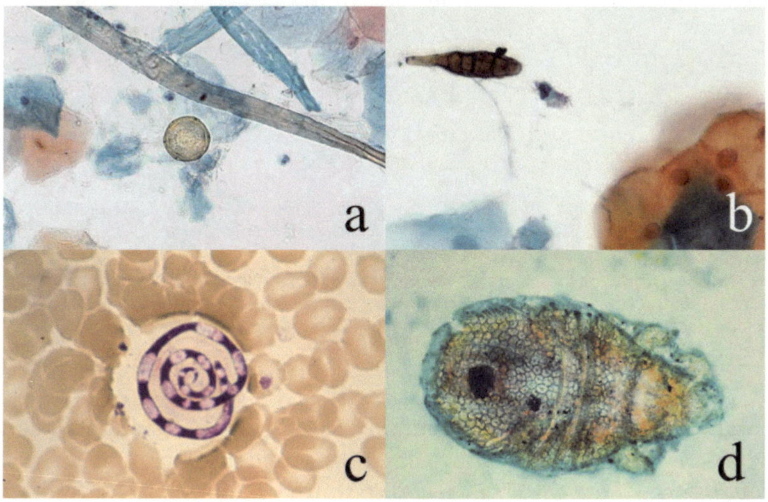

Figure 60.
Incidental airborne contamination during specimen sampling (Papanicolaou [a,b&d], Giemsa [c], a: Pollen, b: Alternaria alternata, *c:* Helicosporium, *d: Mite). A variety of living bodies floating in the air may attach onto cytology specimens rich in sticky mucinous material. A pollen is seen in the cervical smear (a). The shape, color and size of pollen depend upon the kind of flowers and blossoms. The brown-colored conidia of* A. alternata *in the cervical smear show short breaks (b). Hyphae of* Helicosporium *in the blood sample should not be confused with microfilaria (c). Mites living in house dust have four pairs of short legs (d).*

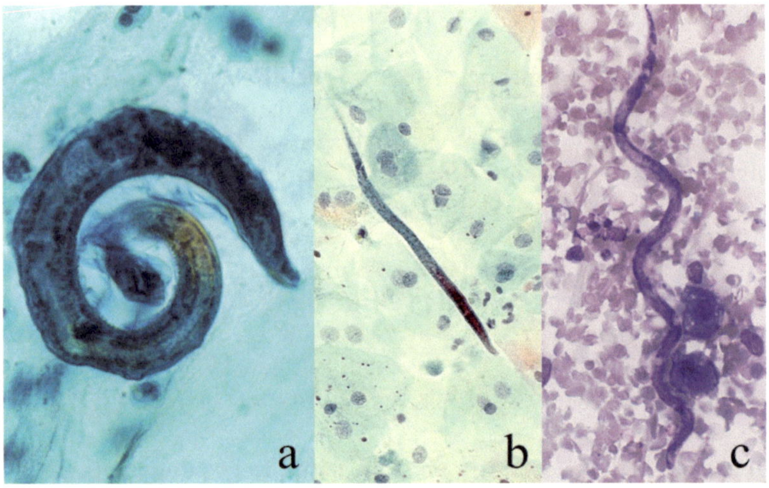

Figure 61.
Incidental contamination of nematode larvae (Papanicolaou [a&b], Giemsa [c], a: Sputum, b: Urine, c: Scraping cytology from severe-degree athlete foot). Cytology specimens may contain a single larval nematode. The larva is indistinguishable from pathogenic S. stercoralis *in disseminated strongyloidiasis. The patient remains asymptomatic. The larva of non-pathogenic free-living nematodes in the soil appears in the cytology specimen from children. The larva in the sputum (a) may have hatched from the inhaled egg in the dust, while larval nematodes are directly contaminated from the soil to the urine (b) and eroded and hemorrhagic skin lesion of the toe (c).*

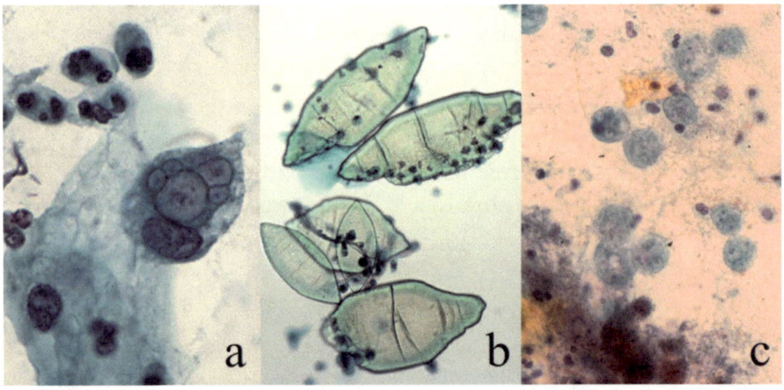

Figure 62.
Structures confusing with pathogenic microbes (Papanicolaou, a: Cervical smear, b: Urine sediment, c: Sputum). Certain microscopic structures may be confused with pathogenic microbes. Sharp-margined vacuoles formed in the cytoplasm of uterine cervical columnar/metaplastic cells resemble chlamydial cytoplasmic inclusions (a). Calcium urate crystals in the sediment of acidic urine resemble S. haematobium *ova (b). Note the size variation, thinness and the lack of miracidium in order for avoiding confusion.* Entamoeba gingivalis *aspirated into the airway must not be confused with pathogenic protozoa (c). The patient commonly suffers from severe periodontitis with bad breath.*

cervical columnar/metaplastic cells should be distinguished from chlamydial inclusions. Calcium urate crystals in the sediment of acidic urine may be confusing with *S. haematobium* ova. Note the size variation, thinness and the lack of miracidium to avoid confusion. Starch granules in sputum cytology may resemble *Paragonimus* eggs [148].

Aspirated *Entamoeba gingivalis* may be observed in sputum cytology specimens [149]. The non-pathogenic protozoa are especially plentiful in the mouth of patients with periodontitis and bad breath. Characteristically, they phagocytize neutrophils, as shown in **Figure 30**.

Airway aspiration of food residue may contain pieces of mushroom. Mushrooms in sputum cytology microscopically consist of parallel-arranged lamified hyphae similar to pathogenic *Aspergillus*. The presence of clamp connection at the site of septum is characteristic of mushroom cells [150]. Co-aspirated food-derived plant cells are often seen in the background (**Figure 63**).

9.4 Non-pathogenic fungi in sputum

Aspiration of air-floating non-pathogenic fungal conidia (spores) may induce growth of hypha-forming fungi in the sputum. Little cellular response is seen. Four different kinds of such fungi are presented in **Figure 64**: *Penicillium* spp., *Ductylaria (Ochroconis) gallopava*, *Petriellidium (Allescheria) boydii* and unknown fungus with a beaded appearance. From a clinical point of view, the appropriate recognition of non-pathogenic microorganisms in cytology specimens is requested. In other words, pathogenic hypha-forming fungi belong to either Aspergillus, Mucor or Candida. It should be noted, however, that these fungi may cause pneumonia in immunocompromised patients [151–153].

9.5 Myospherulosis

In myospherulosis (or spherulocytosis), macrophages contain clustered small globular material (endobody) in the cytoplasm, suggesting infection of yeast-form fungi such as cryptococcosis and coccidioidomycosis. They accumulate in aspirated fluid of cystic lesions in the paranasal cavity or in the breast [154, 155]. PAS and Grocott stains are negative, and they may represent hemolytic red cells or fat droplets phagocytized by the macrophages (**Figure 65**). The term myospherulosis comes from small globular structures seen in a cystic lesion formed in the striated muscle of the neck [156].

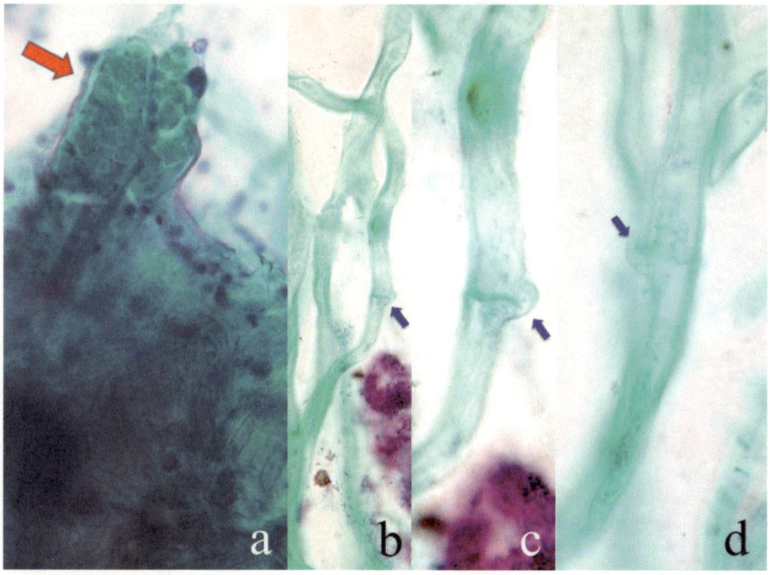

Figure 63.
Aspirated mushroom in the sputum (Papanicolaou, a: Aspirated food debris, b-d: Clamp connection). Airway aspiration of food residues may contain pieces of mushroom. Co-aspirated food-derived plant cells (red arrow) are often seen in the background (a). Mushroom cells in sputum cytology microscopically consist of parallel-arranged, lamified hyphae similar to pathogenic Aspergillus. *The presence of clamp connection (blue arrows) at the site of septum is characteristic of the mushroom cells (b-d). By courtesy of Mr. Tomohiro Watanabe, Chuken Kumamoto, Japan.*

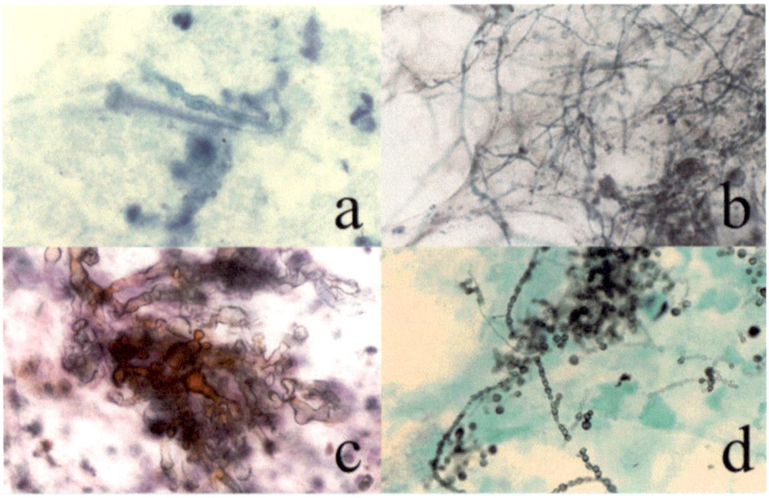

Figure 64.
Non-pathogenic fungi in sputum (Papanicolaou [a-c], Grocott [d], a: Penicillium *spp., b:* Ductylaria (Ochroconis) gallopava, *c:* Petriellidium (Allescheria) boydii, *d: Unknown fungus with a beaded appearance. Aspiration of air-floating non-pathogenic fungal conidia (spores) have induced growth of hypha-forming fungi in the sputum. Four different kinds of fungi are presented. Neutrophilic reaction is scarcely seen. Cytopathologists are requested to appropriately recognize non-pathogenic microorganisms in the sputum.*

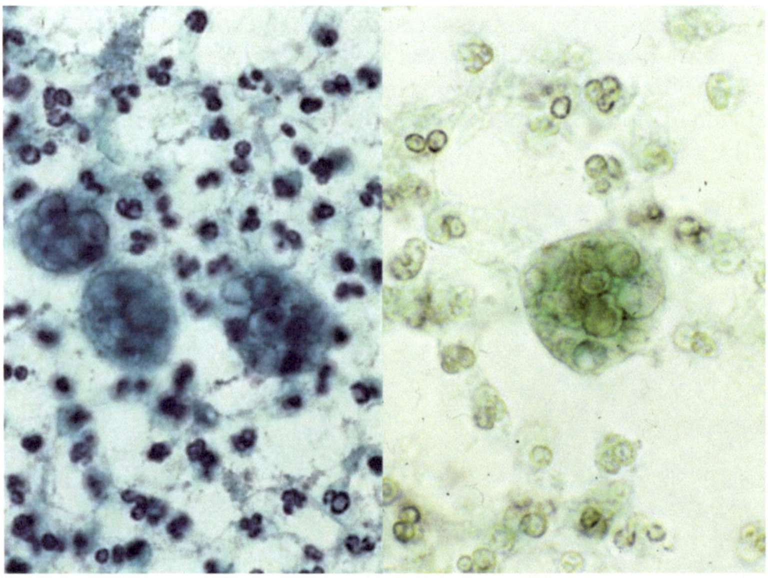

Figure 65.
Myospherulosis (aspirate from paranasal cavity, left: Papanicolaou, right: Grocott). In the aspirated cystic fluid, macrophages contain clustered small globular material (endobody) in the cytoplasm (left). Myospherulosis may resemble infection of yeast-form fungi such as cryptococcosis and coccidioidomycosis. Grocott stain is negative (right). The macrophages have phagocytized hemolytic red cells or fat droplets.

9.6 Grocott stain-positive structures confusing with cryptococcal yeasts or *Pneumocystis jirovecii*

It should be noted that Grocott methenamine silver staining stains not only fungi but also some microorganisms. Grocott may stain *Strongyloides stercoralis*, CMV and *Mycobacterium tuberculosis* [157]. Neutrophils and mucin granules are also

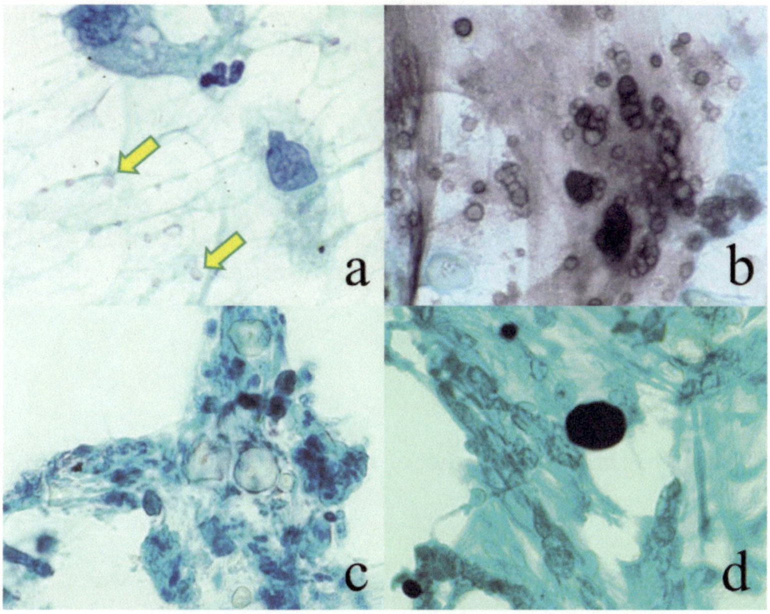

Figure 66.
Grocott stain-positive structures confusing with cryptococcal yeasts or Pneumocystis jirovecii *(Papanicolaou [a&c], Grocott [b&d], destroyed goblet cell mucin [a&b] and starch grains contaminated from rubber gloves [c&d]). In bronchial brushing (scraping) cytology specimens, mucin granules released from destroyed goblet cells (a: Arrows) show black granularity with Grocott stain (b). The Grocott-positive granules resemble cryptococcal yeasts or cysts of* P. jirovecii*. Rubber glove-derived starch grains, contaminated in cytology specimens, are also Grocott-positive (d). Size variation and Papanicolaou-stained appearance (navel-forming figure and birefringence) are helpful for the distinction from pathogens (c).*

Grocott-reactive [158]. When bronchial brushing (scraping) cytology specimens are stained with Grocott method, mucin granules released from destroyed goblet cells show black granularity. The black-stained mucin granules resemble cryptococcal yeasts or cysts of *P. jirovecii* (**Figure 66**). For the diagnosis of pneumocystosis, bronchial/alveolar lavage solution should be evaluated, instead of the bronchial brushing cytology specimen.

Starch grains powdered on the surface of rubber gloves may be contaminated in the sputum/bronchial cytology specimens [159]. Starch grains are also Grocott-reactive, and may be confused with yeast-type fungi. It is requested to use gloves without starch powders for preparing cytology specimens. Size variation and Papanicolaou-stained appearance (navel-forming figure and birefringence) make hints for distinction.

10. Conclusive remarks

The present review described varied cytomorphologic features of infection. Inflammatory cellular responses against pathogens are emphasized. Changes of sexual behavior, globalization-based increase of imported infection and the growing application of immunosuppressive therapy accelerate the chance to encounter unexpected or little-known infection. A wide variety of pathogens may cause infectious diseases. It is not easy for cytopathologists to prove the causative pathogen in cytology specimens. We must realize that the exact and prompt pathogenic diagnosis, with the aid of clinical and epidemiological information, may lead the patient to appropriate treatment. Avoidance of avoidable microbial transmission eventually contributes to the safety of the human society. The recognition of the type of background cellular responses helps us make an appropriate cytodiagnosis.

Acknowledgements

The author cordially thanks numbers of cytotechnologists and cytopathologists who contributed to staining and supplying specimens. In particular, Mr. Tomohiro Watanabe, a veteran cytotechnologist, Chuken Kumamoto, Japan, provided the author with precious materials.

Disclosure

The author do not have any conflict of interest or funding sources in reporting the present review.

Author details

Yutaka Tsutsumi
Diagnostic Pathology Clinic, Pathos Tsutsumi, Nagoya, Aichi, Japan

*Address all correspondence to: pathos223@kind.ocn.ne.jp

References

[1] Tsutsumi Y. Application of the immunoperoxidase method for histopathological diagnosis of infectious diseases. *Acta Histochem Cytochem.* 1994; **27**: 547–560.

[2] Tsutsumi Y. Atlas of Infectious Disease Pathology. Bunkodo, Tokyo, 2000 (349 pages). https://pathos223.com/atlas/index.htm (*in Japanese*).

[3] Tsutsumi Y. Microorganisms found in cytology preparations. *Byori-to-Rinsho.* 2002; **20**(Special Issue: Cytology): 56–68 (*in Japanese*).

[4] Tsutsumi Y. Pathology of Infectious Diseases. 2003; https://pathos223.com/en/

[5] Tsutsumi Y. Infectious diseases. *Byori-to-Rinsho.* 2013; **31**(Special Issue: Topics in Cytology): 390–404 (*in Japanese*).

[6] Tsutsumi Y. Cytological diagnosis of infectious diseases. *Iyaku J.* 2018; **54**(5): 1135–1154 (*in Japanese*).

[7] Tsutsumi Y. Low-specificity and high-sensitivity immunostaining for demonstrating pathogens in formalin-fixed, paraffin-embedded sections. *Intechopen* 2019; doi: 10.5772/intechopen.85055

[8] Reynolds HY. Defense mechanisms against infections. *Curr Opin Pulm Med.* 1999; **5**(3): 136–42. doi: 10.1097/00063198-199905000-00003

[9] Accolla RS. Host defense mechanisms against pathogens. *Surg Infect.* 2006; **7**(Suppl 2): S57. doi: 10.1089/sur.2006.7.s2-5

[10] Chen L, Deng H, Cui H, et al. Inflammatory responses and inflammation-associated diseases in organs. *Oncotarget.* 2018; **9**: 7204–7218.

[11] Altan-Bonnet G, Mukherjee R. Cytokine-mediated communication: a quantitative appraisal of immune complexity. *Nature Rev Immunol.* 2019; **19**: 205–217.

[12] Lehrer RI, Bevins CL, Ganz T. Defensins and Other Antimicrobial peptides and proteins. *Mucosal Immunol.* 2005; 95–110. doi: 10.1016/B978-012491543-5/50010-3

[13] Bruni N, Capucchio MT, Biasibetti E, et al. Antimicrobial activity of lactoferrin-related peptides and applications in human and veterinary medicine. *Molecules* (Basel, Switzerland), 2016; **21**(6): 752. doi: 10.3390/molecules21060752

[14] Ellison III RT, Giehl TJ. Killing of gram-negative bacteria by lactoferrin and lysozyme. *J Clin Invest.* 1991; **88**:1080–91.

[15] van Kessel KPM, Bestebroer J, van Strijp JAG. Neutrophil-mediated phagocytosis of *Staphylococcus aureus*. *Front. Immunol.* 2014; doi. 10.3389/fimmu.2014.00467

[16] Borregaard N, Cowland JB. Granules of the human neutrophilic polymorphonuclear leukocyte. *Blood.* 1997; **89**; 3503–3521.

[17] Sinkovics JG, Horvath JC. Human natural killer cells: a comprehensive review. *Int J Oncol.* 2005; **27**(1): 5–47.

[18] Brinkmann V, Reichard U, Goosmann C, et al. Neutrophil extracellular traps kill bacteria. *Science.* 2004; **303**; 1532–1535.

[19] Shiogama K, Onouchi T, Mizutani Y, et al. Visualization of neutrophil extracellular traps and fibrin meshwork in human fibrinopurulent inflammatory lesions: I. Light microscopic study. *Acta Histochem Cytochem.* 2016; **49**: 109–116. doi. 10.1267/ahc.16015

[20] Medzhitov R. Toll-like receptors and innate immunity. *Nat Rev Immunol.* 2001; **1**: 135–145. doi.org/10.1038/35100529

[21] Medzhitov R. TLR-mediated innate immune recognition. *Semin Immunol.* 2007; **19**(1): 1–2. doi: 10.1016/j.smim.2007.02.001

[22] Perez-Lopez, A., Behnsen, J., Nuccio, SP. Raffatellu M. Mucosal immunity to pathogenic intestinal bacteria. *Nat Rev Immunol.* 2016; **16**: 135–148. doi: 10.1038/nri.2015.17

[23] Harada N, Iijima S, Kobayashi K, et al. Human IgGFc binding protein (FcgammaBP) in colonic epithelial cells exhibits mucin-like structure. J Biol Chem. 1997; 272(24): 15232–15241. doi: 10.1074/jbc.272.24.15232

[24] Kobayashi K, Ogata H, Morikawa M, et al. Distribution and partial characterization of IgG Fc binding protein in various mucin producing cells and body fluids. Gut. 2002; 51(2): 169–176. doi: 10.1136/gut.51.2.169

[25] Heinonen KM, Perreault C. Development and functional properties of thymic and extrathymic T lymphocytes. *Crit Rev Immunol.* 2008; **28**(5):441–66. doi: 10.1615/CritRevImmunol.v28.i5.40

[26] Neutra MR, Frey A, Kraehenbuhl J-P. Epithelial M cells: Gateways for mucosal infection and immunization. *Cell.* 1996; **86**: 345–348.

[27] Gatfield J, Ferrari G, Pieters J. Immunity against extracellular pathogens. *Protoplasma.* 2000; **210**, 99–107. doi: 10.1007/BF01276849

[28] Thakur A, Mikkelsen H, Jungersen G. Intracellular pathogens: Host immunity and microbial persistence strategies. *J Immunol Res.* 2019; **2019**:1356540. doi:10.1155/2019/1356540

[29] Rohra P, Yan L, Mir, F, et al. Follicular cystitis in urine cytology: a clinical and cytopathologic study. *J Pathol.* 2019; **47**(11): 1223–1228.

[30] Halford JA. Cytological features of chronic follicular cervicitis in liquid-based specimens: a potential diagnostic pitfall. *Cytopathology.* 2002; **13**(6): 364–370. doi: 10.1046/j.1365-2303.2002.00434.x.

[31] Konttinen YT, Bergroth V, Nordström D, et al. Cellular immunopathology of acute, subacute, and chronic synovitis in rheumatoid arthritis. *Ann Rhem Dis.* 1985; **44**: 549–555.

[32] Clem AS. Fundamentals of vaccine immunology. *J Glob Infect Dis.* 2011; **3** (1):73–78. doi:10.4103/0974-777X.77299

[33] Ross SC, Densen P. Complement deficiency states and infection: epidemiology, pathogenesis and consequences of neisserial and other infections in an immune deficiency. *Medicine (Baltimore)*. 1984; **63**(5): 243–273.

[34] Stebegg M, Kumar SD, Silva-Cayetano1 A, et al. Regulation of the germinal center response. *Front. Immunol.* 2018. doi: 10.3389/fimmu.2018.02469

[35] Kojima M, Nakamura S, Itoh H, et al. Clinical implication of florid reactive follicular hyperplasia in Japanese patients 60 years or older: a study of 46 cases. *Int J Surg Pathol.* 2005; **13**(2):175–180. doi:10.1177/106689690501300208

[36] Le Hir M, Bluethmann H, Kosco-Vilbois MH, et al. Differentiation of follicular dendritic cells and full antibody responses require tumor necrosis factor receptor-1 signaling. *J*

Exp Med. 1996; 183(5): 2367–72. doi: 10.1084/jem.183.5.2367.

[37] Jandl C, King C. Cytokines in the germinal center niche. *Antibodies* (Basel). 2016; **5**(1):5. doi: 10.3390/antib5010005

[38] Mary T, Raghavendra R, Nanda KG. An overview of suppurative granuloma. *Ind J Dermatopathol Diagnost Dermatol.* 2018; **5**(1): 19–26.

[39] Lou TY, Teplitz C. Malakoplakia: pathogenesis and ultrastructural morphogenesis. A problem of altered macrophage (phagolysosomal) response. *Hum Pathol.* 1974; **5**: 191–207.

[40] Ramirez GA, Yacoub MR, Ripa M, et al. Eosinophils from physiology to disease: a comprehensive review. *Biomed Res Int.* 2018; **2018**: 9095275. doi: 10.1155/2018/9095275

[41] Greenberger PA. Allergic bronchopulmonary aspergillosis. *J Allergy Clin Immunol.* 2002; **110**(5): 685–92. doi: 10.1067/mai.2002.130179

[42] Halim WGY. Schistosomiasis: progress and problems. *World J Gastroenterol.* 2000; **6**(1): 12–19. doi: 10.3748/wjg.v6.i1.12

[43] Tsutsumi Y, Fujimoto Y. Early gastric cancer superimposed on infestation of an Anisakis-like larva. A case report. *Tokai J Clin Exp Med.* 1987; **8**(3): 265–273.

[44] Tsutsumi Y. Pathology of gangrene. https://www.intechopen.com/books/pathogenic-bacteria/pathology-of-gangrene. 2020. doi: 10.5772/intechopen.93505

[45] Tsutsumi Y, Shiogama K, Sakurai K, et al. p16 immunostaining can avoid overdiagnosis in postmenopausal cervical cytology. Int Res J Med Med Sci. 2021; **9**(1): 1–8

[46] Khan S, Omar T, Michelow P. Effectiveness of the cell block technique in diagnostic cytopathology. *J Cytol.* 2012; **29**(3): 177–182.

[47] Jain D, Mathur SR, Iyer VK. Cell blocks in cytopathology: a review of preparative methods, utility in diagnosis and role in ancillary studies. *Cytopathology.* 2014; **25**(6): 356–371.

[48] Tsutsumi Y. Large granular lymphocytes in ascitic fluid. *Byori-to-Rinsho.* 2002; **19**(1): 80–81 (*in Japanese*).

[49] Churukian CJ. List of miscellaneous methods and information. Histochemical Stains Methods. 2009. https://www.urmc.rochester.edu/urmc-labs/pathology/stainsmanual/index.html?RESTAININGSLIDES

[50] Tasho T, Mitsui Y, Ishikura M, el al. Study on restaining of discolored cytological slides. *Igaku Kensa.* 2014; **63**(6): 394–699 (*in Japanese*).

[51] Marshall AE, Cramer HM, Wu HH. The usefulness of the cell transfer technique for immunocytochemistry of fine-needle aspirates. *Cancer Cytopathol.* 2014; **122**(12): 898–902.

[52] Itoh H, Miyajima Y, Osamura RY, Tsutsumi Y. A study of quick "cell transfer" technique. *J Jpn Soc Clin Cytol.* 2002; **41**(4): 302–303 (*in Japanese with English abstract*).

[53] Harada E, Miura K, Tsutsumi Y. How to peel off cover slips quickly. *Byori-to-Rinsho.* 2003; **21**(11): 1306–1307 (*in Japanese*).

[54] Capobianco G, Marras V, Wenger JM, et al. P16 immunostaining and HPV testing in histological specimens from the uterine cervix. *Eur J Gynaecol Oncol.* 2013; **34**(3): 227–230.

[55] Kawai K, Tsutsumi Y. How to recover sections on broken glass slides. *Kensa-to-Gijutsu* 1999; **27**(11): 1315–1316 (*in Japanese*).

[56] Hori S, Itoh H, Tsutsumi Y, Osamura RY. Immunoelectron microscopic detection of chlamydial antigens in Papanicolaou-stained vaginal smears. *Acta Cytol.* 1995; **39**(4): 835–837.

[57] Hori S, Kawai K, Tsutsumi Y, Osamura RY. Ultrastructural demonstration of *Chlamydia trachomatis* DNA by *in situ* hybridization using a biotinylated DNA probe, in comparison with immunoelectron microscopy. *Med Sci Res.* 1991; **19**: 429–430.

[58] Clark CG, Diamond LS. Colonization of the uterus by the oral protozoan *Entamoeba gingivalis*. *Am J Trop Med Hyg.* 1992; **46**(2): 158–160. doi: 10.4269/ajtmh.1992.46.158

[59] Okada H, Matsumoto T, Morikawa M, et al. Clinicopathological and cytological study of *Entamoeba gingivalis*. *J Jpn Soc Clin Cytol.* **41**(5): 321–326 (*in Japanese with English abstract*).

[60] Powers CN. Diagnosis of infectious diseases: a cytopathologist's perspective. *Clin Microbiol Rev.* 1998; **11**(2): 341–365.

[61] Lepargneur JP. Protective role of Doderlein flora. *J Gynecol Obstet Biol Reprod.* 2002; **31**(5): 485–494.

[62] Ventolini G. Progresses in vaginal microflora physiology and implications for bacterial vaginosis and candidiasis. *Womens Health (Lond).* 2016; **12**(3): 283–291.

[63] Demirezen S. Review of cytological criteria of bacterial vaginosis: examination of 2,841 Papanicolaou-stained vaginal smears. *Diagn Cytopathol.* 2003; **29**(3): 156–159.

[64] Onderdonk AB, Delaney ML, Fichorova RN. The human microbiome during bacterial vaginosis. *Clin Microbiol Rev.* 2016; **29**(2): 223–238.

[65] Mendling W, Palmeira-de-Oliveira A, Biber S, Prasauskas V. An update on the role of *Atopobium vaginae* in bacterial vaginosis: what to consider when choosing a treatment? A mini review. *Arch Gynecol Obstet.* 2019; **300** (1): 1–6.

[66] Vinette-Leduc D, Yazdi HM, Jessamine P, Peeling RW. Reliability of cytology to detect chlamydial infection in asymptomatic women. *Diagn Cytopathol.* 1997; **17**(4): 258–261.

[67] Hare MJ, Toone E, Taylor-Robinson D, et al. Follicular cervicitis: colposcopic appearances and association with Chlamydia trachomatis. *Br J Obstet Gynaecol.* 1981; **88**(2): 174–80. doi: 10.1111/j.1471-0528.1981.tb00964.x.

[68] Chan PA, Robinette A, Montgomery M, et al. Extragenital infections caused by *Chlamydia trachomatis* and *Neisseria gonorrhoeae*: a review of the literature. *Infect Dis Obstet Gynecol.* 2016; Article ID 5758387, 17 pages.

[69] Harvey HA, Ketterer MR, Preston A, Lubaroff D, Williams R, Apicella MA. Ultrastructural analysis of primary human urethral epithelial cell cultures infected with *Neisseria gonorrhoeae*. *Infect Immun.* 1997;**65**(6): 2420–2427.

[70] Michael CW, Davidson B. Pre-analytical issues in effusion cytology. *Pleura Peritoneum.* 2016; **1**(1): 45–56.

[71] Prins JM. Endotoxin, antibiotics and inflammation in Gram-negative infections. *Endotoxin in Health and Disease* (eds. Brade H, Opal S, Vogel S, Morrison D), Dekker, New York, 1999; 887–897.

[72] O'Driscoll NH, Lamb AJ. Morphological and ultrastructural changes in bacterial cells as an indicator of antibacterial mechanism of action.

Cell Mol Life Sci. 2016; **73**(23): 4471–4492.

[73] Bjarnsholt T. The role of bacterial biofilms in chronic infections. APMIS 2013; 136(Suppl): 1–51.

[74] Flemming HC, Wingender J, Szewzyk U, et al. Biofilms: an emergent form of bacterial life. *Nat Rev Microbiol.* 2016; **14**(9): 563–575.

[75] Sattar SBA, Sharma S. Bacterial pneumonia. In: StatPearls. Treasure Island (FL): StatPearls Publishing; March 6, 2020. PMID: 30020693

[76] Ishiguro T, Yoshii Y, Kanauchi T, et al. Re-evaluation of the etiology and clinical and radiological features of community-acquired lobar pneumonia in adults. *J Infect Chemother.* 2018; **24** (6): 463–469.

[77] Kandi V. Human Nocardia infections: a review of pulmonary nocardiosis. *Cureus.* 2015; 7(8): e304. doi:10.7759/cureus.304

[78] Shariff M, Gunasekaran J. Pulmonary nocardiosis: review of cases and an update. *Can Respir J.* 2016; **2016**: 7494202. doi:10.1155/2016/7494202

[79] Mabeza GF, Macfarlane J. Pulmonary actinomycosis. *Eur Respir J.* 2003; **21**: 545–551.

[80] Wong VK, Turmezei TD, Weston VC. Actinomycosis. *BMJ* 2011; **343**: d6099. doi: 10.1136/bmj.d6099

[81] Tani EM, Schmitt FC, Oliveira ML, et al. Pulmonary cytology in tuberculosis. *Acta Cytol.* 1987; **31**(4): 460–463.

[82] Jain D, Ghosh S, Teixeira L, Mukhopadhyay S. Pathology of pulmonary tuberculosis and non-tuberculous mycobacterial lung disease: facts, misconceptions, and practical tips for pathologists. *Semin Diagn Pathol.* 2017; **34**(6): 518–529.

[83] Tsutsumi Y. Occupational risk of tuberculosis in pathology. 2003; https://pathos223.com/en/review.htm#Occupational

[84] Prasad CSBR, Narasimha A, Kumar MLH. Negative staining of mycobacteria. A clue to the diagnosis in cytological aspirates: two case reports. *Ann Trop Med Public Health.* 2011; **4**(2): 110–112.

[85] Alderwick LJ, Harrison J, Lloyd GS, Birch HL. The mycobacterial cell wall. Peptidoglycan and arabinogalactan. *Cold Spring Harb Perspect Med.* 2015; **5** (8): a021113. doi: 10.1101/cshperspect.a021113

[86] Moon HW, Hur M. Interferone-gamma release assays for the diagnosis of latent tuberculosis infection: an updated review. *Ann Clin Lab Sci.* 2013; **43**(2): 221–229.

[87] Mazur-Melewska K, Mania A, Kemnitz P, et al. Cat-scratch disease: a wide spectrum of clinical pictures. *Postepy Dermatol Alergol.* 2015; **32**(3): 216–220.

[88] Cutler SJ. Relapsing fever: a forgotten disease revealed. *J Appl Microbiol.* 2010; **108**: 1115–1122.

[89] Lion C, Escande F, Burdin JC. *Capnocytophaga canimorsus* infections in human: review of the literature and cases report. *Eur J Epidemiol.* 1996; **12** (5): 521–533.

[90] Hughes JM, Wilson ME, Wertheim HFL, et al, *Streptococcus suis*: an emerging human pathogen. *Clin Infect Dis.* 2009; **48**(5): 617–625.

[91] Shimada J, Hayashi I, Inamatsu T, et al. Clinical trial of in-situ hybridization method for the rapid

diagnosis of sepsis. *J Infect Chemother.* 1999; **5**(1): 21–31.

[92] Fukuyama, H., Yamashiro, S., Kinjo, K. et al. Validation of sputum Gram stain for treatment of community-acquired pneumonia and healthcare-associated pneumonia: a prospective observational study. *BMC Infect Dis.* 2014; **14**, 534. Doi: 10.1186/1471-2334-14-534

[93] Ogawa H, Kitsios GD, Iwata M, Terasawa T. Sputum Gram stain for bacterial pathogen diagnosis in community-acquired pneumonia: a systematic review and Bayesian meta-analysis of diagnostic accuracy and yield. *Clin Infect Dis.* 2020; **71**(3): 499–513.

[94] Ramana KV, Kandi S, Bharatkumar PV, et al. Invasive fungal infections: a comprehensive review. *Am J Infect Dis Microbiol.* 2013; **1**(4): 64–69.

[95] Guarner J, Brandt ME. Histopathologic diagnosis of fungal infections in the 21st century. *Clin Microbiol Rev.* 2011; **24**(2): 247–280.

[96] Achkar JM, Fries BC. Candida infections of the genitourinary tract. *Clin Microbiol Rev.* 2010; **23**: 253–273.

[97] Rodrigues CF, Silva S, Henriques *M. candida glabrata*: a review of its features and resistance. *Eur J Clin Microbiol Infect Dis.* 2014; **33**(5): 673–688.

[98] Guého E, Improvisi L, de Hoog GS, Dupont B. *Trichosporon* on humans: a practical account. *Mycoses.* 1994; **37**(1–2): 3–10.

[99] Setianingrum F, Rautemaa-Richardson R, Denning DW. Pulmonary cryptococcosis: A review of pathobiology and clinical aspects. *Med Mycol.* 2019; **57**(2): 133–150.

[100] McHugh KE, Gersey M, Rhoads, DD, et al. Sensitivity of cerebrospinal fluid cytology for the diagnosis of cryptococcal infections: a 21-year single-institution retrospective review. *Am J Clin Pathol.* 2019; **151**(2): 198–204.

[101] Dahiya S, Mathur SR, Iyer VK, et al. Diagnosis of Pneumocystis pneumonia by bronchoalveolar lavage cytology: experience at a tertiary care centre in India. *Indian J Chest Dis Allied Sci.* 2005; **47**(4): 259–265.

[102] Gilroy SA, Bennett NJ. Pneumocystis pneumonia. *Semin Resp Crit Care Med.* 2011; **32**(6): 775–782.

[103] Fischler DF, Hall GS, Gordon S, et al. Aspergillus in cytology specimens: a review of 45 specimens from 36 patients. *Diagnost Cytol.* 1997; **16**(1): 26–30.

[104] Assante G, Camarda L, Locci R, et al. Isolation and structure of red pigments from *Aspergillus flavus* and related species, grown on a differential medium. *J Agric Food Chem.* 1981, **29**(4): 785–787.

[105] Oda M, Saraya T, Wakayama M, et al. Calcium oxalate crystal deposition in a patient with Aspergilloma due to *Aspergillus niger*. *J Thorac Dis.* 2013; **5** (4): E174-E178.

[106] Naran S, Lallu S, Kenwright D. Pulmonary mucormycosis diagnosed by brushing cytology. A case report. *NZ J Med Lab Sci.* 2008; **62**(2): 32–34.

[107] Raveenthiran V. Gastrointestinal mucormycosis mimicking necrotizing enterocolitis of newborn. *J Neonatal Surg.* 2013; **2**(4):41.

[108] Tsutsumi Y. The shapes of pathogens. *Kagakuryoho-no-Ryoiki* 2017; **33**(10): 1928–1946 (*in Japanese*).

[109] Tsutsumi Y, Kawai K, Hori S, Osamura RY. Ultrastructural visualization of human papillomavirus DNA in verrucous and precancerous

squamous lesions. *Acta Pathol Jpn.* 1991; **41**(10): 757–762.

[110] Thiel MA, Bossart W, Bernauer W. Improved impression cytology techniques for the immunopathological diagnosis of superficial viral infections. *Br J Ophthalmol.* 1997; **81**(11): 984–988.

[111] Naib ZM. Cytological diagnosis of herpes simplex virus infection. *Clin Dermatol.* 1984; **2**(2): 83–89.

[112] Fugl A, Andersen CL. Epstein-Barr virus and its association with disease: a review of relevance to general practice. *BMC Fam Pract.* 2019; **20**(1): 62. doi: 10.1186/s12875-019-0954-3.

[113] Weiss RL, Snow GW, Schumann GB, Hammond ME. Diagnosis of cytomegalovirus pneumonitis on bronchoalveolar lavage fluid: comparison of cytology, immunofluorescence, and in situ hybridization with viral isolation. *Diagn Cytopathol.* 1991; **7**(3): 243–247.

[114] Suwabe H, Yabe H, Tsutsumi Y. Relapsing hemorrhagic varicella. *Pathol Int.* 1996; **46**(8): 605–609.

[115] Singh HK, Bubendorf L, Mihatsch MJ, et al. Urine cytology findings of polyomavirus infections. In: Ahsan N (eds) *Polyomaviruses and Human Diseases. Advances in Experimental Medicine and Biology*, vol 577, 2006; pp. 201–212, Springer, New York, USA.

[116] Imam T. The complexities in the classification of protozoa: a challenge to parasitologists. *Bayero J Pure Appl Sci.* 2011; **2**(2). DOI: 10.4314/bajopas.v2i2.63805

[117] Muller M. Energy metabolism of protozoa without mitochondria. *Ann Rev Microbiol.* 1988; **42**: 465–488.

[118] Leanza V, Carlo Pafumi, Leanza G, D'Agati A. Cytology of *Trichomonas vaginalis*. *EC Microbiol.* 2016; **3**(3): 451–454.

[119] Ortega YR, Adam RD. Giardia: overview and update. *Clin Infect Dis.* 1997; **25**(3): 545–550.

[120] Buret AG. Immunopathology of giardiasis: the role of lymphocytes in intestinal epithelial injury and malfunction. *Mem Inst Oswaldo Cruz.* 2005; **100**(Suppl.1). doi: 10.1590/S0074-02762005000900032

[121] Patel SP, Schaefer JL, Jaber R, et al. The value of cytology smears for Acanthamoeba keratitis. *Case Rep Ophthalmol Med.* 2016; **2016**: 4148968. doi:10.1155/2016/4148968

[122] Kassi M, Tareen I, Qazi A, Kasi PM. Fine-needle aspiration cytology in the diagnosis of cutaneous leishmaniasis. *Ann Saudi Med.* 2004; **24**(2): 93–97.

[123] Moody AH, Chiodini PL. Methods for the detection of blood parasites. *Clin Lab Haematol.* 2000; **22**: 189–201.

[124] Rosenblatt JE, Reller LB, Weinstein MP. Laboratory diagnosis of infections due to blood and tissue parasites, *Clin Infect Dis.* 2009; **49**(7): 1103–1108.

[125] Armah H, Wired EK, Dodoo AK, et al. Cytokines and adhesion molecules expression in the brain in human cerebral malaria. *Int J Environ Res Public Health.* 2005; **2**(1): 123–131. doi: 10.3390/ijerph2005010123

[126] Adeoye GO, Nga IC. Comparison of quantitative buffy coat technique (QBC) with Giemsa-stained thick film (GTF) for diagnosis of malaria. *Parasitol Int.* 2007; **56**(4): 308–312.

[127] Hunfeld KP, Hildebrandt A, Gray JS. Babesiosis: recent insights into an ancient disease. *Int J Parasitol.* 2008; **38**(11): 1219–1237.

[128] Garcia L. Trypanosomiasis, In *Diagnostic Medical Parasitology*, 5th Ed. ASM Press, Washington, DC, 2007, pp. 218–248. doi: 10.1128/9781555816018.ch9

[129] Skappak C, Akierman S, Belga S, et al. Invasive amoebiasis: a review of Entamoeba infections highlighted with case reports. *Can J Gastroenterol Hepatol.* 2014; **28**(7): 355–359.

[130] Checkley W, White AC, Jaganath D, et al. A review of the global burden, novel diagnostics, therapeutics, and vaccine targets for *Cryptosporidium*. *Lancet Infect Dis.* 2015; **15**(1): 85–94.

[131] Swierzy IJ, Muhammad M, Kroll J, et al. Toxoplasma gondii within skeletal muscle cells: a critical interplay for food-borne parasite transmission. *Int J Parasitol.* 2014; **44**(2): 91–98.

[132] Franzen C. Microsporidia: a review of 150 years of research. *Open Parasitol J.* 2008; **2**(1): 1–34.

[133] Keiser PB, Nutman TB. *Strongyloides stercoralis* in the immunocompromised population. *Clin Microbiol Rev.* 2004, **17**(1): 208–217.

[134] Mukaigawara M, Narita M, Shiiki S, et al. Clinical characteristics of disseminated strongyloidiasis, Japan, 1975–2017. *Emerg Infect Dis.* 2020; **26**(3): 401–408.

[135] Taylor MJ, Hoerauf A, Bockarie M. Lymphatic filariasis and onchocerciasis. *Lancet.* 2010; **376**(9747): 1175–1185.

[136] Fujiya K, Ganno H, Ando M, Chong J-M. *Clonorchis sinensis* ova in bile juice cytology from a patient with severe hyperbilirubinemia and portal vein thrombosis. *Diagn. Cytopathol.* 2016; **44**: 223–225.

[137] Gulsen MT, Savas MC, Koruk M, et al. Fascioliasis: a report of five cases presenting with common bile duct obstruction. *Neth J Med.* 2006; **64**(1): 17–19.

[138] Doanh PN, Horii Y, Nawa Y. Paragonimus and paragonimiasis in Vietnam: an update. *Korean J Parasitol.* 2013; **51**: 621–627.

[139] Nmorsi OP, Ukwandu NC, Ogoinja S, et al. Urinary tract pathology in *Schistosoma haematobium* infected rural Nigerians. *Southeast Asian J Trop Med Public Health.* 2007; **38**(1): 32–37.

[140] Tsutsumi Y. Contamination of non-pathogenic microbes. 2003. https://pathos223.com/en/case/case237.htm

[141] Accorsi CA, Mazzanti MB, Forlani L, Rivasi F. Pollen grains in human cytology, *Grana.* 1991; **30**(1): 102–108.

[142] Radio SJ, Rennard SI, Ghafouri MA, Linder J. Cytomorphology of Alternaria in bronchoalveolar lavage specimens. *Acta Cytol.* 1987; **31**(3): 243–248.

[143] Yoon K-H. Peripheral blood smear contamination with *Helicosporium* fungi resembling microfilaria. *Ann Lab Med.* 2015; **35**(1): 169–171.

[144] Farley ML, Mabry LC, Hieger LR. Mites in pulmonary cytology specimens. *Diagn Cytopathol.* 1989; **5**(4): 416–426.

[145] Bechtold E, Staunton CE, Katz SS. Carpet beetle larval parts in cervical cytology specimens. *Acta Cytol.* 1985; **29** (3): 345–352.

[146] Agarwal A, Helliwell TR. Marsh stars in Liverpool. *J Clin Pathol.* 2005; **58** (10): 1104–1106.

[147] Tsutsumi Y. Nematode larva in urine. https://pathos223.com/en/case/case151.htm

[148] Martínez-Girón R, Martínez-Torre C. False Paragonimus eggs in sputum

cytology. *J Cytol.* 2019; **36**(4): 209–210. Starch granules

[149] Dao AH. *Entamoeba gingivalis* in sputum smears. *Acta Cytol.* 1985; **29**(4): 632–633.

[150] Niederpruem DJ, Jersild RA, Lane PL. Direct microscopic studies of clamp connection formation in growing hyphae of *Schizophyllum commune*. I. The dikaryon. *Arch Mikrobiol.* 1971; **78**, 268–280.

[151] Gill MV, Schoch PE, Rinaldi MG, et al. *Penicillium janthinellum* in sputum and bronchoalveolar lavage in an AIDS patient with pneumonia. *Clin Microbiol Infect.* 1997; **3**: 261–264.

[152] Meriden Z, Marr KA, Lederman HM, et al. *Ochroconis gallopava* infection in a patient with chronic granulomatous disease: case report and review of the literature, Med Mycol. 2012; 50(8): 883–889.

[153] Mader JT, Ream RS, Heath PW. Petriellidium boydii (Allescheria boydii) sphenoidal sinusitis. *JAMA.* 1978; **239** (22): 2368–2369.

[154] Hata S, Kanomata N, Kozuka Y, et al. Cytologic appearance of myospherulosis of the breast diagnosed by fine-needle aspirates: a clinical, cytological and immunocytochemical study of 23 cases. *Diagn Cytopathol.* 2011; **39**(3): 177–180.

[155] Phillip V, Becker K, Bajbouj M, Schmid RM. Myospherulosis. *Ann Diagn Pathol.* 2013; **17**(4): 383–389.

[156] McClatchie S, Warambo MW, Bremner AD. Myospherulosis: a previously unreported disease? *Am J Clin Pathol.* 1969; **51**: 699–704.

[157] Wright AM, Mody, DR, Anton, RC, Schwartz, MR. Aberrant staining with Grocott's methenamine silver: utility beyond fungal organisms. *J Am Soc Cytopathol.* 2017; **6**: 223–227.

[158] Agarwal A, Helliwell TR. Marsh stars in Liverpool. *J Clin Pathol.* 2005; **58** (10): 1104–1106.

[159] Sahay K, Mehendiratta M, Rehani S, et al. Cytological artifacts masquerading interpretation. *J Cytol.* 2013; **30**(4): 241–246.

Chapter 4

Immune Dysfunction during Enteric Protozoal Infection: The Current Trends

Renu Kumari Yadav, Shalini Malhotra and Nandini Duggal

Abstract

Enteric protozoa usually cause severe morbidity and mortality in humans. Protozoal infections contribute to the high burden of infectious diseases. Despite recent advances in the epidemiology, diagnostic tool, molecular biology, and treatment of protozoan illnesses, gaps in knowledge still exist; hence, protozoal infections require further research. We are describing here some important enteric protozoal infections along with the immune dysfunction produced by them. Genus-1. Entamoeba; 2. Giardia; 3. Cryptosporidium; 4. Cyclospora; 5. Cystoisospora; 6. Dientamoeba; 7. Blastocystis; 8. Balantidium.

Keywords: parasite, protozoa, trophozoites, cysts, molecular characterization, laboratory diagnosis, treatment, vaccine

1. Introduction

Parasites are living organisms, which live in or upon another organism and derive nutrients directly from it, without giving any benefit to host. There are three main classes of parasites: protozoa, helminthes, and ectoparasites.

In the current chapter, we are going to discuss about pathogenic protozoan parasites responsible for intestinal infections and the immune system disturbance produced by them. Protozoa are unicellular organisms classified as eukaryotes. Protozoa responsible for intestinal infections are *Entamoeba*, *Giardia*, *Cryptosporidium*, *Cyclospora*, *Cystoisospora*, *Dientamoeba*, *Blastocystis*, and *Balantidium*. These enteric protozoa are associated with diarrheal illnesses in humans, with some causing severe debilitating illnesses, especially in immunosuppressed populations [1].

Enteric protozoa have been given much attention in developing countries, because of their poor sanitary conditions and the unavailability of effective water treatment, which provide a suitable environment for their transmission [2–4]. In more industrialized settings, less focus has been placed on the impact of protozoal infection, presumably because of better health standards. As a result of underdiagnosis and lack of monitoring programs, reliable data are not available for the estimation of the protozoal disease burden in developed settings [5]. The major focus in developed countries is on bacterial and viral infections; hence, in operational surveillance systems, only a few or no parasitic protozoa are included [6, 7]. However, evidence suggests that some protozoa, like *Entamoeba* spp., *Cryptosporidium*, and

Giardia, are isolated frequently from diarrheal patients in developing countries (8). Others, such as *Blastocystis* spp. and *Dientamoeba fragilis*, are isolated mainly in developed countries [8, 9]. Despite the lower prevalence of parasitic diseases in developed countries, they may result in a greater economic burden on the country.

Many protozoal diseases appearing and reappearing in developed countries presented as a public health problem in developing countries long before. The factors that influence the emergence and reemergence of the protozoal disease are similar to other diseases like change in the parasite or host that favor increased transmission, environment, and demographic change that favor increased human-parasite contact and increased the recognition of previously existing problem due to availability of more advanced diagnostic techniques.

There is lack of sensitive diagnostic techniques to detect parasite in clinical specimens, and hence, the estimation of parasite prevalence is difficult, and the carrier stages and subclinical infections of parasites are often not diagnosed [10]. So the development of technologies that can simultaneously detect several protozoa in stool is the current need of the hour [11, 12]. For the sensitive, accurate, and simultaneous detection of protozoan parasites, the molecular techniques are the most promising methods in comparison to conventional staining and microscopy methods [12]. Unfortunately, the molecular methods are not used routinely for the detection of parasitic protozoa because these are costly and labor-intensive [13]. Much effort must now be placed in the development of inexpensive molecular tools for routine laboratory applications.

2. Enteric protozoal infections

The enteric protozoa that are considered as a major public health problem are *Entamoeba histolytica*, *Giardia intestinalis*, *Cryptosporidium* spp., *Dientamoeba fragilis*, *Cyclospora cayetanensis*, *Blastocystis* spp., *Cystoisospora belli*, and *Balantidium coli*.

2.1 *Entamoeba* species

2.1.1 Introduction

Genus *Entamoeba* is divided into six species that have been described in humans, including *Entamoeba histolytica*, *Entamoeba dispar*, *Entamoeba moshkovskii*, *Entamoeba polecki*, *Entamoeba coli*, and *Entamoeba hartmanni*. Among these, *E. histolytica* is the only pathogenic species [10, 14, 15]. Infection of *E. histolytica* is mainly seen in immigrants from or travelers to areas of endemicity, male having sex with male (MSM), HIV-infected patients, and institutionalized populations [16]. Millions of people are infected with *E. histolytica,* and more than 55,000 people die each year globally due to amoebic colitis, a leading cause of diarrhea [17].

2.1.2 Epidemiology

Globally, the highest burden of amebiasis is seen in tropical and subtropical areas of developing countries because of their improper hygiene and water sanitation. It was estimated that *E. histolytica* affect millions of people and responsible for the death of more than 50,000 people each year globally [1]. Amebiasis is also seen in the developed world, mostly in returning travelers or immigrants from endemic countries [18].

It has been noticed that the prevalence of *E. histolytica/dispar* has increased to 36.6% in developed world from 2000 to 2015 [19] in comparison to 1986 in which it was 20% as per Boston report [20].

2.1.3 Pathogenesis

Transmission—Amebiasis is acquired by the ingestion of cyst via the fecal-oral route, but now it is found that fecal contact can occur during sexual contact also (MSM–male having sex with men).

Life cycle—Following ingestion cyst converts into trophozoites known as invasive form of *Entamoeba histolytica*. In asymptomatic individuals, cysts and trophozoites pass into the stool without invading intestinal mucosa, whereas in a few cases, it invade intestinal mucosa and lead to invasive manifestations of *Entamoeba histolytica* like dysentery, amoebic colitis, and ameboma. After invading intestinal mucosa, trophozoites can travel to other organ via bloodstream and manifest as brain abscess and liver abscess [21, 22]. Males are more commonly affected from this invasive disease than females (**Figure 1**).

Pathophysiology—The virulence factors of *E. histolytica* are its enzymes and proteases that help in the invasion of the epithelial cells, penetrating the intestinal mucosa and degradation of the extracellular matrix proteins [24, 25]. More recently it is thought that some individuals are genetically resistant to infection, while malnourished children are more susceptible, and a polymorphism in the **leptin receptor** is found to be associated with increased susceptibility to amebiasis [25, 26].

2.1.4 Immune dysfunction

In some cases, *E. histolytica* colonizes the colon by high-affinity binding to MUC2 mucin without producing any symptoms, whereas in some it causes an

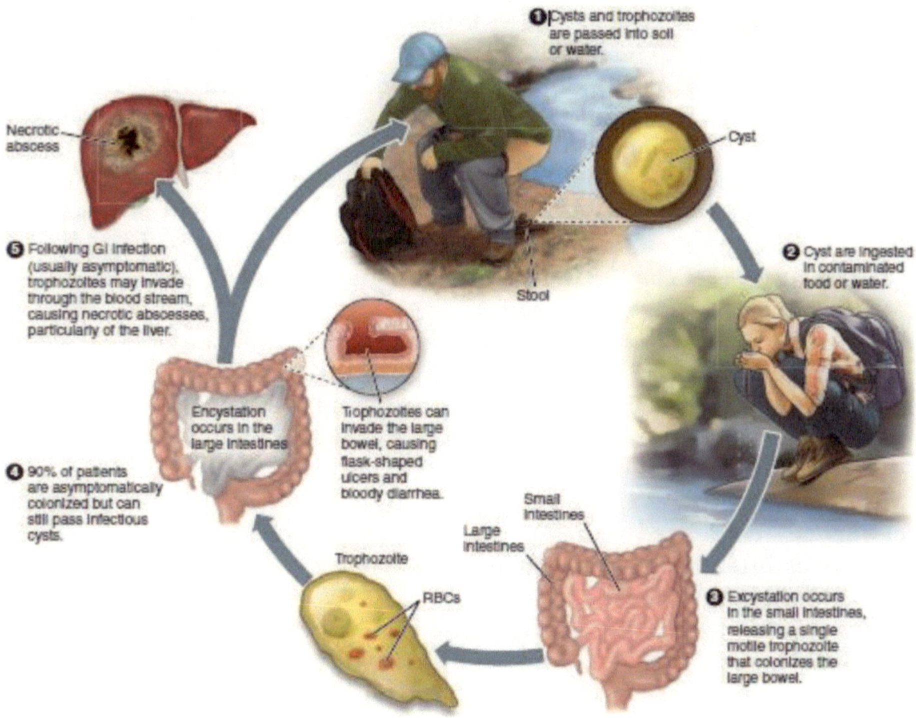

Figure 1.
Life cycle of Entamoeba histolytica. *Courtesy: Ref. [23], Fig. 218–1.*

aggressive inflammatory response upon invasion of the colonic mucosa. The parasite has cysteine protease that cleaves C-terminus of MUC2 that dissolves the mucus layer followed by binding of *E. histolytica* to the mucosal epithelium. As a result, the host mounts a pro-inflammatory response that causes tissue damage and participates in disease pathogenesis. *E. histolytica* escapes the host immune system by mechanisms that are not completely understood. The parasite can destroy effector immune cells by inducing neutrophil apoptosis and suppressing respiratory burst. It also suppresses the production of nitric oxide (NO) from macrophages. Following the adherence of *E. histolytica* to the host cells, multiple cytotoxic effects occur that can promote cell death through phagocytosis and apoptosis, which might play critical roles in immune evasion [27].

2.1.5 Diagnosis

For the early diagnosis of invasive and extraintestinal amebiasis, a high level of clinical suspicion is necessary [28].

Morphological similarities of the pathogenic species *E. histolytica* to the non-pathogenic species *E. dispar* and *E. moshkovskii* make the diagnosis difficult. Hence, microscopy is usually considered insufficient for the differentiation of these species [29]. In microscopic techniques, trichrome-stained smears and wet mount preparations of stool specimens are used in routine for the identification of *Entamoeba* spp. [30]. Other methods for *Entamoeba* identification include culture, antigen-based, and molecular tests.

In culture methods, polyxenic culture media like Balamuth's medium and axenic culture media like diamond's medium are used. A new axenic culture medium **CLUPS** for the growth of *E. histolytica* was also there which is superior to previously used **PEHPS** media [31]. However, cultivation is more sensitive than microscopy, but these methods are time-consuming, not cost-effective, and hence not routinely utilized by most diagnostic laboratories. Cultivation can effectively distinguish between *E. dispar* and *E. histolytica* [32].

Enzyme immunoassay (EI) kits are commercially available that detect *E. histolytica* or differentiate between *E. histolytica* and *E. dispar* by detecting 170 kDa of lectin antigen.

A rapid immunochromatographic testing (ICT) assay is available that detects antigens of *E. histolytica* and *E. dispar* (29 kDa surface antigen) in stool; however, this assay cannot differentiate between *E. histolytica* and *E. dispar* [30]. This assay also detects antigens of *Giardia* (alpha 1 Giardia antigen) and *Cryptosporidium* (protein disulfide isomerase antigen).

For the detection of *E. histolytica* antibodies in human serum serological methods such as latex agglutination, indirect hemagglutination (IHA), immunofluorescence assay (IFA), and enzyme-linked immunosorbent assay (ELISA) are highly sensitive [10, 33]. However, in endemic areas serology is of limited use because of the difficulty in distinguishing between past and present infections [10].

Molecular assays (like PCR) for the detection and differentiation of *Entamoeba* species are now considered as the gold standard for diagnosis of *Entamoeba* [29, 34]. When species differentiation is difficult or not possible than *E. histolytica/E. dispar* complex should be reported.

BioFire FilmArray system is the fully automated PCR system which can detect bacterial, viral, parasitic (*E. histolytica*, *Giardia*, *Cyclospora*, *Cryptosporidium)* diarrheal pathogen by using gastrointestinal panel [35].

For the diagnosis of invasive disease, a combination of microscopy, culture, and serology should be used with a PCR assay or with abdominal imaging (when PCR is unavailable) [33].

2.1.6 Treatment

In asymptomatic carriers of *E. histolytica*, luminal agents should be used to minimize the spread of disease. The treatment of choice is different for intestinal and invasive disease; hence, diagnosis is important before treatment. Metronidazole, a tissue amoebicide, is highly effective and used for invasive amoebic disease along with a luminal agent for the elimination of intestinal colonization [36]. Other agents like tinidazole and ornidazole can also be used for the treatment of invasive disease. For the treatment of intestinal and asymptomatic infections, the luminal agents paromomycin, iodoquinol, and diloxanide furoate are strictly recommended, as they are effective in eliminating cysts from the intestinal tract [37].

Regimen: asymptomatic carriage—luminal agent, Idoquinolone 650 mg tid for 20 days or Paromomycine 500 mg tid for 10 days.

Acute colitis—Metronidazole 750 PO or IV for 5–10 days or tinidazole 2 g/d PO for 3 days Plus luminal agent as above.

Amebic liver abscess—Metronidazole 750 PO or IV for 5–10 days or tinidazole 2 g/d PO once or ornidazole 2 g/d PO once Plus luminal agent as above.

2.1.7 Prevention and control

By avoiding the consumption of contaminated food and water and by treating the asymptomatic person, the infection can be prevented.

Certain **vaccines** against *E. histolytica* are under trial like colonization blocking vaccine targeting SREHP, 170 kDa subunit of lectin antigen, and 29 kDa cysteine-rich protein. Till now no vaccines against *E. histolytica* are licensed for human use [38, 39].

2.2 *Giardia intestinalis*

2.2.1 Introduction

Giardia intestinalis also known as *Giardia duodenalis* and *Giardia lamblia* is a common cause of parasitic diarrhea. It belongs to the flagellate group and first discovered by Leeuwenhoek in 1681 in his own stool.

2.2.2 Epidemiology

Giardia lamblia is distributed worldwide and reported from both temperate and tropical countries of the world. The prevalence of *Giardia* in low- and high-income countries is 4–43% and 1–7%, respectively [40, 41]. It was reported that *Giardia* affect approximately 2% of adults and 8% of children in developed countries [42].

Because of the increasing burden of illness from *Cryptosporidium* spp. and *Giardia* and their ability to impair development and socioeconomic improvements, they were included in the WHO Neglected Diseases initiative in 2004 [14].

2.2.3 Pathogenesis

Transmission—The most important mode of infection is fecal-oral route, but various studies have also found evidence of zoonotic transmission [43]. Giardiasis presents mainly as acute or chronic diarrhea associated with abdominal pain, nausea, malabsorption, and weight loss. *Giardia* infection can lead to growth retardation in malnourished children [44] and zinc deficiency in school-aged children [45]. *Giardia* infection is also common in dogs and cats [46]. They are also a potential source of human infection that may be acquired through handling, sleeping together, licking, and kissing, as the zoonotic genotypes of *Giardia* were isolated from cats and dogs [47]. Giardiasis is also associated with waterborne disease outbreaks and is related to travel-associated diarrhea [48].

Life cycle—Cyst is the infective form of *Giardia* and ingestion of as low as 10 cysts can initiate infection. After ingestion excystation occurs in the small intestine, and its cyst converts into trophozoites. These trophozoites do not disseminate hematogenously and remain in the small bowel. It can either attach to the intestinal epithelium via suckling disc or remain free in the intestinal lumen. Trophozoites convert into cyst form, and both trophozoite and cysts are excreted in feces, but only the cysts are capable of surviving in the outside environment and disease transmission (**Figure 2**).

Pathophysiology—After attachment to the intestinal epithelium, it feed on the mucus secretions and lead to apoptosis of enterocyte; epithelial detection tests are used for screening and are more sensitive than routine microscopic examination. Loss of brush border is also seen.

All of these consequences manifest as lactose intolerance and malabsorption syndrome.

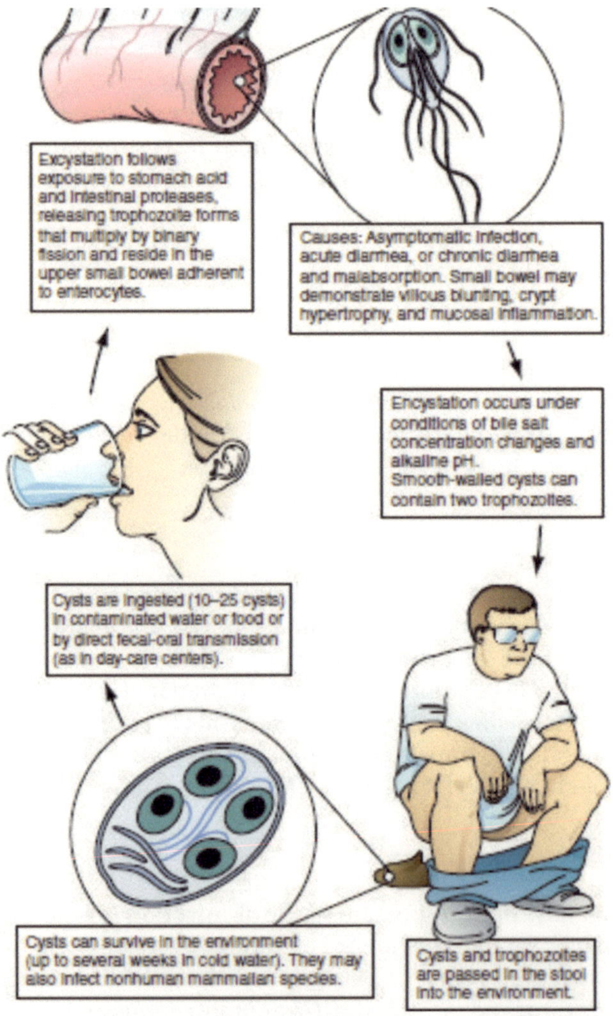

Figure 2.
Life cycle of Giardia. *Courtesy: Ref. [49], Fig. 224–1.*

2.2.4 Immune dysfunction

Both B cell-mediated antibody production and T cell-mediated immune responses play an in important role in protection against *Giardiasis*. Humans with immunodeficiency disorders like common variable immunodeficiency (CVID) and impaired IgA function have an increased risk of developing chronic *Giardia* infections. People living in endemic areas are less prone to infection or reinfection, indicating that acquired immunity exists.

According to a study, cellular immune response to *Giardia* can be understood by murine models. The mice model shows that in the absence of $CD4^+$ T cells, poor control or chronic infection to giardia occurs, indicating that these cells are crucial for the murine defense against *G. lamblia*. Cytokines like tumor necrosis factor alpha (TNF-α) and gamma interferon (IFN-γ) has been shown to be important for determining the parasite load and the duration of an infection.

Cytokines that are secreted by $CD4^+$ T cells were found in the mice infected with *Giardia*, indicating that a range of T_H responses may contribute to the protection against *Giardia*.

In another study, it was found that following infection of *G. lamblia*, there is an upregulation of peroxisome proliferator-activated receptor alpha and interleukin-17A (IL-17A) [50]. IL-17A was also found to be upregulated in another mouse study also, where IL-17A and its receptor were important for defense against the parasite [51]. Hence, on the basis of these studies, IL-17A may be linked to protection against *Giardia*.

Some human studies for *Giardia*-specific immune responses were also there. In one study, IFN-γ was secreted, and cells were proliferating after the stimulation of $CD4^+$ T cells with *Giardia* trophozoites, suggesting that *Giardia*-specific proliferation of $CD4^+$ T cells exists in humans as well [52].

The relative importance of the cytokines IL-17A, IFN-γ, TNF-α, IL-4, and IL-10 in the T cell response to *Giardia* infection in humans has not been determined [53].

2.2.5 Diagnosis

In outline the diagnosis of *Giardia* is based on the microscopic detection of *Giardia* cysts or trophozoites in a stool specimen by using wet mount, iodine mount, and trichrome stain method.

The string test (entro-test) may be useful for revealing *Giardia* trophozoites from duodenal sample [54].

Antigen detection assays that are available for *G. intestinalis* include EIAs, ELISAs, and direct fluorescent-antibody tests [55, 56]. For the cultivation of Giardia axenic media like diamond media can be used [57].

Molecular methods have higher sensitivity than conventional methods; however, many of them are still not commercially available [10, 54]. The use of real-time PCR is increasing for the detection of *Giardia*. Conventional single, nested, and multiplex PCRs have also been developed [10, 54].

2.2.6 Treatment

Metronidazole 250 mg TDS for 5 days or tinidazole 2 g once PO has been used as the therapy of choice. Due to the emergence of resistant isolates of *Giardia*, treatment failures and clinical relapses have been known to occur [44]. Alternatively, nitazoxanide 500 mg BD for 3 days can be used. Other drugs like furazolidone, albendazole, and paromomycin are also effective against giardiasis.

2.2.7 Prevention and control

By improving food and personal hygiene and by treating asymptomatic person, *Giardia* infection can be prevented.

One vaccine (GiardiaVax) has been licensed for dogs and cats in the United States [58, 59]. No human vaccines are currently available [59, 60].

2.3 *Cryptosporidium* species

2.3.1 Introduction

It is the coccidian parasite which causes self-limiting illness in immunocompetent person but can lead to severe disease in immunocompromised (AIDS) person. Two species of *Cryptosporidium*—*C. parvum* and *C. hominis*—found to be associated with human disease.

Epidemiology—The distribution of *Cryptosporidium* spp. is seen worldwide, and it mainly affects or causes severe illness in children and immunocompromised population. The prevalence of cryptosporidiosis in children of developing and developed nations was found to be 20 and 9%, respectively [61]. In adult population the change in prevalence of *Cryptosporidium* spp. was seen in the year 2000; before its prevalence was 3.8% in Los Angeles, and after 2000 it increased to 13% in the developed world [19].

Different clinical manifestations are produced by species of *Cryptosporidium* and subtype families of *Cryptosporidium hominis* [62].

2.3.2 Pathogenesis

Transmission—It is a waterborne disease transmitted via the fecal-oral route, by ingestion of contaminated salads, contaminated water supply, or recreational water, swimming in public pools, and it is observed that it can be transmitted from person to person, men who have sex with men, as well as through zoonotic infections [62]. *Cryptosporidium* spp. have a wide host range including humans, domestic pets, and wildlife [63]. Zoonotic transmission occurs by direct contact with infected animals or their feces or indirectly through the consumption of contaminated water [63, 64].

Life cycle—The environmentally resistant oocysts are produced by *Cryptosporidium*, and they sporulate when excreted in feces, and therefore, these oocysts are immediately infectious [65]. After ingestion of oocyst, excystation occurs in the intestine to liberate sporozoite which further develops into meront (with four merozoite). Meront form converts into macrogamete and microgametes that lead to zygote and finally unsporulated oocyst excreted into the feces.

Pathophysiology—In the small intestine, *Cryptosporidium* present inside the intracellular vacuoles and no characteristic changes were found on biopsy. Infection of *Cryptosporidium* is characterized by self-limiting diarrhea along with abdominal pain, dehydration, and malabsorption. It is an important infection in both immunosuppressed (especially HIV-infected) and in transplant recipients [66] **Figure 3** (Mandell, Douglas, and Bennett's Principles and Practice of Infectious Disease; 8th ed. Philadelphia. Elsevier saunder's 2015 Fig. 284–1. Life cycle of Cryptosporidium, p. 3174).

2.3.3 Immune dysfunction

The occurrence of cryptosporidiosis is closely related to the immune status of its host. It primarily affects the infants and immunocompromised individuals.

In recent years, several studies have highlighted the importance of innate immunity in cryptosporidiosis. Intestinal epithelial cells play a key role in cryptosporidiosis as these are the exclusive host cell for the replication of the parasite and also participate in the protective immune response. Epithelial cells produce chemokines and attract immune cells to the infected area. They also release antimicrobial peptides with parasiticidal activity and induce apoptosis. Intestinal dendritic cells induce adaptive immunity and control *Cryptosporidium parvum* infection in the early stage [68].

Both innate and adaptive immune responses are important for controlling cryptosporidiosis. Innate immune responses are mediated by Toll-like receptor pathways, antimicrobial peptides, prostaglandins, mannose-binding lectin, cytokines, and chemokines. Cell-mediated responses, particularly those involving $CD4^+$ T cells and IFN-γ, play an important role. The parasite has developed several escape mechanisms to slow down these protective mechanisms [69].

2.3.4 Diagnosis

Routinely the diagnosis of cryptosporidiosis is done by the identification of oocysts in stool by microscopic examination. The staining and preservation technique enhance the sensitivity of the test. The staining technique for *Cryptosporidium* spp. includes modified Ziehl-Neelsen technique, Kinyoun's acid-fast staining technique, modified Sheather's flotation technique, and the iron-hematoxylin staining technique. Out of these staining techniques, modified Ziehl-Neelsen technique is widely used [70, 71].

Antigen detection assay is also used widely in the diagnosis of cryptosporidiosis; it is more effective in cases where oocyst numbers are low. These assays are thought to be more sensitive than conventional staining [72, 73]. Other tests like fluorescence

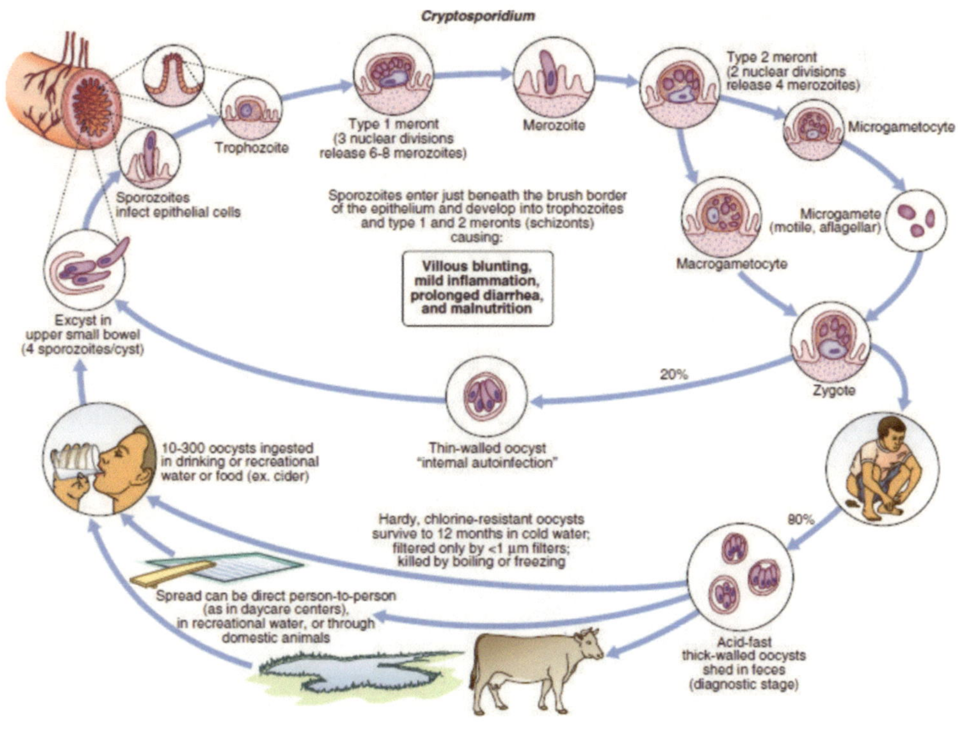

Figure 3.
Life cycle of Cryptosporidium. *Courtesy: Ref. [67], Fig. 284–1.*

microscopy and direct fluorescent-antibody (DFA) assay have been used for the detection of *Cryptosporidium* oocysts with relatively high specificities (96–100%) and sensitivities (98.5–100%) [73, 74]. Molecular methods for the detection and differentiation of *Cryptosporidium* spp. include nested PCR, real-time PCR, multiplex real-time PCR, reverse transcription-quantitative real-time PCR, and multiplex tandem real-time PCR [12, 75]. More recently, an automated multiplex tandem PCR is used; it is based on **a robotic platform** and used to detect *Cryptosporidium* spp. and coinfecting diarrheal pathogens from the human fecal genomic sample [76]. Some new methods such as **reverse line blot (RLB) hybridization**, loop-mediated isothermal amplification method, and nucleic acid sequence-based amplification (NASBA) methods that amplify RNA from either RNA or DNA templates are also used for the diagnosis of *Cryptosporidium* spp. [75, 77]. Another new technique for the detection of *Cryptosporidium* spp. includes **next-generation sequencing, DNA microarray,** etc. which are also available [73] (**Figure 4**).

2.3.5 Treatment

In immunocompetent persons, cryptosporidiosis is usually self-limiting, requiring little or no treatment, but in immunosuppressed person, treatment is required [78]. Nitazoxanide 500 mg BD for 3 days has proven to be effective against cryptosporidiosis. In patients with dehydration, rehydration fluids and nutritional management may be required.

Nitazoxanide is the only licensed drug for the treatment of cryptosporidiosis, and it is given only in persons after the first year of life and with healthy immune systems. Recently trials are going on clofazimine, a drug used for leprosy, to know its effectiveness against *Cryptosporidium* spp. [79].

2.3.6 Prevention and control

Prevention can be done by avoiding exposure to infectious oocyst in human or animal feces, proper hand washing, and improved personal hygiene. No vaccines are available for use against *Cryptosporidium* spp. [79].

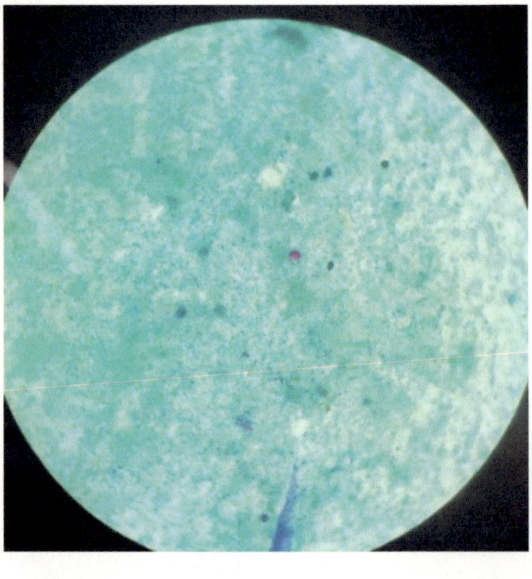

Figure 4.
Oocyst of Cryptosporidium *on modified ZN stain.*

2.4 *Cyclospora*

2.4.1 Introduction

This is a coccidian parasite, and the only species of this genus found in humans is *Cyclospora cayetanensis*. Similar to *Cryptosporidium, Cyclospora* is also associated with severe illness in immunocompromised patients in comparison to immunocompetent patients.

Epidemiology—Human cyclosporiasis is distributed worldwide, and cases are reported both from tropical and subtropical areas of the world. According to a study from China, the prevalence of *Cyclospora* spp. from 2007 to 2009 was 0.70%, whereas in the United Kingdom at the same time, the prevalence was 0.07% [80].

2.4.2 Pathogenesis

Transmission—An important feature of the *C. cayetanensis* is that the direct fecal-oral transmission from fresh stool is not seen because the oocysts require days to weeks outside the host to sporulate and become infectious.

Life cycle and pathophysiology—Similar to *cryptosporidium* spp.

Clinical feature—Infection due to *Cyclospora* manifest as persistent diarrhea, fever, bloating, flatulence, abdominal cramps, constipation, and fatigue, weight loss, etc.

Immune dysfunction—similar to *Cryptosporidium* spp.

2.4.3 Diagnosis

Routinely the diagnosis of *Cyclospora* is done by using microscopic methods. *Cyclospora* oocyst produces autofluorescence (white-blue) under an epifluorescence microscope [81, 82]. By demonstration of *Cyclospora* oocysts, autofluorescence and staining characteristics diagnosis can be made. Oocysts appear as acid-fast round or ovoid very small (8–10 μm) structures. The diameter measurement of oocyst is important to differentiate it from smaller *Cryptosporidium* oocysts, which measure 4–6 μm in size. The procedures used to diagnose *Cyclospora* includes concentration by the formalin-ethyl acetate technique followed by either (i) UV epifluorescence and bright-field microscopy, (ii) examination of a modified acid-fast-stained stool slide, or (iii) examination using a modified safranin-based technique [83].

Molecular techniques have been developed like various PCR tools that target the internal transcribed spacer region, which use primers for the 18S rRNA gene [82]. These tools include conventional PCR, reverse transcriptase PCR in combination with agarose gel electrophoresis, and nested PCR [82]. *Cyclospora* is included in the BioFire FilmArray systems GIT panel; hence, syndromic diagnosis can be made.

2.4.4 Treatment

Laboratory diagnosis is important as the treatment of *Cyclospora* is different from the other protozoa of similar presentation. Trimethoprim-sulfamethoxazole (also known as co-trimoxazole) 160/800 mg BD for 7–10 days is the drug of choice for cyclosporiasis.

2.4.5 Prevention and control

No vaccine is available till now against *Cyclospora*; by avoiding contaminated food and water it can be prevented.

2.5 *Cystoisospora belli* (formerly *Isospora belli*)

2.5.1 Introduction

Cystoisospora belli is responsible for intestinal disease in several mammalian hosts. It is also a coccidian parasite similar to *Cryptosporidium* and *Cyclospora*. It is also found to be associated with more severe illness in immunocompromised host.

2.5.2 Epidemiology

Very little is known about the epidemiology and prevalence of *Cystoisospora belli*. The worldwide reported prevalence of *Cystoisospora belli* was 2.5–26.1%, and it is mostly seen in HIV-positive individuals [18].

2.5.3 Pathogenesis

Transmission—Infection occurs via the ingestion of mature sporulated oocysts present in contaminated food or water.

Life cycle and pathophysiology—Similar to *Cryptosporidium* spp.

Clinical feature—Infection is manifested as watery diarrhea, abdominal cramps, anorexia, and weight loss, and it is almost indistinguishable from cryptosporidiosis. It is more common in AIDS patients and causes traveler's diarrhea in travelers to developing countries. In HIV-infected patients, infection of *Cystoisospora belli* may lead to chronic diarrhea, acalculous cholecystitis cholangiopathy, and extraintestinal infection [84]. Other *Cystoisospora* species causes of diarrhea in domestic animals, like *Cystoisospora suis*, lead to severe diarrheal illness in pigs [85].

Immune dysfunction—Similar to *cryptosporidium* spp.

Diagnosis is made by demonstration of oocyst by direct microscopic examination of the feces, with acid-fast staining or fluorescent staining with auramine and rhodamine. Oocysts of *Cystoisospora belli* are large (20–23 μm × 10–19 μm) and ellipsoidal in shape in comparison to oocyst of *Cryptosporidium* which is round and 4–6 μm in diameter.

Autofluorescence can be seen, but this property is not consistent in *Cystoisospora*.

In molecular techniques PCRs using SSU rRNA primer have shown excellent sensitivity and specificity for the detection of *C. belli* in fecal samples. But these tests are not commercially available [86, 87] (**Figure 5**).

Treatment is done by giving oral co-trimoxazole 160/800 mg BD for 10 days and in case of HIV-infected patients three times daily for 3 weeks. Other drugs like ciprofloxacin, pyrimethamine, and folinic acid are good alternative.

2.5.4 Prevention and control

To date no vaccines are available for *Cystoisospora belli*; however, trials are going on to identify the suitable candidate of parasite for vaccine preparation [88]. Hence, prevention can be done by avoiding contaminated food and water consumption.

2.6 *Dientamoeba fragilis*

2.6.1 Introduction

It was initially considered as amoeba but recently classified under ameboflagellate as it has internal flagellum. It mostly involves the lumen of the cecum and upper colon of the human intestine.

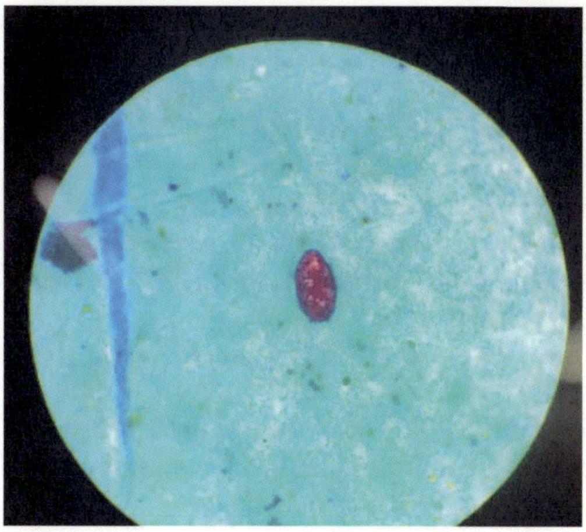

Figure 5.
Oocyst of Cystoisospora *in modified ZN stain.*

Epidemiology—The reported prevalence of *Dientamoeba fragilis* in the year of 2015 was around 1–70% in developed settings [89]. It varies globally as in United States and Australia; the reported prevalence is 5%, and in Netherland it was around 8% [90].

2.6.2 Pathogenesis

Many debates are there for the pathogenicity of *D. fragilis* [91, 92]. Previously it was thought that in *D. fragilis*, cyst stage has not been found [92], but recently both the **cyst and precyst stage** have been identified [93]. The nuclear chromatin of trophozoite of *D. fragilis* is fragmented into 3–5 granules; hence, it is known as fragilis. Although the mode of transmission remains unknown, previously it was thought that infection occurs via the pinworm vector because of its high rates of coinfection with *Enterobius vermicularis* [94]. But according to recent studies, it is found that transmission occurs directly via the fecal-oral route [92].

Clinical symptoms of *Dientamoeba fragilis* infection include abdominal pain, persistent diarrhea, loss of appetite, weight loss, and flatulence. Symptoms of *D. fragilis* are similar to those of irritable bowel syndrome (IBS); hence, it should be considered in the differential diagnosis of IBS [95].

2.6.3 Immune dysfunction

Dientamoeba fragilis is often described as "neglected parasite" [96]. Recently, several studies have occurred on the life cycle and molecular biology of the organism, but the knowledge on immune response against the pathogen is lacking. Because of the close relativity of *Dientamoeba fragilis* to histomonads, the immunological research on *H. meleagridis* can give an indication to the immunological responsiveness of host against *D. fragilis* [97].

2.6.4 Diagnosis

It is based on the microscopic detection of trophozoites in fresh or fixed stool specimens. The trophozoites degrade soon after passing in the stool; hence, the examination should be done immediately. Modified iron-hematoxylin or trichrome staining of fixed smears is considered the gold standard for the diagnosis of

D. fragilis infection [98]. But in comparison to molecular methods, this method is time-consuming and relatively insensitive [99]. The demonstration of the characteristic nuclear structure is needed for the definitive diagnosis of *D. fragilis* which cannot be seen in unstained fecal material. Hence, fixation and staining are necessary for definitive diagnosis of *D. fragilis* [99].

Antigen detection by rapid fecal immunoassays like enzyme immunoassays, fluorescent antibody, and rapid cartridge formats is previously under trial, but now enzyme immunoassay and immunofluorescence tests are available for antigen detection [93].

Antibody detection can be done by using an indirect immunofluorescence method [93].

For the cultivation of *D. fragilis*, no axenic cultures exist, and it will grow well in xenic cultures. For growth of *D. fragilis*, Loeffler's slope medium and modified Earle's balanced salt solution media are used [93].

Recently, the more advanced tests based on various conventional and real-time PCRs are used to detect the small subunit (SSU) rRNA gene of *D. fragilis.* These tests are more rapid, sensitive, and specific for the diagnosis in fresh stools [99]. Molecular methods are more sensitive than microscopy and staining but still not commercially available. *Dientamoeba fragilis* is not included in the gastrointestinal panel of BioFire FilmArray system [93].

2.6.5 Treatment

For the treatment of *D. fragilis* paromomycin, secnidazole, iodoquinol, tetracycline, ornidazole, and metronidazole have been used successfully. The treatment is recommended both in symptomatic patients and asymptomatic family members to prevent reinfection. There are few studies which show emergence of resistance or treatment failures of metronidazole therapy; it suggests that the combination therapy should be used for these protozoa [100]. For the complete eradication of the parasite and in resolution of symptoms, combination therapy is effective [100].

2.6.6 Prevention and control

By improving personal hygiene, *Dientamoeba* infection can be prevented. No vaccines available yet for *Dientamoeba fragilis*.

2.7 *Blastocystis* species

2.7.1 Introduction

Blastocystis spp. are commonly isolated from stools of humans and animals [13]. There are numerous subtypes of *Blastocystis* spp., but it is unclear whether any of these subtypes are specific to humans disease. It was found that *Blastocystis* sp. subtype 3 is most commonly associated with illness in human [13, 101]. It appears in various morphological forms like vacuolar, granular, amoeboid, and cyst forms, and less frequently avacuolar and multivacuolar cells and cells containing filament-like inclusions [102, 103].

Epidemiology—The prevalence of *Blastocystis* spp. in developed countries is 0.5–23% and in developing countries 22–100%. The prevalence varies in different geographical are according to the socioeconomic status of that area [104].

2.7.2 Pathogenesis

Transmission is usually taken place via the fecal-oral route through the consumption of cyst. These cysts may be waterborne or foodborne. There are many zoonotic subtypes of *Blastocystis* spp. which supports the increased potential for zoonotic transmission [105].

Life cycle—Infection is acquired through the consumption of cyst, and it converts to vacuolar, granular, and amoeboid form in human intestine. Cyst form is excreted in the feces and contaminates the food and water ingested by other animals.

Clinical features of *Blastocystis* spp. infection include mostly asymptomatic or nausea, anorexia, abdominal pain, flatulence, and acute or chronic diarrhea. *Blastocystis* spp. have been found to be the most common enteric organisms isolated from diarrheal patients but have been reported as noninfectious pathogens or a member of normal gut microbiota [106].

2.7.3 Immune dysfunction

According to a study, the incubation of *Blastocystis* sp. ST1 cells or culture filtrates induces the production of proinflammatory cytokines interleukin (IL)-8 and granulocyte-macrophage colony-stimulating factor. It suggests that the parasite was able to modulate the host immune response. *Blastocystis* sp. releases cysteine proteases, which induce human colonic epithelial cells (HT84) for the production of IL-8. Cysteine proteases of *Blastocystis* sp. are involved in parasite survival in vivo and represent virulence factors. *Blastocystis* sp. also secretes protease, which is able to cleave human secretory immunoglobulin A. Apart from proteases, hydrolases and protease inhibitors were predicted to be secreted by *Blastocystis* sp. which participates in the blastocystosis physiopathology and its mechanisms for immune evasion [107].

2.7.4 Diagnosis

The diagnosis of infection with *Blastocystis* spp. is done by using wet mount smears, iodine staining, trichrome staining, or iron-hematoxylin staining and detecting the vacuolar, granular, amoebic, or cystic form in stool samples [13, 103]. Trichrome staining of fecal smears or xenic in vitro culture systems has better sensitivity [101]. Subtype 7 which is a slower-growing subtype of *Blastocystis* spp. may be missed with this procedure [101].

For antibody detection in serum ELISA and immunofluorescent assay are also available. ELISA for the antigen detection in the stool is also introduced recently [108].

In molecular techniques, PCR using the SSU rRNA gene is being used increasingly for the detection of *Blastocystis* spp. with increased sensitivity, and it is more commonly used method nowadays [13]. Based on SSU rRNA 17, subtypes of *Blastocystis* spp. were identified out of which ST3 is the commonest worldwide [109] (**Figure 6**).

2.7.5 Treatment

Drugs used for the treatment of pathogenic *Blastocystis* spp. include metronidazole, co-trimoxazole, nitazoxanide, and a combination of paromomycin and metronidazole. Among these metronidazole is the drug of choice, although failures of this drug in eradicating the organism are common [111].

2.7.6 Prevention and control

Prevention can be done by avoiding the contaminated food and water consumption; no vaccines are available for *Blastocystis* spp. till now.

2.8 *Balantidium coli*

2.8.1 Introduction

Balantidium coli is the largest protozoan that infects humans. It belongs to the ciliate group. The whole body of the organism is covered with a row of cilia which is responsible for the rotary motility of *B. coli* (**Figure 7**).

2.8.2 Epidemiology

It mostly occurred in tropical and subtropical areas of developing countries. The estimated prevalence of *B. coli* worldwide is 1% as the human infection with this protozoa is rare [113].

2.8.3 Pathogenesis

Transmission—Pigs are the natural host for *B. coli*. Human acquires infection by ingestion of cysts present in water and undercooked food (fecal-oral route).

Life cycle—After ingestion of cyst, it converts into trophozoite in the large intestine of human. Both cyst and trophozoite are excreted into the feces, but in external environment, trophozoite disintegrates and cyst will survive. It contaminates the water and food, and the life cycle continues.

Clinical feature—Infections of *B. coli* are mostly asymptomatic, but in symptomatic patients, mild diarrhea and abdominal discomfort have been reported. In a few patients fulminating acute balantidiasis, with intestinal perforation, can

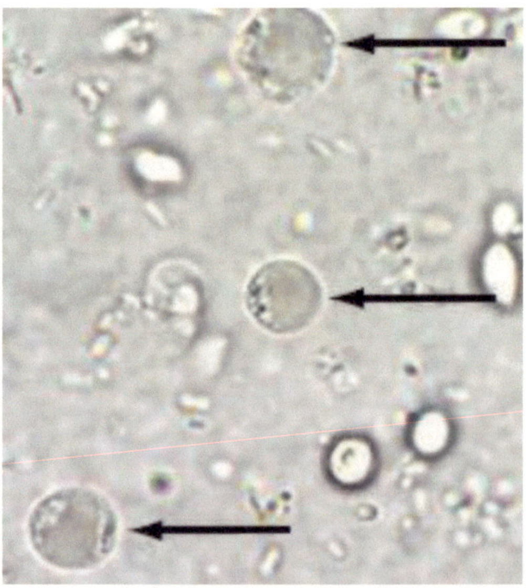

Figure 6.
Cysts of Blastocystis hominis *on wet mount. Courtesy: Ref. [110], Fig. 285–5.*

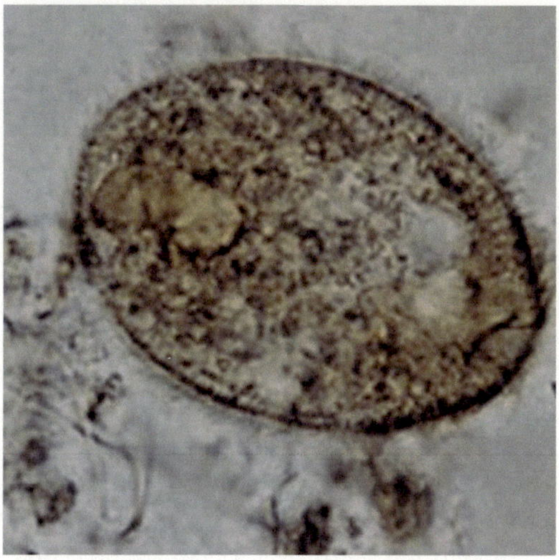

Figure 7.
Trophozoite of B. coli *demonstrating cilia on the surface. Courtesy: Ref. [112], Fig. 285–4.*

develop which has a case fatality rate of about 30%. Fulminating dysentery, resembling amoebic dysentery, can also occur in *B. coli* infection [113].

2.8.4 Immune dysfunction

Immune response in *B. coli* is poorly understood due to the lack of studies available. It was found that infection of *B. coli* is more severe in immunocompromised patients.

2.8.5 Diagnosis

B. coli is diagnosed by its large size cysts (50–70 μm) and trophozoite (30–200 μm × 40–70 μm) and rotatory motility which is demonstrated by wet mount slide preparations. Trophozoites of *B. coli* are visible even with hand lens and, sometimes with the naked eye, in freshly collected diarrheic stools as well as in bronchoalveolar wash fluid [113]. Collection of stool samples should be done for several days because the excretion of parasites can be erratic [113].

Balantidium coli can be cultivated on Boeck and Drbohlav egg serum media and Balamuth media. It can also be cultivated on Pavlova xenic culture media, and the growth of trophozoites can be obtained within 72 h on this media [114].

2.8.6 Treatment

B. coli is treated with tetracycline or metronidazole:

- Tetracycline—500 mg qid for 10 days
- Metronidazole 750 mg tds for 5–7 days

2.8.7 Prevention and control

By avoiding contamination of food and water with pig and human feces and by treating the carrier shedding the cyst, the infection can be prevented. No vaccines are available till now for *Balantidium coli* infection.

3. Conclusion

Enteric protozoa are responsible for severe morbidity and mortality in humans. Protozoal infections contribute to the high burden of infectious diseases. Despite recent advances in the epidemiology, diagnostic tool, molecular biology, and treatment of protozoan illnesses, gaps in knowledge still exist; hence, protozoal infections require further research. Both humoral and cellular immunities play an important role in the protection of host against enteric protozoal infection. Protozoa have some virulence factor or mechanism to evade the immune system of the host, for the production of disease. An already dysfunctional host immune system (immunocompromised conditions) helps the protozoa to cause easy and more serious disease. Among these enteric protozoal infection *Entamoeba histolytica*, *Giardia intestinalis*, *Cryptosporidium*, *Cyclospora*, *Cystoisospora belli*, *Dientamoeba fragilis*, *Blastocystis*, and *Balantidium coli* are common. In the recent years, increasing trends of *Entamoeba histolytica*, *Giardia intestinalis*, and *Cryptosporidium* is seen both in developing and developed countries. Blastocystis is an emerging infection to humans. Mostly enteric protozoal infections are acquired through fecal-oral route and present with abdominal pain and fever mainly. Some of them can lead to anemia and malabsorption syndrome. If not diagnosed or treated timely, enteric protozoal infection can be fatal. Diagnosis for most of them is based on microscopy,but recently many molecular methods are available to diagnose them. By using multiplex PCR, BioFire Gastrointestinal panel, common diarrheal illness-causing organisms can be identified in a shorter time span. ELISA and immunofluorescent assay are also available for some of them, e.g., *Blastocystis* spp. and *Dientamoeba fragilis*. For the treatment of protozoal infection, drugs like metronidazole, paromomycin, diloxanide furoate, nitazoxanide, cotrimoxazole, etc. are used. The infection can be prevented by good personal hygiene and by avoiding food and water contamination. Till date, no vaccines are available for human use against these protozoal infections.

Acknowledgements

- I want to thank our technical staff for their support.
- I also want to thank my juniors for their assistance and support.
- Special thanks to the whole microbiology department, my family, and my teachers for their guidance.

Author details

Renu Kumari Yadav, Shalini Malhotra* and Nandini Duggal
ABVIMS and Dr. RML Hospital, New Delhi, India

*Address all correspondence to: drshalinimalhotra@yahoo.com

References

[1] Kucerova Z, Sokolova O, Demyanov A, Kvac M, Sak B, Kvetonova D, et al. Microsporidiosis and cryptosporidiosis in HIV/AIDS patients in St. Petersburg, Russia: Serological identification of microsporidia and *Cryptosporidium parvum* in sera samples from HIV/AIDS patients. AIDS Research and Human Retroviruses. 2011;**27**(1):13-15

[2] Bakheit MA, Torra D, Palomino LA, Thekisoe OM, Mbati PA, Ongerth J, et al. Sensitive and specific detection of *Cryptosporidium* species in PCR-negative samples by loop-mediated isothermal DNA amplification and confirmation of generated LAMP products by sequencing. Veterinary Parasitology. 2008;**158**(1-2):11-22

[3] Escobedo A, Almirall P, Robertson L, Franco R, Hanevik K, Morch K, et al. Giardiasis: The ever-present threat of a neglected disease. Infectious Disorders-Drug Targets (Formerly Current Drug Targets-Infectious Disorders). 2010;**10**(5):329-348

[4] Mor SM, Tzipori S. Cryptosporidiosis in children in Sub-Saharan Africa: A lingering challenge. Clinical Infectious Diseases. 2008;**47**(7):915-921

[5] Newell DG, Koopmans M, Verhoef L, Duizer E, Aidara-Kane A, Sprong H, et al. Food-borne diseases—The challenges of 20 years ago still persist while new ones continue to emerge. International Journal of Food Microbiology. 2010;**139**:S3-S15

[6] Waldron LS, Ferrari BC, Cheung-Kwok-Sang C, Beggs PJ, Stephens N, Power ML. Molecular epidemiology and spatial distribution of a waterborne cryptosporidiosis outbreak in Australia. Applied and Environmental Microbiology. 2011;**77**(21):7766-7771

[7] Yoder JS, Hlavsa MC, Craun GF, Hill V, Roberts V, Yu PA, et al. Surveillance for waterborne disease and outbreaks associated with recreational water use and other aquatic facility-associated health events–United States, 2005-2006. MMWR Surveillance Summaries. 2008;**57**(9):1-29

[8] Shah N, DuPont HL, Ramsey DJ. Global etiology of travelers' diarrhea: Systematic review from 1973 to the present. The American Journal of Tropical Medicine and Hygiene. 2009;**80**(4):609-614

[9] Fletcher SM, Stark D, Ellis J. Prevalence of gastrointestinal pathogens in Sub-Saharan Africa: Systematic review and meta-analysis. Journal of Public Health in Africa. 2011;**2**(2)

[10] Lebbad M. Molecular Diagnosis and Characterization of Two Intestinal Protozoa: *Entamoeba histolytica* & *Giardia intestinalis*. Institutionen för mikrobiologi, tumör-och cellbiologi/ Department of Microbiology, Tumor and Cell Biology. Karolinska Institutet; 2010

[11] Hove RJ. Molecular Detection of Intestinal Parasites for Clinical Diagnosis and Epidemiology. Doctoral dissertation, Department of Parasitology/Epi-Diag, Faculty of Medicine, Leiden University Medical Center (LUMC), Leiden University; 2009

[12] Stark D, Al-Qassab S, Barratt J, Stanley K, Roberts T, Marriott D, et al. Evaluation of multiplex tandem real-time PCR for detection of *Cryptosporidium* spp., *Dientamoeba fragilis*, *Entamoeba histolytica*, and *Giardia intestinalis* in clinical stool samples. Journal of Clinical Microbiology. 2011;**49**(1):257-262

[13] Roberts T, Barratt J, Harkness J, Ellis J, Stark D. Comparison of microscopy, culture, and conventional polymerase chain reaction for

detection of *Blastocystis* sp. in clinical stool samples. The American Journal of Tropical Medicine and Hygiene. 2011;**84**(2):308-312

[14] Santos HLC, Bandea R, Martins LAF, De Macedo HW, Peralta RHS, Peralta JM, et al. Differential identification of *Entamoeba* spp. based on the analysis of 18S rRNA. Parasitology Research. 2010;**106**(4):883-888

[15] Saleh R, El-zahraa F, Gad M, Ashour A, Soliman M, El-Senousy W, et al. Molecular detection of *Entamoeba histolytica* in fresh vegetables and irrigation. Egyptian Journal of Aquatic Biology and Fisheries. 2019;**22**(5 (Special Issue)):551-561

[16] Gualdieri L, Rinaldi L, Petrullo L, Morgoglione M, Maurelli M, Musella V, et al. Intestinal parasites in immigrants in the city of Naples (southern Italy). Acta Tropica. 2011;**117**(3):196-201

[17] Shirley D-AT, Farr L, Watanabe K, Moonah S. A review of the global burden, new diagnostics, and current therapeutics for amebiasis. In: Open Forum Infectious Diseases. USA: Oxford University Press; 2018

[18] Ghoshal U, Jain V, Tejan N, Kalra SK, Ranjan P, Sinha R, et al. A road less travelled: Clinical comparison of HIV seropositive and seronegative patients with cystoisosporiasis—An 11-year experience from a tertiary care centre in Northern India. Indian Journal of Medical Microbiology. 2018;**36**(4):508

[19] Momba M, Ebdon J, Kamika I, Verbyla M. Using indicators to assess microbial treatment and disinfection efficacy. In: Global Water Pathogen Project. East Lansing, Michigan: Michigan State University, UNESCO; 2019

[20] Walsh JA. Problems in recognition and diagnosis of amebiasis: Estimation of the global magnitude of morbidity and mortality. Reviews of Infectious Diseases. 1986;**8**(2):228-238

[21] Maldonado-Barrera CA, del Rosario C-EM, Muñoz-Fernández L, Victoria-Hernández JA, Campos-Rodríguez R, Talamás-Rohana P, et al. Clinical case of cerebral amebiasis caused by *E. histolytica*. Parasitology Research. 2012;**110**(3):1291-1296

[22] Tamer GS, Öncel S, Gökbulut S, Arisoy ES. A rare case of multilocus brain abscess due to *Entamoeba histolytica* infection in a child. Saudi Medical Journal. 2015;**36**(3):356

[23] Fauci J et al. Harrison's Principles of Internal Medicine. 20th ed. New York, London, Chicago: McGraw Hill Education; 2019. p. 1569

[24] Fletcher SM, Stark D, Harkness J, Ellis J. Enteric protozoa in the developed world: A public health perspective. Clinical Microbiology Reviews. 2012;**25**(3):420-449

[25] Anaya-Velázquez F, Padilla-Vaca F. Virulence of *Entamoeba histolytica*: A challenge for human health research. Future Microbiology. 2011;**6**(3):255-258

[26] Mondal D, Haque R, Sack RB, Kirkpatrick BD, Petri WA Jr. Attribution of malnutrition to cause-specific diarrheal illness: Evidence from a prospective study of preschool children in Mirpur, Dhaka, Bangladesh. The American Journal of Tropical Medicine and Hygiene. 2009;**80**(5):824-826

[27] Begum S, Quach J, Chadee K. Immune evasion mechanisms of *Entamoeba histolytica*: Progression to disease. Frontiers in Microbiology. 2015;**6**:1394

[28] Verma GK, Sharma NL, Shanker V, Mahajan VK, Kaushik R, Verma S, et al. *Amoebiasis cutis*: Clinical suspicion is the key to early diagnosis. The Australasian Journal of Dermatology. 2010;**51**(1):52-55

[29] Stark D. *Entamoeba histolytica*. In: PCR for Clinical Microbiology. Springer; 2010. pp. 363-367

[30] Pritt BS, Clark CG. Amebiasis. In: Mayo Clinic Proceedings; Elsevier. 2008

[31] Gonzalez-Salazar F, Meester I, Guzmán De La Garza F, La Garza-Salinas D, Sampayo-Reyes A, Garza-Gonzalez J, et al. "CLUPS": A new culture medium for the axenic growth of *Entamoeba histolytica*. Journal of Parasitology Research. 2018:133-137. DOI: http://dx.doi.org/10.5772/intechopen.91698

[32] Fotedar R, Stark D, Marriott D, Ellis J, Harkness J. *Entamoeba moshkovskii* infections in Sydney, Australia. European Journal of Clinical Microbiology & Infectious Diseases. 2008;**27**(2):133-137

[33] Fotedar R, Stark D, Beebe N, Marriott D, Ellis J, Harkness J. Laboratory diagnostic techniques for *Entamoeba* species. Clinical Microbiology Reviews. 2007;**20**(3):511-532

[34] Soares NM, Azevedo HC, Pacheco FT, de Souza JN, Del-Rei RP, Teixeira MC, et al. A cross-sectional study of *Entamoeba histolytica*/dispar/moshkovskii complex in Salvador, Bahia, Brazil. BioMed Research International. 2019:1-7

[35] Piralla A, Lunghi G, Ardissino G, Girello A, Premoli M, Bava E, et al. FilmArray™ GI panel performance for the diagnosis of acute gastroenteritis or hemorrhagic diarrhea. BMC Microbiology. 2017;**17**(1):111

[36] Kelly P. Protozoal gastrointestinal infections. Medicine. 2013;**41**(12):705-708

[37] Kappagoda S, Singh U, Blackburn BG. Antiparasitic therapy. In: Mayo Clinic Proceedings; Elsevier. 2011

[38] Quach J, St-Pierre J, Chadee K. The future for vaccine development against *Entamoeba histolytica*. Human Vaccines & Immunotherapeutics. 2014;**10**(6):1514-1521

[39] Roncolato EC, Teixeira JE, Barbosa JE, Ramalho LNZ, Huston CD. Immunization with the *Entamoeba histolytica* surface metalloprotease EhMSP-1 protects hamsters from amebic liver abscess. Infection and Immunity. 2015;**83**(2):713-720

[40] Waldram A, Vivancos R, Hartley C, Lamden K. Prevalence of Giardia infection in households of Giardia cases and risk factors for household transmission. BMC Infectious Diseases. 2017;**17**(1):486

[41] Leder K, Weller DPF. Giardiasis: Epidemiology, clinical manifestations, and diagnosis.. UpToDate. [Accessed on: 15 March 2019]

[42] Dunn N, Juergens AL. Giardiasis. Treasure Island, FL: StatPearls; 2019

[43] Dixon B, Parrington L, Cook A, Pintar K, Pollari F, Kelton D, et al. The potential for zoonotic transmission of *Giardia duodenalis* and *Cryptosporidium* spp. from beef and dairy cattle in Ontario, Canada. Veterinary Parasitology. 2011;**175**(1-2):20-26

[44] Busatti HG, Santos JF, Gomes MA. The old and new therapeutic approaches to the treatment of giardiasis: Where are we? Biologics: Targets & Therapy. 2009;**3**:273

[45] Demirci M, Delibas N, Altuntas I, Oktem F, Yönden Z. Serum iron, zinc and copper levels and lipid peroxidation in children with chronic giardiasis. Journal of Health, Population, and Nutrition. 2003:72-75

[46] Olson ME, Leonard NJ, Strout J. Prevalence and diagnosis of Giardia infection in dogs and cats using a

fecal antigen test and fecal smear. The Canadian Veterinary Journal. 2010;**51**(6):640

[47] Yoshiuchi R, Matsubayashi M, Kimata I, Furuya M, Tani H, Sasai K. Survey and molecular characterization of Cryptosporidium and *Giardia* spp. in owned companion animal, dogs and cats, in Japan. Veterinary Parasitology. 2010;**174**(3-4):313-316

[48] Wilson N, Baker M, Edwards R, Simmons G. Case-case analysis of enteric diseases with routine surveillance data: Potential use and example results. Epidemiologic Perspectives & Innovations. 2008;**5**(1):6

[49] Fauci J et al. Harrison's Principles of Internal Medicine. 20th ed. New York, London, Chicago: McGraw Hill Education; 2019. p. 1616

[50] Dreesen L, De Bosscher K, Grit G, Staels B, Lubberts E, Bauge E, et al. *Giardia muris* infection in mice is associated with a protective interleukin 17A response and induction of peroxisome proliferator-activated receptor alpha. Infection and Immunity. 2014;**82**(8):3333-3340

[51] Dann SM, Manthey CF, Le C, Miyamoto Y, Gima L, Abrahim A, et al. IL-17A promotes protective IgA responses and expression of other potential effectors against the lumen-dwelling enteric parasite Giardia. Experimental Parasitology. 2015;**156**:68-78

[52] Ebert E. Giardia induces proliferation and interferon γ production by intestinal lymphocytes. Gut. 1999;**44**(3):342-346

[53] Saghaug CS, Sørnes S, Peirasmaki D, Svärd S, Langeland N, Hanevik K. Human memory CD4+ T cell immune responses against *Giardia lamblia*. Clinical and Vaccine Immunology. 2016;**23**(1):11-18

[54] Hooshyar H, Rostamkhani P, Arbabi M, Delavari M. *Giardia lamblia* infection: Review of current diagnostic strategies. Gastroenterology and Hepatology from Bed to Bench. 2019;**12**(1):3

[55] Foronda P, Bargues M, Abreu-Acosta N, Periago M, Valero M, Valladares B, et al. Identification of genotypes of *Giardia intestinalis* of human isolates in Egypt. Parasitology Research. 2008;**103**(5):1177-1181

[56] Barbecho JM, Bowman DD, Liotta JL. Comparative performance of reference laboratory tests and in-clinic tests for Giardia in canine feces. Parasites & Vectors. 2018;**11**(1):444

[57] Garcia LS. Diagnostic medical parasitology: An update. Clinical Microbiology Newsletter. 1994;**16**(14):105-110

[58] Meeusen EN, Walker J, Peters A, Pastoret P-P, Jungersen G. Current status of veterinary vaccines. Clinical Microbiology Reviews. 2007;**20**(3):489-510

[59] Serradell MC, Saura A, Rupil LL, Gargantini PR, Faya MI, Furlan PJ, et al. Vaccination of domestic animals with a novel oral vaccine prevents *Giardia* infections, alleviates signs of giardiasis and reduces transmission to humans. NPJ Vaccines. 2016;**1**:16018

[60] Tian H-F, Chen B, Wen J-F. Giardiasis, drug resistance, and new target discovery. Infectious Disorders-Drug Targets (Formerly Current Drug Targets-Infectious Disorders). 2010;**10**(4):295-302

[61] Elwin K, Hadfield SJ, Robinson G, Crouch ND, Chalmers RM. *Cryptosporidium viatorum* n. sp. (Apicomplexa: Cryptosporidiidae) among travellers returning to Great Britain from the Indian subcontinent, 2007-2011. International Journal for Parasitology. 2012;**42**(7):675-682

[62] Xiao L. Molecular epidemiology of cryptosporidiosis: An update. Experimental Parasitology. 2010;**124**(1):80-89

[63] Clarke SH. Reducing the public health risk of cryptosporidiosis by optimizing treatment processes at a military water system. US Army Medical Department Journal. 2011;**12**:50-61

[64] Chalmers RM, Smith RP, Hadfield SJ, Elwin K, Giles M. Zoonotic linkage and variation in *Cryptosporidium parvum* from patients in the United Kingdom. Parasitology Research. 2011;**108**(5):1321-1325

[65] Yoder JS, Harral C, Beach MJ. Control CfD, Prevention. Giardiasis surveillance—United States, 2006-2008. MMWR Surveillance Summaries. 2010;**59**(6):15-25

[66] Areeshi M, Dove W, Papaventsis D, Gatei W, Combe P, Grosjean P, et al. *Cryptosporidium* species causing acute diarrhoea in children in Antananarivo, Madagascar. Annals of Tropical Medicine and Parasitology. 2008;**102**(4):309-315

[67] Mandell, Douglas, and Bennett's Principles and Practice of Infectious Disease. 8th ed. Philadelphia: Elsevier Saunder's; 2015. p. 3174

[68] Laurent F, Lacroix-Lamandé S. Innate immune responses play a key role in controlling infection of the intestinal epithelium by *Cryptosporidium*. International Journal for Parasitology. 2017;**47**(12):711-721

[69] Borad A, Ward H. Human immune responses in cryptosporidiosis. Future Microbiology. 2010;**5**(3):507-519

[70] Khurana S, Chaudhary P. Laboratory diagnosis of cryptosporidiosis. Tropical Parasitology. 2018;**8**(1):2

[71] Mirhashemi ME, Zintl A, Grant T, Lucy FE, Mulcahy G, De Waal T. Comparison of diagnostic techniques for the detection of *Cryptosporidium* oocysts in animal samples. Experimental Parasitology. 2015;**151**:14-20

[72] Quilez J, Sánchez-Acedo C, Clavel A, Del Cacho E, Lopez-Bernad F. Prevalence of *Cryptosporidium* infections in pigs in Aragon (northeastern Spain). Veterinary Parasitology. 1996;**67**(1-2):83-88

[73] Cunha FS, Peralta RHS, Peralta JM. New insights into the detection and molecular characterization of Cryptosporidium with emphasis in Brazilian studies: A review. Revista do Instituto de Medicina Tropical de São Paulo. 2019;**61**:30

[74] Jex A, Smith H, Monis P, Campbell B, Gasser R. Cryptosporidium—Biotechnological advances in the detection, diagnosis and analysis of genetic variation. Biotechnology Advances. 2008;**26**(4):304-317

[75] Skotarczak B. Progress in the molecular methods for the detection and genetic characterization of *Cryptosporidium* in water samples. Annals of Agricultural and Environmental Medicine. 2010;**17**(1):1-8

[76] Jex AR, Stanley KK, Lo W, Littman R, Verweij JJ, Campbell BE, et al. Detection of diarrhoeal pathogens in human faeces using an automated, robotic platform. Molecular and Cellular Probes. 2012;**26**(1):11-15

[77] Plutzer J, Tomor B. The role of aquatic birds in the environmental dissemination of human pathogenic *Giardia duodenalis* cysts and *Cryptosporidium* oocysts in Hungary. Parasitology International. 2009;**58**(3):227-231

[78] Cabada MM, White AC Jr. Treatment of cryptosporidiosis: Do we know what we think we know? Current Opinion in Infectious Diseases. 2010;**23**(5):494-499

[79] Nachipo P, Hermann D, Quinnan G, Gordon MA, Van Voorhis WC, Tam P-YI. Evaluating the safety, tolerability, pharmacokinetics and efficacy of clofazimine in cryptosporidiosis (CRYPTOFAZ): Study protocol for a randomized controlled trial. Trials. 2018;**19**(1):456

[80] Zhou Y, Lv B, Wang Q, Wang R, Jian F, Zhang L, et al. Prevalence and molecular characterization of *Cyclospora cayetanensis*, Henan, China. Emerging Infectious Diseases. 2011;**17**(10):1887

[81] Chacín-Bonilla L. Epidemiology of *Cyclospora cayetanensis*: A review focusing in endemic areas. Acta Tropica. 2010;**115**(3):181-193

[82] Ortega YR, Sanchez R. Update on *Cyclospora cayetanensis*, a food-borne and waterborne parasite. Clinical Microbiology Reviews. 2010;**23**(1):218-234

[83] Smith HV. Cyclospora. In: Foodborne Diseases. Switzerland: Springer Humana Press; 2007. pp. 277-301

[84] Resende DV, Pedrosa AL, CorreiaD,Cabrine-SantosM,Lages-SilvaE, Meira WS, et al. Polymorphisms in the 18S rDNA gene of *Cystoisospora belli* and clinical features of cystoisosporosis in HIV-infected patients. Parasitology Research. 2011;**108**(3):679-685

[85] Tadesse T, Tilahun B, Mengistu S, Alemu S, Zeryehun T, Kefyalew D. Prevalence and species distribution of ectoparasite of domestic dogs in Jimma town, Oromia regional state, southwest Ethiopia. Age. 2019;**93**(24.22):0.770

[86] Murphy SC, Hoogestraat DR, SenGupta DJ, Prentice J, Chakrapani A, Cookson BT. Molecular diagnosis of cystoisosporiasis using extended-range PCR screening. The Journal of Molecular Diagnostics. 2011;**13**(3):359-362

[87] Basyoni MM, Elghobary HAF. GenotypicidentificationofCystoisospora in immunocompromised patients using Tm-variation analysis. The Korean Journal of Parasitology. 2017;**55**(6):601

[88] Palmieri N, Shrestha A, Ruttkowski B, Beck T, Vogl C, Tomley F, et al. The genome of the protozoan parasite *Cystoisospora suis* and a reverse vaccinology approach to identify vaccine candidates. International Journal for Parasitology. 2017;**47**(4):189-202

[89] Ögren J, Dienus O, Löfgren S, Einemo I-M, Iveroth P, Matussek A. *Dientamoeba fragilis* prevalence coincides with gastrointestinal symptoms in children less than 11 years old in Sweden. European Journal of Clinical Microbiology & Infectious Diseases. 2015;**34**(10):1995-1998

[90] Leder K, Weller DPF, Weller PF, Baron EL. *Blastocystis species*. UpToDate. 2019

[91] Stark D, Barratt J, Van Hal S, Marriott D, Harkness J, Ellis J. Clinical significance of enteric protozoa in the immunosuppressed human population. Clinical Microbiology Reviews. 2009;**22**(4):634-650

[92] Barratt JL, Harkness J, Marriott D, Ellis JT, Stark D. A review of *Dientamoeba fragilis* carriage in humans: Several reasons why this organism should be considered in the diagnosis of gastrointestinal illness. Gut Microbes. 2011;**2**(1):3-12

[93] Garcia LS. *Dientamoeba fragilis*, one of the neglected intestinal Protozoa. Journal of Clinical Microbiology. 2016;**54**(9):2243-2250

[94] Barratt JL, Harkness J, Marriott D, Ellis JT, Stark D. The ambiguous life of *Dientamoeba fragilis*: The need to investigate current hypotheses on transmission. Parasitology. 2011;**138**(5):557-572

[95] Stark D, Van Hal S, Marriott D, Ellis J, Harkness J. Irritable bowel syndrome: A review on the role of intestinal protozoa and the importance of their detection and diagnosis. International Journal for Parasitology. 2007;**37**(1):11-20

[96] Stark D, Barratt J, Chan D, Ellis JT. *Dientamoeba fragilis*, the neglected trichomonad of the human bowel. Clinical Microbiology Reviews. 2016;**29**(3):553-580

[97] Mitra T, Kidane FA, Hess M, Liebhart D. Unravelling the immunity of poultry against the extracellular protozoan parasite *Histomonas meleagridis* is a cornerstone for vaccine development: A review. Frontiers in Immunology. 2018;**9**:2518

[98] Fletcher SM. Gastrointestinal illnesses caused by microbes in Sydney. Australia: OPUS; 2013

[99] Stark D, Barratt J, Roberts T, Marriott D, Harkness J, Ellis J. Comparison of microscopy, two xenic culture techniques, conventional and real-time PCR for the detection of *Dientamoeba fragilis* in clinical stool samples. European Journal of Clinical Microbiology & Infectious Diseases. 2010;**29**(4):411-416

[100] Stark D, Barratt J, Roberts T, Marriott D, Harkness J, Ellis J. A review of the clinical presentation of dientamoebiasis. The American Journal of Tropical Medicine and Hygiene. 2010;**82**(4):614-619

[101] Tan KS, Mirza H, Teo JD, Wu B, MacAry PA. Current views on the clinical relevance of *Blastocystis* spp. Current Infectious Disease Reports. 2010;**12**(1):28-35

[102] Poirier P, Wawrzyniak I, Albert A, El Alaoui H, Delbac F, Livrelli V. Development and evaluation of a real-time PCR assay for detection and quantification of Blastocystis parasites in human stool samples: Prospective study of patients with hematological malignancies. Journal of Clinical Microbiology. 2011;**49**(3):975-983

[103] Elghareeb AS, Younis MS, El Fakahany AF, Nagaty IM, Nagib MM. Laboratory diagnosis of *Blastocystis* spp. in diarrheic patients. Tropical Parasitology. 2015;**5**(1):36

[104] Seyer A, Karasartova D, Ruh E, Güreser AS, Turgal E, Imir T, et al. Epidemiology and prevalence of *Blastocystis* spp. in North Cyprus. The American Journal of Tropical Medicine and Hygiene. 2017;**96**(5):1164-1170

[105] Torgerson PR, Macpherson CN. The socioeconomic burden of parasitic zoonoses: Global trends. Veterinary Parasitology. 2011;**182**(1):79-95

[106] Yason JA, Liang YR, Png CW, Zhang Y, Tan KSW. Interactions between a pathogenic Blastocystis subtype and gut microbiota: In vitro and in vivo studies. Microbiome. 2019;**7**(1):30

[107] Wawrzyniak I, Poirier P, Viscogliosi E, Dionigia M, Texier C, Delbac F, et al. Blastocystis, an unrecognized parasite: An overview of pathogenesis and diagnosis. Therapeutic Advances in Infectious Disease. 2013;**1**(5):167-178

[108] Dogruman-Al F, Turk S, Adiyaman-Korkmaz G, Hananel A, Levi L, Kopelowitz J, et al. A novel ELISA test for laboratory diagnosis of *Blastocystis* spp. in human stool specimens. Parasitology Research. 2015;**114**(2):495-500

[109] Zulfa F, Sari IP, Kurniawan A. Association of Blastocystis subtypes with diarrhea in children. Journal of Physics: Conference Series. 2017;**884**:1742-6596

[110] Mandell, Douglas, and Bennett's Principles and Practice of Infectious Disease. 8th ed. Philadelphia: Elsevier Saunder's; 2015. p. 3190

[111] Stensvold CR, Smith HV, Nagel R, Olsen KE, Traub RJ. Eradication of Blastocystis carriage with antimicrobials: Reality or delusion? Journal of Clinical Gastroenterology. 2010;**44**(2):85-90

[112] Mandell, Douglas, and Bennett's Principles and Practice of Infectious Disease. 8th ed. Philadelphia: Elsevier Saunder's; 2015. p. 3189

[113] Schuster FL, Ramirez-Avila L. Current world status of *Balantidium coli*. Clinical Microbiology Reviews. 2008;**21**(4):626-638

[114] Barbosa AS, Cardozo ML, Dib LV, Fonseca ABM, Uchôa CMA, Bastos OMP, et al. Comparative study of three xenic media culture for cultivation of *Balantidium coli* strains. Revista Brasileira de Parasitologia Veterinária. 2018;**27**:20-26

Chapter 5

Toll-Like Receptors, Keys of the Innate Immune System

Alaa Fadhel Hassan

Abstract

Toll-like receptors (TLRs) are members of the integral glycoproteins family, which are consist of intracellular and endoplasmic domains. TLRs are widely distributed in body tissues and expressed by immune and nonimmune cells. They are able to identify pathogens that cause cell injury and distinguish them from harmless microbes, and pathogenic nucleic acids as their binding ligand. Upon binding to their ligands, TLRs first underwent conformational changes; either forming homodimers or heterodimers, starting signaling pathways involve adaptor molecules utilization and then signal transduction through either myeloid differential (MyD)-88 dependent or independent pathways. Ending with activation of several transcription factors (TF) and release of pro-inflammatory cytokines (CK) and Type I interferons (IFN) and initiation of inflammation. TLRs are involved in almost all-inflammatory processes due to underlying disorders and diseases, which made them interesting targets for therapeutic development, via the synthesis of different agonists, antagonists, and even naturalized antibodies.

Keywords: Innate immune response, Toll-like receptors, Myeloid differential88, Cluster differential 14, Lymphocyte antigen 96 & Pro-inflammatory cytokines

1. Introduction

Our start point is that: inflammation is known pathogenesis of different pathophysiological conditions and diseases affecting different body tissues whether acute or chronic. Every inflammation involves an immune response -innate and adaptive- that started with specific receptors called recognition receptors to identify stimuli/damage signal, activation of consequence inflammatory pathway/cascade, the release of inflammatory markers, and recruitment of inflammatory immune cells [1].

The innate immune response is initiated by either endogenous ligands acting as damage signals known as the damage-associated molecular pattern (DAMPs), or exogenous pathogenic ligands-that are accurately portion of the pathogenic microorganism- lead to the same fate; damage signals throughout pathogen-associated molecular patterns (PAMPs) [2]. These patterns alter the body of the cell and cause tissue injuries leading to massive necrosis that release intracellular component into surrounding, these components activate TLRs [3, 4]. These processes, which are both the mechanism and the net results of inflammations, infections, or ischemic injuries cause more, harm than the initial causes itself by improper stimulation of the immune response [3, 4].

TLRs are a family of pattern recognition receptors (PPR), which also involves nucleotide oligomerization domain (NOD)-like receptors (NLR) and retinoic acid-inducible gene I (RIG-I)-like receptors (RLR). They are located on cell membrane/surface and nucleus, are responsible for the detection/recognition of the pathogen or intracellular damaged derived molecular signals to start immune response [1, 2].

These complicated inflammatory processes induced by the immune system are the "Classical typical scenario" involved in the majority of ischemic events, cancers, infectious and inflammatory diseases [4]. For further information about the immune system, Video 1 (https://youtu.be/8mEnyBdsrr8) can be shown on Armando Hasudungan YouTube channel [2] that would explain the innate immunity link with TLRs.

2. Toll-like receptors

TLRs are PRR family involves 13 members that exist in mammals with 10 members detected in the human genome [5, 6], depending on their similar morphology with Toll. Toll is a gene product that participate in both embryonic polarity development and adult fly -antimicrobial response of the species *Drosophila melanogaster* [6, 7]. A 1996 study of this gene product linked the loss/gain of function to the insect's susceptibility and immunologic response to fungal infections; increasing the temptation to seek for the amino acids sequence of the genome. This lead to the final identification of toll-like receptors in 1998 [4].

2.1 Toll-like receptors tissue expression and cellular distribution

TLRs are expressed in almost all body tissues involved in immunologic response as well as those exposed to external environments like the spleen, blood, lung & gastrointestinal tract [4, 8]. The particular cellular expression involves innate and adaptive immunity as well as different nonimmune cells. TLRs cellular expression involves the white blood cells "the sentinel of the innate immune response": microphages (MΦ) & mast cell (MC) "innate immune response keys", dendritic cells (DCs) (primarily pathogenic detector of the adaptive immune response) [4, 6, 8, 9], endothelial cells, epithelial cells, fibroblast, glial cells, astrocytes, oligodendrocytes, etc. [1, 5, 8, 10].

Cellular expression of TLRs family members largely variable and mainly depends on the presence of active infections [8]; according to the same source, as ex., bacterial product & pro-inflammatory cytokines can induce the expression of TL3 while IL-10 blocks TLR4 expression. It has been found that TLR2 expression is more specifically involved in the gram-positive bacteria signaling [8]. TLRs are located either primarily to immune cell plasma membrane phospholipids including TLR 1, 2, 4, 5, 6, & 11 [3, 4, 8]; Or located at the endosomal and lysosomal phospholipids where their extracellular domain (ECD) and its ligand-binding site project into the interior of the organelles like TLR 3, 7, 8, 9, 10 and 13 [2, 3, 10, 11].

2.2 Toll-like receptors biochemistry and Structure

TLRs are a type I integral transmembrane glycoprotein family of very conserved structure [5, 7], consist of 700–1100 amino acids [2, 4]. Their structure, shown in **Figure 1** consist of 2 domains: an ECD that recognize ligands, consist of repetitive motifs rich with leucine and an intracellular domain (ICD) –called cytoplasmic- that maintain inflammatory signal consequence, the last consist of interleukin (IL)-1 receptor region called Toll/IL receptor (TIR) domain [12, 13].

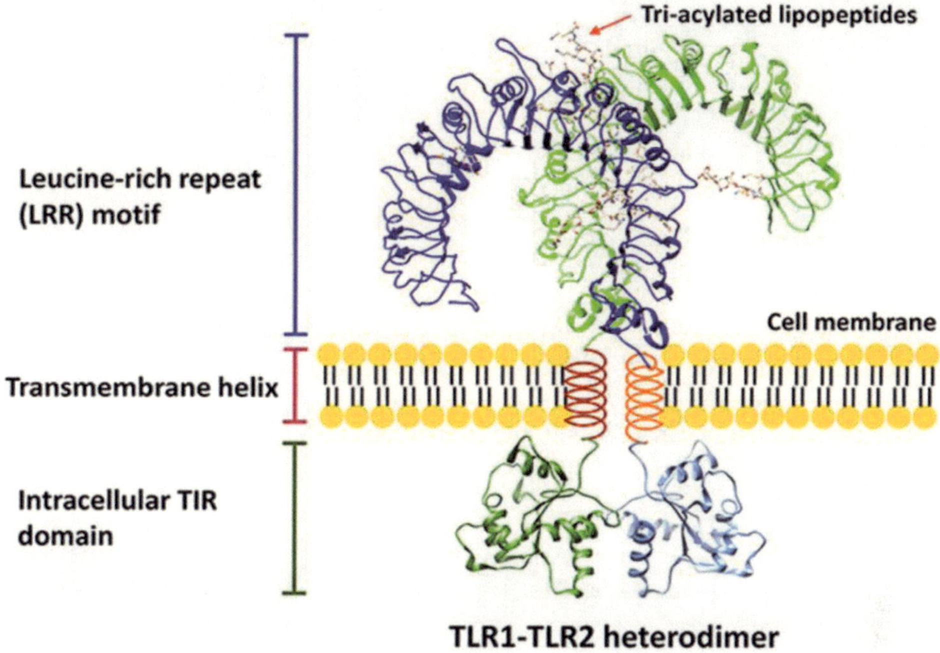

Figure 1.
A representative structure of TLR. The conserved structural features of all TLRs consist of three critical components: (1) leucine-rich repeat (LRR) motif; (2) transmembrane helix; (3) intracellular TIR domain. The LRR structure is based on the model of TLR1-TLR2 heterodimer (Protein Data Bank, PDB, ID: 2z7x) interacting with six triacylated-lipopeptides, Pam3CysSerLys4 (Pam3CSK4), whereas the TIR domain homology model is based on TLR2-TIR structure (PDB ID: 1fyw) [12].

3. Toll-like receptors family members

TLRs involves 13 family members that exist in mammals with 10 members detected in the human genome [5, 6]. Human TLRs amino acids sequence allow a subfamily classification into the TLR2, TLR3, TLR4, TLR5, and TLR9 subfamilies. The TLR2 subfamily involves TLR1, 2, 6, and 10; the TLR9 subfamily involves TLR7, 8, and 9 [14].

TLRs members can form homodimers/heterodimers among their same protein family or associates with an "outside TLR family" protein; both formations contribute to their structural and functional diversity [4]. Homodimers are formed by TLR4 while TLR members 1, 2, and 6 like TLR1/2 or TLR2/6 dimers form heterodimers [2, 3, 15–17]. TLRs members, their dimerization, cellular distribution, ligands, induced signaling pathway, and product are shown in **Table 1**; for further information about TLRs, Video 2 (https://youtu.be/8mEnyBdsrr8) about TLR overview can be shown at Armando Hasudungan YouTube channel [18].

3.1 Toll-like receptors binding ligands

TLRs family members can recognize two types of associated molecular patterns as their ligands, derived from pathogens or damaged organelles damaged structures.

3.1.1 PAMPs

PAMPs derived from pathogen [5, 19]; like gram-negative bacterial lipopolysaccharides (LPS), gram-positive bacterial lipoteichoic acid (LTA) and peptidoglycan

TLRs	Immune Cell Expression	PAMPs	DAMPs	Signal Adaptor	Production
TLR1+ TLR2	Cell surface Mo, MΦ, DC, B	Tri-acylated lipoproteins (Pam3CSK4) Peptidoglycans, Lipopolysaccharides	(TLR2 DAMPs listed below)	TIRAP, MyD88, Mal	IC
TLR2+ TLR6	Cell surface Mo, MΦ, MC, B	Diacylated lipoproteins (FSL-1)	Heat Shock Proteins (HSP 60, 70, Gp96) High mobility group proteins (HMGB1) Proteoglycans (Versican, Hyaluronic Acid fragments)	TIRAP, MyD88, Mal	IC
TLR3	Endosomes B, T, NK, DC	dsRNA (poly (I:C)) tRNA, siRNA	mRNA tRNA	TRIF	IC, type1 IFN
TLR4	Cell surface/ endosomes Mo, MΦ, DC, MC, IE	Lipopolysaccharides (LPS) Paclitaxel	Heat Shock Proteins (HSP22, 60, 70,72, Gp96) High mobility group proteins (HMGB1) Proteoglycans (Versican, Heparin sulfate, Hyaluronic Acid fragments) Fibronectin, Tenascin-C	TRAM, TRIF TIRAP, MyD88 Mal	IC, type1 IFN
TLR5	Cell surface Mo, MΦ, DC, IE	Flagellin		MyD88	IC
TLR7	Endosomes Mo, MΦ, DC. B	ssRNA Imidazoquinolin-es (R848) Guanosine analogues (Loxoribine)	ssRNA	MyD88	IC, type1 IFN
TLR8	Endosomes Mo, MΦ, DC, MC	ssRNA, Imidazoquinolines (R848)	ssRNA	MyD88	IC, type1 IFN
TLR9	Endosomes Mo, MΦ, DC, B, T	CpG DNA CpG ODNs	Chromatin IgG complex	MyD88	IC, type1 IFN
TLR10	Endosomes Mo, MΦ, DC	profilin-like proteins		MyD88	IC

Mo: monocytes, MΦ: macrophages, DC: dendritic cells, MC: Mast cells, B: B cells, T: T cells, IE: Intestinal epithelium, IC: Inflammatory cytokines [2]

Table 1.
TLRs cellular expression, binding ligands, signal adaptor & production [2].

(PGN), mycobacterial lipopeptides, yeast zymosan, viral and bacterial ribonucleic acid (RNA), and unmethylated cytosine phosphate guanine containing- (CpG) deoxyribonucleic acid (DNA) [20, 21].

3.1.2 DAMPs

DAMPs damaged organelles structures, extracellular matrix, cytosolic and nuclear proteins, Heat shock protein-60 (HSP-60) and HSP-70, hyaluronic acid fragments, and free fatty acids (FFA) [5, 22, 23]. They cause activation of the innate and inflammatory immune responses, epithelial regeneration, and sterile inflammation control [6, 24].

4. Toll-like receptors signaling pathway

Upon TLRs recognition and binding to their ligands, they undergo conformational changes, dimerization as well as interaction with adaptor molecules passing series of intracellular signal transduction pathways that involve transcription factors NF-κB, IRFs, and mitogen-activated protein kinase (MAPK) activation. These pathways finally resulting in the secretion of pro-inflammatory mediators including nitric oxide (NO), CK- like tumour necrosis factor-alpha (TNF-α), IL-6 & IL-1β, chemokines (CC), and type I IFN [15, 21, 25, 26]. As shown in **Figure 2**.

4.1 Co-receptors

Co –receptors involved in TLRs signalling include Cluster differential 14 (CD14) and Lymphocyte antigen 96 (MD-2). Both have a major role in TLR4 activation after LPS recognition. CD14 is a glycophosphatidylinositol attached

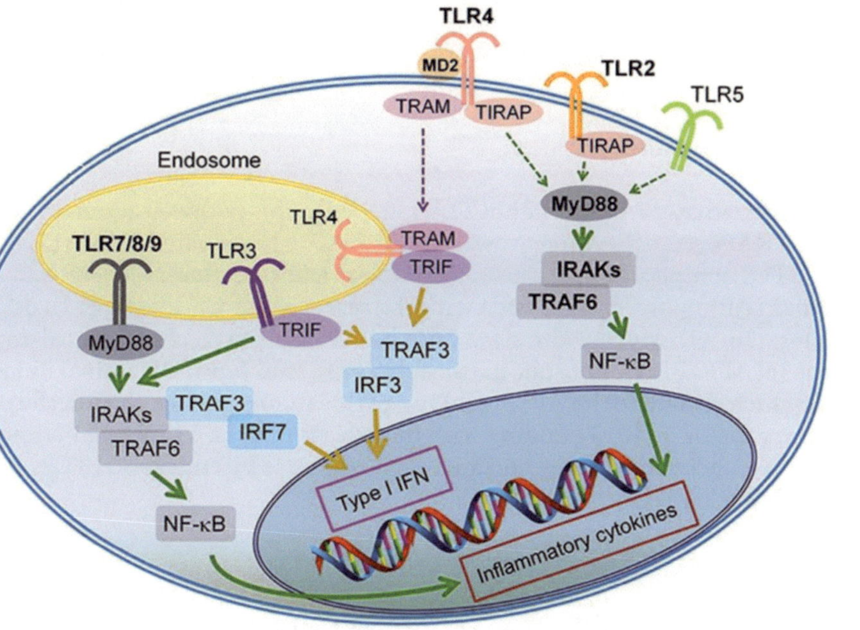

Figure 2.
Signaling pathways of TLR. Surface and endosomal TLRs bind to adaptor molecules and co-receptors. Signal through Myd88 dependent/independent pathway ending with proinflammatory CK or type I IFN [12].

protein expressed on innate immune cells as macrophage and monocytes that function as co-receptor for both cell surface & endosomal expressed TLRs. Lymphocyte antigen 96 (MD-2), which is a cell membrane glycoprotein associated specifically with TLR4 ECD, and expressed at myeloid and endothelial cells [6, 13, 21, 26, 27].

4.2 Adaptor proteins and kinase molecules

TLRs signaling pathways involves four main adaptor protein molecules: MyD88, TIR domain-containing adaptor protein/MyD88 adaptor-like molecules (TIRAP) also called MAL, TIR domain-containing adaptor protein inducing interferon-β (TRIF), and TRIF related adaptor molecule (TRAM) [13, 21, 28]. TLRs signaling pathways involves activation of five TIR containing adaptor kinase molecules, like IL-1 receptor-associated kinase (IRAK) -1 and 4, TNF receptor-associated factor-6 (TRAF6), serine/threonine binding kinase (TBK)-1, MAPK, and inhibitor of kappa-B (IκB) kinase (IKK) [13, 28].

4.3 Transcription factors

There are three transcription factors involved in the TLRs signalling pathway including NF-κB, AP1, and IRF. NF-κB is an intracellular pleiotropic protein complex; it is responsible for gene regulation of proinflammatory CK, CC, adhesion molecules, and cell cycle/survival regulating proteins as cyclin D1 and B cell lymphoma 2 (Bcl-2). AP1 is a dimer of both protein Jun and Fos families; that is associated with cell replication and survival regulation. Finally, the IRFs protein regulating IFNs, are responsible for signal stimulation via MyD88independent/TRIF pathway [6, 13].

4.4 Intracellular signaling pathways

There are two intracellular signalling pathways for TLRs involve MyD88-dependent/& MyD88-independent also called (TRIF-dependant) signal transduction pathway.

4.4.1 MyD88-dependent pathway

It is utilized by all TLRs but not TLR3 [21, 29]. This pathway activates the IRAKs, TRAF6, transforming growth factor (TGF)-β-activated kinase (TAK)-1 and the IKK complex [15]. It causes the nuclear translocation of NF-κB and adaptor protein-1 (AP1) [28, 30], and ends with the secretion of CK like IL-6, IL-10, IL-12 & TNF-α [16, 29]. MyD88 also stimulate the classical extracellular signal-regulated kinases (MAPK/ERK), phosphoinositide-3 (PI3), and Jun (N)terminal kinase (JNK) which stimulate the AP1 signalling pathway, and induce the interferon regulatory factor-7 (IRF7) ending with the release of type-I IFN or co-stimulatory molecules associated with the antimicrobial response by endosomal TLRs 3, 7, 8 and 9 [13, 29, 31, 32].

4.4.2 MyD88-independent pathway

The main pathway of TLR3 and 4, involve TRIF signalling pathway activation which involves TRAF6 activation, results in inositol triphosphate-3 (IP3) phosphorylation and induction of IFN-β gene expression as well as activation of TRAF6 [21, 29].

Surprisingly the same outcome was obtained from plasmatoid dendritic cells (pDCs) stimulated by TLR 7& 9 throughout the activation of the MyD88/IRF7 dependent pathway [15, 33].

4.5 Unique pathways

TLR4 further utilizes TIRAP to activate MyD88 and TRAM to bridge the TRIF activation, which means that TLR4 uniquely utilizes both the MYD88 dependent and independent pathways [11, 21, 29].

As stated by S. Kiziltas et al. "TLR stimulation product is dependent on the nature of PAMPs, the activated TLR, the activated cell and the level of CK. Moreover, the chronically activated signalling pathway would possibly induce transcription of oncogenic factor; adding further complexation to the intracellular signalling for these receptors" [5, 13].

5. Toll-like receptors and pathophysiological disorders

TLRs play an important role in pathophysiological disorders due to their wide tissue distribution, their function as pattern recognition receptors that respond to variable bacterial and damage associated molecules, and involvement in multiple inflammatory signal pathways/& process all render TLRs being a major player in any inflammation-related disorder [4–6, 19, 22, 23, 34]. In addition, analysis of TLRs gene polymorphism in human disorders revealed an increased risk of bacterial infection and sepsis as an example [34]. This section is a shortcut or summary to TLR involvement in different pathophysiological disorders rather than a full description section.

5.1 Toll-like receptors and pathophysiology of inflammatory oxidative stress

Inflammation is a common etiology of many disorders and disease including ischemic injuries, microbial infections, diabetes, arthritis and cancer [3, 4, 35]; still, any inflammatory process is triggered by damage signal recognized by pattern receptors and induce activation of signaling pathways leading to the production of pro-inflammatory markers and activation of immune cells [35]. These processes also induce the release of free radicals (FR) such as reactive oxygen species (ROS) and the activation of hypoxia-inducible transcription factor-1 (HIF1), causing tissue stress and reduced tissue oxygen status, so-called tissue hypoxia. Hypoxia is believed to be a hallmark as well as a key trigger of inflammation itself [35, 36].

Under normal conditions HIF1-α subunit (the inducible form of the heterodimer protein HIF-1 transcription complex) [35], is controlled by hydroxylation of proline residue via prolyl hydroxylase enzyme, and breaking down via proteasome. However under inflammatory conditions LPS activate TLRs that stimulates nicotine amide adenine dinucleotide phosphate (NADPH) oxidase (Nox)-associated cross-talk with the MAPK signaling pathways [36, 37], that causes proinflammatory CK & markers production thus increasing mitochondrial FR release like ROS causing more and more tissue stress. That causes HIF1- α activation; here HIF1- α protein inactivation process will be inhibited due to proline consumption, leading to HIF1-α accumulation in MΦ, DCs and other non-immune cells that exposed to hypoxia/ & non-hypoxic damage signals [38]. Furthermore, this would induce metabolic reprogramming of mitochondrial respiration causing succinate release, and production of IL-1β [35, 38].

In dendritic cells, TLRs cause further stabilization of HIF1-α via release of NF-κB, which would further increase glucose uptake and render shifting of mitochondrial respiration to the anaerobic glycolytic pathway due to the increased oxygen demand versus the decreased supply [35, 36]. Finally result in disruption of the normal function of DCs, the primary pathogenic detector of the adaptive immune response; which undergo cellular maturation upon TLRS activation that results in further expression of co-stimulatory molecules, further production of pro-inflammatory CK & CC, and migration to lymph node so to present antigens to naïve T-cells [4, 35]. All these scenarios would further amplify the existing inflammation and tissue damage [35].

HIF1-α is a transcription factor that responsible for cellular adaptive responses after exposure to injury/stress environment, including maintenance like controlling angiogenesis to improve blood vessel formation, shifting cellular mitochondria respiration to anaerobic glycolysis through improving cellular survival and cellular adhesion in oxidative stress environment's [36]. In addition, it is the major controller of phagocytes bactericidal capacity, and involved in myeloid cell-mediated inflammation, and is an essential factor for inhibition of myeloid cell apoptosis induced by LPS. The last point made it an important factor also in the TLR4 signaling pathway [36, 38]. HIF1-α function as a double-edged sword, that mediate cellular adaptive to stress but progress disease status by the same time [38].

5.2 Toll-like receptors important role in the pathophysiological disorder

5.2.1 Central nervous system

TLRs are expressed in various central nervous system (CNS) cells predominantly in neurons, astrocytes, resident microglia, cerebral microvasculature, plexuses choroid, and leptomeninges. They are associated with the detection of- and regulated by central DAMPs [33]. TLR4 is further upregulated centrally by glutamate via N-methyl-D-aspartate (NMDA) dependent mechanism and peripherally by noradrenaline/β2 receptor, & corticotrophin-releasing factor. TLRs play an important role in restoring central homeostasis, physiology of stress-sensitive behaviour after injuries or diseases as multiple sclerosis, Alzheimer's, and stroke [33].

In the experimental model of CNS, stress exposure revealed mRNA upregulation and activation of TLRs in the brain frontal cortex after the stress is involved in the loss of neuronal plasticity and survival depending on the activation of NF-κB induced ROS production. Also resultant bacterial translocation from the gut to the systemic circulation and other organs such as the liver, spleen, and mesenteric lymph nodes; These circulating gram-negative bacteria are the major source of LPS, which can activate brain TLR4 through multiple pathways, including a neuroinflammatory response. This is partially explained by the theory known as leaky gut [11, 33].

In another experimental model of neurogenesis, TLR3 & 4 were found to act as down regulators, TLR3 deletion/loss of function was also linked to improved cognitive function. The same reference state an opposed case in viral meningitis when TLR3 & 9 recruitment help to decrease neuronal injury and localize infection area and in Alzheimer disease where TLR2, 4, 5, 7 & 9 were suggested to improve disease progression by inhibiting amyloid plaque accumulation [1].

5.2.2 Respiratory system

TLR is thought to play a considerable role in several respiratory disorders starting from allergic rhinitis ending with severe inflammatory disorders like acute respiratory distress syndrome (ARDS), through their activation by the causative inflammations derived by pulmonary oedema, trauma, sepsis & even drug overdose [9, 37]. In allergic rhinitis TLR2, 3, & 4 were found to be both upregulated by- and involved in-the causative inflammation [37].

TLR2 has the mainstay of involvement & determination in respiratory allergic disease due to considerable genetic variation. In asthma, an experimental study shows TLR2 induction by synthetic Pam3Cys triggers immune response & disease severity [37]. While in acute lung injury (ALI) & ARDS, TLR2 was found to be activated by Toll interacting protein (Tollip) [14]. TLR4 was found to increase asthmatics severity & prevalence in paediatrics. TLR4 genetic polymorphism affects cluster differentials (CD)41–251 regulatory T cells (Tregs) which are activated by LPS, the same ligand of TLR4 itself. [2, 3, 37].

5.2.3 Cardiovascular system

An experimental model of doxorubicin and hydrogen peroxide-induced cardiac injury showed TLR2 to be involved in cardio myocytes apoptosis, besides TLR2 targeting suggested to be protective in septic cardiomyopathy [1]. In addition, murine models revealed cardiac tissue expression of TLR4 increased after hypertension, myocardial ischemia, maladaptive left ventricular hypertrophy, and angiotensin II (AngII) infusion participating in vascular remodelling & stiffness, endothelial dysfunction, increase myocardial infarction (MI) size & susceptibility. While Human studies revealed the same in patients with unstable angina, MI, heart failure, atherosclerosis & myocarditis [9, 27, 39–41].

TLR4 expression & signalling was increased in patients' monocytes during attacks of unstable angina & MI [37]. In the experimental model & human vascular inflammation, TLR4 was found to increase the production of CK, CC as well as increase TLR2 expression. In the early stage of the atherosclerotic lesion, TLR4 mRNA protein was detected & MyD88 -the mainstay of TLR signalling pathway- gene deficiency was linked to decrement in CK, CC & lipid content production, as well as in atherosclerotic lesion size. The same reference stated that TLR2 genetic polymorphism was linked to increased coronary artery stenosis, while TLR7 & 8 was involved in cardiac inflammation caused by the Coxsackie virus [37].

5.2.4 Gastrointestinal system

The liver is the major organ that deals with gut-derived endotoxin, exposed by portal circulation [13, 42]. This continuous exposure would trigger frequent activation of the hepatic innate immune system; which contributes to the induction of inflammation in acute hepatic injuries, which means involvement of TLRs in the induction of inflammation [13]. Pathogenic suppression/& inhibition of TLRs found to mediates chronic hepatic injuries/disorders like hepatitis, fibrosis, alcoholic liver injuries, ischemia/reperfusion injury, and carcinoma [13, 28].

In Paracetamol human hepatotoxicity, endogenous chemical injury derives extracellular matrix (ECM) the ligand that activates TLR4 to release TNF-α, induce inducible nitric oxide synthase (iNOS), peroxynitrite, glutathione depletion, so that will amplify immune response, sequestering leukocytes, increase serum hyaluronic acid, causing steatosis, necrosis, and hepatic congestion [16].

Hepatitis viral nucleic acid & proteins are the ligands detected by TLR3, 7, 8, & 9. Starting with hepatitis B virus (HBV), in vitro activation of TLR1, 2, 3, 4, 5, 6, 7, 8, & 9 result in the release of IFN which inhibit HBV DNA replication and RNA transcription. Whilst HBV itself downregulates the expression of TLR1, 2, 4, & 6, this limits their antiviral effect or even renders them nugatory [28]. This downregulation of TLRs is attributed to the presence of HBV e antigen (HBeAg) during acute infection. About hepatitis C (HCV), its core protein activates TLR 1, 2, 4 & 6, which are supposed to produce antiviral IFNs as well as increased hepatic inflammation. The same effect is presumed by TLR 3 & 4 in HBV is achieved here to produce IFN-β [28].

In alcoholic liver disease (ALD), alcohol mainstay effects are to increase gut mucosal permeability to LPS, modification of gut flora, reducing endotoxin clearance rate, and increasing hepatic endotoxin level [16]. These scenarios lead to higher expression of TLR1, 2, 4, 6 & 9 by both parenchymal and non-parenchymal cells, activating their pathway and release of inflammatory mediators, this process observed in the chronic alcohol experimental model [28, 29]. While a patient with cirrhosis expresses a high level of TNF-α, IL-1β, & IL-6, as well as chronic endotoxemia, recurrent bacterial infection [16]. Finally, the process of hepatic regeneration depends on the interplay between the immune system and non-parenchymal cell, which involves activation of TLRs/MyD88 pathway, here the bulky activation of TLRs, would inversely affect the regeneration process, which indicates that the extent of such activation is essential for hepatic regeneration. TLR2, 4 & 9 reported no important role in liver regeneration process [28, 43].

Both human patients and experimental models of diabetes linked the active TLR to the progression of diabetes complication throughout the activation of NF-κB signalling in adipose tissue MΦ due to high level of plasma FFA associated with obesity & diabetes type 2 (T2DM) [44].

In vivo & in-vitro studies performed by Zhang N. et al. revealed that TLR 2 & 4 activation in insulin target tissues as the liver, adipose tissue & immune cells linked them with insulin resistance. The first suggests that high TLRs loss of function or genetic modification protects against high FFA level resulted from large mass adipose tissue secreting non-esterified free fatty acids & reduction of their clearance/oxidation which disturbs gut permeability to LPs [45].

TLR4 resultant inflammation associated with activation IKK, MAPK, JNK, and p38 pathways would further increase insulin receptor substrate-1 (IRS1) serine phosphorylation thus decrease insulin receptor's signal transduction [31, 45]. Furthermore, TLR4-MyD88 signalling pathway activation was suggested throughout developmental researches for several anti-hyperlipidemic medications, while TLR1, 2, 3 & 7 were triggering both host immune defence and/autoimmune response that aggravate diabetic state [37].

5.2.5 Urinary system

TLRs expression in renal tube epithelial lining render their activation to be essential in renal vascular remodelling, endothelial dysfunction in multiple renal disorders like acute kidney injury (AKI), solid organ transplant, glomerulonephritis, ischemic/reperfusion injury (I/R injury) & diabetic renal disorders [27, 44]. Experimental streptozocin induced diabetic model revealed podcytopathy & fibrosis regression after TLR4 knocking out, as they are expressed by podocytes & decreased diabetic nephropathy after TLR2 knocking out [46, 47]. TLR4 gene polymorphism was linked to prostate cancer among gene clusters of TLR1, 6 & 10 [37].

6. Toll-like receptors as therapeutic targets

TLRs, as the primary receptor for many ligands that trigger innate & adaptive immune response, with complex signaling pathways involving many adaptor molecules & co-receptors seem interesting for therapeutic target development. Synthetic agonist, antagonist and even naturalized antibodies could modify TLRs signaling to make them attractive targets for the management of different inflammatory disease. For example at 2013, Savva and Roger enlisted around 32 clinical trials at different phases for TLRs agonist/antagonist agent for the management of sepsis and infectious disease, these trials include even the antimalarial old agent chloroquine [28, 34].

6.1 Toll-like receptors 1 and 2

TLR1/2 heterodimers were found to be increased in patients with atherosclerotic lesions, while administration of TLR1/2 agonist aggravates disease status, also TLR2 inhibition was suggested as diabetes and cardiovascular disorders therapy besides statins & thiazolidinedione by anti-inflammatory action [9]. Pam2/3CSK4 TLR2 ligands covalently linked to CD8+ or B-cell epitopes associated peptides were found to enhance therapeutic response in tumour models, by stimulating TLR2 induced T-cell activation [15]. A 3 component carbohydrate-based cancer vaccine involved TLR2 activator that mediates humoral immune response against tumour-induced glycopeptide antigens by affecting the maturation of cellular component of the innate immune system (DC & natural killer cells), furthermore cancer treatment with chimers of anti-tumour antibodies and small molecule agonist of TLR2 would alleviate disease progression [9].

6.2 Toll-like receptor 3

Since high synovial expression of TLR3 in RA patients was found, one scenario for rheumatoid arthritis and possibly bone malignancy is to inhabit the TLR3 pathway via the RNA synthetic analogue Polyinosine-polycytidylic acid (poly (I:C) that affect monocyte –osteoclast cellular differentiation [9].

6.3 Toll like receptor 4

Various TLR4 antagonist was developed as a therapeutic agent, starting with the peptide P13- an inhibitor of TIR domain signalling pathway- that was found to ameliorate inflammatory response and improve surviving in a TLR4-mediated hepatic injury of murine model [16]. In addition, Lipid A mimetics E5564 and CRX526 bind to TLR4-MD2 complex showing valuable inhibition of pro-inflammatory cytokine IL-1 and TNF-α production in LPS treated animal models as well as septic shock patients in phase III clinical trial [9, 16, 29]. TLR4 inhibition was suggested as the scenario for treatment of thrombosis, atherosclerosis & vascular restenosis throughout coating TLR4 or MyD88 with inhibitory compound, small molecule antagonist, then by giving viral vectors that express antisense gene to TLR4 RNA [9], and finally TLR4/MD2/anti-Human IgG (Fc specific) (IgG-Fc) fusion protein inhibitor of NF-κB and JNK activation provides interesting biologic therapy for liver fibrosis, alcoholic and non-alcoholic steatohepatitis by decreasing IL-6 and monocyte Chemoattractant Protein-1(MCP-1) production [16].

Another TLR4-synergizer Fc/fusion protein and TL4 ligand α-1 acid glycoprotein were found to inhibit LPS-induced activation of hepatic MΦ by blocking the triggering receptor expressed on myeloid cells-1 (TREM1), and boosting the anti-inflammatory immune response. Other theoretically interesting scenarios involving the I.V administration of monophosphoryl lipid A derivatives as 2 adult HBV vaccine in treating viral hepatitis [13, 15, 16].

6.4 Toll-like receptors 5 and 7

One possible scenario for cancer immunotherapy involved TLR5 binding to flagellin that can turn the tolerogenic DCs into active antigen-presenting cells (APC) [9].

Isatoribine, a TLR7 agonist administered I.V was found to decrease viral load with a moderate adverse effect profile in HCV patients. In addition, IGS-9620 that was experimentally assessed on the HBV animal model was found to decrease HBVs antigen (HBsAg) level in serum, HBV viral load as well as IFN-α in dose dependent-manner [15, 29]. Note that some TLR7 targeting therapies were approved by Food and Drug Administration (FDA) like imiquimod, TLR7-immune response modifier that was approved since 1997 for treatment of superficial skin malignant melanoma & genital warts by increasing cellular production of CK like IFN, IL-6 & TNF [9].

6.5 Toll-like receptor 9

Selective TLR9 agonists like 1018 ISS (immunomodulatory sequences) that contain repeated CpG motifs were found to modulate the TLR9 signalling pathway involved in HBV infection and have been tested in phase III clinical trials. Another agonist IMO-2055 was under assessment in 2011 for oncologic disease as well as IMO-2125 which was found to maintain the high level of IFN was under assessment as a possible therapy to HCV patients. The TLR9 intracellular signalling inhibitors ST2825 and RO0884 designed to block IRAK1 &4/MyD88 singling pathway caused inhibition of the NF-κB, IL-1β, and TNF-α activation as well as decreased hepatic IL-6 secretion [9, 15, 29].

7. Conclusions

Medical and pharmacological development is focusing on the molecular level, in all aspects including analytical, physiological, pharmacological and even genetic aspects. Understanding immune response is thus important subject, furthermore, the target receptors which damage signals bind to, their signaling pathways end products will tell what possible immune response happened to human body. Toll-like receptors are those targets, the family of integral transmembrane glycoprotein expressed intracellularly or at cellular surface, considered main component and link between innate and adaptive immune response, which can induce signaling pathways involving four main adaptor molecules that initiate divaricated steps ending with inflammatory cytokines. These pathways could be involved in any inflammatory process/disorders and thus seems interesting targets for pharmacological intervention; all these steps bring us back to the bullet that explodes all these events in the body, the immune system.

Conflict of interest

The author declares no conflict of interest.

Notes/thanks/other declarations

After precious Thanks to Almighty and Merciful GOD, I would like to express my thanks and gratitude to Professor G. H. Majeed who made it possible to me to accomplish this situation and reach you, my readers. Also to the magnificent group who introduced me to the world of toll-like receptors: professor Dr. Abduladheem Y. Abbood Al-Barrak, professor Dr. Bassim I. Mohammad, Dr. Samer Fadhel Hassan, Dr. Asma A. Swadi, and Dr. Huda J. Merza, and sure my own family.

Abbreviations and nomenclature

AKI	Acute kidney injury
ALD	Alcoholic liver disease
ALI	Acute lung injury
AngII	Angiotensin II
AP-1	Adaptor protein-1
APC	Antigen-presenting cell
ARDS	Acute respiratory distress syndrome
Bcl-2	B cell lymphoma 2
CC	Chemokine
CD14	Cluster differential 14
CK	Pro-inflammatory cytokines
CpG	Cytosine phosphate guanine
DAMPs	Damage-associated molecular pattern
DC	Dendritic cell
ECD	Extracellular domain
ECM	Extracellular matrix
FFA	Free fatty acids
FR	Free radicals
HBeAg	HBV e antigen
HBsAg	HBV s antigen
HIF1	Hypoxia-inducible factor-1
HSP	Heat shock protein
I/R injury	Ischemic/reperfusion injury
ICD	Cytoplasmic domain
IFN	Type-I interferon
IgG-Fc	Anti-Human IgG (Fc specific)
IKK	Inhibitor of kappa-B (IκB) kinase
IL	Interleukin
iNOS	Inducible nitric oxide synthase
IP3	Inositol triphosphate-3
IRAK	IL-1 receptor-associated kinase
IRF	Interferon regulatory factor
IκB	Inhibitor of kappa-B
JNK	Jun (N)terminal kinase
LPS	Lipopolysaccharides
LRR	Leucine-rich repeat
LTA	Lipoteichoic acid
MAPK	Mitogen-activated protein kinase
MAPK/ERK	Extracellular signal-regulated kinases
MC	Mast cell
MCP-1	Monocyte Chemoattractant Protein-1
MD-2	Lymphocyte antigen 96
MI	Myocardial infarction
mRNA	Messenger ribonucleic acid
MyD88	Myeloid differential88

MΦ	Macrophage
NADPH	nicotine amide adenine dinucleotide phosphate
NLR	nucleotide oligomerization domain (NOD)-like receptors
NO	Nitric oxide
NOD	nucleotide oligomerization domain
Nox	NADPH oxidase
Pam3CSK4	Pam3CysSerLys4
PAMPs	Pathogen-associated molecular patterns
pDCs	Plasmatoid dendritic cells
PGN	Peptidoglycan
PI3	Phosphoinositide-3
poly(I:C)	Polyinosine-polycytidylic acid
PRRs	Pattern recognition receptors
RIG-I	retinoic acid-inducible gene I
RLR	retinoic acid-inducible gene I (RIG-I)-like receptors
ROS	Reactive oxygen species
T2DM	Diabetes type 2
TAK	Transforming growth factor (TGF)-β-activated kinase
TBK-1	Serine/threonine binding kinase
TF	Transcription factors
TGF	Transforming growth factor
TIR	Toll/IL-receptor
TIRAP/MAL	TIR domain-containing adaptor protein/MyD88 adaptor like
TLRs	Toll-like receptors
TNF-α	Tumour necrosis factor-alpha
Tollip	Toll interacting protein
TRAF6	TNF receptor-associated factor-6
TRAM	TRIF related adaptor molecule
Tregs	Regulatory T cells
TREM1	Triggering receptor expressed on myeloid cells-1
TRIF	TIR domain-containing adaptor protein inducing interferon-β

Video materials

Vedio 1. Immunology-Toll Like Receptors Overview

YouTube video: [3] Armando Hasudungan, Immunology-Toll Like Receptors Overview [Internet. YouTube]. 2014. Available from: https://youtu.be/8mEnyBdsrr8

Vedio 2. Toll Like Receptors Overview

YouTube video: [18] Armando Hasudungan, Immunology - Toll Like Receptors Overview [Internet YouTube]. 2014. Available from: https://youtu.be/8mEnyBdsrr8

Author details

Alaa Fadhel Hassan
Department of Pharmacy, Al-Mahmoudiya General Hospital, Baghdad, Iraq

*Address all correspondence to: alaa.alwazni@yahoo.co.uk

References

[1] Amene Saghazadeh and Nima Rezaei Introductory Chapter: Toll-Like Receptors, Toll-like Receptors. Nima Rezaei, editor. IntechOpen January 15th 2020). DOI: 10.5772/intechopen.88493

[2] Invivogen Reviews Toll-Like Receptors [Internet]. 2012. Available from: http://www.invivogen.com/review-tlr [Accessed: 2021-03-09]

[3] Armando Hasudungan, Immunology-Toll Like Receptors Overview [Internet. YouTube]. 2014. Available from: https://youtu.be/8mEnyBdsrr8

[4] Christmas P. Toll-Like Receptors: Sensors that Detect Infection, Nature Education [Internet]. 2010. Available from: https://www.nature.com/scitable/topicpage/toll-like-receptors-sensors-that-detect-infection-14396559/ [Accessed: 2021-03-09]

[5] Kiziltas S: Toll-like receptors in pathophysiology of liver disease. World Journal of Hepatology 2016;8(32):1354-1369. DOI: 10.4254/wjh.v8.i32.1354

[6] Guo J, and Fridman S: Toll-like receptor 4 signaling in liver injury and hepatic fibrogenesis. Fibrogenesis & Tissue Repair 2010,3:21. DOI: 10.1186/1755-1536-3-21

[7] Matsunaga N, Tsuchimori N, Matsumoto T, and Li M: TAK-242 (Resatorvid), a small-molecule inhibitor of toll-like receptor (TLR) 4 signaling, binds selectively to TLR4 and interferes with interaction between TLR4 and its adaptor molecules. Molecular Pharmacology Journal 1,2011,79(1)34-4. DOI: 10.1124/mol.110.068064

[8] Marta Muzio, Daniela Bosisio, Nadia Polentarutti, Giovanna D'amico, Antonella Stoppacciaro, Roberta Mancinelli, Cornelis van't Veer, Giselle Penton-Rol, Luigi P. Ruco, Paola Allavena, and Alberto Mantovani: Differential Expression and Regulation of Toll-Like Receptors (TLR) in Human Leukocytes: Selective Expression of TLR3 in Dendritic Cells1. The Hournal of Immunology Hune 1,2000, 164(11)5998-6004. DOI: 10.4049/jimmunol.164.11.5998

[9] Jezierska A, Kolosova IA, and Verin AD. Toll Like Receptors Signaling Pathways as a Target for Therapeutic Interventions. Curr Signal Transduct Ther. 2011,6(3):428-440. DOI: 10.2174/157436211797483930.

[10] Kawasaki T, Kawai T. Toll-like receptor signaling pathways. Front Immunol. 2014;5:461. Published 2014 Sep 25. DOI:10.3389/fimmu.2014.00461

[11] Chalasani N, Fontana RJ, Bonkovsky HL, Watkins PB, Davern T, Serrano J, Yang H, Rochon J; Drug Induced Liver Injury Network (DILIN). Causes, clinical features, and outcomes from a prospective study of drug-induced liver injury in the United States. Gastroenterology. 2008 Dec;135(6): 1924-34, 1934.e1-4. DOI: 10.1053/j.gastro.2008.09.011

[12] Gao W, Xiong Y, Li Q and Yang H: Inhibition of Toll-Like Receptor Signaling as a Promising Therapy for Inflammatory Diseases: A Journey from Molecular to Nano Therapeutics.Front. Physiol. 8:508.2017. DOI: 10.3389/fphys.2017.00508

[13] Hassan AF: Eavluation of the effects of GIT 27 and TAK 242 on methotrexate-induced liver injury [thesis] Al-Mustansiriyah University, College of Medicine; 2018. Available from: https://pt.slideshare.net/AlaaAlwazni/evaluation-of-the-effect-of-git-27-tak-242-on-druginduced-liver-injury-149321031

[14] Kiyoshi Takeda, Tsuneyasu Kaisho, Shizuo Akira. Toll-Like Receptors. Annual Review of Immunology 2003 21:1, 335-376. DOI:10.1146/annurev.immunol.21.120601.141126

[15] Quetglas EG, Armuzzi A, Wigge S, Fiorino G, Barnscheid L, Froelich M, and Danese S. Review article: The pharmacokinetics and pharmaco dynamics of drugs used in inflammatory bowel disease treatment. European Journal of Clinical Pharmacology, 2015;71(7):773-99. DOI: doi.org/10.1007/s00228-015-1862-7

[16] Howard SC, McCormick J, Pui CH, Buddington RK, Harvey RD. Preventing and Managing Toxicities of High-Dose Methotrexate. Oncologist. 2016 Dec;21(12):1471-1482. Doi: 10.1634/theoncologist.2015-0164.

[17] Yang C, Cai J, Geng Y. and Gao Y. Multiple-dose and double-dose versus single dose administration of Methotrexate for the treatment of ectopic pregnancy: A systematic review and meta-analysis. Reproductive BioMedicine Online. 2017Apr;34(4):383-391. DOI: 10.1016/j.rbmo.2017.01.004

[18] Armando Hasudungan, Immunology - Toll Like Receptors Overview [Internet YouTube]. 2014. Available from: https://youtu.be/8mEnyBdsrr8

[19] Zhang E, and Lu M: Toll-like receptor (TLR)-mediated innate immune responses in the control of hepatitis B virus (HBV) infection. Medical Microbiology and Immunity 2015;204:11-20. DOI: 10.1007/s00430-014-0370-1

[20] Min HS, Kim JE, Lee MH, Song HK, Lee MJ, Lee JE, Kim HW, Cha JJ, Hyun YY, Han JY, Cha DR and Kang YS: Effects of toll-like receptor antagonist 4,5-dihydro-3-phenyl-5-isoxasole acetic acid on the progression of kidney disease in mice on high-fat diet. Kidney Research and Clinical Practice 33(2014)33-44. DOI: 10.1016/j.krcp.2013.11.022

[21] Takashima K, Matsunaga N, Yokshimatsu M, Hazeki K, Kaisho T, Uekata M, Hazeki O, Akira S, Lizawa Y, and Li M: Analysis of binding site for the novel small-molecule TLR 4 signal transduction inhibitor TAK242 and its therapeutic effect on mouse sepsis model. British Journal of Pharmacology. (2009),157,1250-1262. DOI: 10.1111/j.1476-5381.2009.00297.x

[22] Zhang Y, Peng W, Ao X, Dai H, Yuan L, Hung X, et al: TAK-242, a toll-like receptor 4 antagonist, protect against aldosterone-induced cardiac and renal injury. Plos ONE 2015, 10(11):e0142456. DOI: 10.1371/journal.pone.0142456

[23] Broering R, Lu M, and Schlaak F: Role of toll-like receptors in liver health and disease. Clinical Science Nov 2011,121(10),415-426. DOI: 10.1042/CS20110065

[24] Hadi N, and Jabber H: Potential activity of GIT27 against renal ischemia reperfusion injury: An experimental study in male rats. Pathophysiology of Cell Injury Journal 2016;5(2):87-99. DOI: 10.18081/2378-5225-016-12187-99

[25] Carrascosa J.M., de la Cueva P., Ara M., Puig L., Bordas X., Carretero G. et al. Metotrexato en psoriasis moderadagrave: Revisión de la literature y recomendaciones de expert. Actas Dermosifiliogr. 2016.Apr,107(3): 194-206. DOI: 10.1016/j.ad.2015-10.005

[26] Granucci Francesca, Zanoni Ivan. Role of CD14 in host protection against infections and in metabolism regulation. Frontiers in Cellular and Infection Microbiology 3, 2013 pages 32. DOI=10.3389/fcimb.2013.00032

[27] Li M, Matsunaga N, Hazeki K, Nakamura K, Takashima K, Seya T,

Hazeki O, Kitazaki T, and Lizawa Y. A novel cyclohexene derivative, ethyl(6R)-6-[N(2-choro-4-fluorophenyl) sulfamoyl]cyclohex-1-ene-1-carboxylate (TAK-242), selectively inhibits toll-like receptor 4-mediated cytokine production through suppression of intracellular signaling. Molecular Pharmacology April1,2006,69(4)1288-1295. DOI:10.1124/mol.105.019695

[28] Yee J, and Orchard D. Monitoring recommendations for oral azathioprine, methotrexate and cyclosporine in a pediatric dermatology clinic and literature review. Australasian Journal of Dermatology 2018Feb,59(1):31-40. DOI: 10.1111/ajd.12526

[29] Wan S, Xiang Y, Fang W, and Huang D. The effect of methotrexate in combination with mifepristone on ectopic pregnancy: A meta-analysis. International Journal of Clinical and Experimental Medicine 2016,9(8):14990-15003. Available from:http://www.ijcem.com/files/ijcem0028751.pdf

[30] Coskun M, Steenholdt C, de Boer NK and Nielsen OH. Pharmacology and optimization of thiopurines and methotrexate in inflammatory bowel disease. Clinical pharmacokinetics (2016)55,257-74. DOI: 10.1007/s40262-015-0316-9

[31] Kaplowitz N. (2004) Drug induced liver injury. Clinical Infectious disease 2004Mar:38(Suppl2):S44-8. DOI:10.1086/38446

[32] Bianchi G, Caporali R, Todoerti M, and Mattana P. Methotrexate and rheumatoid arthritis: Current evidence regarding subcutaneous versus oral routes of administration. ADVANCES IN THERAPY 2016;33:369-378. DOI: 10.1007/s12325-016-0295-8

[33] Roth AD and Lee MY. Idiosyncratic drug-induced liver injury (IDILI): Potential mechanisms and protective assays. BioMed Research International 2017,2017:9176937,23pages. DOI: 10.1155/2017/9176937

[34] Savva A, and Roger T. Targeting toll-like receptors: promising therapeutic strategies for the management of sepsis-associated pathology and infectious diseases. Front Immunol. 2013 Nov 18;4:387. DOI:10.3389/fimmu.2013.00387

[35] Spirig R, Djafarzadeh S, Regueira T, Shaw SG, von Garnier C, et al. Effects of TLR Agonists on the Hypoxia-Regulated Transcription Factor HIF-1α and Dendritic Cell Maturation under Normoxic Conditions. PLOS ONE 2010 5(6): e10983. DOI: 10.1371/journal.pone.0010983

[36] Wood, E.G., Macdougall, C.E., Blythe, H. et al. HIF1α activation in dendritic cells under sterile conditions promotes an anti-inflammatory phenotype through accumulation of intracellular lipids. Sci Rep **10**, 20825 (2020). DOI:10.1038/s41598-020-77793-6

[37] Pandey S, Agrawal DK. Immunobiology of Toll-like receptors: emerging trends. Immunol Cell Biol. 2006 Aug;84(4):333-41. DOI: 10.1111/j.1440-1711.2006.01444.x.

[38] Nicholas S, and Sumbayev, V. The involvement of hypoxia-inducible factor 1 alpha in Toll-like receptor 7/8-mediated inflammatory response. Cell Res 19, 973-983 (2009). DOI:10.1038/cr.2009.44

[39] A Licata, MG Minissale, V Calvaruso, A Craxì A focus on epidemiology of drug-induced liver injury: analysis of a prospective cohort. Eur Rev Med Pharmacol Sci 2017,Vol. 21 - N. 1 Suppl, Pages: 112-121. Available from: https://www.europeanreview.org/article/12449

[40] Luu B, Rodway GW. Does low-dose methotrexate deserve more respect from

clinicians? JAAPA. 2017 May;30(5):12-15. DOI: 10.1097/01.JAA.0000515554.91731.82.

[41] Campbell JM, Bateman E, Stephenson MD, Bowen JM, Keefe DM, and Peters MD. Methotrexate-induced toxicity pharmacogenetics: An umbrella review of systematic reviews and meta analysis. Cancer Chemotherapy and Pharmacology 2016 Jul;78(1):27-39. DOI: 10.1007/s00280-016-3043-5

[42] Singh D, Cho WC, and Upadhyay G. Drug-induced liver toxicity and prevention by herbal antioxidants: An overview. Frontiers In Physiology 1.6:363. DOI: 10.3389/fphys.2015-00363

[43] Olayinka ET, Ore A, Adeyemo OA, and Ola OS. Ameliorative effect of gallic acid on methotrexate-induced hepatotoxicity in rat. Journal of xenobiotics 2016;6:6092. DOI: doi.org/10.408/xeno.2016,6092

[44] Fagone P, Mangano K, Mammana S, Pesce A, Pesce A, Caltabino R and Nicoletti F, et al. Identification of novel targets for the diagnosis and treatment of liver fibrosis. International Journal of Molecular Medicine 36:747-752,2015. DOI: 10.3892/ijmm.2015.2264

[45] Zhang N, Liang H, Farese RV, Li J, Musi N, Hussey SE. Pharmacological TLR 4 inhibition protects against acute and chronic fat-induced insulin resistance in rats. Plos ONE 10(7):e013575. DOI: 10.1371/journal.pone.013575

[46] Privitera G, Agarwal B, and Jalan R. Acute Liver Failure: Pathophysiologic basis, and the current and emerging therapies. EMJ Hepatol. 2014;1:99-107. Available from: https://emj.emg-health.com/wp-content/uploads/sites/2/2018/02/Acute-Liver-Failure-Pathophysiologic-Basis-And-The-Current-And-Emerging-Therapies.pdf

[47] Cao L, Quan XB, Zeng WJ, Yang XO, Wang MJ. Mechanisms of hepatocyte apoptosis. Journal of cell death 2016:919-26. DOI:10.4137/JCD.S39824

Chapter 6

Human Herpetic Viruses and Immune Profiles

Marwa Mohammed Ali Jassim,
Majid Mohammed Mahmood and Murtada Hafedh Hussein

Abstract

Herpesviruses are large, spherical, enveloped viral particles with linear double-stranded DNA genome. Herpesvirus virion consists of an icosahedral capsid containing viral DNA, surrounded by a protein layer called tegument, and enclosed by an envelope consisting of a lipid bilayer with various glycoproteins. Herpesviruses persist lifelong in their hosts after primary infection by establishing a latent infection interrupted recurrently by reactivations. The Herpesviridae family is divided into three subfamilies; α-herpesviruses, β-herpesviruses, and γ-herpesviruses based on the genome organization, sequence homology, and biological properties. There are eight human herpes viruses: Herpes simplex virus type 1 and 2 (HSV-1, −2) and-Varicella-zoster virus (VZV), which belong to the α-herpesvirus subfamily; Human cytomegalovirus (HCMV), and Human herpesvirus type 6 and 7 (HHV-6,HHV-7), which belong to the β-herpesvirus subfamily; and Epstein–Barr virus (EBV) and Kaposi's sarcoma-associated herpesvirus (KSHV) or Human herpesvirus 8 (HHV-8), which belong to the γ-herpesvirus subfamily. Within this chapter, we summarize the current knowledge about EBV and CMV, regarding their genome organization, structural characteristics, mehanisms of latency, types of infections, mechanisms of immune escape and prevention. Epstein–Barr Virus (EBV) genome encodes over 100 proteins, of which only (30) proteins are well characterized, including the proteins expressed during latent infection and lytic cycle proteins. Based on major variation in the EBNA-2 gene sequence, two types of EBV are recognized, EBV type 1 and 2. Epstein–Barr virus types occur worldwide and differ in their geographic distribution depending on the type of virus. **EBV** spreads most commonly through bodily fluids, especially saliva. However, EBV can also spread through blood, blood transfusions, and organ transplantations. The EBV is associated with many malignant diseases such as lymphomas, carcinomas, and also more benign such as infectious mononucleosis, chronic active infection. The EBV has also been suggested as a trigger/cofactor for some autoimmune diseases. Overall, 1–1.5% of the cancer burden worldwide is estimated to be attributable to EBV The latently infected human cancer cells express the most powerful monogenic proteins, LMP-1 and LMP-2(Latent Membrane Protein-1,-2), as well as Epstein–Barr Nuclear Antigens (EBNA) and two small RNAs called Epstein–Barr Encoded Small RNAs (EBERs). The EBV can evade the immune system by its gene products that interfering with both innate and adaptive immunity, these include EBV-encoded proteins as well as small noncoding RNAs with immune-evasive properties. Currently no vaccine is available, although there are few candidates under evaluation. Human cytomegalovirus (HCMV) is a ubiquitous beta herpesvirus type 5 with seroprevalence ranges between 60 to 100% in developing countries. CMV is spread from one person to another, usually by direct and prolonged contact with bodily fluids, mainly saliva, but it can be transmitted by genital secretions, blood transfusion and organ transplantation. In addition, CMV can be transmitted vertically from mother to child. CMV infection can result in severe disease for babies, people who receive solid organ transplants or bone marrow/stem cell transplants and people with severe immune suppression such as advanced human immunodeficiency virus (HIV) infection. The HCMV has several mechanisms of immune system evasion. It interferes with the initiation of adaptive immune responses, as well as prevent CD8+ and CD4+ T cell recognition interfering with the normal cellular MHC Class I and MHC Class II processing and presentation pathways. Challenges in developing a vaccine include adeptness of CMV in evading the immune system. Though several vaccine candidates are under investigation.

Keywords: human cytomegalovirus, Epstein-Barr virus, mononucleosis, transplantation, immune evasion, oncogenesis

1. Introduction

1.1 The Herpesviridae family

Herpesviridae is a large family of double-stranded DNA viruses, which is included in the recently classified order Herpesvirales. This family can be further classified into three distinct subfamilies: *Alphaherpesvirinae, Betaherpesvirinae, and Gammaherpesvirinae*, according to their biology and DNA genomic sequence [1]. The *Alphaherpesvirinae* subfamily includes five distinct genera, *Simplexvirus* and *Varicellovirus* are most important members causes infection to human. The members of Herpes simplex viruses 1 and 2 (HSV-1 and HSV-2) infect almost 85% of the world human population and cause orolabial.herpes and genital herpes; other members of the family include varicella-zoster virus (VZV), which is responsible for chicken-pox and shingles. The *Betaherpesvirinae* subfamily includes four genera, the most important members that infect the human are Human *cytomegalovirus* also known as human herpesvirus 5 (HCMV or HHV-5) and Human herpesviruses 6 and 7 (HHV-6 and HHV-7) [2].

The *Gammaherpesvirinae* subfamily is composed of four distinct genera, *Lymphocryptovirus* (LCV) and *Rhadinovirus* (RDV)infect the human, Taxonomically, the oncogenic Epstein–Barr virus (EBV) is also designated as human herpesvirus 4 (HHV-4) belongs to the genus lymphocryptovirus (LCV) and it is the only human pathogen of this genus. The RDV Kaposi's sarcoma-associated virus (KSHV, also known as HHV-8), another oncogenic herpesvirus, is the only known human RDV [3]. The *Gammaherpesviruses* may promote oncogenic effects and also contribute to the development of malignancies but this is a rare outcome [2]. Altogether herpesviruses can establish latent infection within specific tissues, with immune surveillance evasion. The human herpesviruses and their diseases are summarized in (**Table 1**).

1.2 Epstein–Barr Virus (EBV)

EBV is ubiquitous virus, with a seroprevalence of more than 90% of the adult population worldwide. It was first identified in 1964 by Anthony Epstein's group in a cell line from a Burkitt's lymphoma biopsy [4, 5]. The EBV has also been identified as a B lymphotropic oncogenic virus owing to its capacity to convert resting B lymphocytes in vitro, inducing continuous dissemination of infected B cells and producing lymphoblastic cell lines (LCLs) [6]. This discovery was central to the identification of EBV as the first nominee human tumor virus. Subsequently,

Subfamily	Genus	Species	Disease
Alphaherpesvirinae	*Simplexvirus*	*Human alphaherpesvirus* 1 (Herpes simplex virus 1) *Human alphaherpesvirus* 2 (Herpes simplex virus 2)	Acute Herpetic gingivostomatitis, Keratitis, Conjunctivitis, Encephalitis, Dermal whitlow, Herpes labialis Herpes genitalis
	Varicellovirus	*Human alphaherpesvirus* 3 (*Varicella Zoster* virus)	Chickenpox/ shingles
Betaherpesvirinae	*Cytomegalovirus*	*Human betaherpesvirus* 5 (Human cytomegalovirus)	Congenital abnormalities
	Roseolovirus	*Human betaherpesvirus* 7 (Human herpesvirus 7)	Febrile illnesses
		Human betaherpesvirus 6A, 6B (Human herpesvirus 6A, 6B)	Infant rash Exanthem subitum
Gammaherpesvirinae	*Lymphocryptovirus*	*Human gammaherpesvirus* 4 (Epstein–Barr Virus)	Infectious mononucleosis (Glandular fever), Burkitt's lymphoma, Hodgkin's lymphoma, Nasopharyngeal carcinoma, Oral hairy leukoplakia
	Rhadinovirus	*Human gammaherpesvirus* 8 (Kaposi's sarcoma-associated virus or Human herpesvirus 8)	Kaposi's sarcoma

Table 1.
Taxonomy of Human Herpesviruses [3].

EBV was correlated with a variety of clinical malignancies, including Hodgkin's Lymphoma (HL), post-transplant lymphoproliferative disease (PTLD) and X-linked lymphoproliferative disease (XLPD). The potential to invade other cell types other than B lymphocyte, such as T, natural killer (NK) and epithelial cells, has led to the association of EBV with other malignancies: peripheral T cell, nasal T or NK cell lymphomas, gastric and nasopharyngeal carcinomas (NPC) [2, 7]. However, infection with EBV induces contagious mononucleosis during or after adolescence [8]. Even though EBV exhibits a strong growth transforming capacity, that asymptomatically infects up to 95% of the human population, whereas it is perfectly immune-controlled [2].

1.2.1 The EBV Virion and Genome organization

The virus is 122–180 nm in diameter. Epstein–Barr virion contains a linear, double-stranded DNA genome wrapped on an icosahedral capsid, approximately (100–110) nm in diameter, containing 162 capsomeres with a pore running down the long axis. The protein tegument with viral and cellular proteins including actin, tubulin, and cofilin separates the nucleocapsid from the lipid envelope that coats the virus and contains numerous viral glycoproteins (GP) spikes such as gp350/220, gp42, GH, GB and gp150 on the outer surface. These glycoproteins play an important role in cell tropism and recognition of receptors [8, 9] as shown in the **Figure 1**.

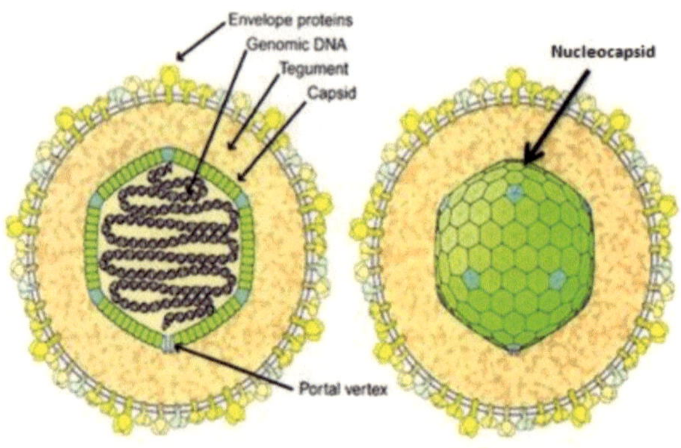

Figure 1.
The structure of EBV viral particle.

The double-stranded DNA (172Kb) linear genome encodes more than 100 proteins as well as non-coding functional RNAs (EBER RNAs, BART miRNAs, and BHRF1 miRNAs). There are some similar tandem terminal repeats (TR) of 0.5 kb at each terminal of the genome [10] and other internal direct repeats of 3 kb (IR) including the latency promoter (Wp) and the special short unique sequence domains (US)and UL (long). The US and UL sequences comprise nearly all of the genome encoding capacity [11] as shown in **Figure 2**. The EBV genome is classified as C genome, which is linear in a virus particle, but distributed as an episome in the nucleus of infected cells; circulating occurs by terminal repeat units (TRs) following B cell infection with EBV [11]. The first cloned and sequenced EBV strain was typing 1 EBV: B95.8, this strain was obtained from an infectious mononucleosis patient's. Sequencing was based on previously generated EcoRI and BamHI restriction fragments (**Figure 3**). B95.8 strain is commonly used in labs around the world; however, a 13.6 kb portion of its genome is incomplete. Subsequently, the missing fragment was sequenced from the Raji strain and a revised EBV consensus genome was released several years later [12].

1.2.2 The EBV classification

Two major types of EBV, type 1 and 2, have been described in humans based on major variations in EBNA-2 gene sequence [11]. Type 1 is dominant throughout most of the world, but the two types are equally prevalent in Africa. The EBNA-2 is the most variable locus in the EBV genome which is characterized by 70% identity at the level of nucleotide sequence whereas only 56% similarity at the amino acid level between these two types (3). In addition, the variation between type 1 and type 2 is also linked to the sequence variation in the viral latent genes EBNA-3A, EBNA-3B, EBNA-3C and EBNA-LP [13].

1.2.3 Epstein Barr virus life cycle

1.2.3.1 Cell attachment and viral entry

The initial attachment of EBV is mainly regulated by the association between its envelope protein (gp350/220) and the cellular complement component receptor 2 (CR2/CD21) protein located on the B cell surface. This association activates Cluster of differentiation (CD21) receptor aggregation in the plasma membrane and also a tyrosine kinase signal transduction through CD19 that contributes to Nuclear factor-kappa B (NF- kB) activation and cell cycle entry [14].

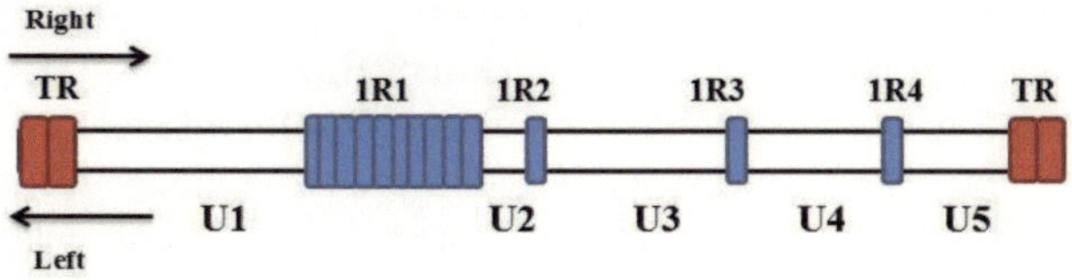

Figure 2.
Linear Organization of the EBV genome.

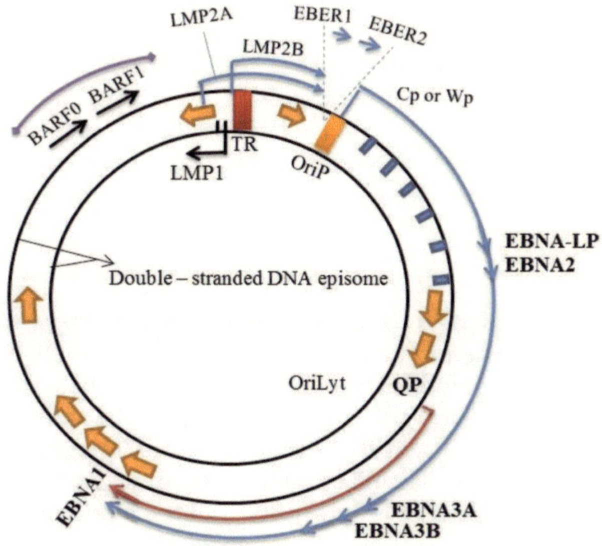

Figure 3.
Circular Organization of the EBV genome.

The attachment of the second viral glycoprotein gp42 to the human leukocyte antigen class II receptor (HLA class II) activates the viral envelope fusion with the membrane of the cell and the viral entrance in a cycle that relies on the glycoprotein complex GH/GL and also on GB [15]. The GH/GL complex is supposed to serve as a receptor that activates GB-mediated fusion after gp42 binding to HLA Class II molecules. Thereafter, virion nucleocapsids are released into the cytoplasm and transported to nuclear pores on microtubules [13]. As a result, the viral linear genome is transferred to the nucleus of B lymphocyte and the viral genome is then retained in the nucleus as a covalently locked extrachromosomal episome [16].

For epithelial cells, as there are no CD21 or HLA class II molecules on their surface, the entrance of EBV does not involve gp350/220 and gp42. Viral BMRF2 protein can mediates interaction with cellular β1 integrins [14]. The fusion of viral envelope is activated by the attachment of the viral gH/gL complex to 5–007vβ6/8 integrins, which is confirmed by the effectiveness of infection in virions missing gp350/220 glycoproteins. The EBV virion expresses three- gH / gL/ gp42 and two- gH/gL glycoprotein complexes that grant the capacity to invade either B cells or epithelial cells [13].

The virus is endocytosed into a low pH vesicle where fusion occurs after the interplay of EBV glycoprotein gp350 and receptor type 2 (CR2). Glycoprotein gp42 is bound directly to GH and transforms dimeric GHGL in a trimeric gHgLgp42, modifying the conformation of gp42 to cause its attachment to the human leukocyte antigen (HLA) class II molecule. It will allow the central fusion machine to support effective B cell infection. Besides, GH can bind cellular components [15].

The epithelial cells do not constitutively express HLA class II, which makes gp42 useless in the process of fusion. The interaction of dimeric GHGL complexes with integrins, however, replaces the cell fusion caused by the interaction between gp42 and HLA class II. The use of dimeric GHGL complexes to cause epithelial cell fusion and gHgLgp42 trimeric complexes to contribute to B cell fusion was expected that the virus would trigger B cells and epithelial cells to alter the viral tropism: The gp42 spike in epithelial viral particles makes it 100 times more infectious than the virus produced from B-cells. The opposite is not so dramatic: the B-virus is five times more contagious for the epithelial cell than the epithelial virus [17]. After binding to the primary B cell, most virions do not internalize with the epithelial cell and the infection can be significantly increased by co-culturing with EBV negative B cells. Such virions stay on the surface of the cell B and can then be passed via the formation of the intracellular synapse to CR2-negative epithelial cells. This transfer technique involves the interaction between gp350-CR2 and GH and GB viral glycoproteins. This mechanism has been suggested to allow EBV to enter both lymphoid and epithelial cells simultaneously [18].

1.2.3.2 EBV Lytic Infection

The lytic infection is characterized by the active release of new contagious virus particles, either infecting new human hosts or infecting other naive B cells in the same host. The lytic cycle is divided into three stages: Immediate-early (IE), Early (E), and Late (L). The expression of immediate early BZLF1 and BRLF1 genes included in the activation of the lytic process is activated by signal transduction by the B cell receptor (BCR) [15]. The BZLF1 is a viral transactivator protein responsible for activating the production of lytic genes and the repression of latent genes, resulting in cells' death and the release of contagious virons. The signal transduction of BCR initiates BRLF1 development and also improves its production allowing the transition from latency to lytic cycle [14]. The BZLF1 protein is a bZIP-specific transcription factor close to c-FOS and C/EBP. The BZFL1 and BRLF1 motivate functions of early genes, such as viral DNA polymerase (BALF5) and thymidine kinase, to initialize viral DNA replication from the lytic origin of replication (OriLyt) in tandem with other direct and early gene products [13]. Late lytic genes encode viral structural proteins, including tegument proteins, glycoproteins, and BcLF1 main capsid proteins. Newly synthesized viral DNAs are packed into nucleocapsids in the nucleus of the cell, which moves across the nuclear membrane to the cytoplasm, creating vesicles carrying virons with an envelope. The vesicles fuse with the plasma cell membrane and the virus particles exocytose [19].

1.2.3.3 Latency

Herpesviruses are distinguished by their ability to establish and sustain a latent infection in their hosts. Latent EBV expresses its genes in one of three latency systems: Latency I, II, or III variations in either of these systems assist in the development of a distinct series of viral RNAs and proteins [20] **Table 2** and **Figure 4**. This chronic infection is characterized by inhibition of viral replication and viral dormancy, and immune evasion in the host. The EBV determines latency in the B cell pool which is the long-term reservoir for the virus in vivo. Naive B cells infected with EBV in the Waldeyer ring proliferate as activated B bursts, which are close to antigen-activated B lymphocytes in terms of the structure and morphology of their cell surface [24].

Opposite to the lytic infection, replication of the viral genome in latent infection occurs through host DNA polymerase and from a separate source, Orig of

EBV latent protein	Function related to B-cell lymphomagenesis
EBNA1	Regulation of viral DNA replication and transcription of many viral and cellular genes; It facilitates p53 disintegration and thus promotes tumorigenesis
EBNA2	One of the main viral transcription factors; In combination with EBNALP, EBNA2 regulates the transcription of many of viral and cellular genes; Fundamental for B cell transformation
EBNALP	EBNA2-mediated transcription activator, both for viral and cellular genes; Bypassing the innate immune response of cells; Fundamental for B cell transformation
EBNA3A	Together with EBNA3C, it suppresses the genetic transcription of BIM, p14, p15, p16 and p18 through epigenetic regulation; Prevents differentiation of B cells into plasma; Essential for B cell transformation
EBAN3B	Virus-encoded tumor suppressor protein
EBNA3C	Together with EBNA3A, it suppresses the genetic transcription of BIM, p14, p15, p16 and p18 through epigenetic regulation; Assists G1-S and G2-M transformations of the cell cycle; Ubiquitin-proteasome pathway; Supresses p53-, E3F1- and Bim-mediated apoptosis; Activates autophagy necessary for B cell transformation
LMP1	Functionally mimics CD40 signaling pathway; one of the major transcriptional regulators; Mainly activates NF-kB, JAK/STAT, ERK MAPK, IRF and Wn't signaling pathways; Induces BCL-2 and a20 expression to prevent apoptosis; Essential for B cell transformation
LMP2A	Functionally mimics BCR signaling pathway; prevents apoptosis; EBV latency regulation
LMP2B	Regulates LMP2A functions
EBERs	Most of the non-coding viral RNA is found in all forms of latency programs; Affects the innate immune response and gene expression; Inhibits apoptosis dependent on activated dsRNA dependent protein kinase (PKR)
miRNAs	Transcribed from BART and BHRF1 loci; maintains latently infected B cells through blocking cellular apoptosis

Table 2.
Impact of latent antigens in EBV on B-cell transformation and subsequent development of lymphoma [21–23].

replication. During latent infection, the viral genome is present as a closed circular, extrachromosomal plasmid or episome. The viral DNA is wrapped with host histone molecules and replicates steadily once throughout the cell cycle together with the host genome [25], this enabled EBV infected B blasts during proliferation to express all latent EBV genes which are known as latency III or growth-program that play important role in cell activation and proliferation.

This is achieved by the expression of two viral latent membrane proteins (LMPs), LMP-1, and LMP-2A, which constitute a functional homolog of the CD40 receptor in B lymphocytes and often mimic the constitutively active BCR, respectively [26]. B cell migrates to nearby primary follicles to form germ centers and the viral transcription system switches to latency II or a default system to enable the B cells to differentiate into memory B cells, Latency II is characterized by the expression of LMPs and EBNA-1 protein.. In the absence of antigen-mediated signals, LMPs are necessary to provide cell survival signals needed to prevent apoptosis of latently infected B cells [26].

Epstein Barr-virus nuclear antigen-1(EBNA-1) protein is important for EBV DNA replication and for preservation of viral genome in the cells [20]. The memory B cells lately infected reach peripheral circulation and represent viral persistence reservoir; [27]. Such latently infected memory B cells with EBV are distinguished by a silence of the expression of viral protein in a program called latency 0 or latency-program which is intended to permit immune evasion and therefore lifelong

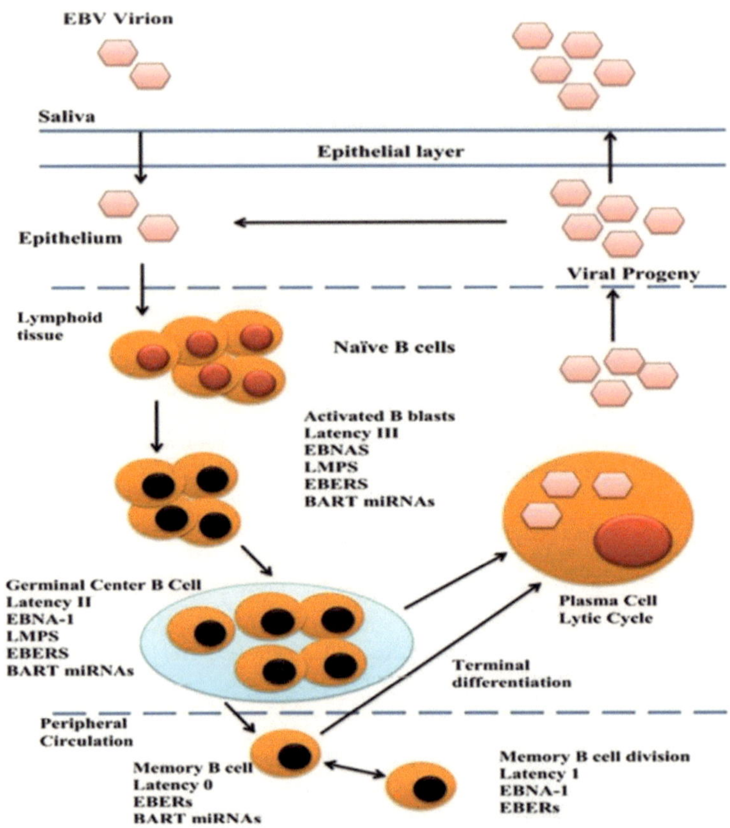

The EBV transmitted by saliva, infects human naïve B cells in tonsils and viral latency III is established. Infected B cells proliferate and differentiate into B blasts, and migrate to germinal centers where they differentiate into memory B cells.Then latency II is established to offer signals of survival which permit the cell to leave the germinal center as a memory B cell. The viral expression latency-program in resting memory B cells is established and no viral proteins are expressed, which permits the viral to elude revealing by ,· or latency only program and -\-the immune system and persist in the host. When these memory B cells sometimes divide they express the EBNA move to the tonsils and differentiate into plasma cells to initiate the lytic cycle and triggers viral replication. The viral progeny may be .shed into the saliva for dissemination into new hosts or may infect new B cells

Figure 4.
Model for the establishment of EBV persistent latent infection.

persistence on the host. The expression EBNA-1 is enabled and allows the division of the viral genome in the cells carrying the virus. This is known as the transcription program Latency I or EBNA-1 only program [26]. From peripheral circulation, latently infected memory B cells migrate into oropharynx and tonsils and then differentiate into plasma antibody-producing cells. Reactivation of the virus is triggered and infectious viruses are created as they bear the virus. Therefore, these viral particles will infect additional hosts with new naive B cells [27].

1.2.4 Transmission

The oral route is the primary route of the EBV transmission commonly through bodily fluids, especially saliva [28]. However, it has been reported that EBV infection can also be transmitted after the transfusion of a large volume of fresh blood [29]. Although EBV has been detected in cervical secretions of 8% -28%, of women, it is still controversy on whether EBV is transmitted through sexual contact [30]. Possible spread via organ transplantation can occur which is of particular concern in association with subsequent infection by EBV [25]. Transmission by milk is also a possible route, but is a non-significant mode of EBV transmission [31].

1.2.5 EBV epidemiology

Epstein–Barr virus types occur worldwide, but they differ in their geographic distribution. For instance, Type 1 is prevalent in population from Europe, America, China, and South Asia, while Type 2 is less prevalent in these populations and is more observed in African and Papua New Guinean populations. Over 90–100% of adults have been infected with EBV, and the infection is most commonly affecting those patients aged 2 to 4 years and those aged 15 years. Epstein–Barr virus causes approximately 90% of the cases of infectious mononucleosis, which is commonly seen in both the community and the hospital setting. Among infants and young children who are primarily infected with EBV, in Africa, and where Burkitt lymphoma is common, 50% of them are infected with this virus before their 1 year of age. About 70% of cases of PTLD are associated with Epstein–Barr virus (EBV), especially in cases that occur early after transplantation [32]. Recent studies from Kenya reported a striking overlap between increased incidence of malaria transmission and Burkitt lymphoma [33]. Furthermore, various studies have demonstrated the presence of 8: 14 translocation in both the endemic African Burkitt lymphoma and in the non-endemic tumor type (Europe, America, and Japan).

1.2.6 EBV clinical features

The EBV is associated with many malignant diseases such as lymphomas, carcinomas, and also more benign such as infectious mononucleosis, chronic active infection. The EBV has also been suggested as a trigger/cofactor for some autoimmune diseases. Overall, 1–1.5% of the cancer burden worldwide is estimated to be attributable to EBV [3].

1.2.6.1 Primary EBV infection

The primary EBV infections of infants and children are often asymptomatic or have nonspecific symptoms, but infections of adolescents and adults frequently result in infectious mononucleosis (IM). Around 80% of infected adults mostly experience symptoms, including sore throat, cervical lymphadenopathy, weakness, upper respiratory infection, headache, reduced appetite, fever, and myalgia (muscle aches). It is characterized by a large number of lymphocytes, mainly CD8 + T-cells, which, as opposed to healthy individuals, can reach five to ten times more numbers in the blood. The causes of this expansion of T-cells in IM are not clear, but factors such as failure of natural immune control by natural killer (NK) cell, memory CD8+ T cells of memory of EBV or genetic background have been suggested [34].

The severity of symptoms in primary EBV infection is associated with age and immune system of the patients. The complications of the disease include splenomegaly, and/or chronic hepatitis, pneumonia and lymphadenitis. Less common are complications, such as hemolytic/aplastic anemia, myocarditis, Guillain–Barré syndrome, encephalitis, and meningitis [35].

Chronic active EBV infection is a rare disorder characterized by the presence of severe illness of more than six months' duration, high virus-specific antibody titers and organ disease with the demonstration of EBV antigens or EBV DNA in tissue [35].

1.2.6.2 EBV reactivation and EBV associated diseases

The reactivation of latent EBV infection has been shown to occur following impairment of the cellular immune response which is important in the long-term control of persistent EBV infection. Chronic uncontrolled EBV reactivation may result in the development of carcinoma. The followings are diseases and cancers associated with EBV infection [2, 36–39]:

1. X-linked Lymphoproliferative Syndrome-XLP (Duncan's disease). It is an inherited disease of males, characterized by an uncontrolled increase in the number of cytotoxic T and NK lymphocytes where their activity is directed against normal cells of various organs. Subsequent studies showed that the disease has a variety of phenotypes after primary EBV infection, such as fatal infectious mononucleosis, hypogammaglobulinemia, and malignant lymphoma.

2. Post-transplant Lymphoproliferative Disorders (PTLD). They are a heterogeneous group of diseases characterized by uncontrolled proliferation of B cells (90%), T cells (9%), and NK cells (0.5%). 90% of PTLD lymphomas are EBV positive and they are typically limited to the lymph nodes. PTDL often occurs as a complication of both solid organ transplant (SOT) and hematologic stem cell transplant (HSCT) patients.

3. Leukoplakia Hairy Mouth. It is a benign, mild proliferative change in epithelial cells of the tongue and the nasopharynx. It is caused by EBV and occur usually in persons who are immunocompromised, most frequently seen in those infected with HIV.

4. Burkitt's Lymphoma (BL). It is a rapidly growing malignant tumor composed of large D-type lymphoblasts and mainly affects facial bones, mandibles, and jaws. EBV-associated BL has an incidence of 5–10/100 000 children and accounts for up to 74% of childhood malignancies in the African equatorial belt.

5. Hodgkin's Lymphoma (HL). It is a proliferative disease of the lymphatic system with a peak incidence in those between 20–30 years and after 50 years of their age, the risk of developing HL is four times higher in those infected with EBV as compared to the general population.

6. Nasopharyngeal Carcinomas (NPC). It is the most common cancer originating in the nasopharynx, most commonly in the postero-lateral nasopharynx or pharyngeal recess which can extend to the base of the skull, palate, nasal cavity, or oropharynx [40]. Most NPC is associated with EBV infection, mainly in middle-aged patients and is more common in men than women [41, 42]. The International Agency for Research on Cancer (IARC) classified the EBV virus into the first group of carcinogens due to this association [43].

7. Sinonasal Carcinoma.

About 90% of head and neck cancers are squamous cell carcinoma (SCC), they originate from the mucosal lining that, causes tumor development in the nasal cavity and mouth, nasopharynx, larynx, esophagus and paranasal sinuses [44]. The International Agency for Research on Cancer (IARC) estimated that 16% of total new cancers, as well as 20% of deaths caused by cancers worldwide, were due to infections with EBV [2, 7]. Sinonasal carcinoma is a rare tumor comprised of about 1% of all cancers and 3% of all head and neck cancers [44, 45].

1.2.6.3 C. Autoimmune diseases.

EBV has variously been linked with a number of autoimmune diseases including multiple sclerosis (MS), systemic lupus erythematosus (SLE), Sjögren's syndrome (SS) and rheumatoid arthritis (RA). The EBV links with these diseases include raised titers of EBV antibodies, decreased T cell response to EBV and elevated EBV viral load. It has been suggested that EBV triggers the activation state of the immune system by inducing the development of pro-inflammatory mediators, which may play a role in autoimmune pathogenesis [46].

1.2.7 EBV oncogenesis and association of latency type

Oncogenesis is a cytological, genetic, and cellular transformation process that results in malignant tumors. Discovery of viral oncogenes and the discovery that they are derived from cellular genes called protooncogenes led to the understanding that c-onc genes have roles in different tumor types. The activation of viral oncogenes requires genetic changes in cellular protooncogenesby 3 genetic mechanisms: (a) Mutation (b) Gene amplification (c) Chromosome rearrangements. These mechanisms result in either an increase in protooncogene expression or a change in protooncogene structure [47]. The EBV-mediated B-cell change is associated with a global improvement in viral and cellular expression of genes. The biologic characteristics of the virus were instantly fascinating, as it was shown that cell lines could be determined from samples of Burkitt's lymphoma (BL) and could propagate a virus that could strike primary B cells with EBV and turn them into immortalized cell lines [48]. This study of molecular phenotype led to the discovery of viral proteins that are necessary for latent infection and needed for cell transformation [49]. The mechanisms of EBV oncogenesis include:

1.2.7.1 Alteration of host cellular signaling pathways

EBV encoding oncogenes induce the changes in the host cellular signaling pathways that control proliferation, differentiation, cell death, genomic integrity, and recognition by the immune system.

LMP1, LMP2A, and LMP2B, latent membrane proteins are generated of the common viral locus with converging and interfering primary transcripts [50]. The LMP1 is one of the main EBV-encoded oncoproteins and it is a constitutively active mimic of cellular CD40 receptor. It is critically important for the EBV-induced B-cell transformation via the activation of NF-κB, c-Jun N-terminal kinase (JNK), and p38 cascades [21]. LMP1 also regulates cellular apoptosis by triggering the NF-κB pathway by increasing the antiapoptotic expression of Bcl2 via IRAK1 and TRAF6 whereIRAK1 is necessary for both p38 and p65/RelA phosphorylation [50]. Also, LMP1-stimulated proapoptotic polycomb complex protein (Bmi-1) is further been recruiting by EBNA3C for the transcriptional funnel of other genes. LMP2A acts as a functional homolog of the B-cell receptor (BCR) and thus promotes the survival of B-cells. Likewise, it is essential for the growth transformation of germinal center-derived B-cells which are BCR negative [21]. LMP2B negatively regulates LMP2A functions and transition from latent to lytic activation by depleting LMP2A-mediated BCR cross-linking and restoring $Ca2+$ mobilization [51].

EBNA1 is important for the DNA replication and maintenance of the viral latent genome. It binds to the viral episomal replication origin (OriP) and simultaneously to the host cell hromosomes that enable viral genome duplication during each cell cycle [22]. Through promotor selection, combined with comprehensive epigenetic control, EBNA1 can organize the shift between different latency programs, and EBNA1 can produce transcripts for different cells and help improve the control of telomeres on cell chromosomes [52]. The p53 and Mdm2 affected by the EBNA1 binding with ubiquitin-specified protease 7 (USP7), contributes to antiapoptotic activity control, likely by promoting survivin expression levels [23, 53].

EBNA2 and EBNALPare the first latent genes expressed following B-cell infection. EBNA2 is the main viral transcription factor responsible for activation of the expression of the entire repertoire of latent transcripts along with several host genes, utilizing cell transcription factors, RBP-J and EBF1 [22]. At the same time, EBNALP supports transcriptional regulation by EBNA2 via blocking off the activity of NCoR and RBP-J [52]. The EBNA2 contributes most strongly to the proliferation of B-cells through the activation of about 300 cell genes, such as the transcription of MYC and RUNX3 [52].

EBNA3 protein family consisting of EBNA3A, -3B, and -3C are transcription factors that precisely regulate host gene transcription and the proliferation of B-cells, particularly in the immunosuppression environment. Also, EBNA3B knockout virus-induced tumors demonstrated a lack of T-cell infiltrate and related CXCL10 chemokine activation [53]. In comparison, EBNA3A and EBNA3C cooperate as predominant viral oncoproteins by controlling the transcription of the cellular gene. This phenomenon is also true for EBNA3A [54].

The EBNA3A and EBNA3C have been demonstrated to react with a long list of cellular proteins and transcription factors involved in the regulation of multiple cell signaling pathways [55]. Interactive partners for EBNA3C involve transcription factors, chromatin modulators (both histone deacetylase and histone acetylase enzymes), cell-cycle proteins including G1-S and G2-M transitions, metastases suppressors, post-translational modifiers, E3-ubiquitin ligase, ubiquitin-specific proteases, unfolded protein response (UPR) regulators, cell tumor suppressors, and oncoproteins [56]. The EBNA3C has been shown to form a complex with Chk2 and thus manipulates the G2/M step of the cell cycle [54]. Overall, the B-cell transformation and B-cell lymphoma are directly affected by the EBNA3 protectors by targeting main cell signaling cascades including cell cycle, apoptosis, and autophagy [56].

- Noncoding viral transcripts.

A variety of noncoding RNAs (ncRNAs) in EBV infected B cells can be expressed, known as the EBV-encoded non-polyadenylated RNAs (EBER1 and EBER2) and numerous miRNAs [57]. Such ncRNAs are not necessary for the transformation of B-cells, but they are associated with immune evasion, and demonstrated in various forms of latency systems. In addition, EBER in situ hybridization is the most reliable and sensitive method to detect EBV infection in tissues of various EBV-related malignancies. EBER expression promotes the growth of B cells by blocking of PKR phosphorylation and inhibition of translational initiation factor eIF-2a and alpha-interferon (IFN-α)-induced apoptosis [56]. EBER can interact with ribosomal protein L22 that regulates protein translation, gene expression and PKR dependent apoptosis [58]. The EBER2 directly recruits PAX5 for the control of LMP2A expression, which has been verified by the usage of the EBER2 mutant virus with lower LMP2A expression [58]. EBV encodes more than 40 mature miRNAs, which are encoded at 2 different loci in the EBV genome:

BHRF1 locus (BHRF1 miRNAs) and BART locus (BART miRNAs). The expression of various EBV miRNA is different among different cells. Viral miRNAs can either target other EBV transcripts or transcripts of host cells. BHRF1 miRNAs exhibit expression that is restricted to latency 3 whereas the BART miRNAs are expressed in all latency types. The expression of BHRF1 miRNAs in infected B lymphocytes, target multiple tumor suppressor proteins such as PTEN and P27KIP1 for the B-cell transformation. Viral miRNAs also inhibit the expression of several tumor suppressor genes, including DICE1, PUMA, PTEN and BCL2L11 to promote the survival of an epithelial cell [59].

Table 2 and **Figure 5** explain the key latent transcription mechanisms of EBV.

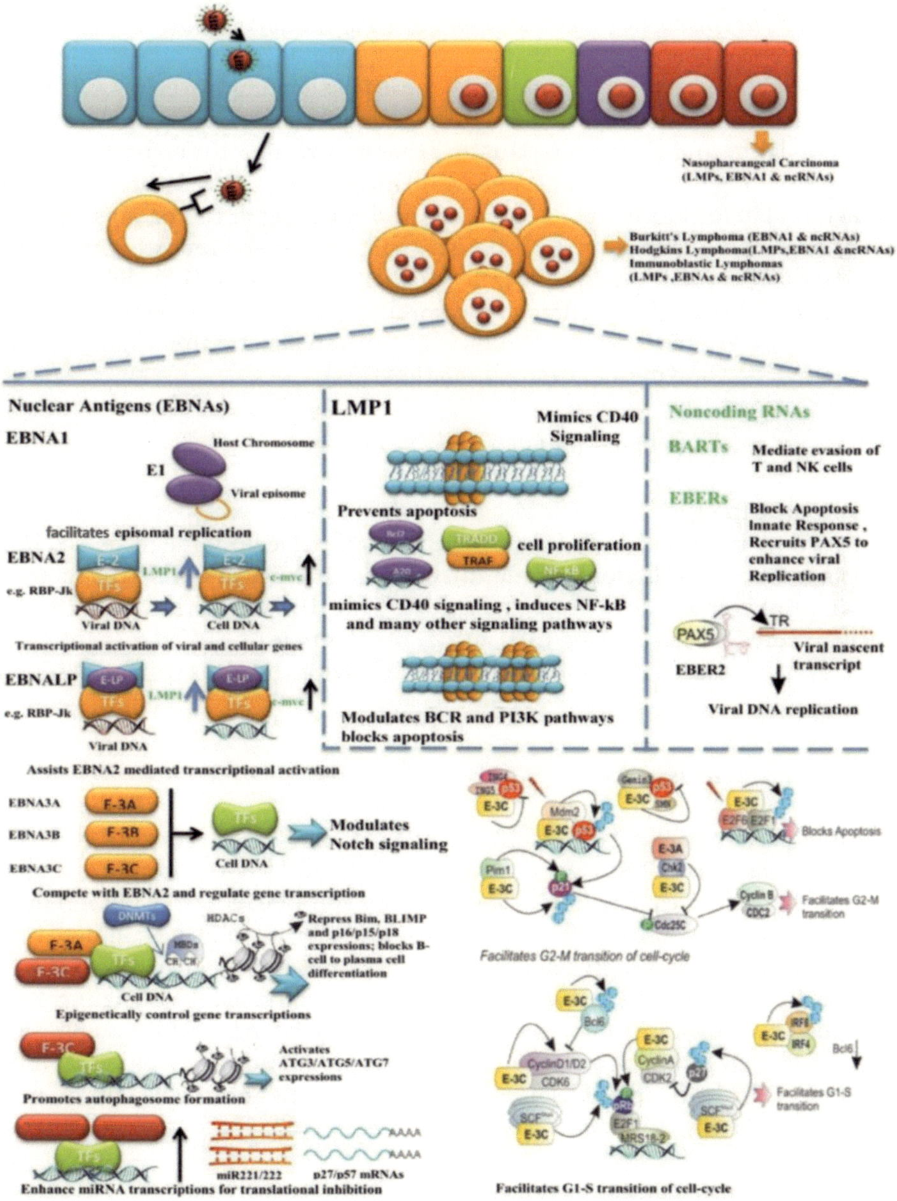

Figure 5.
Special features of EBV latent transcripts during B-cell transformation associated in the development of B-cell lymphoma.

1.2.7.2 Suppression, escape, and modulation of host immunity

Multiple genes have been reported to suppress antigen presentation. EBNA1 contains a Gly/Ala repeat sequence, through which proteasomal degradation and antigen presentation of the protein is impaired, while BNLF2A targets the transporter associated with antigen processing and blocks antigen presentation. BGLF5 represses HLA class I synthesis, whereas BILF1 downregulates cell surface expression of the molecule. It is highly likely that at least LMP1 and LMP2A, the viral functional mimic of CD40 and BCR, have tactfully evolved to modify those processes in the germinal center, and thus, these EBV gene products can deregulate the immune system for survival [60]. See section 1.1.8.

1.2.7.3 Genetic or epigenetic background/alteration of the host genome

Virally-induced epigenetic alterations of the host genome are evident in EBV-associated cancers, which are the result of genetic mutations, changes involving DNA methylation and chromatin structure that in turn alter the expression of growth promoting or suppressing genes. Enhanced Ig-Myc translocation has been demonstrated in Burkitt lymphoma [34, 35] and silencing of tumor suppressor genes (e.g. p16INK4A) in many EBV-positive cancers. Several reports have demonstrated that the predisposition of individual HLA allele significantly affects the morbidity of EBV-positive proliferative disorders, particularly in NPC and Hodgkin lymphoma [37–42]. LMP1 seems to induce genetic/epigenetic alterations by DNA hypermethylation and chromatin modifications [51].

1.2.8 Immune evasion mechanisms by Epstein–Barr virus

In response to primary EBV infection, both innate and adaptive antiviral responses have been activated. Despite the very effective immune response, the virus is not cleared.. A lifelong, latent infection is established within the memory B-cell and EBV genomes are propagated during the division of the transformed, latently infected B cells. During this period, broad range of EBV early gene products interfere with immune response which helps the virus to persist and to reactivate [60].

1.2.8.1 Evasion of innate immunity

Various pattern recognition receptors (PRRs) including cell surfaces and endosomal Toll- receptors (TLRs) and cytoplasmic DNA and RNA sensors are capable of detecting EBV particles. The EBV is identified by Toll-like receptors (TLRs) on the cell surface and in endosomes [61]. Virus-derived or virus-induced components can be detected by RNA and DNA sensors as well as by inflammasomes within the cytosol. TLRs, RNA and DNA sensors trigger a cascade of intracellular signaling events that enable the activation of the interferon regulatory factors (IRFs) and NF-κB. As a result, activated gene transcriptions induce the production of cytokines and type I interferon (IFN I). Different levels of these PRR signals are attacked by latent and/or lytic EBV proteins or EBV miRNA as recently seen [62, 63].

1.2.8.1.1 Reduction of Toll-like receptor expression

EBV can inhibit the synthesis of cellular proteins in infected cells through global mRNA destabilization. This mechanism is via the EBV DNase (alkaline exonuclease) BGLF5 regulated by the supplementary BGLF 5 RNase feature, which is expressed during the active period of infection and uses the same catalytic position as DNase action, but the substrate-bindings site is only partly shared by DNA and RNA substrates [62].

1.2.8.1.2 Modulation of IRF signaling and Type I interferon production

A significant number of lytic EBV proteins interact with host IRFs, which are the transcription factors that stimulate the synthesis of type IFN. The immediate-early EBV transactivator BZLF1, BRLF1 and tegument protein LF2 interact with IRF7 andIRF3, and inhibit its transcriptional activity on IFNα4 and IFNβ promoters resulting in the suppression of antiviral state induction. In addition, EBV protein kinase BGLF4 phosphorylates and inhibits IRF3 transcriptional activity, thus decreasing IFNβ expression [64].

1.2.8.1.3 Interference with NF-κB and inflammatory pathways

The EBV infection is linked to the decreased NF-κB--dependent gene expression. The expression of viral BZLF1 and cellular NF-κB is reciprocally inhibited. The higher levels of the NF-κB in the absence of BZLF1 instigate EBV latency, while increased expression of the BZLF1 after the induction of the lytic cycle overwhelms the minimal amount of the NF-κB [65].

1.2.8.1.4 Interference with innate effector molecules

Several EBV gene products have an impact on the function of effector molecules of innate immunity. The host cytokine colony-stimulating factor 1 (CSF-1) activates the differentiation of the macrophage, and the secretion of the IFN-α, EBV encodes the soluble form of the CSF-1 receptor BARF1, which neutralizes the effects of the host CSF, leading to the reduction of the IFN-secretion of EBV infected mononuclear cells. EBV BZLF1 counteracts intrinsic effector molecules in a variety of ways [62]. First, BZLF1 decreases TNF5–007 and IFNΔ receptors to minimize cellular susceptibility to these cytokines; second, BZLF1 induces SOCS3-signaling cytokine suppressor, which inhibits JAK/STAT signaling and thus promotes IFN-responsiveness Type I state; Third, BZLF1 triggers TGFβ immunosuppressive cytokine expression and disrupts the development of Promyelocytic leukemia bodies (PML-bodies) that may have antiviral activity [66].

1.2.8.2 Evasion of adaptive immunity

The EBV compromises the activation of both CD8 + and CD4 + T cells by interfering in different stages of HLA Class I and Class II antigen presentation pathways, especially during the productive phase of infection [67].

1.2.8.2.1 Evasion of CD8+ T Lymphocytes

The EBV encodes at least three proteins that independently interfere with antigen presentation through deregulation of the surface expression of HLA I in many ways to prevent EBV-specific (memory) T cell recognition [68]. BGLF5 induces degradation of HLA I-encoding mRNA and reduces the presence of peptide at the cell surface which inhibits T-cell recognition. Ut has been suggested that BNLF2adeplete peptides from the ER (HLA I loading compartment) and inhibits the importation of peptides by the antigen- transporter (TAP). BILF1, encoding a constitutively active G protein-coupled receptors (GPCR) which reduce the transportation of HLA I from the trans-Golgi network. In addition, cell surface turnover is increased and subsequently i degradation by lysosomal proteases. These proteins are expressed during the replicative process of EBV and function in tandem with the prevention of CD8 + T cells being recognized [67].

1.2.8.2.2 Evasion of CD4+ T Lymphocytes

The EBV has adopted several strategies for immune evasion that interfere with CD4 + T-cell immunity. The EBV receptor Gp42 can bind to the B-cell HLA class II molecules. The HLA Class II/peptide complex relationship blocks T-cell receptor (TCR)—class II interactions and prevents CD4 + T cell activation. Besides, protein GP42/gH/gLdecreases the HLA II cell surface expression by the HLA II mRNA degradation. Inhibiting the activities of CIITA promoters and, as a result, lowering the HLA II surface levels, EBV also encodes a viral IL-10 homolog (BCRF1) that has been identified as impairing the IFNβ signal. The IL-10 is an anti-inflammatory cytokine that can inhibit CD4 + priming and effector functions and modulates them; BCRF1 was suggested to inhibit CD4 + T-cell antiviral response similar to IL-10 [69].

1.2.8.2.3 Immune evasion during latency

The EBV severely restricts latent infection viral protein expression to prevent host immune recognition. Different latency forms represent different stages from primary B cell infection to the transformation of the growth. Thus, in latency III cells, EBNA1inhibits its translation and proteasomal degradation. This strategy ensures adequate levels of EBNA1 to preserve the viral genome while decreasing the turnover of proteins to minimize the appearance of viral antigens to CD8 + T cells. LMP1 and 2 mediate NF-κB activation and decrease the TLR9 surface expression and accelerate the turnover of IFN receptors, resulting in a decrease in the incidence of IFN receptors. During latency II, the expression is limited to EBNA1 and LMP1, and 2. Latency I only contains an expression of EBNA1, and latency 0 occurs without any expression of EBV protein [70]. The EBV encodes different types of non-coding RNAs, including two EBV-encoded small RNAs (EBERs) that inhibit PKR activity and miRNAs that de-regulate T-cell attracting CXCL-11 chemokines and de-regulates T-bet and IFNΔ transcriptional regulator [71].

1.2.9 EBV prevention and vaccine

In the vast majority of individuals, EBV is a harmless passenger, controlled easily by immune defenses, but in some individuals, EBV drives a broad range of diseases that can cause significant morbidity and mortality.

A vaccine is currently unavailable. A prophylactic vaccine which prevents acute disease, the most beneficial using the humoral immune approach, vaccines expressing the major viral envelop protein, gp350 have been developed. Most recently live recombinant vaccinia vaccine expressing gp350 protected two- thirds of the vaccinated infants [20]. as Also, therapeutic vaccines are investigated. These vaccines are based on direct peptide immunization approaches. The use of immunodominant HLA Class I and II epitopes of LMP1, LMP2, and EBNA1 may induce a strong and sustained T-cell response, which was demonstrated with some success primarily in reactivating CD4+ and CD8+ cell in vitro [72, 73].

The use of antiviral therapy in EBV infection is limited. Antiviral therapy can be used as preemptive therapy of PTLD in EBV- organ transplant recipients. These agents can block EBV replication in donor B-cells and infection of recipient B-cells. Prophylactic intravenous ganciclovir after liver transplantation lead to decreasing of PTLD incidence, which may be due to a reduction in the number of latently-infected B-lymphocytes [32].

1.3 Human Cytomegalovirus (HCMV)

Human Cytomegalovirus (HCMV) or human herpesvirus 5 (HHV-5) [74], is ubiquitous in human populations and was first isolated and cultivated in 1956 [56]. The HCMV derives its name from the Greek cyto-, "cell", and -megalo-, "large", because of the enlargement of virus infected cells, (cytomegaly), [75]. Human cytomegalovirus is a leading - cause of congenital viral infections with numerous consequences such as birth defects including intrauterine growth restriction, still-birth, low birth weight, preterm birth, microcephaly, neurodevelopmental delay, cerebral palsy, hematological disorders, pneumonitis, blindness, and sensorineural hearing loss [76]. HCMV infection is typically clinically silent in immunocompetent hosts, with few mild symptoms like fever, myalgia andcervical lymphadenopathy. Individuals with weakened immunity – neonates, diagnosed with HIV/AIDS, and those on long-term immunosuppressive treatments, such as transplant recipients – HCMV infection often results in clinically severe diseases. The worst outcomes, including mortality and long-term morbidity, are with congenital infection [56].

1.3.1 The virion and genome organization

A complete virus particle consists of a DNA core with a protein coat and external envelopes representing the extracellular infective form of a virus called virion [77]. The average size of viral particle varies between 200–300 nanometers and has linear double-stranded (235-kb) DNA genome which is enclosed by an icosahedral capsid. The capsid's diameter is (110–130 nm) and made of 162 capsomeres (12 pentons and 150 hexons).. HCMV has three kinds of capsids: A capsid (only capsid shell), B capsid (capsid shell and assembled proteins), and C capsid (a mature capsid containing the viral genome). These three capsids represented in three various stages of capsid maturation that takes place in the nucleus of infected cells capsid is composed of at least 5 proteins, i- Major Capsid Protein (UL86), ii- Minor Capsid Protein (UL85), iii- Smallest Capsid Protein (SCP, UL48–49), iv- Assembly protein (Fragments of UL80) and v- Minor Capsid Binding Protein (MCP, UL46) [78].

The widest layer inside the virion structure is the tegument layer that closely surrounds the capsid; anchoring the envelope to the tégumented capsid is believed to be essential. The layer of tegument is composed of several proteins like pp65/ppUL83, pp71/ppUL82, pp150/pUL32, and pp28/pUL99 which play main roles during the entry of virus (un-coating), intracellular capsid transportation and assembly [79]. The tegument is covered by a lipid bilayer called the envelope that keeps the entire virion intact. It interacts with the host cell membrane in target cells and thus plays a significant role in the attachment and entry of viruses. Lipid bilayer envelope is derived from cellular ER or endosomal membranes and associated with 23 viral glycoproteins. The viral glycoproteins gpUL55 (gB), gpUL73 (gN), gpUL74 (gO), gpUL75 (gH), UL100 (gM), gpUL115 (gL) and the pentameric complex consist of gL, gH and UL128–131, are known to play crucial roles in viral entry, cell-to-cell spread and virion maturation [80] (**Figure 6**).

Among herpesviruses, HCMV has the largest genome consisting of a linear dsDNA ranging from 220 to 240 kbp and comprising up to 250 opening reading frames (ORFs) [81]. The herpesvirus genomes are categorized into A-F sections depending on the organization of the genome segments (**Figure 7**). The genome of

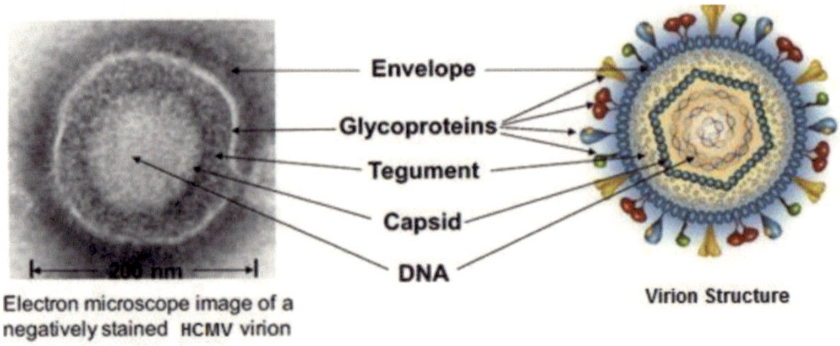

Figure 6.
Structure of an HCMV Virion.

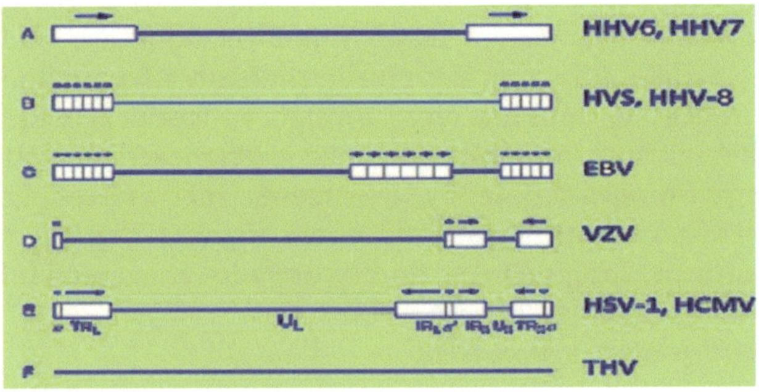

Figure 7.
The six classes of herpesvirus genomes.

HCMV is classified as an E genome. The sequencing analysis of the HCMV genome has shown that it has a very complex structure. Generally, the genome is organized into two parts: The single long regions (UL) and unique small regions (US) flanked by terminal repeats (TR) and internal repeats (IR), UL-area ORFs and US-region ORFs are classified according to their location [82] (**Figure 8**). More than 70 viral proteins have been identified from the purified virions [83]. Only 50 proteins are important for viral replication, while the vast majority of HCMV proteins are involved in host immunomodulation via their interference with cellular signals [84]. HCMV encodes for at least 4 long polyad-encoded RNAs and 26 microRNAs which have major functions during host-virus and virus replication interactions [85].

1.3.2 Life cycle of human Cytomegalovirus

1.3.2.1 Entry

Cytomegalovirus virus (CMV) enters host cells through membrane fusion as shown in **Figure 9**. The viral entry involves the binding of viral glycoproteins on the surface of the viral lipid envelope and the specific receptors on the external membrane of the host cell [86]. This initial interaction makes the cell susceptible to further interactions that fuse the membranes and eventually disassemble and release the viral genomic DNA into the host cell. Many tegument proteins are thought to mediate the delivery of the DNA-containing nucleocapsid to the nuclear pore complex and the release of the viral DNA into the nucleus [87].

HCMV-gB mediates attachment to cells via binding to cellular receptors that include heparan sulfate proteoglycan, integrins, and epidermal growth factor receptor (EGFR) to promote the entrance process [9]. The heparan sulfate proteoglycan

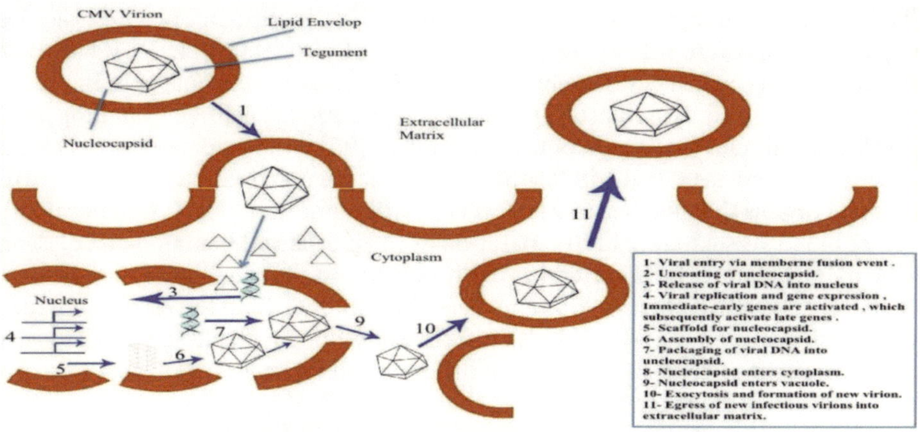

Figure 8.
Illustration of the CMV life cycle from viral entry to egress of new infectious virions.

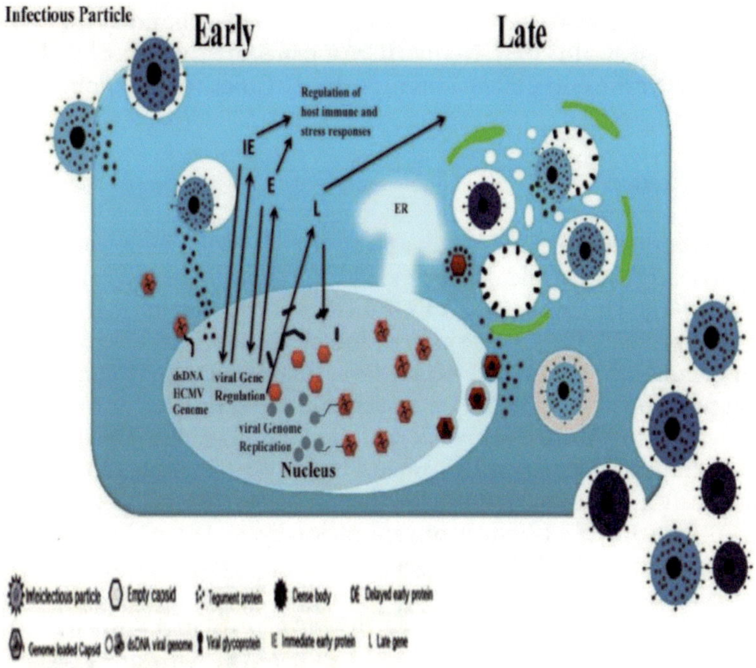

Figure 9.
Overview of the human cytomegalovirus life cycle.

molecule permits the HCMV particle to attach closer to the cell membrane where viral glycoproteins bind to more specific receptors. This is mediated by gB and gH or their complexes, and many other viral protein complexes may also mediate the process of entry; a homodimer of gB known as gC-I, and a heterodimer composed of gM and gN which form gC-II and a heterotrimer of gH, gL and gO which form the gC-III complex. HCMV has two different routes of entry in different cell types [88]. Virus entry in fibroblasts is mediated by viral envelope glycoprotein complexes gH/gL-gB and gH/gL/gO via direct fusion at the plasma membrane. Aminopeptidase N (CD13) and Annexin-II may also serve as receptors to promote entry at the plasma membrane. The entry of CMV into other relevant cell types, such as endothelial cells, follows an endocytic route, clathrin is the major constituent of coated vesicles and plays an important role in the endocytic entry of viruses [89]. HCMV has a wide cell tropism and can infect different cell types [34], such as neuronal cells in the brain and retina [90], fibroblasts, epithelial and endothelial cells (EC) in lung and gastrointestinal tract [91], hepatocytes in the liver [90] and peripheral blood mononuclear lymphocytes (PBML) [92]. Infected monocytes can release infectious viruses into target organs through tissue-macrophage differentiation [93]. Efficient infection of EC of blood vessel may lead to separation of these cells and hematogenic dissemination may, therefore, be initiated [94]. Fibroblasts, on the other hand, are likely to contribute to the efficient development of HCMV and will help the establishment of the primary infection. HCMV can also infect various kinds of cells in vitro. However, the replication of viruses varies between different cell types. In fibroblasts, smooth muscle cells, endothelial cells, and epithelial cells, HCMV inducesproductive infection, while in poorly differentiated cells, such as myeloid-linear progenitor cells, viral replication is limited [91]. Fibroblasts are the most commonly used cells for HCMV cultivation in the laboratory because HCMV binds to fibroblasts with the efficiency of 2500–3000 particles per cell [94].

HCMV strains show broad variations in the relative pentamer and trimer rates of viral glycoproteins incorporated into virions, which correspond to the cell tropism differences between strains [93]. Many HCMV genes are capable of influencing viral cell tropism at the entry stage and most likely function by composite effects of gH/gL complexes.

1.3.2.2 Lytic replication

HCMV infection of the cell leads to an active replication with the production of new viral particles that are released by exocytosis of the infected cells, known as the lytic phase [95]. The replication of viral DNA is dependent on the expression of certain viral proteins [96]. Gene expression of viral DNA takes place in three stages: immediate early genes (IE) expression, followed by early genes (E) and late genes (L) expressions [95]. The IE gene products act as transcription factors as well as trans-activators to regulate expression of the E and L genes. Two predominant nuclear phosphoproteins, IE1-p72 (Immediate Early 1 – Protein 72KD) and IE2-p86 (Immediate Early 2- Protein 86KD) have the key roles., IE1-p72 is expressed from the UL123 ORF during the IE phase of replication cycle. IE1-p72 transactivates the promoters of numerous HCMV early genes including gene products that facilitate the replication process. It also interacts with the p107 protein through a domain at the N-terminus of IE1-p72 and increase the p107- mediated repression of E2F promoters leading to the inhibition of p107- mediated growth suppression [97]. Therefore, it appears that IE1-p72 can induce E2F activity. The expression of IE1-p72 can promote S phase entry only in cells lacking p53 or p21 [96]. Controversial to that, IE1- p72 expression causes wild-type cells to arrest, most likely in G1 due to increased levels of p53 protein, which results in a p53-dependent induction of p21 expression and subsequent growth arrest [96, 97]. IE2-p86 is expressed from the UL122 ORF during the IE phase of the replication cycle and is essential for HCMV replication [98]. The protein IE2-p86 specifically interacts with pRb through more than one domain and induce pRb mediated repression of E2F promoters. The IE2-p86 disruption of pRb-E2F complexes enable E2Fs transactivation of its target genes [99]. However, IE2-p86 induces cells to enter S phase, an effect that could be attributed in part to IE2-p86 transactivation of the cyclin E promoter and induction of E2F activity [100].

Phosphoprotein 65 (pp65) is a tegument phosphoprotein that exhibits kinase activity. This protein may affect the activity of a specific subset of cytotoxic

T –Lymphocytes (CTLs) by modification of IE1- p72, possibly through its phosphorylation and may interfere with its processing and/or degradation [82].

Following peak expression of IE regulatory proteins, early genes (E genes) become transcriptionally active (**Figure 9**). These proteins regulate replication process of HCMV DNA, such as a DNA-polymerase (UL54) and DNA primase (UL70) which sustain an efficient production of new viral progeny [89]. The L proteins, which are mainly structural components, are essential for virion assembly and egress. HCMV genome contains a cis-acting lytic origin of DNA replication (OriLyt) element to initiate bidirectional DNA replication (theta form of replication), followed by a rolling circle mode of replicationof viral DNA molecules and their incorporation into new virus particles [80]. The entire replication cycle for HCMV takes approximately 72 h and the mature new virions infect the new cells either by their release from the infected cells or by the spread via cell-to-cell mechanisms [80]. During latency, only selected IE gene transcription and translation of viral proteins occur and when conditions are favorable, the virus may be reactivated into lytic replication and new viral progeny occurs [101].

1.3.2.3 Virus assembly

The newly synthetized DNA is inserted into an immature B capsid after the pre-capsid assembly stage and becomes a fully mature C capsid. This DNA packed capsid, egresses through the nuclear membrane from the nucleus, through an envelope and de-envelopment cycle [102]. The mechanism of assembly of tegument proteins is still unclear. It has been suggested that tegument proteins are added to nucleocapsids sequentially starting in the nucleus and continuing in the cytoplasm, which provide stability during nucleocapsid translocation from the nucleus to the cytoplasm.. The final envelopment of tegumented particle occurs at ER/endosomal membranes. By transporting Rab3 secretory vesicles, mature particles are released by fusion of the vacuole with the plasma membrane and shed out by exocytosis [103] (**Figure 9**).

1.3.2.4 Latency and reactivation

The establishment of latency is one of the major biological characteristics of herpesviruses. Primary HCMV infection is often asymptomatic in a healthy person (immune-competent host) and leads to latent and recurrent infection [104].

The MIE gene acts as a transactivator for transcribing the majority of encoded HCMV genes and is necessary for the virus replication and the lytic process of infection. During latency, cellular factors transcripts control the MIE promoter. These factors also suppress the chromatin around the MIE gene, which prevents the lytic cycle stage and is a part of preserving the latent stage of the infection [105].

People who have had an organ or bone marrow transplant and those with AIDS can develop serious illness caused by CMV. Typically, latent virus from a previous infection (the primary CMV infection may have occurred many years earlier) becomes active again because the person's immune system is compromised [106].

To establish viral latency and maintenance, the latency- gene product UL138 is required. HCMV-LUNA and UL138 are generated during HCMV latency and lytic infection and activate CD4 + T cells, resulting in the development of IL-10 and IFN-π; one with immunosuppressive effects and one with immune-activating effects that facilitate the replication and reactivation of latent HCMV. The produced IFN-β can promote macrophage differentiation, which can lead to reactivation and enhanced HCMV replication. UL138 also downgrades protein-1, MRP, which can lead to lower exports of C4 cell leukotriene, preventing DCs from entering lymph nodes and impairing an HCMV-specific immune response [105].

The expression of the UL 111a gene, which encodes a functional IL-10 homolog with strong immunosuppressive effects, also offers HCMV strategy in viral latency to suppress the immune system. Through the latent infection process, UL111a undergoes alternative splicing, which results in the expression of a latency related transcript cmvIL-10, and the production of a protein that mimics the function of human immunosuppressive cytokine IL-10. This favors the infected cells not to be recognized by the immune system and to avoid clearance. Also, US28 and UL144 are expressed during latency redirect the immune response or block the immune recognition [107].

Latent HCMV infection also modulate the cell expression of MHC molecules class I in order to evade the immune response. At least four proteins encoded by (US) region of HCMV genome involved in inhibition of MHC class I expression, either by directly acting on MHC class I moleculeor acting on MHC class I-associated proteins, such as TAP (transporter associated with antigen processing) and tapas in [101]. These loci of HCMV DNA encodes several distinct IE proteins, pUL36, pUL37, and pUL37_1, which appear to stop engagement of the apoptotic signals associated with the tumor necrosis factor (TNF) family of receptors including Fas. As a result, the activation of Fas-associated death domain (FADD) is inhibited, which prevents activation of procaspase-8 FLICE (FADD-like interleukin-1 beta-converting enzyme) and ultimately prevent the active caspase-8 to cause the subsequent activation of downstream effector caspases and prevent apoptosis of the cells [107, 108].

1.3.3 Transmission

The spreading of HCMV from one person to another primarily occurs through infected bodily secretions such as saliva, blood, tears, milk, and urine [109]. Close or intimate person to person contact usually is necessary for viral transmission. Accordingly, sexual transmission has been implicated in the spread of CMV. Seropositivity is higher among persons with multiple sexual partners and histories of sexually transmitted diseases [110].

Cytomegalovirus (CMV) may be transmited from mother to infant before, during or after birth. During pregnancy, vertical CMV transmission occurs via either passage of virions from maternal blood to the fetus and subsequent infection of the placenta or via the entry of infected maternal leukocytes, endometrial, or cervical cells into the fetal circulation. The infection ascending from the genital tract may also be possible. During birth, the infection of neonate occurs via the contact with infected maternal genital secretions. After birth, breastfeeding is the most frequent route of CMV transmission to the neonate [111]. Cytomegalovirus transmission also occurs naturally after receipt of an organ containing latent virus where these transplant recipients (who are undergoing immunosuppressive treatment) are at risk of contracting the disease [112].

1.3.4 CMV clinical features

1.3.4.1 CMV infections in the immunocompetent host

Cytomegalovirus (CMV) infection is common among patients of all age groups, but it has traditionally been considered as a problem in neonatal and immunosupressed patients. Cytomegalovirus infection in immunocompetent patients is usually asymptomatic or subclinical. Symptomatic disease usually results in mononucleosis-like syndrome. The symptoms are similar to classic mononucleosis, caused by the Epstein–Barr Virus. However, the mononucleosis syndrome associated with CMV typically lacks signs of enlarged cervical lymph nodes and splenomegaly [113]. Rarely, CMV infection in immunocompetent persons is associated with severe diseases such as enteritis, thrombotic disease, hemolytic anemia, thrombocytopenia, encephalitis, Guillain–Barré syndrome and myocarditis. Guillain–Barré syndrome is immune-mediated peripheral neuropathy characterized by neuromuscular paralysis. The first case of Guillain-Barré Syndrome associated with cytomegalovirus (CMV) infection was reported in a renal transplant recipient in 1970 [114].

1.3.4.2 CMV infections in the immunocompromised patients

CMV infection is one of the most important infectious complications of solid-organ transplantation and is responsible for serious, life-threatening diseases in patients infected with human immunodeficiency virus (HIV) and other immunodeficiencies [115]. CMV disease manifests in the vast majority of transplant recipients as a viral syndrome that includes fever, malaise, myalgia, or headache (sometimes called *CMV syndrome*) or more severe organ-specific diseases such as pneumonitis, gastrointestinal lesions, hepatitis, retinitis, pancreatitis, myocarditis, and rarely, encephalitis or peripheral neuropathy. In solid organ transplant (SOT) recipients, primary HCMV infection has been consistently linked with dysfunction of the transplanted organ. In HIV-infected patients, retinitis is the single most common manifestation of CMV disease, accounting for 85% of all cases [116]. In developing countries, CMV retinitis is still the most frequent cause of visual loss in HIV-infected patients [115].

1.3.4.3 Congenital infection and diseases

Congenital infection refers to a condition where cytomegalovirus is transmitted in the prenatal period. Worldwide, approximately 1 in 100 to 500 babies are born with congenital CMV. Approximately 1 in 3000 will show symptoms and 1 in 7000 will die [117]. Congenital HCMV infection occurs after primary infection (or reactivation) during pregnancy. Congenital infections are less common in poorer communities with high seropositivity of, women in childbearing age.. In industrialized countries, up to 8% of HCMV seronegative mothers acquire primary HCMV infection during pregnancy, of which roughly 50% will transmit CMV to the fetus [118]. Between 10–15% of infected fetuses are born with the symptoms of congenital CMV disease, [119] which may include pneumonia, gastrointestinal, retinal, and neurological diseases [120].

1.3.5 Epidemiology

Human CMV is an ancient virus that is ubiquitous in human populations, reaching a prevalence of 100% in Africa and Asia, and approximately 80% in Europe and the USA, depending on socioeconomic status [121]. Seroprevalence rates of HCMV vary depending on age (higher rates are observed among older persons), geography (higher rates in developing countries), and socioeconomic status (higher rates in economically depressed regions) **Figure 10** [122].

A comparison of literature estimates shows that congenital CMV-related disabilities are as common among newborns and children as other better known diseases such as Down syndrome, fetal alcohol syndrome, or spina bifida [123]. Of the approximately 30,000 United States babies born with an infection with CMV every year almost 20 percent are born with or experience permanent sequelae such as hearing loss, eyesight loss, brain damage or cognitive impairment.

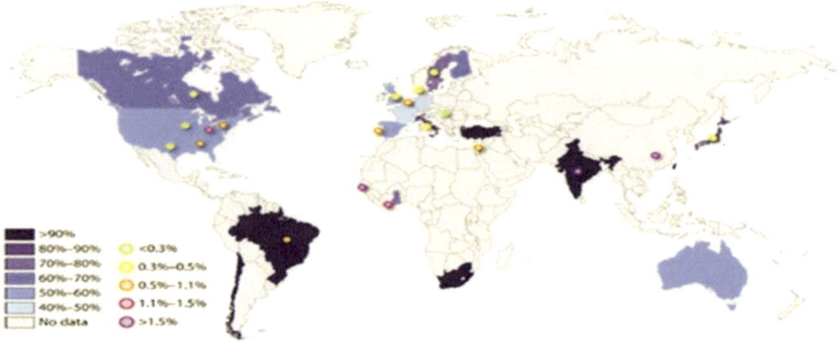

Figure 10.
Globally, rates of CMV seroprevalence in women of reproductive age and the incidence of congenital CMV infection. Studies have been performed in Australia, Belgium, Brazil, Canada, Chile, England, Finland, France, Germany, Ghana, India, Israel, Italy, Japan, South Africa, Spain, Sweden, Taiwan, Turkey, and the USA.

The contribution of congenital CMV infection to childhood hearing loss is of particular importance, with approximately 20% of moderate to profound bilateral sensorineural hearing impairment occurring due to congenital CMV infection [124]. Approximately 90% of newborns with congenital infection have no symptoms at birth; If symptoms arc present, they are often nonspecific [125]. When disabilities such as hearing loss appear, often months or years later, it is usually too late to make a retrospective diagnosis that identifies congenital CMV infection as the cause. For an individual woman, the greatest risk of having a baby with congenital infection comes from the mother's primary infection during pregnancy [126]. Consequently, babies of women who are CMV negative prior to pregnancy are particularly vulnerable to poor outcomes if the mother becomes infected during pregnancy [127].

In a nationally representative survey it has been stated that between 30% and 50% of United States women under 45 years of age are seronegative for CMV and that as many as a half a million US women of childbearing age experience primary CMV infections every year [128].

1.3.6 Immune evasion by Cytomegalovirus

HCMV persistence is correlated with the interaction between the immune response of the host and the virus evasion mechanisms, where HCMV interferes with both adaptive immune responses and immune effectors. A variety of evasion strategies have been developed by HCMV to prevent selected dendritic cell functions. The differentiation of CMV infected monocyte into macrophages and CD1 a-positive Dendritic cells (DCs) is inhibited, which does not require viral replication [129].

1.3.6.1 Evasion of Innate immunity

The innate immune system is essential in driving an efficacious acquired immune response. This includes the induction of type I interferons, activation of professional antigen presenting cells, and recruitment and activation of NK cells which themselves promote more efficient activation of antigen presenting cells and T cells. The binding and entry of HCMV into the cell initiates several pathways leading to the upregulation of NFkB and interferon regulatory factor 3 (IRF3) which can ultimately lead to the production of type I interferons and certain inflammatory cytokines. This innate cellular response to the initial stages of infection is mediated by Toll like receptor 2 (TLR2) signaling, which has been shown to recognize the viral surface glycoproteins gB and gH [130].

1.3.6.1.1 Natural Killer (NK) cells and immune evasion mechanisms

The importance of NK in the innate immune response to HCMV is suggested by the extensive studies of mechanism of HCMV prevention of the activation of NK cells. NK cells are inhibited by signals sent via inhibitory receptors that interact with class I MHC molecules on the surface of target cell. Low surface levels of Class I MHC on HCMV-infected cells may also reduce the inhibitory signaling of NK. This could make the infected cells susceptible to NK cell cytotoxicity [131]. Two mechanisms describing HCMV-mediated inhibitory receptor signaling have been reported. Firstly, HCMV uses the host HLA-E pathway to inhibit NK cells via the CD94/NKG2 heterodimeric inhibitory receptor by promoting cell surface HLA-E expression. Viral UL40 protein contains a nonomeric peptide which binds HLA-E promoting its cell surface expression. Secondly, HCMV expresses a viral homologue of cellular MHC Class I, UL18. UL18 is trafficked to the cell surface where it binds to the inhibitory NK cell receptor, LILRB1 (LIR-1) with higher affinity than MHC Class I inhibiting LILRB1+ NK cell activation. HCMV encodes five genes controlling NK cell activation and cytotoxicity by the provision of inhibitory signals and suppression of activating signals. The pp65 tegument protein (UL83) dissociates CD3_ signaling from NKp30, whilst intracellular retention of CD155 and CD112 by UL141 prevents activation of NK cells via receptors CD226 and CD96, the remaining viral proteins interfere with a major receptor on all human NK cells (NKG2D) that mediated NK cell activation [132].

1.3.6.2 Evasion of adaptive immunity

Primary infection of HCMV provokes the production of antibodies specific for many HCMV proteins including structural tegument proteins such as (pp65 and pp150), envelope glycoproteins predominantly (gB and gH) as well as non-structural proteins such as the Immediate Early 1 protein (IE1, UL123).

1.3.6.2.1 CD8+ T cell responses and MHC Class I downregulation

In human bone marrow transplantation studies where HCMV infection can cause significant morbidity, it was evident that there was a strong correlation between recovery of the CD8+ T cell population and protection from CMV disease. HCMV employs several mechanisms to interfere with the normal cellular MHC Class I processing and presentation pathways to prevent CD8+ T cell recognition. HCMV encodes at least four related glycoproteins, each with a unique mechanism to prevent antigen presentation. HCMV viral genes US2 and US11 degrade newly synthesized MHC class I heavy chains, US3 retains MHC-I in the endoplasmic reticulum by interfering with chaperone-controlled peptide loading and US6 inhibits the translocation of viral and host peptides across the endoplasmic reticulum membrane by the dedicated peptide transporter TAP (transporter associated with antigen processing) [133]. The action of these genes may not completely protect cells from CD8+ T cell recognition dependent on the T cell-antigen specificity. Whereas HCMV-infected cells expressing US2–11 prevent any presentation of IE antigen to human T cells, pp65 peptides were still presented. CMV-specific T cells are also marked by the lack of expression of the costimulatory receptors CD27 and CD28, which are otherwise commonly expressed on naïve T cells [134].

1.3.6.2.2 CD4+ T cell responses and MHC Class II downregulation

During primary infection, HCMV specific CD4+ T cells can be detected 7 days after the detection of HCMV DNA in peripheral blood in response to same ORFs asCD8 + T cells and pp65 and IE genes gB and gH. These cells produce T helper type 1 (Th1) cytokines IFNγand TNF. A large number of HCMV- encoded gene products target the MHC class I antigen presentation pathway in an effort to avoid recognition by CD8+ T cell [134].

Human cytomegalovirus also avoids the CD4+ T cell response by several ways. Disrupting IFN- induced, upregulation of MHC class II molecules to the cell surfaces by preventing the expression of Janus kinase 1 and suppression of Class II transactivator mRNA. The virally-encoded gene product of US2 also suppress MHC class II presentation to CD4+ T cells by redirecting the HLA-DR-and HLA-DM-chains to the cytosol where they are degraded. HCMV re-programs human hematopoietic progenitor cells (HPCs) into immune-suppressive monocytes that express IL-10 in a process requiring US28. Recently a truncated transcript to UL111A, a viral homolog of the immunomodulatory cytokine IL-10, which is expressed during latency (cmvLA IL-10) has been shown to downregulate expression of MHC class I and class II molecules, inhibit both proliferation of mononuclear cells and the production of inflammatory cytokines [135].

1.3.7 CMV prevention and vaccines

The lack of specific and effective treatments for HCMV infection has highlighted the need to understand HCMV host-cell interactions, including viral entry and host immune responses against this virus [136]. The HCMV vaccine is designed to be used to prevent infection or to prevent its re-activation in those infected already [137]. To eliminate and eradicate congenital HCMV infection, vaccination would be a priority. Several researchers have attempted to develop vaccines against HCMV, such as live attenuated vaccines, recombinant chimeric vaccines, and subunit vaccines such as glycoprotein B and tegument protein pp65 vaccines [138]. However, none of these vaccines exhibited effective protection, as well as, and to this day, no licensed HCMV vaccine is available.

Strategies to reduce Congenital CMV disease burden may be implemented at different stages, and include prevention of maternal infection, prevention of mother-to-child transmission, early detection and intervention by neonatal screening, and neonatal antiviral therapy. The principal sources of exposure for women of child-bearing age are sexual contacts and children secreting the virus [136]. Pregnant women also should be advised to avoid close contact with individuals likely to shed CMV such as adults with symptoms consistent with mononucleosis and toddlers attending group day care. Preemptive therapy is an approach in which patients are monitored for early replication (ie, viremia by polymerase chain reaction [PCR]), and the antiviral drug is administered only when CMV replication is detected to prevent its progression to higher-grade viremia and CMV disease. Small, noncomparative studies have shown lower rates of CMV disease (especially delayed-onset disease) with preemptive compared with the use of antiviral prophylaxis in CMV-seronegative liver transplant recipients with seropositive donors [139].

Author details

Marwa Mohammed Ali Jassim[1], Majid Mohammed Mahmood[2]*
and Murtada Hafedh Hussein[3]

1 Studies and Planning Department, University of Baghdad, Baghdad, Iraq

2 College of Science, Mustansiriyah University, Baghdad, Iraq

3 College of Science, Thi-Qar University, Thi-Qar, Iraq

*Address all correspondence to: majidmahmood93@yahoo.com

References

[1] Palser, A. L., Grayson, N. E., White, R. E., Corton, C., Correia, S., Watson, S. J., ... & Rickinson, A. B. (2015). Genome diversity of Epstein-Barr virus from multiple tumour types and normal infection. Journal of virology, 89(10), 5222-5237.

[2] Kliszczewska, E., Jarzyński, A., Boguszewska, A., Pasternak, J., & Polz-Dacewicz, M. (2017). Epstein-Barr Virus–pathogenesis, latency and cancers. Journal of Pre-Clinical and Clinical Research, 11(2), 142-146.

[3] Lei, H., Li, T., Li, B., Tsai, S., Biggar, R. J., Nkrumah, F., ... & Bhatia, K. (2015). Epstein-Barr virus from Burkitt Lymphoma biopsies from Africa and South America share novel LMP-1 promoter and gene variations. Scientific reports, 5 (1), 16706-16711.

[4] Epstein, M. A. (1964). Virus particles in cultured lymphoblasts from Burkitt's lymphoma. Lancet, 1 (28), 702-703.

[5] Tsao, S. W., Tsang, C. M., & Lo, K. W. (2017). Epstein–Barr virus infection and nasopharyngeal carcinoma. Philosophical Transactions of the Royal Society B: Biological Sciences, 372(1732), 216-270.

[6] Harrison, S.C. (2015). Viral membrane fusion. Virology, 479 (43), 498-507.

[7] Jassim, M.M.A.; Mahmood, M.M.; Ali, S. H. (2019). Localization of Herpetic Viruses Patterns in Sinonasal and Nasopharyngeal Malignant and Benign Tumors. Online Journal of Otolaryngology and Rhinology, 1(5):1-6.

[8] Halldorsson, S., Li, S., Li, M., Harlos, K., Bowden, T. A., & Huiskonen, J. T. (2018). Shielding and activation of a viral membrane fusion protein. Nature Communications, 9(1), 349-354.

[9] Tomaszewski-Flick, M. J. (2008). Functional Analysis and Characterization of Epstein Barr Virus Latent Membrane Protein 2b (Doctoral dissertation, University of Pittsburgh).

[10] Correia, P.J.B. (2017): Mechanism of superior B cell transformation by type 1 Epstein-Barr virus. (Doctoral dissertation, University of Imperial College London).

[11] Chen, H. L., Peng, J., Zhu, X. B., Gao, J., Xue, J. L., Wang, M. W., & Xia, H. S. (2010). Detection of EBV in nasopharyngeal carcinoma by quantum dot fluorescent in situ hybridization. Experimental and molecular pathology, 89(3), 367-371.

[12] Weiss, E. R., Lamers, S. L., Henderson, J. L., Melnikov, A., Somasundaran, M., Garber, M., ... & Luzuriaga, K. (2018). Early Epstein-Barr virus genomic diversity and convergence toward the B95. 8 genome in primary infection. Journal of virology, 92(2), 1466-1475.

[13] Agnieszka, S. (2016). Genetic analysis of the role of Epstein-Barr virus nuclear antigen leader protein (EBNA-LP) in B cell transformation. (Doctoral dissertation, University of Imperial College London).

[14] Yoshida, M., Watanabe, T., Narita, Y., Sato, Y., Goshima, F., Kimura, H., & Murata, T. (2017). The Epstein-Barr virus BRRF1 gene is dispensable for viral replication in HEK293 cells and transformation. Scientific Reports, 7(1), 6044-6048.

[15] Chesnokova, L.S.; Hutt-Fletcher, L.M. (2014). Epstein-Barr virus infection mechanisms. Chin. J. Cancer. 33:545-548.

[16] Chesnokova, L. S., & Hutt-Fletcher, L. M. (2014). Epstein-Barr virus infection mechanisms. Chinese journal of cancer, 33(11), 545-551.

[17] Tsai, M.-H.; Raykova, A.; Klinke, O.; Bernhardt, K. K. Delecluse, H. J. (2013). Spontaneous lytic replication and epitheliotropism define an Epstein-Barr virus strain found in carcinomas. Cell Rep. 5:458-470.

[18] Neuhierl, B.; Feederle, R. W.; Hammerschmidt, H.J. (2002). Glycoprotein gp110 of Epstein-Barr virus determines viral tropism and efficiency of infection. Proc. Natl. Acad. Sci. U. S. A. 99:15036-15041.

[19] Murata, T., & Tsurumi, T. (2014). Switching of EBV cycles between latent and lytic states. Reviews in medical virology, 24(3), 142-153.

[20] Bazot, Q., Paschos, K., & Allday, M. J. (2018). Epstein-Barr virus latent protein EBNA3A directly targets and silences the kinase STK39 in B cells infected by EBV. Journal of virology, 92(7). 1918-1927.

[21] Styles, C.T.; Bazot, Q. Parker, G.A.; White, R.E.; Paschos, K.; Allday, M.J. (2017). EBV epigenetically suppresses the B cell-to-plasma cell differentiation pathway while establishing long-term latency. PLoS Biol, 15:e2001992.

[22] Portal, D.; Zhou, H.; Zhao, B.; Kieff, E. (2013). Epstein-Barr virus nuclear antigen leader protein localizes to promoters and enhancers with cell transcription factors and EBNA2. Proc Natl Acad Sci U S A, 110:18537-18542.

[23] Kalchschmidt , J.S.; Gillman, A.C.; Paschos, K.;Bazot, Q.; Kempkes, B. (2016). EBNA3C directs recruitment of RBPJ (CBF1) to chromatin during the process of gene repression in EBV infected B cells. PLoS Pathog, 12:e1005383.

[24] Moss, W. N., & Steitz, J. A. (2013). Genome-wide analyses of Epstein-Barr virus reveal conserved RNA structures and a novel stable intronic sequence RNA. BMC genomics, 14(1), 543-551.

[25] Le, J., Durand, C. M., Agha, I., & Brennan, D. C. (2017). Epstein–Barr virus and renal transplantation. Transplantation Reviews, 31(1), 55-60.

[26] Teow, S. Y., Liew, K., Khoo, A. S. B., & Peh, S. C. (2017). Pathogenic role of exosomes in epstein-barr virus (ebv)-associated cancers. International Journal Biological Science, 13(10), 1276-1281.

[27] Navari, M., Etebari, M., Ibrahimi, M., Leoncini, L., & Piccaluga, P. (2018). Pathobiologic roles of Epstein–Barr virus-encoded MicroRNAs in human lymphomas. International journal of molecular sciences, 19(4), 1168-1173.

[28] Tyring, S., Moore, A. Y., & Lupi, O. (Eds.). (2016). Mucocutaneous Manifestations of Viral Diseases: An Illustrated Guide to Diagnosis and Management. *CRC press*, p, 132-149.

[29] Abbott, R. J., Pachnio, A., Pedroza-Pacheco, I., Leese, A. M., Begum, J., Long, H. M., ... & Borrow, P. (2017). Asymptomatic primary infection with Epstein-Barr virus: observations on young adult cases. Journal of virology, 91(21), 382-393.

[30] Din, S. S. (2017). Frequency of Epstein–Barr Virus Genotypes in Pakistani Trangender SexWorkers. *In* Open forum infectious diseases, 4, (1), 313-319.

[31] Rochford, R. (2016). Epstein-Barr virus infection of infants: implications of early age of infection on viral control and risk for Burkitt lymphoma. Boletín Médico Del Hospital Infantil de México (English Edition), 73(1), 41-46.

[32] Salehiniya, H., Mohammadian, M., Mohammadian-Hafshejani, A., & Mahdavifar, N. (2018). Nasopharyngeal Cancer in the World: Epidemiology, Incidence, Mortality and Risk Factors. World Cancer Research Journal, 5(1).45-53.

[33] Rainey, J.J.;Mwanda, W.O.; Wairiumu, P (2007). Spatial

distribution of Burkitt's lymphoma in Kenya and association with malaria risk. Trop Med Int Health. 12(8):936-943.

[34] Balfour, H.H.; Odumade, O.A.; Schmeling, D.O.; Hogquist, K.A. (2013). Behavioral, virologic, and immunologic factors associated with acquisition and severity of primary epstein-barr virus infection in university students. J. Infect. Dis. 207:80-88.

[35] Chijioke, O., A.; Müller, R.; Feederle, M.H.M.; Barros, C.; Krieg, V. ; Antsiferova, V.; Landtwing, W.; Münz, C. (2013). Human natural killer cells prevent infectious mononucleosis features by targeting lytic epstein-barr virus infection. Cell Rep. 5:1489-1498.

[36] Acharya, S., Ekalaksananan, T., Vatanasapt, P., Loyha, K., Phusingha, P., Promthet, S., ... & Pientong, C. (2015). Association of Epstein-Barr virus infection with oral squamous cell carcinoma in a case–control study. Journal of Oral Pathology & Medicine, 44(4), 252-257.

[37] Sand, L., & Jalouli, J. (2014). Viruses and oral cancer. Is there a link?. Microbes and infection, 16(5), 371-378.

[38] Kikuchi, K., Inoue, H., Miyazaki, Y., Ide, F., Kojima, M., & Kusama, K. (2017). Epstein–Barr virus (EBV)-associated epithelial and non-epithelial lesions of the oral cavity. Japanese Dental Science Review, 53(3), 95-109.

[39] Jalouli, J., Ibrahim, S. O., Sapkota, D., Jalouli, M. M., Vasstrand, E. N., Hirsch, J. M., & Larsson, P. A. (2010). Presence of human papilloma virus, herpes simplex virus and Epstein–Barr virus DNA in oral biopsies from Sudanese patients with regard to toombak use. Journal of oral pathology & medicine, 39(8), 599-604.

[40] Jalouli, J., Jalouli, M. M., Sapkota, D., Ibrahim, S. O., Larsson, P. A., & Sand, L. (2012). Human papilloma virus, herpes simplex virus and epstein barr virus in oral squamous cell carcinoma from eight different countries. Anticancer research, 32(2), 571-580.

[41] Mimi, C. Y., & Yuan, J. M. (2002). Epidemiology of nasopharyngeal carcinoma. *In* Seminars in cancer biology, 12 (6):421-429.

[42] Nawaz, I., Moumad, K., Martorelli, D., Ennaji, M. M., Zhou, X., Zhang, Z., ... & Hu, L. F. (2015). Detection of nasopharyngeal carcinoma in Morocco (North Africa) using a multiplex methylation-specific PCR biomarker assay. Clinical epigenetics, 7(1), 89-96.

[43] Edris, A., Mohamed, M. A., Mohamed, N. S., & Siddig, E. E. (2016). Molecular detection of Epstein-Barr virus in nasopharyngeal carcinoma among Sudanese population. Infectious agents and cancer, 11(1), 55-61.

[44] Ali Jassim, M.M.; Mahmood, M.M.; Mohammed Ali, S.H.(2019). Intervene of EBV and HCMV in Relation to p27 and p73-Cell Regulatory Proteins in Histopathological Changes of Sinonasal and Nasopharyngeal Carcinomas. Virology & Immunology Journal,3(3):1-7.

[45] Jha, H. C., Banerjee, S., & Robertson, E. S. (2016). The role of gammaherpesviruses in cancer pathogenesis. Pathogens, 5(1), 18-29.

[46] Draborg, A.H.; Duus, K.; Houen. G. (2013). Epstein-Barr virus in systemic autoimmune diseases. Clin. Dev. Immunol. 2013:535738.

[47] White, R.E.; Ramer, P.C.; Naresh, K.N.; Meixlsperger, S.; Pinaud, L.; Rooney, C.; Allday M.J. (2012). EBNA3B-deficient EBV promotes B cell lymphomagenesis in humanized mice and is found in human tumors. J Clin Invest, 122:1487-1502.

[48] Pei, Y.; Banerjee, S.; Jha, H.C.; Sun, Z,; (2017). An essential EBV latent

antigen 3C binds Bcl6 for targeted degradation and cell proliferation. PLoS Pathog, 13:e1006500.

[49] Pei, Y.; Banerjee, S.; Sun, Z.; Jha, H.C.; Saha, A.; Robertson, E.S. (2016). EBV nuclear antigen 3C mediates regulation of E2F6 to inhibit E2F1 transcription and promote cell proliferation. PLoS Pathog, 12:e1005844.

[50] Skalska, L.; White, R.E.; Parker, G.A.;S inclair, A.J.; Paschos, K.; Allday, M.J.(2013). Induction of p16 (INK4a) is the major barrier to proliferation when Epstein–Barr virus (EBV) transforms primary B cells into lymphoblastoid cell lines. PLoS Pathogens 9: e1003187.

[51] Lu, F.; Chen, H.S.; Kossenkov, A.V.; DeWispeleare, K.; (2016). EBNA2 drives formation of new chromosome binding sites and target genes for B cell master regulatory transcription factors RBPjkappa and EBF1. PLoS Pathog, 12:e1005339.

[52] Carbone, A.; Volpi, C.C.; Gualeni, A.V.; Gloghini,A.(2017). Epstein-Barr virus associated lymphomas in people with HIV. Curr Opin HIV AIDS, 12:39-46.

[53] Abhik, S. Erle, S. (2019): Mechanisms of B-Cell Oncogenesis Induced by Epstein-Barr Virus. Journal of Virology, 93(13): e00238–e00219.

[54] Zhao, B.; Barrera, L.A.; Ersing, I.; Willox, B.; Gewurz, B.E. (2014). The NF-kappaB genomic landscape in lymphoblastoid B cells. Cell Rep, 8:1595-1606.

[55] Hurwitz, S.N.; Cheerathodi, M.R.; Nkosi, D.; Meckes, D.G.(2018). Tetraspanin CD63 bridges autophagic and endosomal processes to regulate exosomal secretion and intracellular signaling of Epstein-Barr virus LMP1. J. Virol., 92.

[56] Chen, H.S.; Martin, K.A.; Lu, F.; Tempera I. (2014). Epigenetic deregulation of the LMP1/LMP2 locus of Epstein-Barr virus by mutation of a single CTCF-cohesin binding site. J. Virol, 88:1703-1713.

[57] Wu, R,; Sattarzadeh, A.; Rutgers, B.; Diepstra,A. (2016). The microenvironment of classical Hodgkin lymphoma: heterogeneity by Epstein-Barr virus presence and location within the tumor. Blood Cancer J., 6:e417.

[58] Skalsky, R.L.; Cullen, B.R. (2015). EBV noncoding RNAs. Curr Top Microbiol Immunol , 391:181-217.

[59] Larsson, A.(2017). The need for research infrastructures: a narrative review of large-scale research infrastructures in biobanking. Biopreserv Biobank , 15: 375-383.

[60] Wirtz, T.; Weber, T.; Kracker, S.; Yasuda, T. (2016). Mouse model for acute Epstein-Barr virus infection. Proc Natl Acad Sci U S A, 113:13821-13826.

[61] Ariza, M.E.; Rivailler, P.; Glaser, R.; Chen, M.; Williams, M.V. (2013). Epstein-Barr virus encoded dUTPase containing exosomes modulate innate and adaptive immune responses in human dendritic cells and peripheral blood mononuclear cells. PLoS ONE, 8:e69827

[62] Takeuchi, O.; Akira, S. (2009). Innate immunity to virus infection. *Immunol Rev* ., 227:75-86.

[63] van Gent, M,; Gram, A,M,; Boer, I. (2015) Silencing the shutoff protein of Epstein-Barr virus in productively infected B cells points to (innate) targets for immune evasion. *J Gen Virol* ., 96(4):858-865

[64] Dunmire, S.K.; Odumade, O.A.; Porter, J.L.; Balfour, H.H. (2014) Primary EBV infection induces an expression profile distinct from other viruses but similar to hemophagocytic syndromes. PLoS One, 9:e85422

[65] van Gent, M.; Braem, S.G.E.; de Jong, A.; Wiertz, E.J. (2014) Epstein-Barr virus large tegument protein BPLF1 contributes to innate immune evasion through interference with toll-like receptor signaling. PLoS Pathog , 10:e1003960

[66] Ohashi, M.; Fogg, M.H.; Orlova, N.; Quink, C.; Wang, F. (2012). An Epstein-Barr virus encoded inhibitor of colony stimulating factor-1 signaling is an important determinant for acute and persistent EBV infection. PLoS Pathog , 8:e1003095.

[67] Chaigne-Delalande, B.; Li, F.Y.; O'Connor, G.M.; Lukacs, M.J. (2013). Mg2+ regulates cytotoxic functions of NK and CD8 T cells in chronic EBV infection through NKG2D. Science, 341:186-191

[68] Quinn, L.L.; Zuo, J.; Abbott, R.J.M.; Rowe, M. (2014). Cooperation between Epstein-Barr virus immune evasion proteins spreads protection from CD8+ T cell recognition across all three phases of the Lytic cycle. PLoS Pathog , 10:e1004322

[69] Zuo, J.; Thomas, W.A.; Haigh, T.A.; Fitzsimmons, L.; Long, H.M.; Hislop, A.D.; Taylor, G.S.;Rowe, M. (2011) Epstein-Barr virus evades CD4+ T cell responses in lytic cycle through BZLF1-mediated downregulation of CD74 and the cooperation of vBcl-2. PLoS Pathog 7:e1002455

[70] Huang, W.T.; Lin, C.W. (2014). EBV-encoded miR-BART20-5p and miR-BART8 inhibit the IFNgamma-STAT1 pathway associated with disease progression in nasal NK-cell lymphoma. Am *J. Pathol*, 184:1185-1197.

[71] Ambrosio, M.R.; Navari, M.; Gomez, G. (2014).The Epstein Barr-encoded BART-6-3p microRNA affects regulation of cell growth and immuno response in Burkitt lymphoma. Infect Agent Cancer, 9:12

[72] Ruiss, R., Jochum, S., Wanner, G., Reisbach, G., Hammerschmidt, W., & Zeidler, R. (2011). A VLP-based Epstein-Barr Virus vaccine. *Journal of virology*, 5598-5603.

[73] Cohen, J. I. (2015). Epstein–barr virus vaccines. Clinical & translational immunology, 4(1), 32-39.

[74] Sackman, A. M., Pfeifer, S. P., Kowalik, T. F., & Jensen, J. D. (2018). On the Demographic and Selective Forces Shaping Patterns of Human Cytomegalovirus Variation within Hosts. Journal Pathology, 7(1), 16-23.

[75] Ryan, K.J. and Ray, C.G. (Eds). (2004) Sherris medical microbiology. 4th ed. Chapter 37, McGraw Hill, Pp: 551-552.

[76] Wallace, P. R., Janet, W. H., Samuel, W., Horace, C. T., & Robert, J. H. (1956). Cytopathogenic agent resembling human salivary gland virus recovered from tissue cultures of human adenoids. Proceedings of the Society for Experimental Biology and Medicine, 92(2), 418-424.

[77] Gibson, W. (2008). Structure and Formation of the Cytomegalovirus Virion. In *Human Cytomegalovirus*, pp. 187-204. Edited by T. Shenk & M. Stinski: Springer Berlin Heidelberg.

[78] Tomtishen, J. P. (2012). Human cytomegalovirus tegument proteins (pp65, pp71, pp150, pp28). Virol J 9.

[79] Ryckman, B. J., Rainish, B. L., Chase, M. C., Borton, J. A., Nelson, J. A., Jarvis, M. A. & Johnson, D. C. (2008). Characterization of the human cytomegalovirus gH/gL/UL128-131 complex that mediates entry into epithelial and endothelial cells. J Virol 82, 60-70.

[80] Ali, M. M.; Mahmood, M. M.; Ali, S. H. M. (2019). Co-Molecular Roles of Human Cytomegalovirus and p73

gene in late events of Nasopharyngeal and Sinonasal carncinogenesis. Indian Journal of Forensic Medicine & Toxicology, *13*(3), 473-479.

[81] Jiang, X. J., Zhang, J., Xiong, Y., Jahn, G., Xiong, H. R., Yang, Z. Q., & Liu, Y. Y. (2017). Human cytomegalovirus glycoprotein polymorphisms and increasing viral load in AIDS patients. PloS one, 12(5), e0176160.

[82] Hage, E., Wilkie, G. S., Linnenweber-Held, S., Dhingra, A., Suárez, N. M., Schmidt, J. J., ... & Schulz, T. F. (2017). Characterization of human cytomegalovirus genome diversity in immunocompromised hosts by whole-genome sequencing directly from clinical specimens. The Journal of infectious diseases, 215(11), 1673-1683.

[83] Corcoran, K., Sherrod, C. J., Perkowski, E. F., Texier, J., Li, F., Wang, I. M., ... & Dittmer, D. P. (2017). Genome sequences of diverse human cytomegalovirus strains with utility in drug screening and vaccine evaluation. Genome announcements, 5(3), 1433-1436.

[84] Davison, A. J., Eberle, R., Ehlers, B., Hayward, G. S., McGeoch, D. J., Minson, A. C., & Thiry, E. (2009). The order herpesvirales. Archives of virology, 154(1), 171-177.

[85] Ostermann, E., Spohn, M., Indenbirken, D., & Brune, W. (2016). Complete genome sequence of a human cytomegalovirus strain AD169 bacterial artificial chromosome clone. Genome announcements, 4(2), 91-116.

[86] Kabanova, A., Marcandalli, J., Zhou, T., Bianchi, S., Baxa, U., Tsybovsky, Y., Lilleri, D., Silacci-Fregni, C., Foglierini, M., Fernandez-Rodriguez, B. M., Druz, A., Zhang, B., Geiger, R., Pagani, M., Sallusto, F., Kwong, P. D., Corti, D., Lanzavecchia, A. & Perez, L. (2016). Platelet-derived growth factor-alpha receptor is the cellular receptor for human cytomegalovirus gHgLgO trimer. *Nat Microbiol*.

[87] Krummenacher, C., Carfí, A., Eisenberg, R. J. & Cohen, G. H. (2013). Entry of herpesviruses into cells: the enigma variations. In *Viral Entry into Host Cells*, pp. 178-195: Springer.

[88] Krishna, B. A., Lau, B., Jackson, S. E., Wills, M. R., Sinclair, J. H., & Poole, E. (2016). Transient activation of human cytomegalovirus lytic gene expression during latency allows cytotoxic T cell killing of latently infected cells. Scientific reports, 19(6), 24674.24681.

[89] Van Damme, E., & Van Loock, M. (2014). Functional annotation of human cytomegalovirus gene products: an update. Frontiers in microbiology, 19(5), 218-224.

[90] Revello, M.G. ; Gerna , G.(2010). Human cytomegalovirus tropism for endothelial/epithelial cells:scientific background and clinical implications. Rev Med Virol, 20(3): 136-155.

[91] Calo, S.; Cortese, M.; Ciferri, C.; Bruno, L.; Gerrein, R.; Benucci, B.; Monda, G.; Gentile, M.; Kessler, T.; Uematsu, Y.(2016).. The Human Cytomegalovirus UL116 Gene Encodes an Envelope Glycoprotein Forming a Complex with gH Independently from gL. J. Virol. , 90, 4926-4938.

[92] Si, Z.; Zhang, J.; Shivakoti, S.; Atanasov, I.; Tao, C.L.; Hui, W.H.; Zhou, K.; Yu, X.; Li, W.; Luo, M.(2018). Different functional states of fusion protein gB revealed on human cytomegalovirus by cryo electron tomography with Volta phase plate. PLOS Pathog. 14, e1007452.

[93] Zhou, M.; Lanchy, J.M.; Ryckman, B.J. (2015). Human Cytomegalovirus gH/gL/gO Promotes the Fusion Step of Entry into All Cell Types, whereas gH/

gL/UL128-131 Broadens Virus Tropism through a Distinct Mechanism. J. Virol. 89, 8999-9009.

[94] Zhou, M.; Yu, Q.;Wechsler, A.; Ryckman, B.J.(2013). Comparative analysis of gO isoforms reveals that strains of human cytomegalovirus differ in the ratio of gH/gL/gO and gH/gL/UL128-131 in the virion envelope. J. Virol., 87, 9680-9690.

[95] Arend, K. C., Lenarcic, E. M., & Moorman, N. J. (2018). The 5′ Untranslated Region of the Major Immediate Early mRNA is Necessary for Efficient Human Cytomegalovirus Replication. Journal of virology, 14(7), 92-99.

[96] Boward, B., Wu, T., & Dalton, S. (2016). Concise review: control of cell fate through cell cycle and pluripotency networks. Stem cells, 34(6), 1427-1436.

[97] Jassim, M. M. A.; Mahmood, M. M.; Ali, S. H. M.; Kamal, M. S. (2019). Interplay between EBERS and P27 Tumor Suppressor Proteins in Molecular Transformation of Nasopharyngeal and Sinonasal Carcinomas. Indian Journal of Public Health Research & Development, *10*(6), 894-900.

[98] Kim, J. E., Kim, Y. E., Stinski, M. F., Ahn, J. H., & Song, Y. J. (2018). Corrigendum: Human Cytomegalovirus IE2 86 kDa Protein Induces STING Degradation and Inhibits cGAMP-Mediated IFN-β Induction. Frontiers in Microbiology, 26 (9), 350-356.

[99] Biolatti, M., Gugliesi, F., Dell'Oste, V. & Landolfo, S. (2018). Modulation of the innate immune response by human cytomegalovirus. Infection Genetic Evolution, 64 (9), 105-114.

[100] Hansen, T.H., & Bouvier, M.(2009). MHC class I antigen presentation: learning from viral evasion strategies. National Review Immunology, 9(7),503-513.

[101] Procter, D. J., Banerjee, A., Nukui, M., Kruse, K., Gaponenko, V., Murphy, E. A., ... & Walsh, D. (2018). The HCMV Assembly Compartment is a dynamic Golgi-derived MTOC that controls nuclear rotation and virus spread. Developmental cell, 45(1), 83-100.

[102] Nagarakanti, S., Bishburg, E., & Bapat, A. (2018). Adenovirus, herpes simplex virus and cytomegalovirus infection in a lung transplant recipient. ID Cases, 2(11), 91-93.

[103] Kamtchum-Tatuene, J., Al-Bayati, Z., Mwandumba, H. C., Solomon, T., Christmas, S. E., & Benjamin, L. A. (2018). Serum concentration of anti-Cytomegalovirus IgG and ischaemic stroke in patients with advanced HIV infection in Malawi. PloS one, 13(11), e0208040.

[104] Jean Beltran, P. M., & Cristea, I. M. (2014). The life cycle and pathogenesis of human cytomegalovirus infection: lessons from proteomics. Expert review of proteomics, 11(6), 697-711.

[105] Weekes, M. P.; Tan, S. Y.; Poole, E.; Talbot, S.; Antrobus, R.; Smith, D. L.; Montag, C.; Gygi, S. P.; Sinclair, J. H. ; Lehner, P. J. (2013). Latency-associated degradation of the MRP1 drug transporter during latent human cytomegalovirus infection. Science. 340, 199-202.

[106] Poole, E.; Walther, A.; Raven, K.; Benedict, C. A.; Mason, G. M. ; Sinclair, J. (2013). The myeloid transcription factor GATA-2 regulates the viral UL144 gene during human cytomegalovirus latency in an isolate-specific manner. Journal of virology 87, 4261-4271.

[107] Roy, S., Liu, F., & Arav-Boger, R. (2015). Human Cytomegalovirus Inhibits the PARsylation Activity of Tankyrase—A Potential Strategy for Suppression of the Want Pathway. Viruses, 8(1), 8-16.

[108] McStay, G. P., & Green, D. R. (2014). Measuring apoptosis: caspase

inhibitors and activity assays. *Cold Spring Harbor Protocols*, (8) 1, 799-806.

[109] Smith, T. & revello, R., (2017). Herpesviruses. In J. Cohen, W.G. Powderly, & S.M. Opal, eds. *Infectious Diseases. Elsevier*, Pp. 1

[110] Burrell, C.J., Howard, C.R., & Murphy, F.A. (2017). Herpesviruses, In Fenner & White, eds. Medical Virology. *Academic Press*, Pp. 237-261.

[111] Itell, H. L., Nelson, C. S., Martinez, D. R., & Permar, S. R. (2017). Maternal immune correlates of protection against placental transmission of cytomegalovirus. Placenta, 60(1), S73-S79.

[112] Azevedo, L., Pierrotti, L., Abdala, E., Costa, S., Strabelli, T., Campos, S., Ramos, J.F., Latif, A.Z., Litvinov, N., Maluf, N.Z., Caiaffa Filho, H.H., Pannuti, C.S., Lopes, M.H., Santos, V.A., Linardi, C.C., Yasuda, M.A., & Marques, H. (2015). Cytomegalovirus infection in transplant recipients. Clinics, 70(7), 515-523.

[113] Drachman, D. A.; Paterson, P. Y.;Berlin, B. S.; Roguska, J.(1970). Immunosuppression and the Guillain-Barré Syndrome. JAMA Neurology, 23(5):385-393.

[114] Navarro, D. (2016). "Expanding role of cytomegalovirus as a human pathogen". Journal of Medical Virology, 88 (7): 1103-1112.

[115] Richardson, A. K.; Currie, M. J.; Robinson, B. A.; Morrin, H.; Phung, Y.; Pearson, J. F.;Walker, L. C. (2015). Cytomegalovirus and Epstein-Barr virus in breast cancer. PloS one, 10(2), e0118989.

[116] Hume, A. J.; Finkel, J. S.; Kamil, J. P.; Coen, D. M.; Culbertson, M. R.; Kalejta, R. F. (2008). Phosphorylation of retinoblastoma protein by viral protein with cyclin-dependent kinase function. Science, 320(5877), 797-799.

[117] Pasquereau, S.; Al Moussawi, F.; Karam, W.; Assaf, M. D.; Kumar, A.; Herbein, G. (2017). Cytomegalovirus, macrophages and breast cancer. The open virology journal, 11(1), 15-27.

[118] Herbein, G. (2018). The Human Cytomegalovirus, from Oncomodulation to Oncogenesis. Viruses. 10:E408. 10.3390/v10080408.

[119] Barry, S.; Zuckerman, A. J.; Banatvala, J. E.; Griffiths, P. E. (2004). "Chapter 2C Cytomegalovirus". Principles and Practice of Clinical Virology. *Chichester: John Wiley & Sons*. 85-122.

[120] Vancíková, Z.; Dvorák, P. (2001). "Cytomegalovirus infection in immunocompetent and immunocompromised individuals--a review". Curr. *Drug Targets Immune Endocr. Metabol. Disord*. 1 (2): 179-187.

[121] Cannon, M,J. (2009). Congenital cytomegalovirus (CMV) epidemiology and awareness. J. Clin. Virol. 46 (4): S6 –S10.

[122] Essays, UK. (November 2018). CMV Infection Epidemiology and Pathogenesis. Retrieved from https://www.ukessays.com/essays/sciences/cmv-infection-epidemiology-9342.php?vref=1

[123] Hyde, T.B.;' Schmid, D.S.; Cannon, M.J. (2010). Cytomegalovirus seroconversion rates and risk factors: implications for congenital CMV. Rev. Med. Virol. 20:311-326.

[124] Grosse, S.D.; Ross, D.S.; Dollard, S.C.(2008). Congenital cytomegalovirus (CMV) infection as a cause of permanent bilateral hearing loss: a quantitative assessment. J Clin Virol., 41(2):57-62.

[125] Noyola, D.E.; Demmler, G.J.; Williamson, W.D.; Griesser, C.; Sellers, S.; Llorente, A. (2000). Cytomegalovirus urinary excretion and long term outcome in children with congenital cytomegalovirus infection. Congenital

CMV Longitudinal Study Group. *Pediatr Infect. Dis. J.*, 19(6):505-510

[126] Kenneson, A. Cannon, M.J. (2007).Review and meta-analysis of the epidemiology of congenital cytomegalovirus (CMV) infection. *Rev, Med. Virol.*, 17(4):253-76.

[127] Stockdale, L., Nash, S., Nalwoga, A., Painter, H., Asiki, G., Fletcher, H., & Newton, R. (2018). Human cytomegalovirus epidemiology and relationship to tuberculosis and cardiovascular disease risk factors in a rural Ugandan cohort. PloS one, 13(2), e0192086.

[128] Ljungman, P., Boeckh, M., Hirsch, H. H., Josephson, F., Lundgren, J., Nichols, G., (2017). Definitions of cytomegalovirus infection and disease in transplant patients for use in clinical trials. *Clinical Infectious Diseases*, 1, 64(1),87-91-99.

[129] Gredmark, S.; Soderberg-Naucler, C. (2003). Human cytomegalovirus inhibits differentiation of monocytes into dendritic cells with the consequence of depressed immunological functions. J. Virol. 77(20), 10943-10956 .

[130] Isaacson, M.K., Juckem, L.K., Compton, T. (2008). Virus entry and innate immune activation. Curr. Top. *Microbiol. Immunol*. 325, 85-100.

[131] Prod'homme, V., Sugrue, D.M., Stanton, R.J., Nomoto, A., Davies, J., Rickards, C.R., Cochrane, D., Moore, M., Wilkinson, G.W., Tomasec, P. (2010). Human cytomegalovirus UL141 promotes efficient downregulation of the natural killer cell activating ligand CD112. J. Gen. Virol.

[132] Ashiru, O., Bennett, N.J., Boyle, L.H., Thomas, M., Trowsdale, J., Wills, M.R.(2009). NKG2D ligand MICA is retained in the cis-Golgi apparatus by human cytomegalovirus protein UL142. J. *Virol*. 83 (23), 12345-12354.

[133] Osman, D., J. Mullberg, C. L. Sutherland, W. Chin, R. Armitage, W. Fanslow, M. (2001). ULBPs, novel MHC class I-related molecules, bind to CMV glycoprotein UL16 and stimulate NK cytotoxicity through the NKG2D receptor. Immunity 14:123-133.

[134] Day, E.K., Carmichael, A.J., Ten Berge, I.J., Waller, E.C., Sissons, J.G., Wills, M.R. (2007). Rapid CD8+ T cell repertoire focusing and selection of high-affinity clones into memory following primary infection with a persistent human virus: human cytomegalovirus. J. *Immunol*. 179 (5), 3203-3213.

[135] Gamadia, L.E., Remmerswaal, E.B., Weel, J.F., Bemelman, F., van Lier, R.A., Ten Berge, I.J. (2003). Primary immune responses to human CMV: a critical role for IFNgamma- producing CD4+ T cells in protection against CMV disease. Blood , 101 (7), 2686-2692.

[136] Staley, A.B. & Suresh, B.B.(2018). Vaccination against the human cytomegalovirus. Vaccine, 2(89),1-6.

[137] Verma, Y., Gupta, E., Kumar, N., Hasnain, N., Bhadoria, A. S., Pamecha, V., & Khanna, R. (2018). Earlier and higher rates of cytomegalovirus infection in pediatric liver transplant recipients as compared to adults: An observational study. Journal of laboratory physicians, 10(2), 221-230.

[138] John, S., Yuzhakov, O., Woods, A., Deterling, J., Hassett, K., Shaw, C. A., & Ciaramella, G. (2018). Multi-antigenic human cytomegalovirus mRNA vaccines that elicit potent humoral and cell-mediated immunity. Vaccine, 36(12), 1689-1699.

[139] Nina Singh, M.D.; Drew. J.; Winston, M.D; et al. (2020). Effect of preemptive therapy vs antiviral prophylaxis on cytomegalovirus disease in seronegative liver transplant recipients with seropositive donors: A randomized clinical trial. JAMA., 323:1378.

Chapter 7

Plasticity in Interferon Responses Modulates T-Cell Immunity in Parasitic Infections: Periphery to Thymus

Lovlesh Thakur, Nadeem Akhtar, Aklank Jain, Hridayesh Parkash and Manju Jain

Abstract

Parasitic infections are the major threat prevalent in tropical and subtropical regions throughout the world. Different parasitic infections take a huge toll on mortality and morbidity at global level. Different parasites invade the host system, multiply inside host cells of their choice and sabotage defense mechanisms to overpower the host. T-cell immunity is majorly affected in different parasitic diseases such that the peripheral T-cell immune response is altered along with lesser explored thymic changes. Direct and/or indirect effect of parasitic infection leads to alterations in T-cell development, differentiation and activation resulting in deregulated T-cell immune mechanisms. Cytokines of interferon family play a significant role in determining the disease outcome and severity. Therefore, in this chapter, we here provide a detailed overview of the functional role played by IFNs during parasitic diseases in terms of their influence on peripheral T-cell activation and tolerance along with lesser explored impact on developing T cells in the thymus with altered microenvironmental niches.

Keywords: parasitic diseases, periphery, IFN, T cells, thymus, immunomodulation, disease outcome

1. Introduction

Parasitism is a relationship among species, in which one organism, the parasite, sustains on the host organism. Parasitic diseases can affect almost all living organisms. Parasites are dependent on the host organisms for their own survival. Not all parasites are harmful but some cause severe pathology to the host, such as *Leishmania*, *Plasmodium*, *Trypanosoma*, etc. Parasites known to affect humans are divided into three classes: protozoans, helminths and ectoparasites [1]. Parasite invasion triggers the innate, inflammatory and adaptive immune responses inside the mammalian host. Innate immunity recognizes the non-self and activates the T-cell–mediated adaptive immune system in order to eliminate the invader. Removal or recruitment of parasite is dependent on the production of distinct pattern of cytokines from specific T cells. T cells are formed through an intricate developmental process with the dynamic stage-specific changes in the developing lymphocytes. T-cell development takes place in multiple steps originating from bone marrow to progenitors of T cell maturation in the thymus. It has been known that "thymus" plays the main role in the production of a self-tolerant adaptive immune response that is critical against the pathogen's threat [2]. A variety of infectious agents like protozoans mainly *Trypanosoma* spp., *Plasmodium* spp. and *Leishmania* spp. alter thymic structure and function. Thymic atrophy reflected by lymphocyte depletion is considered as a common feature in response to pathogens but the consequences on thymic function may differ significantly in different infections. Together with structural and functional changes induced by the parasite in the thymic microenvironmental niches, the development of thymocytes and thus the altered thymic output have direct implications in peripheral T-cell response. T-cell–based immune responses are further modulated via different types of cytokines viz. interferons (IFNs), tumor necrosis factors (TNFs) and interleukins (ILs) with implication in the disease outcome and progression. Different cytokines work independently or in collaboration as determinants of the disease establishment and progression. Host peripheral immune response influencing the disease outcome during parasitic infection is substantially studied *in vitro* on cell lines, *in vivo* in experimental models and in human subjects. A heterogeneous T-cell response marks the disease pathogenesis. A Th1 cell-mediated immune response is predominated by pro-inflammatory cytokines such as IFN-γ and TNF-α and plays a key role in arresting the disease by limiting the parasite replication. Contrary to this, a shift toward Th2 immune response, represented by increased expression of the anti-inflammatory IL-4, is associated with exacerbation of infection and uncontrolled parasite replication. This response is due to suppressive effects of Th2 cytokines on Th1 immunity. The important role played by the crucial IFN family of cytokines during parasitic diseases is emphasized in peripheral circulation as well as with regard to thymus-centric modulation of T-cell–based immunity in different infectious diseases.

2. A brief note on interferons

IFNs are the key soluble immune molecules belonging to the IFN family with specific structural and functional characteristics. They are divided into three main groups based on the structural details and functional contribution toward modulating the immune response during parasitic infections: IFN-I, II and III. The IFN-I family includes IFN-α and IFN-β. IFN-I signaling is mediated through a common cell surface receptor, (IFNAR). IFN-I production by a wide variety of cells mediates autocrine and paracrine signaling pathways upon viral infections. The IFN-II family represents IFN-γ. Its response is mediated by IFN-γ receptor (IFNGR). IFN-II plays a role in defense against intracellular pathogens by modulating diverse cellular functions. The third IFN-III family, or IFN-λ, comprises four different subtypes: IFN-λ1, λ2, λ3 and λ4. IFN-III is not well studied but has a role similar to IFN-I. The expression of IFN-λR receptor is mainly restricted to cells of epithelial origin [3].

3. Protozoan diseases

Protozoan parasitic infections are among the most common life-threatening infectious diseases. They can enter into the human body generally by a bite from an insect vector or through fecal-oral route. Protozoan parasites are responsible for serious infections. *Plasmodium falciparum* (*P. falciparum*), *Toxoplasma gondii*

(T. gondii), *Leishmania donovani (L. donovani)*, *Trypanosoma cruzi* (*T. cruzi)*, *Trypanosoma brucei* (*T. brucei*) and *Giardia intestinalis* (*G. intestinalis*) are among the most common protozoan pathogenic parasites and cause malaria, toxoplasmosis, leishmaniasis, Chagas disease, sleeping sickness and giardiasis, respectively. The three pathogenic parasitic diseases viz. Chagas disease caused by *T. cruzi*, malaria caused by *Plasmodium* spp. and leishmaniasis caused by *Leishmania spp*. will be discussed in length in relation to discrete T-cell–associated quantitative and qualitative alterations, reported in all these protozoan diseases. Disease-specific cytokine milieu with distinct role of IFNs is implicated in modulating disease progression and outcome. A snapshot of IFN-associated T-cell immune modulation in context to each of the protozoan disease is discussed in subsequent sections.

4. Chagas disease caused by *T. cruzi*

Chagas disease or American trypanosomiasis is caused by the parasite *T. cruzi*, transmitted to mammals by insect vectors. It is a hemoflagellate protozoan belonging to the kingdom Protista, phylum euglenozoa and class Zoomastigophora. It is a multi-host parasite transmitted by insect triatomines and is also called as "assassin bug" or "kissing bug." It is common in parts of Mexico and Central and South America [4].

4.1 Disease transmission

Transmission of *T. cruzi* parasite to humans is understood based on the route of infection as primary or secondary. The primary route is the most frequent route, and infection occurs through insect bite, blood transfusion or congenital and oral route. The secondary route is less frequent such as accidental infection during animal handling or infected organ transplant. The most common transmission route in the Brazilian region is oral transmission and the second is through contaminated food/beverage, whereas in Argentina, Bolivia, Colombia, Ecuador, French Guinea and Venezuela, contaminated food consumption is the main reason of infection [4].

4.2 Disease severity and diagnosis

Chagas disease has two phases of infection: an acute and a chronic phase. Acute form is mild. The parasite remains in the blood circulation for a long time (few weeks to months). Acute phase is followed by prolonged asymptomatic "chronic phase," marked by very few or negligible parasite in blood. Chronic Chagas disease symptoms include dilated colon or esophagus and different heart rhythm abnormalities. Diagnosis of acute phase of infection is marked by the presence of parasite in peripheral blood circulation and can be observed by microscopic examination of stained blood smear. Diagnosis of chronic Chagas disease is generally made by testing blood for parasite-specific antibodies [4].

4.3 Peripheral T-cell response in *T. cruzi* infection

Chagas disease is associated with several immunological alterations due to change in the expression pattern of cytokines that play a fundamental role in regulating the functionality of almost all cell types. *T. cruzi* infection triggers the nitric oxide (NO) production and may exert protective or toxic effects on the host immune system. NO can induce oxidative stress via damaging the host tissues. Inducible NO synthase pathway gets activated upon parasitic invasion, produces NO and is highly responsible for macrophage-mediated intracellular *T. cruzi* elimination within infected cells. The cytokines such as IFN-γ, TNF-α and chemokines are produced in large amounts during *T. cruzi* acute infection and are potent inducers of NOS [5]. Along with NO synthase (NOS), several potent effector mechanisms such as T-cell–mediated immunity involving both $CD4^+$ and $CD8^+$ T-cell compartments are essentially involved in defense against *T. cruzi* invasion and replication in mammalian host. Relevance of T cells is well documented in experimental *T. cruzi* mice model where mice lacking T-cell subsets develop disease with high parasite load in tissues and periphery [6, 7]. These findings support the important role of T-cell populations in dealing with acute and chronic phase of *T. cruzi* infection in humans. Double positive (DP) $CD4^+CD8^+$ T-cell population was found to be increased in number with increased expression of activation markers (CD38 and HLA-DR) during chronic Chagas disease demonstrating that these DP T cells contribute to immune response against *T. cruzi* infection [8]. Cardiac inflammatory infiltrate of DP T cells in patients who have undergone cardiac transplant suggests that their performance in controlling the cardiac disease in humans is worth considering [8].

4.4 IFN-I–associated immune changes in *T. cruzi* infection

IFN-I has an important role in inhibiting the parasite multiplication. Induction/production of IFN-I in response to *T. cruzi* is stage specific [9] and primarily dependent on the dose/amount of parasite, and route of infection. Exogenous IFN-I treatment in *T. cruzi*–infected mice showed that mice develop increased resistance to infection by stimulating natural killer and T-cell activities [10]. Protective action of IFN-I associated with LRG-47 [IFN-inducible p47GTPase] is well documented in experimental mice models. LRG-47 regulates host resistance against intracellular pathogens in a comparative study with wild-type (WT) and knockout (KO) mice, where LRG-47 KO mice exhibit severe anemia, thrombocytopenia and atrophy of thymus, in contrast to WT counterparts. Similar to *in vivo* model, IFN-I–induced *in vitro* stimulation of LRG-47 KO macrophages also display a defect in intracellular killing of amastigotes [11]. IFN-I is reported to play a dual role during disease: protection from disease and establishing pathology. Disease-exacerbating role of IFN-I has been reported in WT and $IFNAR^{-/-}$ mice model such that $IFNAR^{-/-}$ mice were able to restrict the parasite growth and survive, while the WT mice failed to resist the infection [12]. It suggests that under the conditions of increased parasite load, IFN-I contributes to the pathogenesis of infection [12].

4.5 IFN-II (IFN-γ)–associated immune changes in *T. cruzi* infection

IFN-γ has a central role in Chagas disease cardiomyopathy. The disease is characterized by increased production of IFN-γ in the periphery [13]. Several cytokines, including IFN-γ, IL-1-α, IL-6 and TNF-α, modulate the expression of immune cells and contribute to the inflammatory process by recruiting the T cells into the inflammatory sites. Conversely, IL-4, TGF-β and IL-10 negatively regulate NO production and downregulate the intracellular control of *T. cruzi* infection by IFN-γ–activated macrophages [14]. In humans, IFN-γ was detected as a predominant cytokine in circulation during *T. cruzi* infection [15]. IFN-γ regulates the expression of several genes, transcription factors, inflammatory cytokines such as TNF-α, chemokines, and other pathogen-resistance genes including inducible nitric oxide synthase 2 (iNOS or NOS2) [16]. Higher amount of IFN-γ along with TNF-α leads to an efficient parasite killing and enhanced function of memory T cells [17]. It is evident from the fact that mice deficient in IL-12 that is necessary for IFN-γ production exhibit severe tissue and systemic parasitism suggesting the importance of the IFN-γ in controlling intracellular parasitism [18]. It is known that T-cell–mediated control of the disease is dependent on the duration of infection and tissue damage. The detection of IFN-γ and TNF-α during early phase of chronic *T. cruzi* infection is associated with IL-10 production by $CD4^+$ T cells [17]. IFN-γ production is higher in chronic Chagas cardiomyopathy compared to asymptomatic patients, wherein IL-10 is reported to be highly expressed [16]. IL-10 has a counter effect on Th1 responses via downmodulating IFN-γ response, which, if sustained for long time, may have harmful effects on the host [19]. Altered cytokine profile either quantitatively or qualitatively can be a major cause of chronic Chagas disease. IFN-γ is well known as a protective lymphokine against *T. cruzi*, but there are many reports stating the dual role (antiparasitic, protective and pathogenic) of IFN-γ in Chagas disease [13, 14]. Several reports suggest myocarditis and heart failure in patients with Chagas disease, possibly due to continuous production of IFN-γ by T cells [16]. IFN-γ controls infection through NOS production and activating ROS through induction of NADP oxidases, while resistance of *T. cruzi* to ROS induces serious alterations in heart function. The detrimental role of overexpression of IFN-γ has been proven in experiment with transgenic mice, where it results in TNF-α–dependent murine myocarditis and cardiomyopathy [20].

4.6 IFN-III–associated immune changes in *T. cruzi* infection

Type III IFNs serve as regulatory cytokines by reducing the damage caused by pro-inflammatory cytokines or by retaining the more potent IFN-I for times when immune responses are inadequate [21]. This subtype is poorly recognized and has not been studied in Chagas disease.

4.7 Thymic alterations in *T. cruzi* infection

Circulation of T cells in response to parasitic infection is securely controlled as various cytokines and chemokines influence the disease outcome. It has been reported that in *T. cruzi* infection, both Th1 and Th2 cytokines are associated with resistance and susceptibility to disease, respectively. Influence of cytokines on thymus function is not much studied. Pérez et al. [22] made a detailed evaluation of the effect of pro-inflammatory (IFN-γ, IL-12 and iNOS) and anti-inflammatory cytokines (IL-4 and IL-10) on thymus in a study with experimental C57BL/6 murine model. Uninfected knock out mice for both pro-inflammatory and anti-inflammatory cytokines showed thymocyte cellularity similar to wild-type mice, although apoptotic loss of DP thymocytes was seen in infected mice group, showing that thymic atrophy is independent of IFN-γ or iNOS [22]. However, in another study, it is shown that upon *T. cruzi* infection in C57BL/6 mice, IL-10 and IFN-γ play a role in controlling thymic T-cell activation via altering the thymic cell function, but the extent of immunological disturbances was not clearly described [23]. In experimental Chagas disease, it was observed that the increased level of extracellular matrix (ECM) in thymus favors the export of immature thymocytes from thymus. Increased migration of thymocytes in response to fibronectin leads to a high number of DP T cells' migration from the thymus to peripheral lymphoid organs. The frequency of peripheral $CD4^+CD8^+$ DP T cells is increased in acute *T. cruzi* infection up to 16 times in subcutaneous lymph nodes [24, 25]. Thymic atrophy is an acute phenomenon observed in the infected mice, accompanied by alteration in the thymic structure. The mechanism is understood in terms of hormonal dysregulation induced under infection condition. The production of pro-inflammatory cytokines, IL-1, IL-6 and TNF-α increases during the infection and

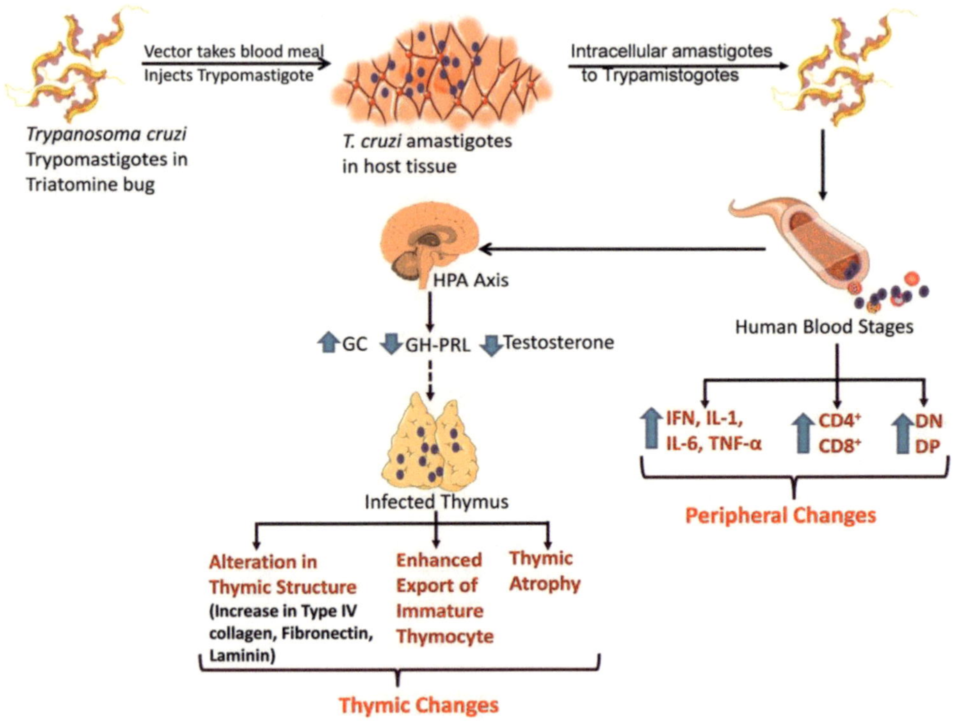

Figure 1.
Peripheral and thymic changes induced in host organism (mice and/or human) upon T. cruzi *infection.*

activates the HPA axis causing the release of glucocorticoids (GCs) [26]. GCs are steroidal hormones that lead to thymus atrophy with depletion of immature cortical thymocytes. Thymocyte depletion is seen to be directly proportional to increased TNF-α levels. However, this depletion is attributed to TNF-induced glucocorticoids rather than TNF-α directly such that it is not the cytokines, but the downstream molecules induced by them that lead to the observed thymic changes [27, 28]. It has been reported that during *T. cruzi* infection, prolactin (PRL) has a significant role in homeostatic balance of thymic corticosterone [29, 30]. Under stressful conditions, PRL balances the negative effects of GC by increasing thymocytes and thymic epithelial cell (TEC) proliferation. PRL rescues these cells from apoptosis in opposition to GC, which inhibits thymocyte growth. Recent reports show that PRL secretion is also altered along with GC secretion with decreased level of PRL paralleling increased GC levels during acute *T. cruzi* infection, causing an imbalanced cross talk that may correlate with the thymic involution [31, 32]. Thus, it can be said that GC and PRL are responsible for the loss of thymocytes, which leads to thymic atrophy. Thus, a dysregulated immuno-endocrine axis leads to profound effects on the thymus and disease outcome during *T. cruzi* infection. Peripheral and thymic changes associated with *Trypanosoma* infection are depicted in **Figure 1**.

5. Malaria caused by *Plasmodium* spp.

Malaria is a deadly disease caused by *Plasmodium* parasite belonging to the family Plasmodiidae. The disease is transmitted to humans by the bite of infected female Anopheles mosquito. Parasite species *P. falciparum* and *P. vivax* cause malaria in humans. Based on WHO reports, malaria is prevalent in 87 countries throughout the world, with estimated 219 million cases and 435,000 estimated death reports [33]. An estimated 91% of all deaths due to malaria occur in Africa.

5.1 Disease transmission

Malaria is generally transmitted through the bite of *Anopheles* mosquitoes with high activity between dusk and dawn. Disease transmission is dependent on factors such as climatic conditions comprising rainfall patterns, temperature and humidity, host immunity, parasite species and the vector involved [33].

5.2 Disease severity and diagnosis

Malaria can be fatal if not treated. Disease outcome is determined by the parasite species and host immunity. Complications may arise in the form of cerebral malaria (CM) wherein the parasite infects the brain and leads to serious damage including seizures and coma accompanied with breathing problems, organ failure and low blood sugar. Early detection and disease treatment are important to reduce the risk of disease severity. Staining-based microscopic parasite diagnosis methods or malaria rapid diagnostic tests (RDTs) are widely used for preliminary diagnosis of the disease. RDTs detect specific antigen produced by malaria parasite in human blood using a dye-labeled capture antibodies providing an evidence of malaria infection [33].

5.3 Peripheral T-cell immune response associated with *Plasmodium* infection

Host immune response against *Plasmodium* parasites *in vitro* and *in vivo* is well studied in murine models (*P. yoelii*, *P. vinckei*, *P. chabaudi* and *P. berghei*) and humans (*P. malariae*, *P. vivax*, *P. falciparum*, *P. ovale* and *P. knowlesi*) [3]. The parasite stimulates multifaceted immune responses, including antibodies, NK and NKT cells, $CD4^+$ and $CD8^+$ T cells [34]. T cells play a major role in protection against *Plasmodium*. Both Th1 and Th2 subsets of $CD4^+$ T cells are the major players to control the systemic infections [35]. $CD4^+$ T cells stimulate $CD8^+$ T-cell cytotoxic activity, inhibit the development of liver stages and prevent the infection of red blood cells [36]. Thus, a balance between the cytokines and other immune molecules produced by different cell types is critical in determining the outcome of the infection.

5.4 IFN-I–associated immune changes in *Plasmodium* infection

IFN-γ is the most widely studied in malaria and has a versatile effect on the host. It may exert a protective or destructive effect, depending on the stage of the infection or the species of *Plasmodium* involved. Disease-protective phenomenon was observed in mice infected with *P. berghei*, where post-IFN-β treatment survival of mice is prolonged compared to non-treated counterparts [37]. Protection to disease is driven by a sensory mechanism against *Plasmodium* in the liver that mediates a functional antiparasite response driven by type I IFN. IFN-I is known to be active during the late phase of the liver stage infection. It is evident by the fact that treatment of *P. yoelii–infected* mice with recombinant IFN-α does not alter the hepatic parasite burden. This results in partially limiting the parasite growth in the liver and influences the commencement of erythrocytic stage infection. Leukocytes are recruited around the liver-stage of the parasite leading to reduced parasitemia [38]. Blood transcriptional profile of mild and severe malaria infection cases revealed that a specific set of genes was significantly associated with a mild form as compared to their expression pattern in severe form of malaria. Studies on malaria-infected individuals from Malawi region revealed that genes responsible for IFN-I signaling pathway have an important role in the development of protective immune response against malaria. This is proved by molecular studies wherein mutations within IFN-α receptor (IFN-αR) lead to disease susceptibility and severe disease in Malawian population [39]. In contrast to the protective effects discussed above, a pathogenic role for IFN-I in *Plasmodium* infections has also been described. This has been reported in murine models, where the absence of IFN-I signaling in *P. berghei*–infected mice led to reduced parasite load and resistance against CM. The development of CM occurs as a result of detrimental brain injuries due to damaging inflammatory host immune response [40, 41]. Expression analysis of $CD4^+$ T cells from *P. berghei* ANKA (PbA)-infected mice revealed that $CD4^+$ cells showed dominance of IFN-I and IFN-γ signaling pathway-related genes. Mice deficient in IFN-I signaling had reduced parasite burden and displayed no CM-related symptoms. IFN-I suppressed IFN-γ production *via* inhibiting $CD4^+$ T-bet^+ T-cell derived IFN-γ production and hampered protective Th1-mediated control of parasitemia in *P. chabaudi*–infected mice [41]. Progression to CM can be modified by host genetic factors. A robust association between IFNAR1 and CM protection is well documented in experimental CM in $IFNAR1^{-/-}$ mice infected with *P. berghei* [40]. It is reported that splenic $CD8^+$ T cells from $IFNAR1^{-/-}$ mice got activated functionally but were unable to mediate any damage to brain tissue and cause CM development. This proves that IFNAR1 signaling promotes $CD8^+$ effector activity, which is mandatory for CM, in both humans and mouse [40]. There are controversial reports stating the IFN-I–mediated suppression on IFN-γ activity. During early stage of *P. chabaudi* infection, IFN-I induced by the infection plays a disease-exacerbating role by suppressing IFN-γ producing $CD4^+$ T cells in C57BL/6 mice [41]; however, in 129 Sv/Ev mice, IFN-I has minor roles in controlling the disease pathology [42]. Similar instance is observed in humans where polymorphism in human gene encoding for IFNARI strongly supports protection from the disease [43]. These controversial reports suggest that duration of activity and levels of IFN-I are important in regulating immune response against parasite growth [3].

5.5 IFN-II (IFN-γ)–associated immune changes in *Plasmodium* infection

IFN-γ regulates various components of the host immune system such as defense against intracellular pathogens by antigen presentation, antimicrobial mechanism, leukocyte development and immune cell trafficking. The protective role of IFN-γ is evident from the *in vitro* and *in vivo* studies, where inhibitory effect of IFN-γ on parasite multiplication was observed in *P. berghei* sporozoites-infected murine hepatocytes and/or human hepatic HEPG2 cells upon treatment with human recombinant IFN-γ [44–47]. IFN-γ helps in controlling the parasitism by activating macrophages and promoting phagocytosis of circulating parasites and plays a crucial protective role during blood-stage infection. *P. chabaudi* AS-infected mice treated with monoclonal antibody against IFN-γ had less control of parasite multiplication, suggesting that IFN-γ is essential for limiting parasite multiplication. Similar effects were evident in *P. chabaudi* AS-infected mice that were lacking IFN-γ receptor. These mice had lower survival rates as compared to the WT controls [48]. This suggests that IFN-γ production at different stages during infection could alter parasite survival and hence disease outcome. In *P. berghei* infection, IFN-γ along with TNF-α also plays a protective role by parasite removal activity [49]. The natural resistance to *Plasmodium* infection is reported in humans from tribes in Mali where resistance was correlated with increased levels of IFN-γ [50], suggesting a protective role for IFN-γ against malaria. IFN-γ is essential in both protective immunity and pathogenesis of the diseases. During malaria infection, elevated levels of IgE antibodies are also observed. IgE containing immune complexes are pathogenic and not protective as they are involved in overproduction of TNF-α. TNF-α acts as a major pathogenic factor in malaria and poses an increased risk of severe disease or death due to *P. falciparum* infection [35]. IFN-γ promotes migration of leukocytes and pathogenic $CD8^+$ T cells to the brain during infection induced by *P. berghei* ANKA in WT 129P2Sv/Ev mice compared to IFN-γ R1-deficient mice. The production of elevated levels of IFN-γ during parasite blood-stage is associated with susceptibility to severe CM malaria [51]. Its protective or harmful effect on the host depends on the stage of infection and target organ [44].

5.6 IFN-III–associated immune changes in *Plasmodium* infection

Type I and type II interferons [IFNs] are critical to govern the disease outcome; however, reports on the involvement of recently identified IFN-III humans during malaria infection are scarce [3].

5.7 Thymic alterations in *Plasmodium* infection

Malarial infection results in increased levels of IFN-γ and TNF-α in human serum. Both these cytokines have been shown to be involved in double positive T-cell death [52–54]. However, neutralizing the effect of IFN-γ and TNF-α did not alter the apoptosis-inducing capacity of the serum [28]. Conversely, TNF-α neutralization helps in the reduction of DP T-cell count due to increased apoptosis, stating that TNF-α exerts a protective rather than a destructive role in malaria-induced thymic atrophy [28]. Studies done on BALB/c mice model show a high level of apoptosis and premature migration of thymocytes in mice upon *Plasmodium* infection along with overexpression of TNF-α associated with thymic atrophy [55, 56]. Reports on direct effect of IFN family on thymic changes during malaria

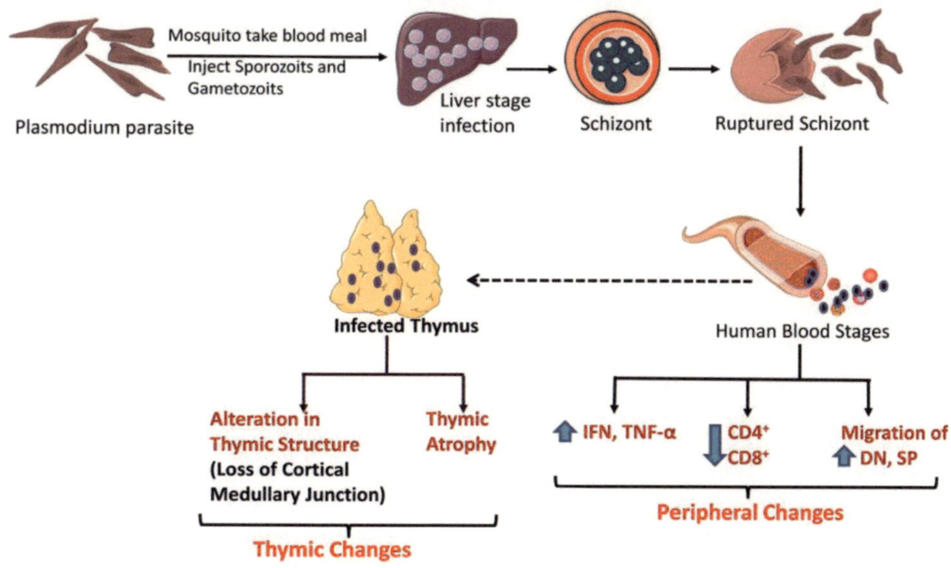

Figure 2.
Peripheral and thymic changes induced in host organism (mice or human) upon Plasmodium *infection.*

infection are scarce. In *P. berghei*–infected mice, changes in the thymic microenvironment alter the thymocytes' migration pattern with the direct implication in the export of immature cells to the periphery [57]. It is modeled that in *Plasmodium*-infected mice, the number of $CD4^+$ T cells decreases due to the destruction or reduced production of $CD4^+$ T cells and the number of $CD8^+$ T cells increases due to peripheral expansion or redistribution of preexisting cytotoxic T cells or due to an increase in thymic output [58]. Thymus atrophy seen in *Plasmodium* infection is accompanied by alterations in thymus architecture with loss of cortical-medullar delimitation [56]. The atrophy of thymus starts with an early stage of infection and the thymus weight is reduced markedly [59]. *Plasmodium* infection interferes with the positive and negative selection process of thymocytes resulting in apoptosis of thymocytes and thymic atrophy. This is evident from *P. chabaudi* non-lethal malaria model where thymic atrophy is reported to occur due to depletion of single positive $CD4^+$ and $CD8^+$ T cells [56, 60]. Changes in the thymic microenvironment, altered expression of the ECM proteins and chemokines observed in *P. berghei*–infected mice result in an altered intrathymic thymocyte migration pattern and defective thymocyte development [57]. Thus, a dysregulation in thymic immune cross talk comprising IFNs results in thymic structural and functional changes as depicted in **Figure 2**.

6. Leishmaniasis caused by *Leishmania* spp.

Leishmania is a tropical protozoan parasite belonging to the family Trypanosomatidae. The parasite is transmitted by the bite of the female phlebotomine fly species in old world countries and by Lutzomyia species in new world countries. More than 20 *Leishmania* species are known to circulate in endemic foci in Africa, Asia, the Middle East, the Mediterranean region, Central-South America, and southern Europe. The *L. donovani* and *L. infantum/chagasi* complex is responsible for VL; the *L. major, L. tropica*, *L. aethiopica* and *L. mexicana* complex causes CL; and the subgenus *L. Viannia* complex causes CL and MCL as per the classical association of specific parasite species with distinct clinical outcomes. The disease has a wide geographical occurrence covering 97 countries and territories with endemic foci for each of the different clinical manifestations [61].

6.1 Disease transmission

Female phlebotomine sandflies transmit the Leishmania parasite during blood meal. Disease transmission is dependent on the parasite or sandfly species, environmental conditions, host immunity and animal reservoir [62].

6.2 Disease severity and diagnosis

There are three main clinical forms of leishmaniasis: Visceral leishmaniasis (VL) or kala-azar is characterized by hepatosplenomegaly, fever and anemia. Cutaneous leishmaniasis (CL) is the most common form of the disease, characterized by skin lesions on exposed body parts, scars on the body and societal stigma. Mucocutaneous leishmaniasis (MCL) manifestation involves mucous membranes of the nose, mouth and throat. Diagnosis is generally based on microscopic examinations of *Leishmania* amastigotes in skin lesions in case of CL and rapid diagnostic recombinant K39 tests in case of VL with recent complementation with parasite-specific molecular diagnostics [62–64].

6.3 Peripheral T-cell response associated with *Leishmania* infection

There is a mixed Th1 and Th2 immune response during *Leishmania* infection with discrete quantitative and qualitative changes in T-cell subsets and the associated cytokines. Numerous reports explain counter-regulatory effects of T-cell subset–specific cytokines both at transcriptional and at translational levels. *Leishmania*-infected host exhibits a dynamic peripheral Th1/Th2 immune environment such that Th1 immune-activation is associated with IL-2, IFN-γ and TNF-α, which leads to macrophage activation and disease resolution, while Th2 response is associated with IL-4, IL-5 and IL-13 that supports disease progression [65]. Treg cells that produce IL-4 and IL-10 cytokines are also involved in regulating Th2/Th1 balance toward disease outcome. In mice infected with *L. donovani*, $CD4^+$ T cells are activated on the first day of infection and proliferate several folds resulting in splenomegaly [66]. $CD8^+$ T cells also produce cytokines and chemokines, which enhance immunity to pathogens [67]. So along with $CD4^+$ T-cell response, $CD8^+$ T cells also provide a level of control through production of IFN-γ and contribute to disease outcome. In contrast to protection mechanism, $CD8^+$ T cells induce cytotoxicity in *L. braziliensis* infection [68]. Thus, in the acute phase of *Leishmania* infection, $CD8^+$ T cells are protective because they produce IFN-γ, while in the chronic phase, they promote pathology because of cytotoxicity. In *L. infantum*-infected murine model, alterations in the number of peripheral $CD4^+$ and $CD8^+$ T cells are observed, wherein increase in peripheral $CD8^+$ T cells is responsible for the control of *L. infantum* infection with a slight decrease in the number of $CD4^+$ T cells [69].

6.4 IFN-I–associated immune changes in *Leishmania* infection

IFN-II is considered the main player in cell-mediated immune responses against infections, but recently, IFN-I is also being reported to play a role in Leishmaniasis pathology outcome. Activated macrophages initiate the parasite elimination via the production of iNOS. Deficiency of this enzyme in mice promotes susceptibility to *L. major* infection [70]. The protective role of IFN-I was studied in vivo where neutralizing IFN-I in mice experimentally infected with *L. major* made them more vulnerable to infection and increase in parasite load due to enhanced parasite multiplication. Blocking IFN-I function led to dissolution of iNOS activity and reduced cytotoxicity at early stages of infection [71]. The stage of parasitic infection and the dose of IFN-I play a significant role in predicting the consequences of the disease [12, 37]. Mattner et al. [72] revealed that IFN-I acts in a dose-dependent manner, where a low dose against a high dose of IFN-I protected the *L. major*-infected BALB/c mice from progressive leishmaniasis. IFN-I treatment aids in IFN-γ production via STAT4-dependent pathway [72].

6.5 IFN-II (IFN-γ)-associated immune changes in *Leishmania* infection

Leishmania immunity is mostly mediated by T lymphocytes and immune response is shown to be dependent on host genotype. This is evident from the fact that some inbred strains of mouse are susceptible, while others are resistant to *Leishmania* infection. In the human body, IFN-γ is not secreted alone, but other cytokines mainly IL-12, IL-10 and IL-4 influence the IFN-γ both at the level of induction and at the level of effector function. This further determines the course of infection [73, 74]. Several *in vivo* and *in vitro* experiments have shown that IFN-γ hinders the activation/expansion of $CD4^+$ Th2 cells, resulting in the preferential expression of Th1 immune response and Th1 immunity. IFN-γ expression pattern is well documented for correlation with protection against the parasitic diseases in old and new world *Leishmania* infection model. The absence of IFN-γ or IFN-γ receptor leads to expansion of Th2-type cellular response in C57BL/6 mice making the host highly vulnerable to *L. major* or *L. amazonensis* infection [75]. IFN-γ-mediated immune protection against *Leishmania* infection is also evident where CXCL10-treated, *L. donovani*-infected BALB/c mice display generation of perforins and granzyme B via $CD8^+$ T-cell–dependent strong host-protective Th1 response, accompanied by significant downregulation in Th2- and Treg-associated cytokines [76, 77]. Pretreatment of macrophages obtained from BALB/c, C57BL/6 and C3H/HeJ mice significantly reduces *L. amazonensis* load via an NO-mediated mechanism of IFN-γ production in the presence of recombinant CXCL10 [78]. Immune response against leishmaniasis is also dependent on *Leishmania* spp. involved in infection. Several comparative studies conducted with crude antigen extracts of *L. braziliensis* and *L. amazonensis* reported that the extracts of *L. braziliensis* are more potent over *L. amazonensis* in stimulating CXCL10 production correlating to IFN-γ-positivity and multi-functional $CD4^+$ T cells in CL patients. Therefore, in agreement with the findings from murine infection models, CXCR3 and CXCL10 chemokines are also involved in protection and disease pathogenesis in leishmaniasis [79, 80]. Human VL is generally known to be predominated by Th2-type response. Anti-leishmanial drug treatments induce a significant Th1-type response in cured patients marked by the production of IFN-γ and IL-4 in the viscera. Contrary to VL, CL patients show a disease-healing response dominated by IFN-γ. IL-4, a Th2 cytokine, is rarely detected in CL cases [81]. In patients with active VL, the expression of IFN-γ is increased in the periphery, but it may be possible that the effect is not enough to overcome the parasite multiplication or there is unresponsiveness to *L. donovani* antigen. This may be due to elevated level of immunosuppressive Th2-specific cytokines in active VL patients [82].

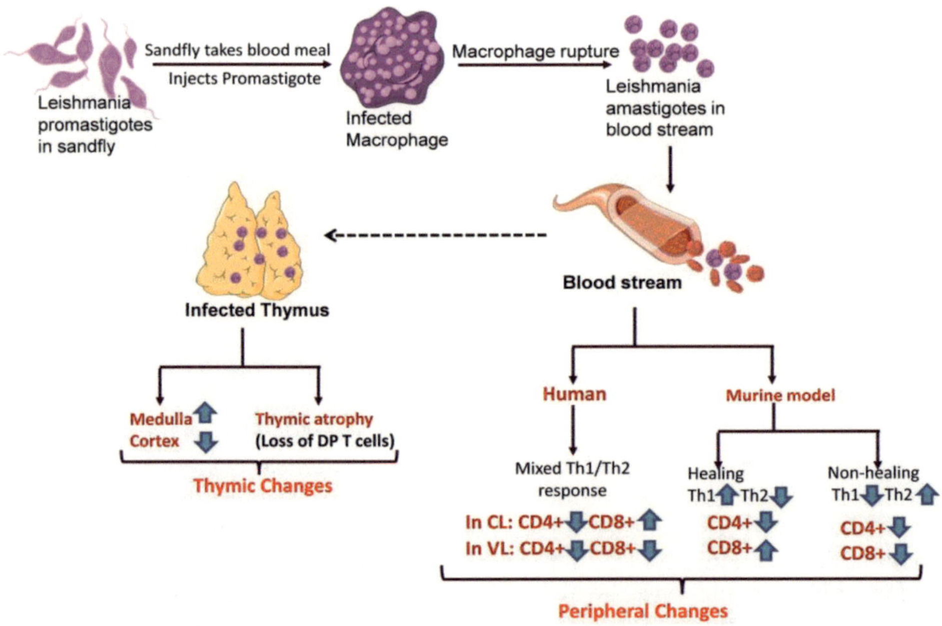

Figure 3.
Peripheral and thymic changes induced in host organism (mice or human) upon Leishmania *infection.*

6.6 Thymic alterations in *Leishmania* infection

Thymus is the least studied in context of *Leishmania* infection. A recent report demonstrates a decrease in thymic cellularity and concomitant thymic atrophy with severely compromised thymic microenvironment in a murine model co-conditioned with protein malnutrition and *L. infantum* infection. These mice exhibited a significant reduction of the thymic corticomedullary ratio [83]. Similar studies done in our laboratory with *L. donovani* infected VL murine model demonstrate that the parasite homes to thymus and lead to expansion of medullary regions when compared to control uninfected mice (unpublished data). *L. infantum* infection in protein-malnourished mice causes thymic atrophy due to a decrease of DP thymocytes and alters thymic chemotactic factors by diminishing CCL5, IGF1, CXCL9, 10 and 12 with significantly increased levels of IL-1α and IL-10 [83, 84]. It has been observed that due to *Leishmania* infection in mice, positively selected $CD8^+$ or $CD4^+$ T cells upregulate CCR7 and migrate to the medulla in response to CCL19/CCL21. CCR7 knockout mice were associated with cortical accumulation of SP thymocytes and decreased medullary $CD4^+$SP and $CD8^+$SP T cells. The migration of T cell is decreased in protein-malnourished infected mice as the components of extracellular matrix and adhesion molecules are altered that compromise the migratory capabilities necessary for adequate lymphocyte proliferation, intrathymic maturation and extrathymic activation [85]. Peripheral and thymic changes associated with *Leishmania* infection are depicted in **Figure 3**.

7. Conclusions

In conclusion, the role of IFN family in both immune-protective and immune-pathogenic processes in parasitic infections makes it a key set of molecules to be studied in depth (**Figure 4**). Modulatory effect of IFNs on T cells and downstream

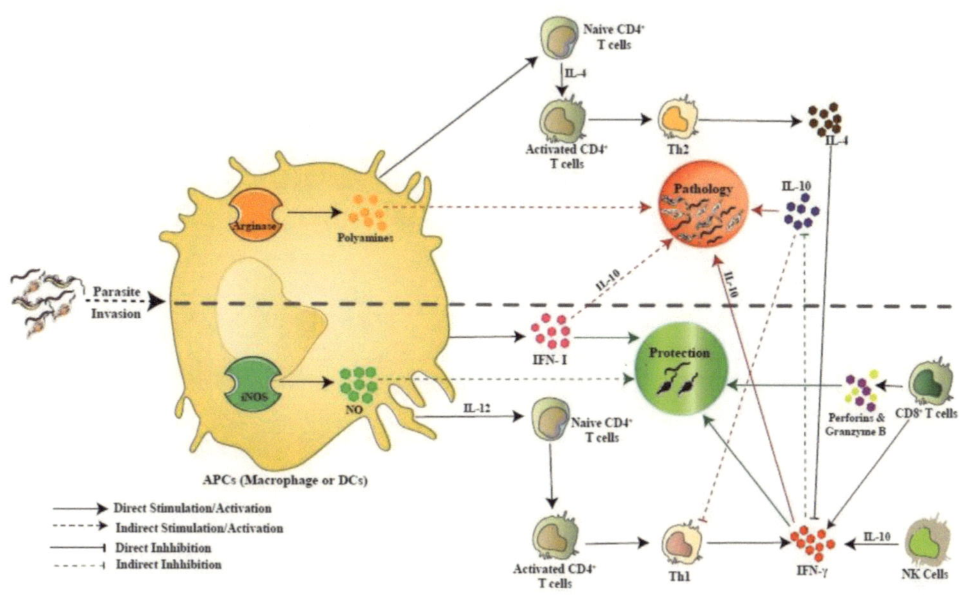

Figure 4.
Mechanistic of IFNs (IFN-I and IFN-γ) upon parasitic invasion in host organism. IFNs play a dual role in disease progression and/or protection, depending on the type and expression levels in relation to other secretory molecules and cytokines (Th1/Th2/Treg) and cause-effect cross talk between different IFN producers and effector immune cells.

effector function of T cells along with their complex cross-network functionality in other circulating blood and tissue-resident immune cells warrant further understanding on their role in disease manifestation and outcome. IFNs as the modulators of thymic structure and function are an interesting dimension of the immune-regulatory capabilities of these soluble immune molecules in infectious diseases. IFNs work as double-edged sword to modulate immune effector mechanisms determined by parasite and host components. This family of important cytokines can be tailored to be used as immunomodulators and/or immunotherapeutic molecules.

Acknowledgements

Aklank Jain and Manju Jain would like to acknowledge Central University of Punjab, Bathinda, India, for providing Research Seed Money Grant.

Author details

Lovlesh Thakur[1], Nadeem Akhtar[2], Aklank Jain[1], Hridayesh Parkash[3] and Manju Jain[2]*

1 Department of Zoology, Central University of Punjab, Bathinda, India

2 Department of Biochemistry, Central University of Punjab, Bathinda, India

3 Amity Institute of Virology and Immunology, Amity University, Noida, India

*Address all correspondence to: manjujainmda@gmail.com

References

[1] Centers for Disease Control and Prevention. Parasites. 2016. Available from: https://www.cdc.gov/parasites/about.html

[2] Savino W, de Moraes MC, Barbosa SD, Da Fonseca EC, De Almeida VC, Hontebeyrie-Joscowicz M. Is the thymus a target organ in infectious diseases? Memórias do Instituto Oswaldo Cruz. 1992;**87**(Suppl 5):73-78

[3] Silva-Barrios S, Stäger S. Protozoan parasites and type I IFNs. Frontiers in Immunology. 2017;**8**:14

[4] Centers for Disease Control and Prevention. Parasites—American Trypanosomiasis (also Known as Chagas Disease). 2019. Available from: https://www.cdc.gov/parasites/chagas/

[5] Gutierrez FR, Mineo TW, Pavanelli WR, Guedes PM, Silva JS. The effects of nitric oxide on the immune system during *T. cruzi* infection. Memórias do Instituto Oswaldo Cruz. 2009;**104**:236-245

[6] Tarleton RL, Sun J, Zhang L, Postan M. Depletion of T-cell subpopulations results in exacerbation of myocarditis and parasitism in experimental Chagas' disease. Infection and Immunity. 1994;**62**(5):1820-1829

[7] Rottenberg M, Bakhiet M, Olsson T, Kristensson K, Mak T, Wigzell H, et al. Differential susceptibilities of mice genomically deleted of CD4 and CD8 to infections with *T. cruzi* or Trypanosoma brucei. Infection and Immunity. 1993;**61**(12):5129-5133

[8] Giraldo NA, Bolaños NI, Cuellar A, Guzman F, Uribe AM, Bedoya A, et al. Increased CD4+/CD8+ double-positive T cells in chronic Chagasic patients. PLoS Neglected Tropical Diseases. 2011;**5**(8):e1294

[9] Chessler A-DC, Ferreira LR, Chang T-H, Fitzgerald KA, Burleigh BA. A novel IFN regulatory factor 3-dependent pathway activated by trypanosomes triggers IFN-β in macrophages and fibroblasts. The Journal of Immunology. 2008;**181**(11):7917-7924

[10] Rottenberg M, Cardoni R, Andersson R, Segura E, Örn A. Role of T helper/inducer cells as well as natural killer cells in resistance to *T. cruzi* infection. Scandinavian Journal of Immunology. 1988;**28**(5):573-582

[11] Santiago HC, Feng CG, Bafica A, Roffe E, Arantes RM, Cheever A, et al. Mice deficient in LRG-47 display enhanced susceptibility to *T. cruzi* infection associated with defective hemopoiesis and intracellular control of parasite growth. The Journal of Immunology. 2005;**175**(12):8165-8172

[12] Chessler A-DC, Caradonna KL, Da'dara A, Burleigh BA. Type I interferons increase host susceptibility to *T. cruzi* infection. Infection and Immunity. 2011;**79**(5):2112-2119

[13] Ferreira LRP, Frade AF, Baron MA, Navarro IC, Kalil J, Chevillard C, et al. Interferon-γ and other inflammatory mediators in cardiomyocyte signaling during Chagas disease cardiomyopathy. World Journal of Cardiology. 2014;**6**(8):782-790

[14] Laucella S, Rottenberg M. Role of cytokines in resistance and pathology in *T. cruzi* infection. Revista Argentina de microbiologia. 1996;**28**(2):99-109

[15] Albareda MC, Laucella SA. Modulation of *T. cruzi*-specific T-cell responses after chemotherapy for chronic Chagas disease. Memórias do Instituto Oswaldo Cruz. 2015;**110**(3):414-421

[16] Chevillard C, Nunes JPS, Frade AF, Almeida RR, Pandey RP,

Nascimento MS, et al. Disease tolerance and pathogen resistance genes may underlie *T. cruzi* evasion, parasite persistence and differential progression to Chagas disease cardiomyopathy. Frontiers in Immunology. 2018;**9**:2791

[17] Sathler-Avelar R, Vitelli-Avelar D, Massara R, Borges J, Lana M, Teixeira-Carvalho A, et al. Benznidazole treatment during early-indeterminate Chagas' disease shifted the cytokine expression by innate and adaptive immunity cells toward a type 1-modulated immune profile. Scandinavian Journal of Immunology. 2006;**64**(5):554-563

[18] Michailowsky V, Silva NM, Rocha CD, Vieira LQ, Lannes-Vieira J, Gazzinelli RT. Pivotal role of interleukin-12 and interferon-γ axis in controlling tissue parasitism and inflammation in the heart and central nervous system during *T. cruzi* infection. The American Journal of Pathology. 2001;**159**(5):1723-1733

[19] Dutra WO, Gollob KJ. Current concepts in immunoregulation and pathology of human Chagas disease. Current Opinion in Infectious Diseases. 2008;**21**(3):287

[20] Torzewski M, Wenzel P, Kleinert H, Becker C, El-Masri J, Wiese E, et al. Chronic inflammatory cardiomyopathy of interferon γ–overexpressing transgenic mice is mediated by tumor necrosis factor-α. The American Journal of Pathology. 2012;**180**(1):73-81

[21] Lazear HM, Schoggins JW, Diamond MS. Shared and distinct functions of type I and type III interferons. Immunity. 2019;**50**(4):907-923

[22] Pérez AR, Silva-Barbosa SD, Roggero E, Calmon-Hamaty F, Villar SR, Gutierrez FR, et al. Immunoendocrinology of the thymus in Chagas disease. Neuroimmunomodulation. 2011;**18**(5):328-338

[23] raes d M, ML, Minoprio P, Dy M, Dardenne M, Savino W, Hontebeyrie-Joskowicz M. Endogenous IL-10 and IFN-γ production controls thymic cell proliferation in mice acutely infected by *T. cruzi*. Scandinavian Journal of Immunology. 1994;**39**(1):51-58

[24] Cotta-de-Almeida V, Bonomo A, Mendes-da-Cruz DA, Riederer I, is DM, J, Lima-Quaresma KR, et al. *T. cruzi* infection modulates intrathymic contents of extracellular matrix ligands and receptors and alters thymocyte migration. European Journal of Immunology. 2003;**33**(9):2439-2448

[25] Mendes-da-Cruz DA, eis d M, J, Cotta-de-Almeida V, Savino W. Experimental *T. cruzi* infection alters the shaping of the central and peripheral T-cell repertoire. Microbes and Infection. 2003;**5**(10):825-832

[26] Roggero E, Perez AR, Tamae-Kakazu M, Piazzon I, Nepomnaschy I, Besedovsky HO, et al. Endogenous glucocorticoids cause thymus atrophy but are protective during acute *T. cruzi* infection. Journal of Endocrinology. 2006;**190**(2):495-503

[27] Cohen O, Kfir-Erenfeld S, Spokoini R, Zilberman Y, Yefenof E, Sionov RV. Nitric oxide cooperates with glucocorticoids in thymic epithelial cell-mediated apoptosis of double positive thymocytes. International Immunology. 2009;**21**(10):1113-1123

[28] Khanam S, Sharma S, Pathak S. Lethal and nonlethal murine malarial infections differentially affect apoptosis, proliferation, and CD 8 expression on thymic T cells. Parasite Immunology. 2015;**37**(7):349-361

[29] Lepletier A, de Carvalho VF, Rodrigues e Silva PM, Villar S, Perez AR,

Savino W, et al. *T. cruzi* disrupts thymic homeostasis by altering intrathymic and systemic stress-related endocrine circuitries. PLoS Neglected Tropical Diseases. 2013;**7**(11):e2470

[30] Lepletier A, de Almeida L, Santos L, da Silva Sampaio L, Paredes B, Gonzalez FB, et al. Early double-negative thymocyte export in *T. cruzi* infection is restricted by sphingosine receptors and associated with human Chagas disease. PLoS Neglected Tropical Diseases. 2014;**8**(10):e3203

[31] Lepletier A, de Frias Carvalho V, Morrot A, Savino W. Thymic atrophy in acute experimental Chagas disease is associated with an imbalance of stress hormones. Annals of the New York Academy of Sciences. 2012;**1262**:45-50

[32] Pérez AR, Morrot A, Carvalho VF, de Meis J, Savino W. Role of hormonal circuitry upon T cell development in Chagas disease: Possible implications on T cell dysfunctions. Frontiers in Endocrinology. 2018;**9**:334

[33] World Health Organization. Fact Sheets, Malaria. 2019. Available from: https://www.who.int/news-room/fact-sheets/detail/malaria

[34] Tsuji M, Zavala F. T cells as mediators of protective immunity against liver stages of *Plasmodium*. Trends in Parasitology. 2003;**19**(2):88-93

[35] Perlmann P, Troye-Blomberg M. Malaria blood-stage infection and its control by the immune system. Folia Biologica. 2000;**46**(6):210-218

[36] Morrot A, Zavala F. Regulation of the CD8+ T cell responses against *Plasmodium* liver stages in mice. International Journal of Parasitology. 2004;**34**(13-14):1529-1534

[37] Morrell CN, Srivastava K, Swaim AM, Lee MT, Chen J, Nagineni C, et al. Beta interferon suppresses the development of experimental cerebral malaria. Infection and Immunity. 2011;**79**(4):1750-1758

[38] Liehl P, Zuzarte-Luís V, Chan J, Zillinger T, Baptista F, Carapau D, et al. Host-cell sensors for *Plasmodium* activate innate immunity against liver-stage infection. Nature Medicine. 2014;**20**(1):47-53

[39] Krupka M, Seydel K, Feintuch CM, Yee K, Kim R, Lin C-Y, et al. Mild *P. falciparum* malaria following an episode of severe malaria is associated with induction of the interferon pathway in Malawian children, Infection and Immunity. 2012;**80**(3):1150-1155

[40] Ball EA, Sambo MR, Martins M, Trovoada MJ, Benchimol C, Costa J, et al. IFNAR1 controls progression to cerebral malaria in children and CD8+ T cell brain pathology in *Plasmodium* berghei-infected mice. Journal of Immunology. 2013;**190**(10):5118-5127

[41] Haque A, Best SE, Ammerdorffer A, Desbarrieres L, de Oca MM, Amante FH, et al. Type I interferons suppress $CD4^+$ T-cell-dependent parasite control during blood-stage *Plasmodium* infection. European Journal of Immunology. 2011;**41**(9):2688-2698

[42] Voisine C, Mastelic B, Sponaas AM, Langhorne J. Classical CD11c+ dendritic cells, not plasmacytoid dendritic cells, induce T cell responses to *Plasmodium* chabaudi malaria. International Journal of Parasitology. 2010;**40**(6):711-719

[43] Aucan C, Walley AJ, Hennig BJ, Fitness J, Frodsham A, Zhang L, et al. Interferon-alpha receptor-1 (IFNAR1) variants are associated with protection against cerebral malaria in the Gambia. Genes and Immunity. 2003;**4**(4):275-282

[44] Gun SY, Claser C, Tan KSW, Rénia L. Interferons and interferon

regulatory factors in Malaria. Mediators of Inflammation. 2014;**2014**:243713

[45] Schofield L, Villaquiran J, Ferreira A, Schellekens H, Nussenzweig R, Nussenzweig V. Gamma interferon, CD8+ T cells and antibodies required for immunity to malaria sporozoites. Nature. 1987;**330**(6149):664-666

[46] Mellouk S, Green SJ, Nacy CA, Hoffman SL. IFN-gamma inhibits development of *Plasmodium* berghei exoerythrocytic stages in hepatocytes by an L-arginine-dependent effector mechanism. Journal of Immunology. 1991;**146**(11):3971-3976

[47] Doolan DL, Sedegah M, Hedstrom RC, Hobart P, Charoenvit Y, Hoffman SL. Circumventing genetic restriction of protection against malaria with multigene DNA immunization: CD8+ cell-, interferon gamma-, and nitric oxide-dependent immunity. Journal of Experimental Medicine. 1996;**183**(4):1739-1746

[48] Zhong S, Stevenson MM. Central role of endogenous gamma interferon in protective immunity against blood-stage *Plasmodium* chabaudi AS infection. Infection and Immunity. 2000;**68**(8):4399-4406

[49] Taylor E, Onditi F, Maina N, Ozwara H. Immunization of mice with soluble lysate of interferon gamma expressing *Plasmodium berghei* ANKA induces high IFN-γ production, Tropical Diseases, Travel Medicine and Vaccines. 2017;**3**(1):11

[50] McCall MBB, Hopman J, Daou M, Maiga B, Dara V, Ploemen I, et al. Early interferon-γ response against *P. falciparum* correlates with interethnic differences in susceptibility to Parasitemia between sympatric Fulani and Dogon in Mali. The Journal of Infectious Diseases. 2010;**201**(1):142-152

[51] Belnoue E, Potter SM, Rosa DS, Mauduit M, Grüner AC, Kayibanda M, et al. Control of pathogenic CD8+ T cell migration to the brain by IFN-gamma during experimental cerebral malaria. Parasite Immunology. 2008;**30**(10):544-553

[52] Guevara Patiño JA, Marino MW, Ivanov VN, Nicolić-Žugić J. Sex steroids induce apoptosis of CD8+ CD4+ double-positive thymocytes via TNF-α. European Journal of Immunology. 2000;**30**(9):2586-2592

[53] Pérez AR, Roggero E, Nicora A, Palazzi J, Besedovsky HO, del Rey A, et al. Thymus atrophy during *T. cruzi* infection is caused by an immuno-endocrine imbalance. Brain, Behavior, and Immunity. 2007;**21**(7):890-900

[54] Fayad R, Sennello JA, Kim SH, Pini M, Dinarello CA, Fantuzzi G. Induction of thymocyte apoptosis by systemic administration of concanavalin A in mice: Role of TNF-α, IFN-γ and glucocorticoids. European Journal of Immunology. 2005;**35**(8):2304-2312

[55] Liepinsh DJ, Kruglov AA, Galimov AR, Shakhov AN, Shebzukhov YV, Kuchmiy AA, et al. Accelerated thymic atrophy as a result of elevated homeostatic expression of the genes encoded by the TNF/lymphotoxin cytokine locus. European Journal of Immunology. 2009;**39**(10):2906-2915

[56] Francelin C, Paulino LC, Gameiro J, Verinaud L. Effects of *Plasmodium* berghei on thymus: High levels of apoptosis and premature egress of CD4(+)CD8(+) thymocytes in experimentally infected mice. Immunobiology. 2011;**216**(10):1148-1154

[57] Gameiro J, Nagib PR, Andrade CF, Villa-Verde DM, Silva-Barbosa SD, Savino W, et al. Changes in cell migration-related molecules expressed by thymic microenvironment

during experimental *Plasmodium* berghei infection: Consequences on thymocyte development. Immunology. 2010;**129**(2):248-256

[58] Gravely SM, Hamburger J, Kreier JP. T and B cell population changes in young and in adult rats infected with *Plasmodium* berghei. Infection and Immunity. 1976;**14**(1):178-183

[59] Andrade CF, Gameiro J, Nagib PR, Carvalho BO, Talaisys RL, Costa FT, et al. Thymic alterations in *Plasmodium* berghei-infected mice. Cellular Immunology. 2008;**253**(1-2):1-4

[60] Seixas E, Ostler D. *Plasmodium* chabaudi chabaudi (AS): Differential cellular responses to infection in resistant and susceptible mice. Experimental Parasitology. 2005;**110**(4):394-405

[61] Thakur L, Singh KK, Shanker V, Negi A, Jain A, Matlashewski G, et al. Atypical leishmaniasis: A global perspective with emphasis on the Indian subcontinent. PLoS Neglected Tropical Diseases. 2018;**12**(9):e0006659

[62] World Health Organization. Fact sheets, Details, Leishmaniasis. 2019. Available from: https://www.who.int/news-room/fact-sheets/detail/Leishmaniasis

[63] Salotra P, Sreenivas G, Pogue GP, Lee N, Nakhasi HL, Ramesh V, et al. Development of a species-specific PCR assay for detection of *Leishmania donovani* in clinical samples from patients with kala-azar and post-kala-azar dermal leishmaniasis. Journal of Clinical Microbiology. 2001;**39**(3):849-854

[64] El Tai N, Osman O, El Fari M, Presber W, Schönian G. Genetic heterogeneity of ribosomal internal transcribed spacer in clinical samples of *Leishmania donovani* spotted on filter paper as revealed by single-strand conformation polymorphisms and sequencing. Transactions of the Royal Society of Tropical Medicine and Hygiene. 2000;**94**(5):575-579

[65] Loeuillet C, Banuls AL, Hide M. Study of *Leishmania* pathogenesis in mice: Experimental considerations. Parasites and Vectors. 2016;**9**:144

[66] Sarween N, Chodos A, Raykundalia C, Khan M, Abbas AK, Walker LS. CD4+CD25+ cells controlling a pathogenic CD4 response inhibit cytokine differentiation, CXCR-3 expression, and tissue invasion. Journal of Immunology. 2004;**173**(5):2942-2951

[67] Novais FO, Scott P. CD8+ T cells in cutaneous Leishmaniasis: The good, the bad, and the ugly. Seminars in Immunopathology. 2015;**37**(3):251-259

[68] Bousoffara T, Louzir H, Ben Salah A, Dellagi K. Analysis of granzyme B activity as a surrogate marker of *Leishmania*-specific cell-mediated cytotoxicity in zoonotic cutaneous Leishmaniasis. Journal of Infectious Diseases. 2004;**189**(7):1265-1273

[69] Daoudaki M, Diakou A, Frydas S, Fouzas I, Karagouni E, Vavatsi N, et al. Vaccination with *Trichinella spirallis* antigens increases CD8+ peripheral T cells and enhances the Th2 immune response in *Leishmania infantum* challenged mice. International Journal of Immunopathology and Pharmacology. 2009;**22**(1):169-174

[70] Olekhnovitch R, Ryffel B, Müller AJ, Bousso P. Collective nitric oxide production provides tissue-wide immunity during *Leishmania* infection. The Journal of Clinical Investigation. 2014;**124**(4):1711-1722

[71] Diefenbach A, Schindler H, Donhauser N, Lorenz E, Laskay T, MacMicking J, et al. Type 1 interferon

(IFNα/β) and type 2 nitric oxide synthase regulate the innate immune response to a protozoan parasite. Immunity. 1998;**8**(1):77-87

[72] Mattner J, Wandersee-Steinhäuser A, Pahl A, Röllinghoff M, Majeau GR, Hochman PS, et al. Protection against progressive Leishmaniasis by IFN-β. The Journal of Immunology. 2004;**172**(12):7574-7582

[73] Alexander J, Brombacher F. T helper1/t helper2 cells and resistance/susceptibility to *Leishmania* infection: Is this paradigm still relevant? Frontiers in Immunology. 2012;**3**:80

[74] Kumar R, Singh N, Gautam S, Singh OP, Gidwani K, Rai M, et al. *Leishmania* specific CD4 T cells release IFNγ that limits parasite replication in patients with visceral leishmaniasis. PLoS Neglected Tropical Diseases. 2014;**8**(10)

[75] Pinheiro RO, Rossi-Bergmann B. Interferon-gamma is required for the late but not early control of *Leishmania amazonensis* infection in C57Bl/6 mice. Memórias do Instituto Oswaldo Cruz. 2007;**102**(1):79-82

[76] Gupta G, Majumdar S, Adhikari A, Bhattacharya P, Mukherjee AK, Majumdar SB, et al. Treatment with IP-10 induces host-protective immune response by regulating the T regulatory cell functioning in *Leishmania donovani*-infected mice. Medical Microbiology and Immunology. 2011;**200**(4):241-253

[77] Majumder S, Bhattacharjee S, Chowdhury BP, Majumdar S. CXCL10 is critical for the generation of protective CD8 T cell response induced by antigen pulsed CpG-ODN activated dendritic cells. PLoS One. 2012;**7**(11):e48727

[78] Vasquez RE, Xin L, Soong L. Effects of CXCL10 on dendritic cell and CD4+ T-cell functions during *Leishmania amazonensis* infection. Infection and Immunity. 2008;**76**(1):161-169

[79] Macedo ABB, Sánchez-Arcila JC, Schubach AO, Mendonça SCF, Marins-Dos-Santos A, de Fatima Madeira M, et al. Multifunctional CD4+T cells in patients with American cutaneous Leishmaniasis. Clinical and Experimental Immunology. 2012;**167**(3):505-513

[80] Schnorr D, Muniz AC, Passos S, Guimaraes LH, Lago EL, Bacellar O, et al. IFN-γ production to *Leishmania* antigen supplements the *Leishmania* skin test in identifying exposure to *L. braziliensis* infection. PLoS Neglected Tropical Diseases. 2012;**6**(12):e1947

[81] Dayakar A, Chandrasekaran S, Kuchipudi SV, Kalangi SK. Cytokines: Key determinants of resistance or disease progression in visceral Leishmaniasis: Opportunities for novel diagnostics and immunotherapy. Frontiers in Immunology. 2019;**10**:670

[82] Andargie TE, Ejara ED. Pro- and anti-inflammatory cytokines in visceral Leishmaniasis. Journal of Cell Science & Therapy. 2015;**6**(3):1

[83] Losada-Barragán M, Umaña-Pérez A, Durães J, Cuervo-Escobar SA, Rodríguez-Vega A, Ribeiro-Gomes FL, et al. Thymic microenvironment is modified by malnutrition and *Leishmania infantum* infection. Frontiers in Cellular and Infection Microbiology. 2019;**9**:252

[84] Losada-Barragan M, Umana-Perez A, Cuervo-Escobar S, Berbert LR, Porrozzi R, Morgado FN, et al. Protein malnutrition promotes dysregulation of molecules involved in T cell migration in the thymus of mice infected with *Leishmania* infantum. Scientific Reports. 2017;**7**:45991

[85] Cuervo-Escobar S, Losada-Barragan M, Umana-Perez A, Porrozzi R, Saboia-Vahia L, Miranda LH, et al. T-cell populations and cytokine expression are impaired in thymus and spleen of protein malnourished BALB/c mice infected with *Leishmania* infantum. PLoS One. 2014;**9**(12):e114584

Chapter 8

Typhoid Fever and Its Nervous System Involvement

Atif Iqbal Ahmed Shaikh and Appasamy Thirumal Prabhakar

Abstract

Typhoid fever is a common cause of febrile illness. The causative organism S. Typhi uses special mechanisms to invade the intestines and then disseminates to the reticuloendothelial system. Thereafter, using the immune mechanism to its own advantage, it can reach the nervous system. The nervous system involvement usually occurs around the second week of fever. It usually occurs when the patient has severe sepsis. Neuropsychiatric manifestations are common, and fatigue is out of proportion to the fever. Diagnosis is often delayed, due to lack of diagnostic facilities in developing nations where it is common. In developed nations diagnosis is delayed as well, as often it is not suspected. Antibiotic therapy usually is effective, unless resistance is present, which is gradually becoming common. Early diagnosis and treatment usually leads to complete resolution of symptoms.

Keywords: Typhoid, brain, CNS

1. Introduction

The label 'enteric fever' suggests a febrile illness arising out of infection of the gut, however it is restricted to infections with *Salmonella enterica* typhi and paratyphi [*Salmonella enterica* serovar typhi and paratyphi] [1] rather than the family-enterobacteriacae itself. The nosology of 'enteric' suggests early and predominant involvement of the ileum and other parts of the gastrointestinal and biliary system in typhoid and paratyphoid fever [2]. The term typhoid is derived from the [3] Greek word *typhus*, in 1829 by a French Pathologist Louis Pierre [4]. He wanted to describe the disease by one of its prominent manifestations in those days. Typhus means 'hazy' or 'smoky' and typhoid means 'Typhus like', differentiating it from the typhus group of fevers. The 'hazy' could have been a reference to the CNS manifestation where the patient is often delirious.

Clinical manifestations: Typhoid fever to this day institutes a very significant proportion of diarrheal and febrile illnesses, especially in the developing world [5]. It has been reduced to very low levels in countries with good sanitation and accessibility to hygienic drinking water [6]. Cases in developed nations often are patients who have traveled to nations where enteric fever is endemic and these patients may have a drug resistance pattern similar to the originating country or locality [7–9].

Classical clinical features in the form of fever in a step ladder pattern, rose-spots and relative bradycardia may be less commonly seen and recognized. Typical fever patterns and classical signs may help to consider *Salmonella typhi* infection early in the disease, however, they are often not recognized [3, 10]. Enteric manifestations are common, with diarrhea, vomiting, abdominal pain and abdominal tenderness

IntechOpen

are present in most patients. Non-enteric manifestations are common as well, and central nervous system manifestations are discussed in this chapter.

Nervous system manifestations: The nervous system manifestations usually occur later in the disease, usually by the second week. CNS manifestations are often associated with severe disease and other toxic manifestation including septicemia and septic shock. Fever with any central manifestations, always raises the possibility of meningitis, if the febrile illness has not been diagnosed earlier. Empirical antibiotics given early in the disease, impacts culture results.

Microbiological diagnosis becomes imperative to decide choice of antibiotics. Appropriate diagnostic facilities are often not available in areas where typhoid infection is common. These patients are often treated with empirical antibiotics [11]. If the choice, duration and doses are incorrect, patients are likely to develop multi-drug resistant infection, and often present late when they become toxic and have signs of severe disease. Antibiotic resistance to fluoroquinolones has become common. Resistance to third generation cephalosporins is rising as well [12, 13]. Inappropriate antibiotic usage also interferes with cultures, hence microbiological diagnosis becomes incorrect, and hence all the more risk of drug resistance and poorer clinical outcomes [14]. Incidence of multi-drug and extensively drug resistant salmonella infections, that includes resistance to extended spectrum penicillin, and carbapenems like meropenem has been reported [15, 16]. This makes it necessary for clinicians to be alert to the possibility of patients presenting with typhoid fever in all its manifestations including the eponymous one.

The earliest possible description which possibly can be typhoid fever was made by the historian Thucydides [17]. He describes 'the plague of Athens' in his writings of the Peloponnesian war in 430 BC and probably later again. The description is that of a slowly rising fever, weakness, diarrhea, muscle pain, rash of flat spots, and in extreme illness, intestinal bleeding, memory loss, and confused behavior.

Clinical studies with large patient cohorts have been described since the end of the nineteenth century and recently as well. Descriptions of neuropsychiatric manifestations, behavior and association with fever along with other classic manifestations have been described. Quite a few classic authors in the field of neurology have dabbled with typhoid and its CNS involvement. In their classic book "On peripheral neuritis- A treatise" James Ross and Judson S Bury describe 'Paralysis after typhoid fever" in detail, amongst other infective and non-infective causes of neuropathies [18]. Adolphe Gubler of Millard-Gubler syndrome fame has described patients with palatal paresis and limb paresis after presumably a bout of typhoid fever. Hermann Nothnagel has described patients with typhoid fever developing ulnar nerve palsies [19]. Even William Osler has attempted to describe and name a syndrome called 'Typhoid spine' which apparently had been used until the 1980s, however, has fallen out of use [20]. Wallenberg had reported 4 cases out of 160 cases of hemiplegia were secondary to typhoid fever. Hemiplegia was also reported by Smithies and Osler [21, 22].

Many large descriptive cohort studies have been conducted, since the beginning of the twentieth century, and later in the post-antibiotic era. CNS manifestations range from 5 to 35% in various studies [10, 23–26].

Psychiatric manifestations in the form of delirium, altered behavior are the commonest CNS manifestations. It may be difficult to differentiate from encephalopathy of sepsis and may have similar pathogenesis. Older authors describe 'scared' 'frightful' patients who worsen to become comatose and develop focal deficits. Aggressive behavior is less commonly described [19]. Memory disturbances are common and may remain persistent after the acute encephalopathy wanes off. Behavioral disturbances generally improve, however often do not resolve completely. Hallucinations, delusions and other psychotic symptoms have been described, and are less common.

Low mood and fatigue out of proportion to fever is common, and has been demonstrated in chimpanzees, who develop excessive fatigue without fever, hinting that fever and fatigue may have different pathogeneses [27].

Other less common manifestations that have been reported are stroke like presentations, cerebellar involvement, reversible extrapyramidal syndrome, myelopathy and optic neuropathy. These presentations are more commonly late presentations. Other peripheral nervous system involvement in the form of Guillain Barre syndrome has been reported. Cases of *Salmonella* brain abscess have been reported [28–30].

Diagnosis of typhoid fever, when CNS involvement is significant, can be delayed. In Mozambique, an outbreak of fever with often patients developing neurological complications was reported. Typhoid was not suspected immediately, however, later investigations revealed it to be the cause. Around 13% of patients had some neurological abnormality, with similar profile as other series, with addition. Vitamin B6 deficiency was also described to be low, however, direct comparison with patients without neurological manifestations was not done [25].

Patients who develop salmonella infection-related CNS complications generally have a more severe form of illness. Older patients, dehydration, lung involvement, thrombocytopenia and low blood counts have been found to be high risk factors for development of encephalopathy. Widal levels were found to be higher in patients with encephalopathy as compared to patients without CNS involvement [26].

Neurological manifestations have largely been reported to occur in the second week of fever, however, can be seen early in the disease as well. Early and empirical use of antibiotics may be one reason for reduced CNS manifestations of typhoid fever. However, cultures must be taken, and an appropriate microbiological diagnosis must be made.

CNS manifestations generally wane with the initiation of early antibiotics. However, some studies have noted features of encephalopathy after the patient has been treated with antibiotics. The mechanism proposed for this is endotoxin release, however, it has not been substantiated adequately. Another possible mechanism is immune mediated disease. Encephalopathy occurring after initiating antibiotics may confound the treating physician. Treatment consists of steroids as discussed later.

Non-typhoidal salmonellosis can cause CNS involvement in the form of encephalopathy and meningitis as well. This has even been reported from developed nations. Minimal brain edema, microvesicular fatty liver and severe enterocolitis was seen in the patients that expired. Focal encephalitis with seizures with frontal intermittent rhythmic delta activity (FIRDA) on electro-encephalography (EEG) has been described. Rare manifestations like cerebellar ataxia, cranial nerve palsies and Guillain Barre Syndrome have been reported [31, 32]. Non-typhoidal *Salmonella* associated meningitis has a very high mortality of around 60%.

2. Immune response to salmonellae

Innate immune response is the first wall in the defense at places where the body is exposed to microorganisms. One important such site is the gastroenteric tract where a balance has to be maintained to enable absorption of nutrients and protection from plethora of organisms residing or invasive organisms [33].

The acidic pH is the one of the initial steps to reduce infections arising in the gastro-intestinal tract. Thus, conditions in which acidic pH is impaired can give rise to higher risk of contracting salmonella or severe salmonella. This includes achlorhydria, proton pump inhibitor therapy or previous gastrectomy. Additionally, salmonellae also show increased acid tolerance, especially if previously exposed to a moderately acidic environment such as ponds or soils [34].

After passing the stomach salmonellae arrive in the intestine. The intestinal flora is one of the luminal barriers to prevent invasion. Antibiotic usage, especially ones that destroy the intestinal flora, increase chances of infection as well as give rise to a risk of severe infection [35]. Also, during an outbreak, patients who developed salmonella infection were significantly more likely to have taken antibiotics in the last one month as compared to individuals who had not taken [36, 37]. The intestinal flora provide competition for nutrients, as well as reduce pH in the intestines

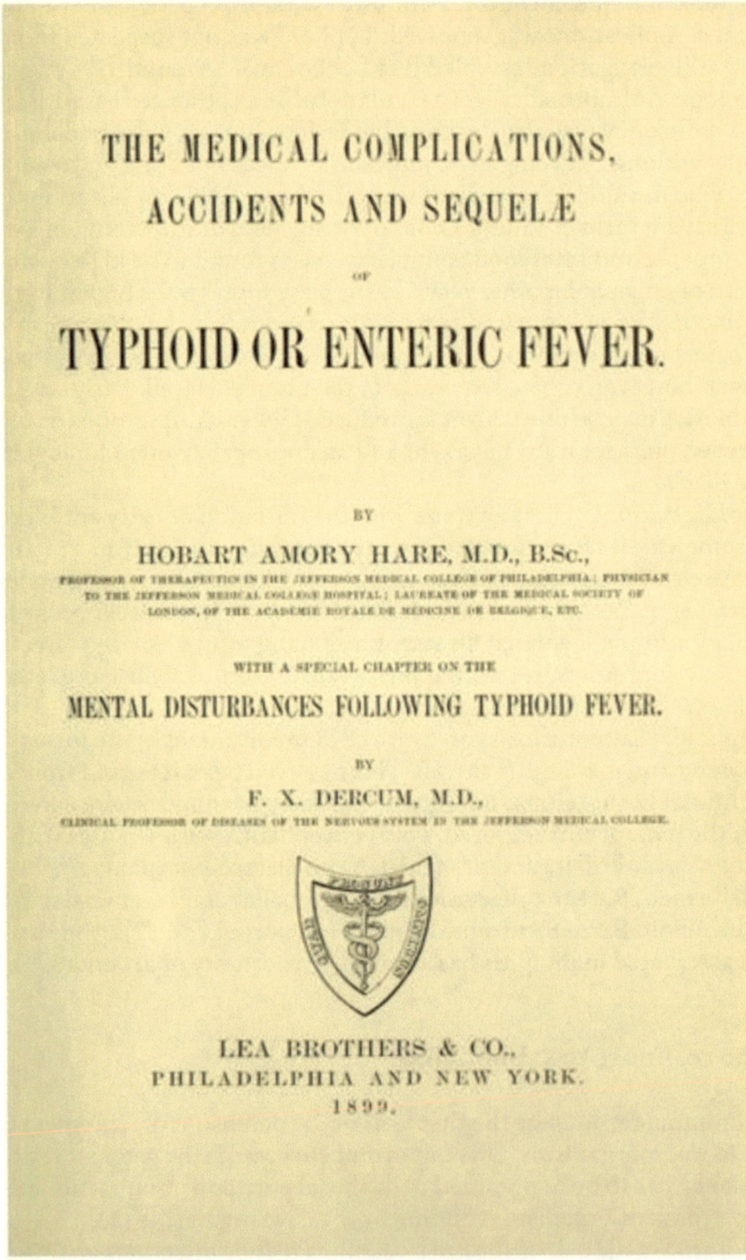

THE MEDICAL COMPLICATIONS,
ACCIDENTS AND SEQUELÆ
OF
TYPHOID OR ENTERIC FEVER.

BY
HOBART AMORY HARE, M.D., B.Sc.,
PROFESSOR OF THERAPEUTICS IN THE JEFFERSON MEDICAL COLLEGE OF PHILADELPHIA; PHYSICIAN TO THE JEFFERSON MEDICAL COLLEGE HOSPITAL; LAUREATE OF THE MEDICAL SOCIETY OF LONDON, OF THE ACADÉMIE ROYALE DE MÉDECINE DE BELGIQUE, ETC.

WITH A SPECIAL CHAPTER ON THE
MENTAL DISTURBANCES FOLLOWING TYPHOID FEVER.

BY
F. X. DERCUM, M.D.,
CLINICAL PROFESSOR OF DISEASES OF THE NERVOUS SYSTEM IN THE JEFFERSON MEDICAL COLLEGE.

LEA BROTHERS & CO.,
PHILADELPHIA AND NEW YORK.
1899.

Figure 1.
A page from the classic book on typhoid fever and its manifestations called 'The Medical Complications, Accidents and Sequelae of Typhoid or Enteric Fever' by Hobart Amory Hare and F.X. Dercum. [5].

and increase volatile fatty acids. However, salmonella use a specific nutrient called ethanolamine, which is especially present when the intestines are inflamed. Therefore, a mild inflammation induced by salmonella increased chances of further infection. Additionally, inflammatory bowel diseases like Crohn's disease provide an additional risk factor for development of salmonella infection.

Salmonella enters the gut by means of antigen presenting cells or direct invasion of the epithelial cells [38, 39]. Entry is partially gained via the cystic fibrosis transmembrane conductance regulator protein, that is abnormal is cystic fibrosis. Indeed, homozygous and heterozygous mutations in this gene may have some protection against salmonella infections.

Salmonella can survive and replicate intracellularly, with a capacity to divide in different tissues. Entry into the enteric cells and later survival in macrophages and that too especially macrophages in the liver and spleen may be essential for it to gain virulence as well as seed to different organs, including the central nervous system. This stage of reticuloendothelial infection gives rise to the common clinical findings of rising fever after complaints of loose stools, and mild hepatosplenomegaly [40].

Salmonella virulence in the gut has been identified in studies that it arises from the *Salmonella* pathogenicity island loci [SPI] placed on the *Salmonella* chromosome. They produce amongst others a protein called Type-III secretion system (TTSS-1) that modulates the intracellular milieu to aid *Salmonella* uptake into the cells and replication [41, 42]. Whether the same mechanism is enough or essential is a topic of many studies. Salmonellae evade the initial response as they can survive in the macrophages. This is mediated by the Salmonella pathogenicity Island – 2 [SPI-2] encoding type three secretion system that blocks movement of

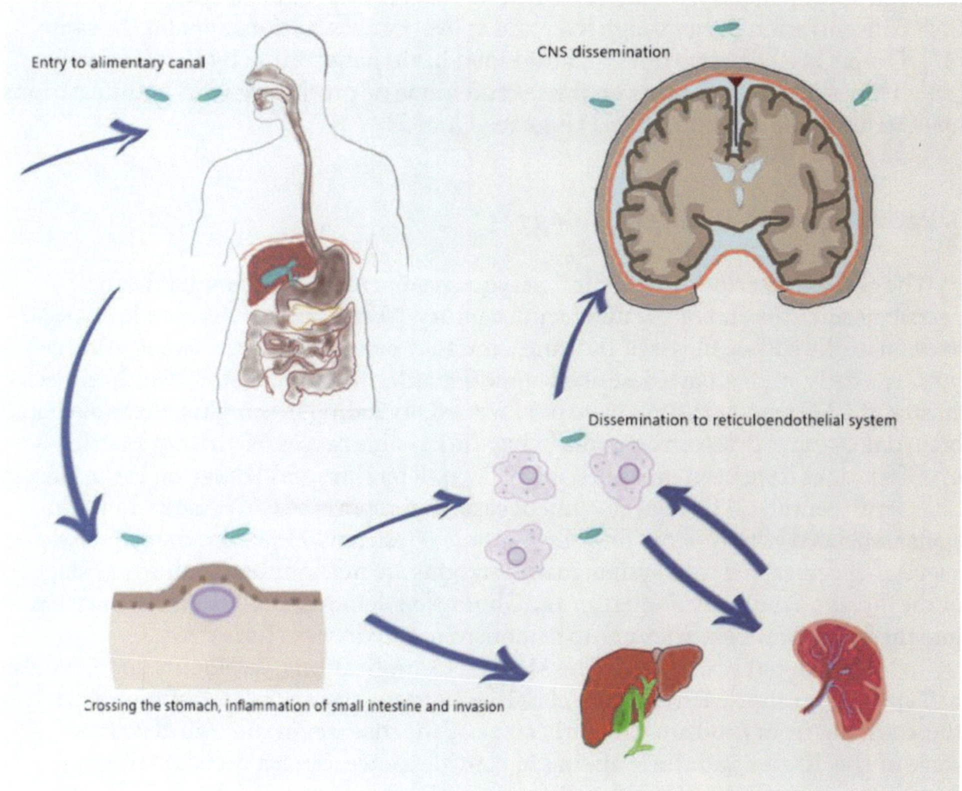

Figure 2.
Basic pathway of salmonella infection to reach the central nervous system.

reactive oxygen and nitrogen species inside the phagosome, where the salmonella survives. This initial survival against the innate immune responses is the key to systemic infection. Once the salmonellae survive in the macrophages, they can travel and multiply in other sites [33, 43]. Developing a mouse model and demonstrating *Salmonella* bacteria in the brain after oral feeding, was amongst the initial steps to study brain infections [44]. Crossing the blood–brain barrier is difficult, however eventual invasion of the neurons and later multiplication of the bacteria in the CNS is a relatively rapid process as has been demonstrated *in vitro* models by Debolina et al. [45, 46].

In the central nervous system, the presence of blood brain barrier poses an extra line of defense to most organisms including *Salmonella*. How *S. typhi* get around this is not clearly known. Most of the studies of central nervous system infection with *Salmonella* have been done with *S typhimurium*, as *S. Typhi* itself is an exclusive human pathogen. Van Sorge et al. used human brain microvascular endothelial cells to demonstrate binding and intracellular uptake of *Salmonella typhimurium* as a substitute to blood brain-barrier [45]. They showed that entry via blood–brain barrier may have additional pathogenic factors apart from the SPI associated proteins. Additionally, it has been studied in septicemia secondary to gram-negative bacteria there are structural and functional deficits in the blood–brain barrier which may give rise to sepsis-associated encephalopathy. This is a complex state with multiple players, including circulatory and microcirculatory dysfunction, cytokine storm, free radical release and oxidative damage amongst other metabolic derangements.

When salmonella interact with the blood brain barrier increased chemokines and a neutrophilic response is noted. Interactions with outer membrane protein A [often a constituent protein of cell membrane of gram negative bacteria] may be involved as ompA deficient strains when injected intraperitoneally have reduced CNS concentration even though liver and spleen concentrations remain the same [47]. OmpA has been incriminated for blood brain penetration by *E. coli* as well [48]. This additionally brings up the second mode of producing CNS manifestations mediated via non-specific toxins (**Figures 1** and **2**).

3. Pathology and pathophysiology

The clinical description of central nervous system manifestations has been described since the end of the nineteenth century. Numerous studies have been done to evaluate the pathogeneses of the same, and they continue into the twenty-first century. The early studies have described clinical manifestations in detail. Multiple mechanisms of CNS manifestations were put forward, including meningitis, toxic-mediated brain damage and dyselectrolytemia being the possible causes. Neuropsychiatric manifestations have been described in good detail by Hare and Foulerton in the late nineteenth century [31]. They document cases and reports of early mania and delusions associated with typhoid fever by multiple physicians. However, they conclude suggesting central nervous system manifestations are not common in the early stages of the disease. Studies by Foulerton and Thompson demonstrated bacterial invasion into the brain, however, attempts to demonstrate a toxin were futile [49].

In their seminal book titled 'The Medical Complications, Accidents and Sequelae of Typhoid or Enteric Fever', Hare classifies nervous system related symptoms into those arising from prodrome or early stage of the disease, in the well-developed stage of the disease and those arising in convalescence. Such a division of manifestations pushes one to consider different pathogenic mechanisms for different manifestations. These are described in **Table 1**.

Period of infection	Manifestations
Early infection or prodrome [First few days]	Confusional state, delusions, mania, meningismus often progress rapidly to coma
Well developed stage [After more than a week]	Delirium- restlessness, insomnia, confusion, apathy, delirious, vivid dreams, delusions, impaired sensorium, retrograde amnesia, melancholy, visual hallucinations, convulsions, meningitis +/− Kernigs sign with pathologically confirmed purulent meningitis involving both dura and leptomeninges, lead-pipe rigidity, cranial nerve palsies, cortical venous thrombosis, bulbar palsy, coma
Convalescence	Paralysis-LMN type- peripheral neuropathy, UMN type- Myelopathy, hemiplegia secondary to cortical involvement, thrombosis, hemorrhage, meningo-encephalitis, aphasia, cerebellar involvement, chorea

Table 1.
CNS manifestations of Salmonella *infection.*

4. Hypotheses and evaluation of pathogenic mechanisms

CNS manifestations have been suspected to result out of one or more of the following

1. *Salmonella* meningitis or meningoencephalitis: For this, the organism if entering the body orally, requires to be absorbed into the gut, get disseminated, cross the blood–brain barrier and finally cause infection of the cells of the CNS. Direct damage to the cells then causes the clinical syndrome.

2. Toxin mediated damage: where chemicals are released locally or systemically which impair the neuronal function. The toxin could be *Salmonella* specific or non-specific, for example, cytokine storm following any gram-negative septicemia.

   The role of toxins in pathogenesis of clinical manifestations in Salmonella has been suspected for more than a century. However, the discovery of toxin and mechanism of actions is yet under investigation. A reason for the inability to demonstrate toxins is that toxins are produced intracellularly and causes severe disease only in humans [27, 50]. The toxins have variable effect on humans and chimpanzees. In chimpanzees, despite there being a higher concentration of bacteria, symptoms are only mild, with manifestations being severe in humans. This has been attributed to the differences in sialoglycans in humans and chimpanzees. Additionally, even though the toxin produced only a mild fever in primates, it did produce severe malaise and fatigue that is seen in patients with typhoid fever. As the malaise occurs without the fever, it is expected to probably be a CNS manifestation rather than secondary to a systemic infection [27]. A recent study indicates that typhoid toxin is not essential for typhoid infection nor may be responsible for early manifestations. The study was not adequate to negate attribution to severe manifestations or chronic disease [51]

3. Electrolyte disturbances: These could arise in typhoid commonly as a result of loose stools or vomiting. Renal impairment in severe infections causing multiple organ dysfunction syndromes can also cause metabolic derangements.

4. Immune mediated damage: Many reported syndromes can best be explained by immunological involvement, for example, Guillain Barre syndrome,

acute disseminated encephalomyelitis like presentation, cerebellitis and late manifestations.

Immune-mediated clinical syndromes associated with typhoid are often suspected and reported [52]. Another reason to suspect immune basis is that the symptoms begin well after the fever subsides, in which case autoimmune mechanisms become very likely. Resolution of symptoms with steroids helps the case further, however, specific antigens have not been conclusively described. A well-known complication of severe typhoid fever is macrophage activation syndrome. In all the series that have studied CNS manifestations, discussion regarding macrophage/microglial activation is surprisingly inadequate. This is one complication that needs further evaluation as a means of encephalopathy. This is especially important as treatment may mandate high dose steroids and carefully follow up till the macrophage activation syndrome reverses.

5. Micronutrient deficiency: Possibly postulated as secondary to intestinal involvement giving rise to difficulty in the absorption of some nutrients, and anorexia as well along with increased metabolic demand.

5. Diagnosis and evaluation

Clinical evaluation: Like any febrile illness, evaluation begins with a history and circumstantial knowledge. In an area of a high incidence of typhoid fever, there is a chance that the diagnosis will be suspected at the outset. History of any contacts with patients who had salmonella may be beneficial. An astute clinical examination in all patients presenting with acute febrile illnesses should include looking for dehydration, coated tongue, Rose spots, splenomegaly and relative bradycardia. A detailed CNS examination is a must especially in cases of manifestations, including brief cognition testing, evaluation for focal motor, or sensory deficits and neck stiffness, at the bare minimum.

Laboratory evaluation: All patients with an acute febrile illness usually have total blood counts done which may be normal or mildly reduced. Thrombocytopenia may be present. Blood cultures in all patients is mandatory before starting antibiotics. Bone marrow culture is more likely to yield a positive result; however, it is invasive in nature [14]. Stool and urine cultures may be helpful as well in the second and third weeks of fever. Serological tests like Widal or Typhidot can be done in the second week if the fever persists. In addition to getting cultures, it is imperative to check for sensitivity patterns, in view of ever-rising drug resistance. CSF is almost always done when CNS manifestations are present, importantly to rule out bacterial meningitis secondary to common organisms. CSF studies are often found to be normal. Cultures may occasionally show the presence of *Salmonella*.

Other evaluations for encephalopathy require checking of hydration status, and metabolic disorders including dys-electrolytemias, renal failure, liver disease, or acidosis.

Imaging: In the absence of florid findings on the CSF, other means to chase CNS involvement in typhoid fever is imaging. MRI findings have been described in multiple cases. They include focal white matter edema suggestive of cerebritis or diffuse vasogenic edema. There may be other focal signs in the form of single or multiple *Salmonella* associated abscesses. Reversible diffusion restriction in the white matter has been described [53–55]. Often findings of swelling of the splenium of corpus callosum have been reported, however it is a finding that can be seen when imaging is done immediately after seizures.

EEG generally shows nonspecific findings. There can be focal slowing, focal spikes, and one case showing features of FIRDA [56].

Serological testing with WIDAL is commonly conducted, as it is widely available, inexpensive and easy to conduct. A rise in titers may be more important than a single study. Although it is non-specific, patients with CNS involvement often have very high titers of antibodies, as compared to patients without CNS involvement. Also, in the same spirit, patients are often sicker, have systemic inflammatory response syndrome, low blood counts, thrombocytopenia and more likely to have pulmonary or hepatic complications [26].

6. Management

Early diagnosis is essential in cases of salmonella infections which can be difficult especially in a non-endemic clinical setting. In endemic areas blood-cultures commercial kits are often not available or may not be affordable. As the incubation period is around one to two weeks, a history of recent travel must be documented in patients. This has shown to be as much as 60 days. The classic clinical presentations of step ladder pattern fever, with loose stools, and abdominal pain, relative bradycardia, rose spots over the abdomen and occasionally chest, and coated tongue should be carefully evaluated, however, they often may be absent. Soft splenomegaly may be present. Peripheral white blood cell counts are generally on the lower side of normal, and thrombocytopenia is often present.

Definitive diagnosis however entails blood culture, especially in the first week. It can take a long time to show results. Immunological tests including WIDAL become positive only after a period of one week. Imaging and CSF analysis are often done for evaluation of differential diagnosis.

Antibiotic therapy: Antibiotic therapy is generally not as urgent as in other gram-negative septicemia. However as soon as the diagnosis is made, therapy should be initiated with help from a local microbiological guide regarding sensitivity to drugs. Empirical antibiotic therapy is to be generally avoided, however, must be started early in cases when the patient has SIRS or CNS manifestations.

Antibiotic therapy is generally advised for a period of two to three weeks. Ceftriaxone becomes a good choice of therapy especially as it has good CNS penetration. As more often *Salmonella* resistance is being reported, it is imperative to have a local drug sensitivity profile. Fluoroquinolones were sensitive, and possibly sensitivity to ciprofloxacin may still be present, however gradually increasing NARST [Nalidixic acid-resistant *S Typhi*] strains may necessitate higher doses or change in therapy. Extensively drug-resistant strains have been reported with resistance to Chloramphenicol, Trimethoprim-sulbactam, ampicillin, third-generation cephalosporin, and fluoroquinolones [57]. Resistance to Azithromycin has been reported in a single case of *S. paratyphi* who was treated with ceftriaxone [58]. However, if the patient is septic, or is in shock, injectable carbapenems like meropenem may be the treatment of choice at the outset. Therapy can be adjusted later based on drug sensitivity testing.

Steroids: Steroids are indicated in the treatment of patients with severe disease [59, 60]. The two common and most important indications are CNS disease and shock. CNS involvement in the form of encephalopathy, psychiatric manifestations, cerebellar involvement, extrapyramidal involvement, myelopathy or seizures are indications of starting steroids. Hydrocortisone, dexamethasone and pulse doses of methylprednisolone have been tried, with case reports suggesting good outcomes [61]. CNS manifestations have been recorded often to revert quickly with steroids, however, there are only case reports present. Additionally, mechanisms of how steroids help is not explored, as also if steroids have caused poor outcomes. This aspect requires further study.

Steroids are often helpful in presumed immunologically mediated syndromes, for example, ADEM and cerebellitis. Intravenous immunoglobulin (IvIg) may be required in some cases, especially post infectious Guillain Barre syndrome.

Supportive therapy: This includes rehydration and fluid resuscitations. Dyselectrolytemia needs to be corrected promptly. If the patient has had seizures, then anti-epileptic therapy is warranted. The choice of anti-epileptic will vary from patient to patient. Phenytoin is to be avoided if cerebellar signs are present. Valproate and phenytoin are to be avoided if hepatic dysfunction is present and levetiracetam is to be avoided or doses need to be adjusted if renal failure is present.

Early and aggressive treatment with close monitoring is required to avoid long-lasting complications. Long-lasting problems known to persist are psychosis, delusions and spasticity. Extrapyramidal and cerebellar involvement generally reverts completely. Patients can have memory deficits and behavioral symptoms and hence follow up is essential.

7. Public health issues

Salmonella infection being an orally acquired infection, has major public health issues. Any such infection should assume that there is a break in the path for fecal waste disposal or contamination. CNS manifestations are assumed to be severe manifestations and hence they have a special regard.

Severe infections are common in patients who have taken antibiotics in the past one month. Widespread illogical use of broad-spectrum antibiotics should hence be controlled. With a single dose of streptomycin, the bacterial flora of the gut gets altered and leads to higher risks of infection. Additionally, drug resistant infections, especially fluoroquinolone and third generation cephalosporin resistant infections are on the rise and drug resistance to carbapenems has been reported as well. This is a huge implication in terms of management of typhoid fever.

Use of oral vaccine has been slow, but gradually increasing. The protection offered by these vaccines should reduce severe typhoid infections, and additionally so the CNS manifestations, however, this is not clearly known.

Author details

Atif Iqbal Ahmed Shaikh* and Appasamy Thirumal Prabhakar
Christian Medical College, Vellore, India

*Address all correspondence to: dr.atifshaikh@gmail.com

References

[1] Stoesser N, Eyre D, Basnyat B, Parry C. Treatment of enteric fever (typhoid and paratyphoid fever) with third and fourth generation cephalosporins. Cochrane Database Syst Rev [Internet]. 2013 [cited 2019 Sep 24];(3). Available from: https://www.cochranelibrary.com/cdsr/doi/10.1002/14651858.CD010452/full

[2] Physical Examination - Royal College Surgeons in Ireland [Internet]. [cited 2019 Sep 24]. Available from: https://www.rcsi.ie/index.jsp?p=1452&n=1459

[3] Habte L, Tadesse E, Ferede G, Amsalu A. Typhoid fever: clinical presentation and associated factors in febrile patients visiting Shashemene Referral Hospital, southern Ethiopia. BMC Res Notes [Internet]. 2018 Aug 22 [cited 2019 Oct 25];11. Available from: https://www.ncbi.nlm.nih.gov/pmc/articles/PMC6103867/

[4] Barnett R. Typhoid fever. The Lancet. 2016 Nov 19;388(10059):2467.

[5] Stanaway JD, Reiner RC, Blacker BF, Goldberg EM, Khalil IA, Troeger CE, et al. The global burden of typhoid and paratyphoid fevers: a systematic analysis for the Global Burden of Disease Study 2017. Lancet Infect Dis. 2019 Apr 1;19(4):369-381.

[6] Typhoid vaccines: WHO position paper. Releve Epidemiol Hebd. 2008 Feb 8;83(6):49-59.

[7] Ackers M-L, Puhr ND, Tauxe RV, Mintz ED. Laboratory-Based Surveillance of *Salmonella* Serotype Typhi Infections in the United States: Antimicrobial Resistance on the Rise. JAMA. 2000 May 24;283(20): 2668-2673.

[8] Steinberg EB, Bishop R, Haber P, Dempsey AF, Hoekstra RM, Nelson JM, et al. Typhoid Fever in Travelers: Who Should Be Targeted for Prevention? Clin Infect Dis. 2004 Jul 15;39(2):186-191.

[9] Ahmedullah H, Khan FY, Al Maslamani M, Al Soub H, Chacko K, Abu Khattab M, et al. Epidemiological and clinical features of *Salmonellatyphi* infection among adult patients in Qatar: A hospital-based study. Oman Med J. 2018 Nov;33(6):468-472.

[10] Klotz SA, Jorgensen JH, Buckwold FJ, Craven PC. Typhoid fever. An epidemic with remarkably few clinical signs and symptoms. Arch Intern Med. 1984 Mar;144(3):533-537.

[11] Ilić K, Jakovljević E, Škodrić-Trifunović V. Social-economic factors and irrational antibiotic use as reasons for antibiotic resistance of bacteria causing common childhood infections in primary healthcare. Eur J Pediatr. 2012 May 1;171(5):767-777.

[12] Shrestha S, Yadav RS, Deo SK. Burgeoning irrational antibiotics use in primary health care in Nepal. J Nepal Health Res Counc. 2019 Jan 28;16(41):473-475.

[13] Dixit A, Kumar N, Kumar S, Trigun V. Antimicrobial resistance: Progress in the decade since emergence of New Delhi metallo-β-lactamase in India. Indian J Community Med. 2019 Jan 1;44(1):4.

[14] Mogasale V, Ramani E, Mogasale VV, Park J. What proportion of *Salmonella Typhi* cases are detected by blood culture? A systematic literature review. Ann Clin Microbiol Antimicrob [Internet]. 2016 May 17 [cited 2019 Oct 26];15. Available from: https://www.ncbi.nlm.nih.gov/pmc/articles/PMC4869319/

[15] Britto CD, Wong VK, Dougan G, Pollard AJ. A systematic review of antimicrobial resistance in *Salmonella*

enterica serovar *Typhi*, the etiological agent of typhoid. PLoSNegl Trop Dis. 2018 Oct 11;12(10):e0006779.

[16] Meher Rizvi. Rising prevalence of enteric fever due to multi-drug resistant *Salmonella*: an epidemio logical study. J Med Microbiol. 57((2008)):1247-1250.

[17] Papagrigorakis MJ, Yapijakis C, Synodinos PN, Baziotopoulou-Valavani E. DNA examination of ancient dental pulp incriminates typhoid fever as a probable cause of the Plague of Athens. Int J Infect Dis. 2006 May;10(3):206-14. doi: 10.1016/j.ijid.2005.09.001. Epub 2006 Jan 18.

[18] Ross J, Bury JS. On peripheral neuritis. 434 p.

[19] Hare HA. The Medical Complications, Accidents and Sequelae of Typhoid or Enteric Fever.

[20] The typhoid spine.JAMA, JAMANetwork1907;XLVIII(1):53. doi:10.1001/jama.1907.02520270059005 Available from: https://jamanetwork.com/journals/jama/article-abstract/461781

[21] Osler W. Hemiplegia in typhoid fever. J NervMent Dis. 1896 May;21(5):295.

[22] Smithies F. Hemiplegia as a complication in typhoid fever,with report of a case. J Am Med Assoc. 1907 Aug 3;XLIX(5):389-95.

[23] Panda PK, Panda K. Study of clinico-epidemiological risk factors associated with enteric encephalopathy in children. Int J Contemp Pediatr. 2018 Aug 24;5(5):1971-1974.

[24] Osuntokun BO, Bademosi O, Ogunremi K, Wright SG. Neuropsychiatric manifestations of typhoid fever in 959 patients. Arch Neurol. 1972 Jul 1;27(1):7-13.

[25] Neurologic Manifestations Associated with an Outbreak of Typhoid Fever, Malawi - Mozambique, 2009: An epidemiologic investigation [Internet]. [cited 2019 Jul 22]. Available from: https://www.ncbi.nlm.nih.gov/pmc/articles/PMC3513310/

[26] Leung DT, Bogetz J, Itoh M, Ganapathi L, Pietroni MAC, Ryan ET, et al. Factors associated with encephalopathy in patients with *Salmonella enterica serotype Typhi* bacteremia presenting to a diarrheal hospital in Dhaka, Bangladesh. Am J Trop Med Hyg. 2012 Apr;86(4):698-702.

[27] Galán JE. Typhoid toxin provides a window into typhoid fever and the biology of <em>Salmonella</em> Typhi. Proc Natl Acad Sci. 2016 Jun 7;113(23):6338.

[28] Chakrabarti. Guillain-Barre syndrome in a case of typhoid fever: A less common scenario [Internet]. [cited 2019 Jul 22]. Available from: http://www.cjhr.org/article.asp?issn=2348-3334;year=2015;volume=2;issue=2;spage=176;epage=177;aulast=Chakrabarti

[29] Berger JR, Ayyar DR, Kaszovitz B. Guillain-Barré syndrome complicating typhoid fever. Ann Neurol. 1986 Nov;20(5):649-650.

[30] Hanel RA, Araújo JC, Antoniuk A, da Silva Ditzel LF, Flenik Martins LT, Linhares MN. Multiple brain abscesses caused by *Salmonella typhi*: case report. Surg Neurol. 2000 Jan;53(1):86-90.

[31] Hare HA. Varieties of onset, Symptoms of onset connected with central nervous system. In: The Medical Complications, Accidents and Sequelae of Typhoid or Enteric Fever. p. 52-60.

[32] Sawhney IM, Prabhakar S, Dhand UK, Chopra JS. Acute cerebellar ataxia in enteric fever. Trans R Soc Trop Med Hyg. 1986;80(1):85-86.

[33] Daniel Hurley, Matthew P. McCusker, Séamus Fanning and Marta Martins. Salmonella–Host Interactions – Modulation of the Host Innate Immune System. Front. Immunol., 07 October 2014 | https://doi.org/10.3389/fimmu.2014.00481

[34] J W Foster. Low pH adaptation and the acid tolerance response of Salmonella typhimurium, Crit Rev Microbiol. 1995;21(4):215-237. doi: 10.3109/10408419509113541.

[35] Bohnhoff M, Miller CP, Martin WR, resistance of the mouse's intestinal tract to experimental salmonella infection. I. Factors which interfere with the initiation of infection by oral inoculation.J Exp Med. 1964;120:805.

[36] Tannock GW, Savage DC, Indigenous microorganisms prevent reduction in cecal size induced by Salmonella typhimurium in vaccinated gnotobiotic mice. Infect Immun. 1976;13(1):172.

[37] Ryan CA, Nickels MK, Hargrett-Bean NT, Potter ME, Endo T, Mayer L, Langkop CW, Gibson C, McDonald RC, Kenney RT JAMA. 1987;258(22):3269. Massive outbreak of antimicrobial-resistant salmonellosis traced to pasteurized milk.

[38] Kops SK, Lowe DK, Bement WM, West AB. Migration of Salmonella typhi through intestinal epithelial monolayers: an in vitro study. Microbiol Immunol. 1996;40(11):799.

[39] Mills SD, Finlay BB Comparison of Salmonella typhi and Salmonella typhimurium invasion, intracellular growth and lo,calization in cultured human epithelial cells. Microb Pathog. 1994;17(6):409.

[40] Giannella RA. Salmonella. In: Baron S, editor. Medical Microbiology [Internet]. 4th ed. Galveston (TX): University of Texas Medical Branch at Galveston; 1996 [cited 2019 Oct 26]. Available from: http://www.ncbi.nlm.nih.gov/books/NBK8435/

[41] Singh Y, Saxena A, Kumar R, KumarSaxena M. Virulence System of *Salmonella* with Special Reference to *Salmonella enterica*. *Salmonella* - Re-EmergPathog [Internet]. 2018 Nov 5 [cited 2019 Oct 27]; Available from: https://www.intechopen.com/books/salmonella-a-re-emerging-pathogen/virulence-system-of-i-salmonella-i-with-special-reference-to-i-salmonella-enterica-i--

[42] Ibarra JA, Steele-Mortimer O. *Salmonella* – the ultimate insider. *Salmonella* virulence factors that modulate intracellular survival. Cell Microbiol. 2009 Nov;11(11): 1579-1586.

[43] Eric AHughesJorge EGalán. Immune Response to *Salmonella*: Location, Location, Location? Immunity; Volume 16, Issue 3, March 2002, Pages 325-328

[44] Mathur R, Oh H, Zhang D, Park S-G, Seo J, Koblansky A, et al. A mouse model of *Salmonella typhi* infection. Cell. 2012 Oct 26;151(3):590-602.

[45] van Sorge NM, Zialcita PA, Browne SH, Quach D, Guiney DG, Doran KS. Penetration and activation of brain endothelium by *Salmonella enterica serovar Typhimurium*. J Infect Dis. 2011 Feb 1;203(3):401-405.

[46] Chaudhuri D, Roy Chowdhury A, Biswas B, Chakravortty D. *Salmonella Typhimurium* Infection Leads to Colonization of the Mouse Brain and Is Not Completely Cured With Antibiotics. Front Microbiol. 2018;9:1632.

[47] Aslam MS, Akhter M, Rasheed R, Samra ZQ, Gull I, Athar MA. Identification and purification of antigenic 34 kDa outer membrane protein of *Salmonella typhi*. Clin Lab. 2012;58(9-10):1071-1077.

[48] Prasadarao NV. Identification of *Escherichia coli* Outer Membrane Protein A Receptor on Human Brain Microvascular Endothelial Cells. Infect Immun. 2002 Aug 1;70(8):4556-4563.

[49] Foulerton Alexander GR, Campbell Thomson H. On the causation of nervous symptoms in typhoid fever,: with an experimental study of the action of typhoid toxins on the ganglion cells of the central nervous system. The Lancet. 1900 Apr 21;155(3999): 1121-1125.

[50] Parkhill J, Dougan G, James KD, Thomson NR, Pickard D, Wain J, et al. Complete genome sequence of a multiple drug resistant Salmonella enterica serovar Typhi CT18. Nature. 2001 Oct 25;413(6858):848-852.

[51] Gibani MM, Jones E, Barton A, Jin C, Meek J, Camara S, et al. Investigation of the role of typhoid toxin in acute typhoid fever in a human challenge model. Nat Med. 2019 Jul;25(7):1082-1088.

[52] Ktsoyan Z, Budaghyan L, Agababova M, Mnatsakanyan A, Arakelova K, Gevorgyan Z, et al. Potential involvement of *Salmonella*infection in autoimmunity. Pathogens [Internet]. 2019 Jul 3 [cited 2019 Oct 27];8(3). Available from: https://www.ncbi.nlm.nih.gov/pmc/articles/PMC6789781/

[53] Ahmed M, Sureka J, Mathew V, Jakkani R, Abhilash KPP. Magnetic resonance imaging findings in a fatal case of *Salmonella typhi*-associated encephalopathy: A case report and literature review. Neurol India. 2011;59(2):270.

[54] Chidhara S, Rangasami R, Chandrasekharan A. Magnetic resonance imaging and magnetic resonance spectroscopy in *Salmonella* meningoencephalitis. J PediatrNeurosci. 2016;11(1):88-90.

[55] Kobuchi N, Tsukahara H, Kawamura Y, Ishimori Y, Ohshima Y, Hiraoka M, et al. Reversible Diffusion-Weighted MR Findings of *Salmonella enteritidis*-Associated Encephalopathy. Eur Neurol. 2003;49(3):182-184.

[56] Uysal H, Karademir A, Kılıç M, Ertürk Ö. *Salmonella* Encephalopathy with Seizure and Frontal Intermittent Rhythmic Delta Activity. Infection. 2001 Apr 1;29(2):103-106.

[57] Huang L-T, Ko S-F, Lui C-C. *Salmonella* meningitis: clinical experience of third-generation cephalosporins. Acta Paediatr. 1997 Oct 1;86(10):1056-1058.

[58] Molloy A, Nair S, Cooke FJ, Wain J, Farrington M, Lehner PJ, et al. First report of *Salmonella enterica serotype paratyphi A* azithromycin resistance leading to treatment failure. J Clin Microbiol. 2010 Dec;48(12):4655-4657.

[59] Chisti MJ, Bardhan PK, Huq S, Khan WA, Khan AM, Sharifuzzaman null, et al. High-dose intravenous dexamethasone in the management of diarrheal patients with enteric fever and encephalopathy. Southeast Asian J Trop Med Public Health. 2009 Sep;40(5):1065-1073.

[60] Hoffman SL, Punjabi NH, Kumala S, Moechtar MA, Pulungsih SP, Rivai AR, et al. Reduction of mortality in chloramphenicol-treated severe typhoid fever by high-dose dexamethasone. N Engl J Med. 1984 Jan 12;310(2):82-88.

[61] Jain A, Baheti G. Steroid Pulse Therapy in the Management of Neuro-psychiatric Manifestations in an Atypical Presentation of Typhoid Fever. J Med Res Innov. 2019 Jul 1;e000178.

Section 2

Innate Immunity in Reproduction

Chapter 9

Innate Immune Defense in the Male Reproductive System and Male Fertility

Fei Wang, Ran Chen and Daishu Han

Abstract

To protect the male germ cells from adverse immune reaction, the male reproductive system adopts special immune environment such as immunoprivileged status. The male genital organs can be infected by various microorganisms via hematogenous dissemination and ascending genitourinary tracts. To overcome the immunoprivileged status, the male genital organs also adopt their own innate defense against microbial infection. The tissue-specific cells in the male reproductive system are well equipped with innate immune machineries, including pattern recognition receptors (PRRs) and their negatively regulatory system. PRR-initiated immune responses must be tightly regulated by the negative regulatory system for the maintenance of immune homeostasis. The immune homeostasis can be disrupted by unrestrictive innate immune response, which may lead to inflammatory conditions in the male genital tracts, an important etiological factor contributing to male infertility. This chapter describes the current understanding of the innate immune responses in the male reproductive system and their effects on male fertility.

Keywords: innate immunity, testis, male fertility, pattern recognition receptor, Tyro3/Axl/Mer receptor tyrosine kinase, orchitis

1. Introduction

Innate immunity is the first line of the body defense against microbial infections. The innate immune responses initiated by pattern recognition receptors (PRRs) play critical roles in building the innate immunity and regulating adaptive immunity [1, 2]. PRR-initiated innate immune responses can lead to acute inflammatory conditions, essential for counteracting invading microbes, which must be tightly restricted by the negative regulatory system for maintaining the immune homeostasis. Unrestricted innate immune responses may result in chronic inflammation that can be harmful to the host self [3].

PRRs not only initiate innate immune responses in the immune cells, but also in the nonimmune epithelial cells of various tissues. In particular, the tissue-specific epithelial cells of organs, such as the intestine, lung, and urogenital tracts, which are frequently infected by microorganisms, abundantly express PRRs, and the PRR-initiated innate immune responses play important roles in the tissue defense against microbial infections. The production of functional sperm is necessary for normal fertility, which requires the close cooperation of a whole reproductive system that is composed of several major organs, including the testes, epididymis, seminal vesicle, and the prostate (**Figure 1**). Sperm cells are produced in the testes, and then mature and are stored in the epididymis. Before ejection, sperm cells are mixed with seminal plasma that is mainly produced by the prostate and seminal vesicle. Sperm is produced post-puberty after the establishment of central immune tolerance. Therefore, sperm production generates immunogenic autoantigens. To avoid detrimental autoimmune responses under physiological conditions, the male productive system adopts a unique immune environment. In particular, the testis is a remarkable immunoprivileged organ. The epididymis also has immunoprivileged properties for protecting sperm during maturation and storage from an immune attack. However, the male reproductive system can be infected by various organisms via ascending genital tracts and hematogenous dissemination, which frequently lead to inflammation of the system and impairment of male fertility. The defined inflammation in the male reproductive system includes urethritis, prostatitis, seminal vesiculitis, epididymitis, and orchitis. Ascending bacterial infections represent frequent etiological factors of inflammation in the male productive system [4]. Epididymitis and orchitis are predominantly caused by hematogenous dissemination of viruses. Moreover, noninfectious epididymitis and orchitis are also frequently observed. In this regard, male germ damage can be a sterile cause of epididymitis. While all of the inflammatory conditions may perturb male fertility, the inflammatory male infertility is mostly caused by epididymitis and orchitis [5].

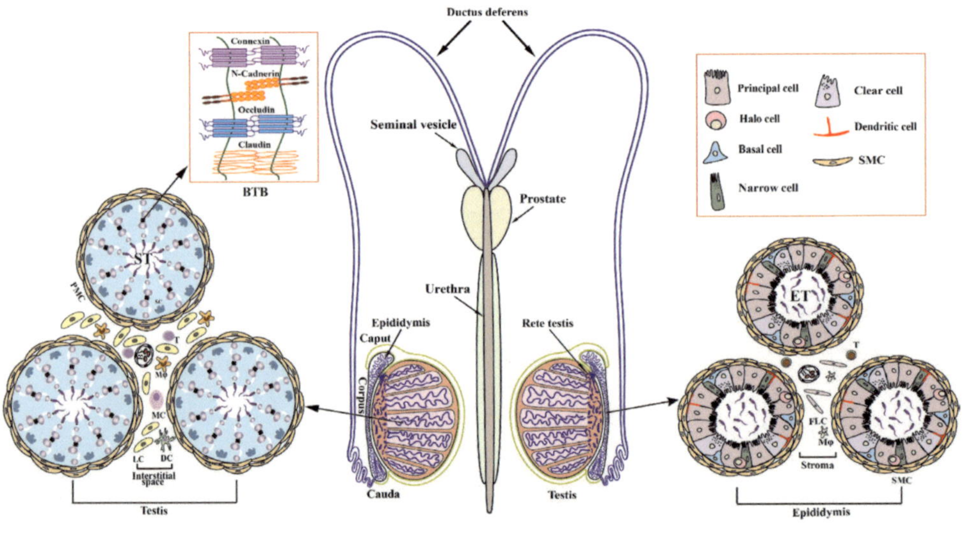

Figure 1.
Schematic of the male reproductive system. The male reproductive system is composed of various organs (middle panel), including genital glands, the testes and epididymides; accessory genital glands, the prostates and seminal vesicles; and genital tracts, the ductus deferens and urethra. Both the testis and epididymis belong to the immunoprivileged organs for protecting male germ cells from adverse immune reaction. The testis histologically contains the seminiferous tubule (ST) and interstitial space (left panel). The ST is surrounded by peritubular myoid cells (PMC) and lined by the seminiferous epithelium that is composed of columnar Sertoli cells (SC) embracing different stages of germ cells. The blood-testis barrier (BTB) is formed between adjacent two SC. In the interstitial spaces of the testis, there are mainly Leydig cells (LC), and various immune cells, including macrophages (Mφ), T lymphocytes (T), mast cells (MC), and dendritic cells (DC). Blood vessels (BV) are located in the interstitial spaces. The epididymis can be divided into the caput, corpus, and cauda epididymides. The caput epididymis is connected with the testis via the rete testis. The epididymis is composed of the coiled epididymal tubules (ET) and stroma among the ET (right panel). The ET is surrounded by smooth muscle cells (SMC) and lined by a pseudostratified epithelium that is composed of principal cells, halo cells, basal cells, narrow cells, clear cells, and DC. The stroma areas contain BV, Mφ, T, and certain fibroblast-like cells (FLC). The corpus epididymis is located between the caput and cauda segments, and the cauda epididymis is connected with the ductus deferens following by the urethra.

Infectious and inflammatory conditions in the male genital tracts represent the most frequent etiological factors contributing to male infertility, which range to 15% in developed countries and 30% in developing countries [6]. In contrast to acute infectious inflammation, which displays evident disease phenotypes, noninfectious chronic inflammatory conditions frequently occur in the male reproductive system and have mild or no visible phenotypes. Therefore, the rates of inflammatory male infertility can be underestimated. The complex inflammatory conditions in the male reproductive system are attributable to its special composition and immune environment. The specificities of the male reproductive system include the following aspects: (1) the male productive system is composed of multiple distinct organs, including the testis, epididymis, prostate, and seminal vesicle; (2) this open system can be frequently infected by sexually transmitted pathogens and pathogenic bacteria; (3) the immunogenic sperm is produced, stored, and transported in the male reproductive system; and (4) tissue-specific cells of the male reproductive system abundantly express PRRs that can be activated by ligands from microbial pathogens and endogenous germ cells. Therefore, PRR-initiated innate immune responses in the tissue-specific cells can play important roles in the defense against microbial infection and the initiation of the inflammatory response in the male reproductive system, thereby being involved in health and disease.

2. Immune environment in the healthy and diseased testis

The mammalian testis possesses a unique immune function for protecting the development of immunogenic germ cells from detrimental immune responses and local defenses against microbial infection. The disruption of testicular immune homeostasis may lead to orchitis, one of the etiological factors contributing to male infertility. Notably, the studies on the mechanisms underlying testicular immune regulation and innate defense against microbial infections are mostly carried out using murine models. How these models inform humans remains to be clarified.

2.1 Testicular immune privilege

The testis shares the remarkable status of immune privilege with the eyes, brain, and uterus [7]. The main goal of immune privilege in the testis is to prevent adverse immune responses against male germ cells. The first round of male germ cell development is only completed after puberty, a long time after the establishment of immune self-tolerance, which occurs during fetal and immediately after birth. Therefore, a majority of male germ cells, particularly the late stages of germ cells that are generated during puberty, are stranger to the immune system and can actively induce immune responses elsewhere extra the testis [8]. These immunogenic male germ cells do not induce immune responses in the testis under physiological conditions due to its immunoprivileged environment.

The histological structure and the local immunosuppressive milieu cooperatively produce the testicular immunoprivileged environment.

2.1.1 Testicular structure favoring immune privilege

The mammalian testis is a complex organ with a highly organized histological structure, many different cell types, and a highly efficient immunosuppressive milieu. The testis is histologically composed of two distinct regions: the seminiferous tubules and the interstitial spaces (**Figure 1**). The seminiferous tubules are surrounded by peritubular myoid cells (PMC). Within the seminiferous tubule, the columnar Sertoli cells encompassing different stages of male germ cells, including spermatogonia, primary spermatocytes, secondary spermatocytes, round spermatids, and elongated spermatids, form the seminiferous epithelium where male germ cells develop to form sperm. The seminiferous epithelium provides a special microenvironment for germ cell development. Notably, the blood-testis barrier (BTB) is formed by various cellular junctions between adjacent Sertoli cells near the basal side of the seminiferous epithelium. The seminiferous epithelium is separated into two compartments, namely basal and adluminal compartments, by the BTB.

While the interstitial spaces constitute a minor region in the testis, there are many cell types in this region. Leydig cells represent a majority of interstitial cells and produce testosterone. Moreover, immune components, including blood vessels and various immune cells, are found in the interstitial spaces. Macrophages are major immune cells. Several other immune cells, including T lymphocytes, mast cells, and dendritic cells have been found in the testicular interstitial spaces. By contrast, B lymphocytes have yet to be observed in these spaces under physiological conditions. The testicular immune privilege confers two properties: the testis tolerates a large number of immunogenic male germ cells and allo-xenografts without immune rejection or that can survive for a prolonged time. The BTB can isolate most germ cells within the adluminal compartments behind the BTB from the immune components in the interstitial spaces. Therefore, the BTB plays a critical role in maintaining the immunoprivileged status in the adluminal compartments. However, the BTB cannot be fully responsible for the immune privilege of the whole testis because the germ cells outside the BTB, including preleptotene spermatocytes and spermatogonia, also produce immunogenic substances [9]. Moreover, the interstitial spaces themselves are also immunoprivileged because the grafts in the interstitial spaces can survive for an extended period. Therefore, other mechanisms must be involved in the maintenance of the testicular immune privilege.

2.1.2 Immunosuppressive milieu

The testis is a highly organized histological structure composed of a great diversity of cell types. In addition to various stages of developing male germ cells, there are many types of somatic cells in the testis. PMC and Sertoli cells in the seminiferous epithelium and Leydig cells in the interstitial spaces are essential for the testicular functions. Most types of immune cells can be found in the interstitial spaces under physiological conditions. The testicular cells secrete a large spectrum of endocrine and paracrine immunoinhibitory factors, which form a dense network that cooperatively contributes to the immunoprivileged environment in the testis (**Figure 2**).

Leydig cells represent major tissue-specific cell types that produce androgen, mainly the testosterone that is essential for male germ cell development and the functions of many target organs extra the testis. Testosterone levels in the testis are 10-fold higher than in the peripheral circulation. Substantial evidence supports its inhibitory effects on the autoimmune response, which contributes to the difference in autoimmune diseases between males and females. Testosterone is also involved in the maintenance of the testicular immunoprivileged environment because it inhibits the induction of experimental autoimmune orchitis [10]. Testosterone should not directly act on immune cells because they do not have its receptor. In fact, only Sertoli cells express androgen receptors. Therefore, testosterone contributes to the immune privilege via regulating Sertoli cell functions. Accordingly, conditional knockout of the androgen receptor in Sertoli cells impairs the BTB and leads to autoimmune orchitis [11].

Testicular cells secrete a large panel of paracrine immunosuppressive factors that inhibit immune responses. In addition to the endocrine function, Leydig

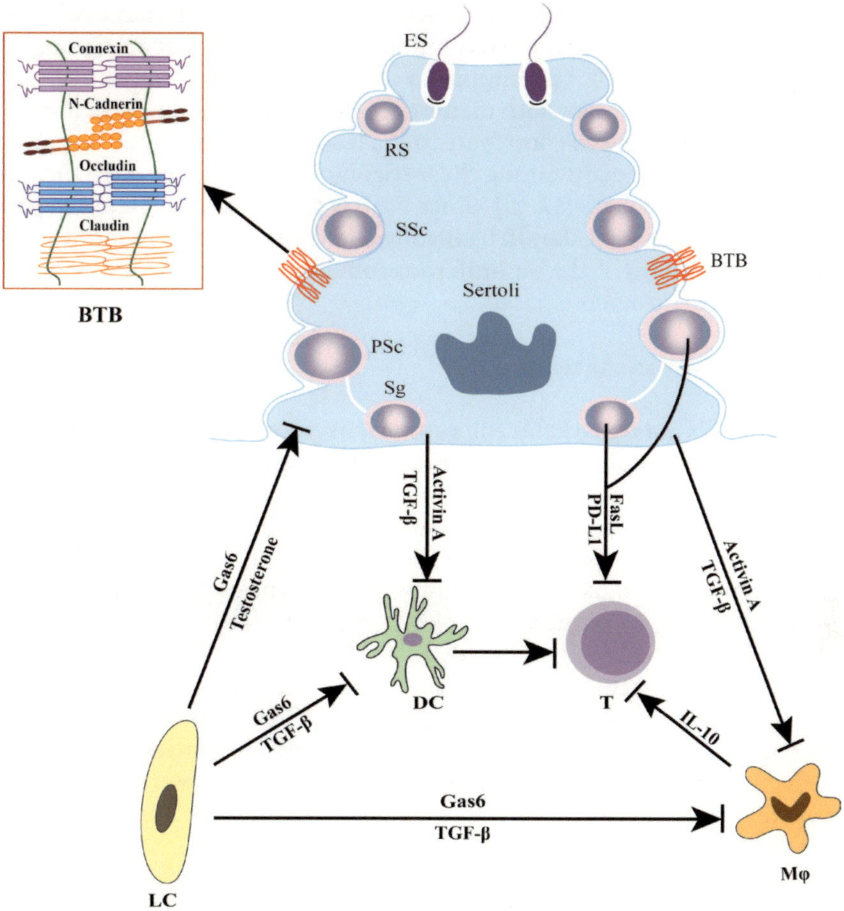

Figure 2.
Immunosuppressive environment in the testis. The systemic immunity in the testis is suppressed by tissue structure and paracrine immunoinhibitory factors. The blood-testis barrier (BTB) sequesters the late stage of germ cells from immune cells in the interstitial spaces. Sertoli cells secrete activin A and transforming growth factor β (TGF-β), and these two factors inhibit immune responses of dendritic cells (DC) and macrophages (Mφ). Leydig cells produce Gas6, TGF-β, and testosterone, which inhibit immune properties of DC, Mφ, and Sertoli cells. Male germ cells produce soluble Fas ligand (FasL) and programmed death ligand 1 (PD-L1). Both FasL and PD-L1 induce apoptosis of T lymphocytes. Mφ exhibit immunosuppressive properties by producing anti-inflammatory cytokine IL-10 to inhibit effective T cell response. DC in the testis predominantly belong to immature subtype and inhibit effector T cells. Sg, spermatogonia; PSc, primary spermatocytes; SSc, secondary spermatocytes; RS, round spermatid; and ES, elongated spermatids.

cells produce growth arrest-specific factor 6 (Gas6). Gas6 inhibits innate immune responses through the activation its receptors Tyro3, Axl, and Mer (TAM) receptor tyrosine kinases not only in immune cells [12], but also in Leydig and Sertoli cells [13]. While testicular macrophages represent considerable populations of interstitial cells, they predominantly exhibit immunosuppressive properties by producing anti-inflammatory factor IL-10 and lacking the innate immune machinery [14]. Therefore, testicular macrophages favor immune privilege, rather than inflammatory conditions in the testis. The seminiferous tubular cells also produce multiple immunoinhibitory molecules contributing to the immunoprivileged status. Most notably, besides the formation of the BTB, Sertoli cells produce various anti-inflammatory factors, such as activin A and TGF-β, that inhibit the activation of immune cells. The paracrine secretions of Sertoli cells make these cells feasible to prevent immune rejection of grafts after co-transplantation [15]. A main function of the immune privilege is the protection of male germ cells from immune surveillance.

The germ cells also have the ability to inhibit the activation of immune cells. Fas ligand (FasL) and PD-L1 are two major immune checkpoints that act through the induction of T cell apoptosis and inhibition of T cell activation, respectively. Male germ cells abundantly express both membrane-bound and soluble FasL and PD-L1. PD-L1 has been confirmed to contribute to the testicular immunoprivileged state [16]. The role of FasL in maintaining the testicular immune privilege in the interstitial spaces is under debate [17, 18]. Since FasL is predominantly found in germ cells behind the BTB [18], it should be involved in the cleanup of immunity in the adluminal compartment of the seminiferous tubules. This speculation or other roles for FasL have yet to be demonstrated.

2.2 Innate immunity in the testis

While the testis is a remarkable immunoprivileged organ, it is not a sterile site. In fact, the testis can be infected by microorganisms from hematogenous dissemination and ascending genital tracts. To overcome the absence of the systemic immune components, the testis adapts its own innate immune defense against microbial infections [19]. Testicular cells express a wide panel of PRRs, which initiate innate immune responses to produce a large number of immunoregulatory factors, including pro-inflammatory cytokines, chemokines, and interferons (IFNs). These factors may activate immune cells to counteract microbial infections or directly restrict microbial replication in the infected cells. PRR-initiated innate immune responses must be negatively regulated because a high level of the immunoregulatory factors for a prolonged period would be harmful to the tissues. Disruption of the innate immune homeostasis may result in orchitis and impair testicular functions. Hereby, we discuss innate immunity in the testis, the cell-to-cell innate immune responses, and their negative regulation in major testicular cells.

2.2.1 Innate immune response signaling pathways

The mechanisms of general innate immune responses are addressed in other chapters. This chapter briefly summarizes PRR-initiated signaling pathways that have been identified in testicular cells (**Figure 3**). Several subfamilies of PRRs, including Toll-like receptors (TLRs), RIG-I-like receptors (RLRs), and cytosolic DNA sensors have been investigated [19]. TLRs, now an established subfamily, were the first PRRs to be identified, which started the field of innate immune receptors and would establish their importance in the initiation of innate immune responses. TLRs exclusively initiate the myeloid differentiation protein 88 (MyD88)-dependent pathways, with the exception of TLR3 and TLR4. TLR3 initiates the Toll/IL-1R-domain-containing adaptor-inducing IFN-β (TRIF)-dependent pathway, whereas TLR4 activation triggers both MyD88- and TRIF-dependent pathways [20]. The MyD88 pathway predominantly activates nuclear factor kappa B (NF-κB), thereby inducing the expression of pro-inflammatory cytokines and chemokines. The TRIF-dependent pathway activates NF-κB and IFN regulatory factor 3 (IRF3), thus leading to the induction of type 1 IFNs (IFN-α and IFN-β) and pro-inflammatory cytokines. These cytokines promote the recruitment and activation of leukocytes, and the expression of IFN-inducible antiviral proteins, thereby counteracting invading microbial pathogens. Moreover, TLR signaling facilitates the maturation of antigen-presenting cells, thereby directing adaptive immunity. RLRs include two functional members, retinoic acid-inducible gene I (RIG-I) and melanoma differentiation-associated protein 5 (MDA5) [21]. RIG-I and MDA5 are cytosolic RNA sensors that recognize cytosolic double-strained RNA (dsRNA). RIG-I and MDA5 activation triggers signaling through an adaptor IFN-β promoter

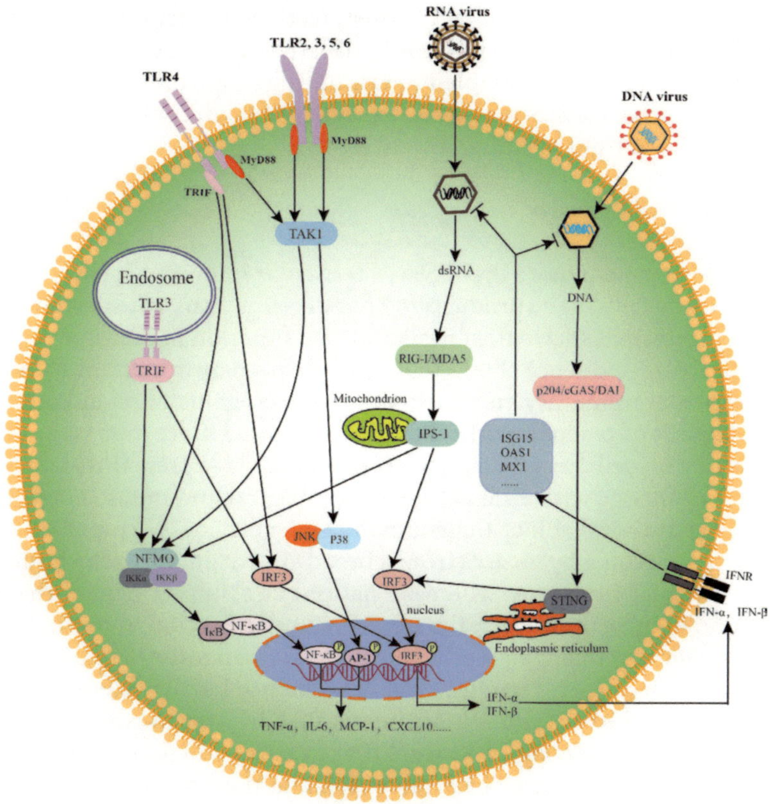

Figure 3.
Schematic of PRR signaling pathways in testicular cells. Expression and function of PRRs in most testicular cells have been revealed. PRR signaling pathways have been intensively investigated in mouse Leydig and Sertoli cells. Various TLRs initiate innate immune responses in Leydig and Sertoli cells using adaptor MyD88 except for TLR3 and TLR4. TLR3 initiates TRIF-dependent pathway, whereas TLR4 triggers innate immune responses through both MyD88- and TRIF-dependent pathways. RNA sensors RIG-I and MDA5 initiate innate antiviral responses using adaptor IPS-1 localized in mitochondria. DNA sensors p204, cGAS, and DAI trigger antiviral response using adaptor STING localized in endoplasmic reticulum. MyD88-, TRIF-, and IPS-1-dependent pathways induce the expression of pro-inflammatory cytokines TNF-α, IL-6 and chemokines MCP-1 and CXCL10 through NF-κB activation, and also induce the expression of IFN-α and IFN-β through IRF3 activation. STING-dependent signaling pathway predominantly induces IFN-α and IFN-β expression. The upregulation of pro-inflammatory cytokines and chemokines facilitate inflammatory conditions in the testis, whereas IFN-α and IFN-β are important for the testicular antiviral defense through the induction of antiviral proteins ISG15, OAS1, and Mx1 in autocrine and paracrine manners.

stimulator-1 (IPS-1) that is localized in the mitochondria [22]. The cytosolic DNA sensor initiates the signaling pathway requiring the stimulator of IFN gene (STING) as an adaptor localized in the endoplasmic reticulum [23]. IPS-1-dependent signaling leading to the activation of IRF3 and NF-κB induces the expression of type 1 IFNs and pro-inflammatory cytokines, whereas the STING-dependent signaling pathway predominantly induces type 1 IFN production through the activation of IRF3. Testicular cells adopt cell-specific PRR-initiated innate immune responses.

2.2.2 Innate immunity testicular macrophages

Under physiological conditions, testicular macrophages represent approximately 20% of total interstitial cells and 80% of immune cells in the testis. It was believed that testicular cells are the front line of innate testicular defense against microbial infections from hematogenous dissemination. However, the response of testicular macrophages to microbial antigen challenges is relatively weak compared to their counterparts in other tissues [24]. By contrast, testicular macrophages predominantly produce anti-inflammatory factor IL-10 after challenge with pathogens. Testicular macrophages show disabled innate immune signaling. This phenotype of testicular macrophages favors the immunoprivileged environment of the testis, but weakens its ability to fight invading microbes.

2.2.3 Innate antiviral defense of Leydig cells

Leydig cells represent more than 75% of testicular interstitial cells. The main function of Leydig cells is the production of androgens, mostly testosterone, which are essential for normal spermatogenesis and also function in multiple organs extra the testis. Increasing evidence shows that Leydig cells play important roles in regulating the testicular immunity and the innate defense against viral infection. The antiviral responses of testicular cells were demonstrated more than two decades ago [25, 26]. Rat Leydig cells, as well as other testicular cells express IFNs and antiviral proteins after challenge with viral antigens. Notably, rat and mouse Leydig cells exhibit stronger antiviral ability than human Leydig cells [27]. This observation may explain the reason why viruses from a broad spectrum infect the human testis and impair testicular functions, whereas a natural viral impairment of the testis has not been found in mice. Moreover, the experimental induction of testicular dysfunction using viruses that frequently impair the human testis has not been successful in wild-type mice. Recent studies reveal that mouse Leydig cells express various PRRs that recognize viruses and initiate innate antivirus responses.

Among PRRs, TLRs among PRRs were first examined in mouse Leydig cells. These cells abundantly express TLR3 and TLR4, and their respective ligands trigger innate immune responses [28]. The TLR3-initiated innate immune response in Leydig cells activates NF-kB and IRF3, thereby inducing the expression of pro-inflammatory cytokines, including TNF-α and IL-6, as well as IFN-α and IFN-β. The activation of both TLR3 and TLR4 in Leydig cells suppresses the synthesis of testosterone. Diminished testosterone production should be caused by the TLR-induced high level of TNF-α and IL-6 because these cytokines inhibit testosterone synthesis [29, 30]. Notably, TLR-initiated innate immune signaling pathways in Leydig cells are negatively regulated by TAM receptors. TAM signaling is an important negative regulatory system of immunity [12]. TAM receptors and their ligand Gas6 are abundantly expressed in the mouse testis [13]. TAM receptors are expressed in Leydig and Sertoli cells, whereas Gas6 is only expressed in Leydig cells. TAM receptors knockout mice develop autoimmune orchitis, suggesting that the Gas6/TAM signaling is essential for the testicular immunoprivileged environment in mice [31]. The roles of the Gas6/TAM signaling in maintaining testicular immune privilege can be attributed to different mechanisms: (1) Gas6 facilitates the phagocytic removal of apoptotic male germ cells through the activation of TAM receptors, which prevents the release of immunogenic male germ cell antigens [32]; (2) TAM receptors favor central immune tolerance to germ cell autoantigens because Axl and Mer knockout mice are susceptible to male germ antigen-induced autoimmune orchitis [33]; and (3) the inhibition of innate immune responses by the Gas6/TAM signaling in testicular cells, such as Leydig cells, contributes to immune privilege in the testis. Therefore, the interplay between the innate immune responses and their negative regulation is important in the testicular defense against microbial infection and the maintenance of the immunoprivileged status.

In addition to TLRs, mouse Leydig cells constitutively express functional RLRs, including RIG-I and MDA5 [34]. Both RIG-I and MDA5 initiate innate antiviral responses through IPS-1 signaling pathway in Leydig cells after challenge with their ligand dsRNA. IPS-1 signaling activates NF-κB and IRF3 in Leydig cells, thereby inducing the pro-inflammatory factors TNF-α and IL-6 as well as IFN-α and IFN-β. The IFNs subsequently induce the expression of several antiviral proteins, including 2′5′-oligoadenylate synthetase (OAS1), MxGTPase1 (Mx1), and IFN-stimulating gene 15 (ISG15), which leads to the degradation of viral RNA, inhibition of viral gene transcription, and amplification of antiviral signaling [35]. RIG-I- and MDA5-initiated IPS-1 signaling in Leydig cells suppresses testosterone synthesis, which may result in the impairment of testicular functions. Therefore, RIG-I/MDA5-initiated innate immune responses are likely to be the mechanism by which RNA viruses, including mumps virus (MuV), human immunodeficiency virus-1, and Zika viruses, frequently induce dysfunction of the testis [36]. In particular, MuV infection frequently induces orchitis, which results in male infertility. MuV can infect most testicular cells and induce immune responses through the activation of RIG-I/MDA5 in Leydig and Sertoli cells [37]. While MuV induces the expression of IFNs and antiviral proteins for antiviral responses, it also upregulates pro-inflammatory factors and chemokines. These cytokines may facilitate inflammatory conditions and the pathogenesis in the testis by inhibiting testosterone synthesis, inducing germ cell apoptosis, and impairing the BTB integrity [38, 39]. RIG-I/MDA5-initiated innate immune responses in testicular cells are a double-edged sword that both counteracts viral infection and impairs the testicular functions.

Mouse Leydig cells constitutively express the cytosolic DNA sensors p204 and STING [40]. The p204/STING signaling can be triggered in Leydig cells after challenge with viral DNA, and induce the expression of IFN-α and IFN-β, as well as antiviral proteins. By contrast, viral DNA induces relatively low levels of pro-inflammatory factors in Leydig cells. Accordingly, the viral DNA sensor-initiated innate immune response in Leydig cells does not inhibit testosterone synthesis and rarely impairs male fertility. Therefore, the DNA sensor/STING signaling seems to be an ideal pathway for preventing viral infection in the testis and protecting the testicular function. The differences in antiviral responses and testicular dysfunction after DNA and RNA viral infections are worthy of further investigation.

2.2.4 TLR-initiated innate immune response in Sertoli cells

TLRs have been well characterized in the murine testis. Functional TLR2 and TLR4 were first demonstrated in mouse Sertoli cells, which opened the study of PRR-initialed innate immune responses in the testis [41]. The expression and function of TLRs in mouse Sertoli cells were subsequently examined in more detail [42–44]. Sertoli cells abundantly express multiple TLRs, including TLR2, TLR3, TLR4, TLR5, and TLR6, and their respective ligands activate TLRs and initiate the innate immune responses. TLR-initiated innate immune responses in Sertoli cells induce the production of TNF-α, IL-6, MCP-1, and IFNs. These cytokines trigger an inflammatory reaction in the testis and facilitate the innate defense against microbial infectious within the seminiferous tubules.

TLR2 and TLR4 in mouse Sertoli cells can be activated by damaged male germ cells, and induce the expression of pro-inflammatory factors and chemokines [45]. These cytokines may impair the BTB integrity, thereby inducing autoimmune orchitis. Male germ cells express high levels of endogenous TLR ligands, such as the high-mobility group box 1 (HMGB1) and several heat shock proteins (HSPs). HMGB1 and HSPs can be released from apoptotic and necrotic male germ cells and subsequently can induce an innate immune response in Sertoli cells, thereby leading to testicular inflammation and dysfunction. Accordingly, extensive apoptosis and necrosis of male germ cells under some pathological conditions, such as physical trauma and cryptorchidism, correlate with autoimmune orchitis [46, 47]. Most male germ cells go to apoptosis before finally developing into sperm. The apoptotic germ cells must be timely removed before necrosis via phagocytosis by Sertoli cells, which is essential for the normal production of sperm [48]. The removal of apoptotic germ cells by phagocytosis contributes to the healthy testicular functions in several ways. One is the removal of autoantigens released from necrotic germ cells. The defective phagocytosis of apoptotic germ cells by Sertoli cells results in autoimmune orchitis [49]. TLR-imitated innate immune responses by endogenous ligands from damaged germ cells are involved in the development of autoimmune orchitis.

The TLR-initiated innate immune response is also negatively regulated by the Gas6/TAM system [50]. TAM receptor knockout mice increase the activation of TLR3 and TLR4 in Sertoli cells after challenge with TLR ligands, thereby overexpressing pro-inflammatory cytokines and type 1 IFNs. By contrast, Gas6 suppresses the TLR-mediated cytokine expression. The inhibition of TLR activation by the Gas6/TAM system is attributable to the induction of suppressor of cytokine signaling 1 and 3 (SOCS1 and SOCS3) proteins because both SOCS1 and SOCS3 can inhibit TLR signaling [51]. Sertoli cells express most of the major TLRs and can be infected by viruses through hematogenous dissemination as well as bacteria via ascending genital tracts. A broad spectrum of microbes may activate multiple TLRs in Sertoli cells to produce high levels of pro-inflammatory cytokines that can perturb the testicular functions. The negative regulation of TLR signaling by TAM receptors in Sertoli cells is important to prevent the impairment of testicular function by increased inflammatory cytokines.

The testis is predominantly infected by hematogenous viruses, although certain microorganisms may reach to the testis via ascending genital tracts [52]. Therefore, the innate antiviral response in the testis is critical for the prevention of virus-impaired testicular function. In addition to the aforementioned Leydig cells, Sertoli cells also possess antiviral ability by producing IFNs in response to viral infection [25]. Mouse Sertoli cells abundantly express TLR3, which recognizes viral dsRNA and initiates innate antiviral responses by inducing IFN expression [44]. However, the expression levels of RLRs and DNA sensor are relatively low in Sertoli cells compared to Leydig cells [34, 40]. Most studies on PRR-initiated innate antiviral response are focused on Leydig cells. These observations suggest that the innate antiviral machinery is better equipped in Leydig cells than Sertoli cells and that Leydig cells are capable of a more efficient antiviral response than Sertoli cells. Accordingly, Leydig cells produce relatively high levels of IFNs compared with Sertoli cells in response to MuV infection [37]. Therefore, MuV replicates more efficiently in Sertoli cells than in Leydig cells [38]. Notably, Sertoli cells produce relatively high levels of pro-inflammatory cytokines compared to Leydig cells after MuV infection. These studies indicate that Leydig cells are mainly responsible for the testicular defense against viral infection, whereas Sertoli cells predominantly contribute to orchitis in response to MuV infection. The different contributions of Sertoli and Leydig cells to the testicular defense against microbial infections and inflammation remain to be further dissected. The clarification of this issue may provide preventive and therapeutic strategies for viral orchitis, an etiological factor contributing to male infertility [30].

2.2.5 Innate defense of male germ cells

Male germ cells represent the largest populations of testicular cells throughout the entire adulthood. The germ cells are generally thought to be protected by the histological structure and testicular somatic cells. In fact, these germ cells can be infected by microbial pathogens with the predominance of viruses. Therefore, male germ cells also adopt their defense against viral infection. The innate defenses of male germ cells are stage-dependent based on their histological locations. TLR3 is expressed in spermatogonia and spermatocytes that are located both outside and inside of the BTB [53]. TLR3 can be activated by its ligand and initiate innate antiviral responses in these early stages of male germ cells. The germ cells express IFNs and antiviral proteins through TLR3 signaling upon encounter with invading viruses. TLR3-initiated innate antiviral responses in spermatogonia and spermatocytes would contribute to the seminiferous epithelial defense against viral infection. Besides spermatogonia and spermatocytes, round and elongating spermatids express TLR11 [54], which can be activated by its ligands in these late stages of germ cells that resided in the adluminal compartments behind the BTB. TLR 11 recognizes *Toxoplasma gondii* (*T. gondii*) and uropathogenic *Escherichia coli* (UPEC), two major pathogens that can reach the testis via ascending genital tracts. TLR11 activation induces the production of interleukin 12 (IL12) and IFN-γ in addition to pro-inflammatory cytokines and chemokines in spermatids. IFN-γ favors the immune defense against *T. gondii* [55]. Therefore, the TLR11-initiated innate immune response in spermatids is involved in the testicular defense against *T. gondii* and UPEC infection via ascending genital tracts in mice. However, the functional TLR11 is absent in human beings, which could be a reason why UPEC and *T. gondii* infect and impair the human testis, whereas natural UPEC and *T. gondii* infections in the murine testis have not been found. It has been demonstrated that TLR11 prevents UPEC infection in mice [56]. In addition to TLR11, MDA5 is also constitutively in spermatids. A viral dsRNA analog induces the expression of IFNs and antiviral proteins through MDA5 in spermatids. The expression levels of pro-inflammatory factors, IFNs, and antiviral proteins in male germ cells are relatively low compared to testicular somatic cells. These studies indicate that male germ cells are also equipped with innate immune machinery. Considering their large numbers, male germ cells should be significantly involved in the testicular defense against microbial infections.

In addition to PRR-initiated immune responses, male germ cells also adopt their own specific defense ability, i.e., autophagy [38]. Autophagy is a conserved intracellular degradation pathway that is tightly controlled by a series of regulatory proteins. By fusion with lysosomes, autophagy can break down dysfunctional organelles and large protein aggregates that cannot be degraded by ubiquitination, which plays various important biological roles under pathophysiological conditions [57]. Autophagy is also an important intracellular innate defense system against invading viruses, bacteria, and protozoa by directly uptaking and degrading microorganisms [58]. Mouse male germ cells are abundantly equipped with autophagic machinery [38]. MuV can be internalized into male germ cells, but fails to replicate in vitro, which is in contrast with what happens in the testicular somatic cells. The presence of an inhibitor of autophagy in culture remarkably increases MuV replication in male germ cells. This observation indicates that autophagy plays a critical role in restricting MuV replication and eliminating invading viruses in male germ cells. The antiviral activity of the autophagic pathway does not up-regulate the expression of pro-inflammatory cytokines that can be harmful for the testis to function at a high level. Therefore, autophagy is an ideal solution for the problem of defending male germ cells against microbial infections. The efficient antiviral defense of male germ cells is particularly important for the prevention of not only viral-caused testicular dysfunction, but also sexual viral transmission. While autophagy is an important mechanism underlying the intracellular antiviral response, certain viruses may escape other antiviral mechanisms by hijacking autophagy [59]. Notably, the Zika virus (ZIKV) can be sexually transmitted in humans and impair testicular function and male fertility in mice after experimental infection [60, 61]. ZIKV is detected in the semen and male germ cells of humans and mice for a prolonged period [62, 63]. Viable ZIKV can be isolated from the spermatozoa of patients with acute infection [64]. These observations suggest that male germ cells can be a reservoir for ZIKV. Previous studies detected human immunodeficiency virus 1 (HIV-1) and hepatitis virus B, typical sexual transmitted viruses, in spermatozoa [65, 66]. Taken together, spermatozoa can be vectors for sexual transmission of certain viruses. In general, while virus families covering a broad spectrum have been detected in the testis and semen [52], only a few of them have been confirmed in male germ cells. Understanding the mechanisms underlying antiviral responses and viral storage in male germ cells is particularly important because it may provide novel clues into the preventive and therapeutic strategies for virus-impaired male fertility and the sexual transmission of pathogens.

3. Innate immunity in the epididymis and epididymitis

The epididymis is the organ where spermatozoa mature and are stored. The epididymis also adopts special structural and immune environments for the protection of spermatozoa maturation from adverse immune responses.

3.1 Histological structure of the epididymis

The epididymis is composed of a highly coiled tubule of several meters in length and the stroma, which can be divided into three segments, namely the caput, corpus, and cauda epididymis. The caput epididymis connects to the rete testis via the efferent ducts and receives sperm from the testis. The sperm passes through the caput segment and is stored in the corpus and cauda epididymis before ejaculation, which is an essential step for sperm maturation and motility. The epididymal tubule is formed by a pseudostratified epithelium consisting of various cell types, including peritubular myoid cells that surround the tubular epithelium, a majority of principal cells, as well as minor narrow cells, clear cells, and basal cells. The epididymal stroma comprises connective tissue containing certain fibroblasts and blood vessels. The three segments of the epididymis are distinct in morphology and cell compositions. The peritubular myoid cell layer increases, whereas the epithelium decreases, in thickness from the caput to the cauda segments. Various immune cells can be found and are differently distributed in the three segments.

3.2 Epididymal immune environment

3.2.1 Distribution of immune cells in the epididymis

The mammalian epididymis is also considered to be immunoprivileged due to its tolerance to sperm, which undergoes maturation and is stored in this organ. However, the immune privilege in the epididymis is not as typical as in the testis. While the blood-epididymis barrier (BEB) is formed between epithelial principal cells near the lumen of the epididymal tubule, it is not as effective as the BTB because certain leukocytes may pass through BEB [67]. Moreover, various leukocytes with immunoregulative properties reside in the epididymis (**Figure 4**). Dendritic cells (DCs) and macrophages are among the major leukocyte populations and show distinct distribution in different segments of the epididymis with abundance in the caput segment and in the basal region of the epididymal epithelium [68]. The protrusions of DCs may reach the lumen via gaps between principal cells. By contrast, DCs are much fewer in the cauda epididymis. A dense network of DCs in the caput epididymis is believed to regulate immune tolerance to antigenic sperm.

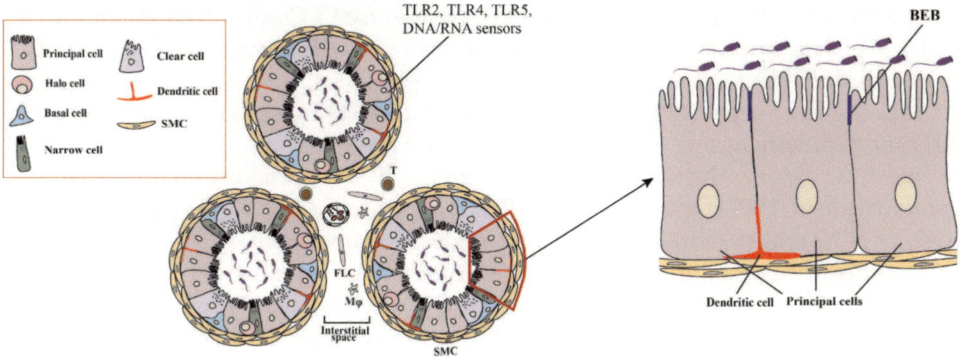

Figure 4.
Schematic of immune environment in the epididymis. Epididymal immune environment comprises local innate immune defense (left panels) and immune privilege (right panel). Dendritic cells and macrophages (Mφ) are major leukocytes. Dendritic cells are resided in the basal region of the epididymal tubule and their protrusions can reach the lumen side. Mφ are predominantly distributed in the stroma. Minor T lymphocytes (T) can be also focused in the stroma areas. Various PRRs, including TLR2, TLR4, TLR5, DNA, and RNA sensors are expressed in principal cells, and the principal cells also produce defensins to counteract microbial infections. The immunoprivileged status of the epididymis is mainly maintained by the blood-epididymis barrier (BEB).

Interestingly, the DCs distribution and the fact that inflammation is more frequent in the cauda than caput segments seem to be two related observations. However, while the distribution of DCs and macrophages in the epididymis has been intensively examined, their functions under pathophysiological conditions are mostly speculated and remain to be elucidated [69]. In addition to DCs and macrophages in the basal region of the epididymal tubules, a minor T lymphocyte subset and mast cells can be found in the stroma [70]. The function of immune cells in the stroma remains unknown.

3.2.2 Innate immune response in epididymal epithelial cells (EECs)

The epididymis can be frequently infected by ascending microbes, which may result in epididymitis, the most important etiological factor of male infertility. Since major leukocytes in the epididymis predominantly adopt properties contributing to tolerance to sperm, the majority of tissue-specific cells should be responsible for the defense against microbial infections and inflammatory conditions in the epididymis. EECs express a large number of defensins that have potent activities against microbial infection and are important for sperm function [71]. Much like testicular cells, EECs also express a broad spectrum of functional PRRs. The activation of PRRs in EECs initiates the innate immune response, thereby producing pro-inflammatory cytokines, chemokines, and IFNs. These cytokines should play roles in the epididymal defense against invading microbes by impairing microbial survival and the development of inflammatory conditions by recruiting leukocytes from circulation. TLR2 and TLR4 are expressed in the rat EECs [72]. Challenges of EECs with *Staphylococcus aureus* in vitro induce the expression of nitric oxide synthase and production of nitro oxide and TNF-α, which is attributable to the activation of NF-κB and MAPKs. Using gene knockout mice, substantial evidence was gathered showing that TLR4 and TLR5 cooperatively initiate innate immune responses in EECs after infection with UPEC [73]. UPEC induces the expression of TNF-α, IL-6, MCP-1 in EECs, and the TLR4- and TLR5-mediated NF-κB activation. Moreover, UPEC also induces the expression of type 1 IFNs in mouse EECs through the activation of IRF3 in vitro and in vivo, which should be involved in the innate defense against UPEC and inflammatory responses in the epididymis.

In addition to PRRs that recognize bacteria, mouse EECs are also well equipped with PRRs that initiate innate antiviral responses [74]. EECs express TLR3, RIG-I, and cytosolic DNA sensors. The synthetic analogs of viral RNA and DNA induce the expression of type 1 IFNs through the IPS-1 and STING signaling pathways. The IFNs subsequently induce expression of antiviral proteins OAS1, ISG15, and Mx1 in an autocrine manner. Notably, viral RNA induces IFN expression and a remarkable upregulation of TNF-α and MCP-1 in EECs. However, viral DNA induces the high levels of IFNs and antiviral proteins, but moderately upregulates TNF-α and MCP-1. These observations indicate that viral RNA significantly induces both antiviral and inflammatory responses, whereas viral DNA predominantly induces innate antiviral responses in EECs. These phenotypes are also observed in testicular cells. Considering that the high level of TNF-α and MCP-1 may impair sperm survival and favor inflammatory conditions, the IPS-1 and STING antiviral signaling pathways should differentially impact pathophysiological conditions in the testis and epididymis, thus affecting male fertility. This speculation is supported by the fact that RNA viruses such as MuV and HIV-1 frequently cause inflammation in the testis and epididymis, thereby impairing male fertility. By contrast, DNA viral orchitis and epididymitis are not evident. Mechanisms underlying different effects of RNA and DNA viruses on male fertility are interesting issues that are worthy of further examination, which may provide novel clues to develop preventive and therapeutic strategies for virus-impaired male fertility.

3.3 Epididymitis

Epididymitis is the most common male genital tract inflammation with scrotal swelling and pain or asymptomatic depending on acute or chronic phenotypes [4]. While acute epididymitis is mostly caused by bacterial infections via the ascending genital tracts, chronic epididymitis is mainly the result of noninfectious stimuli that are associated with various risk factors, including physical trauma, vasectomy, testicular cancer, post systemic or genital tract infection, and adverse effects of certain medications. UPEC, *C. trachomatis*, and *N. gonorrhoeae* are common pathogens responsible for acute infectious epididymitis. The mechanisms underlying the acute epididymitis of rodent models after experimental bacterial infection have been intensively investigated [6]. Epididymitis can be induced by injection of clinically relevant bacteria into the local or mimicking retrograde routes. Similar to observations in clinical patients, the experimental epididymitis in small animals displays reddening, swelling, and enlargement of the scrotum. Massive infiltrations of leukocytes with the predominance of lymphocytes and neutrophils are observed in the experimental epididymitis. These models are valuable to investigate the duration of infection, the mechanisms of disease, and the efficiency of treatments [75].

While antibiotic therapy is effective for acute epididymitis, the treatment of chronic epididymitis has been less successful [76]. Mechanisms underlying chronic epididymitis remain largely unknown, and their further understanding is essential for developing an effective therapy. Although it has been known that chronic epididymitis is associated with various noninfectious risk factors, the pathogenic antigens resulting in noninfectious epididymitis are less understood. Recently, two studies using mouse models showed that damaged germ cells due to different causes induced noninfectious epididymitis characterized by massive macrophage infiltration [77]. Whether the male germ cell damage is a common etiological factor contributing to infectious epididymitis associated with multiple risk factors is the most urgent issue worthy of clarification. Most male germ cells are produced during puberty, when central immune tolerance has been already established. Therefore, male germ cells express a large number of new autoantigens that can induce innate immune responses. These antigens are not released out and do not induce an immune response in the male reproductive tracts under physiological conditions. However, male germ cell damages produce endogenous TLR agonists that can induce the expression of pro-inflammatory cytokines and chemokines through the activation of TLR2 and TLR4 in Sertoli cells. Both TLR2 and TLR4 are also expressed in mouse EECs, and damaged male germ cells induce the innate immune response in EECs. Whether damaged male germ cells induce epididymitis in human beings should urgently be clarified, as it would be helpful for the development of novel diagnostic and therapeutic approaches for noninfectious epididymitis. In humans, this hypothesis is based on several phenotypes of clinical observations: (1) triggers of the noninfectious epididymitis have not been identified; (2) all known risk factors may damage male germ cells; (3) most cases of noninfectious epididymitis are unilateral and male germ cells might be damaged; and (4) chronic epididymitis is mostly observed in the cauda segment where damaged male germ cells may be stored. Further studies using human samples are required to confirm this hypothesis. In this regard, it is a priority to identify specific germ cell antigens that can induce innate immune responses and inflammation in the epididymis. These antigens in the semen may be used as a diagnostic marker to distinguish the noninfectious and infectious epididymitis, thereby choosing a suitable therapeutic approach. Epididymitis is considered to significantly impair male fertility and may be the most important single factor contributing to male infertility. Epididymitis may impair male fertility through different ways. The inflammatory conditions in the epididymis can impair sperm parameters. Moreover, acute epididymitis may spread to the testis and result in "epididymo-orchitis," and the orchitis can perturb sperm production. Acute epididymitis can be easily found because patients timely visit outpatient due to the typical symptoms. By contrast, most cases of chronic epididymitis are asymptomatic and are mainly diagnosed in outpatient visitors for infertility, suggesting that chronic epididymitis is closely associated with male infertility. The chronic epididymitis may damage the epididymal tubule and result in tissue fibrosis, thereby leading to the obstruction of the excurrent tubules. It is believed that a great portion of obstructive azoospermia and oligozoospermia, important etiological factors of male infertility, can be caused by acute and chronic epididymitis [78]. Due to lack of common diagnostic and therapeutic standards for chronic epididymitis, the effect of epididymitis on male fertility seems to be underestimated. In particular, chronic asymptomatic epididymitis would be one of the etiological factors of idiopathic male infertility, representing about 50% of total male infertility. Therefore, understanding of mechanisms underlying chronic epididymitis and its impact on male fertility can provide clues for the prevention and therapy of male infertility caused by inflammation in male genital tracts.

4. Innate immunity and inflammation in male accessory glands

Healthy male fertility not only requires normal testicular and epididymal functions for sperm development and maturation, but also needs the functions of the male accessory glands, including mainly the prostate and seminal vesicle, which produce majority of seminal plasma essential for sperm function and fertility. As organs downstream of the epididymis, both the prostate and seminal vesicle can be infected before the epididymis by retrograde microorganisms via the ascending urethra. Therefore, infectious prostatitis and seminal vesiculitis are more frequent than the inflammation in the epididymis and testis. While prostatitis and seminal vesiculitis are not considered as important etiological factors of male infertility, these inflammations may impair sperm parameters and cause male subfertility [79–81].

4.1 Prostate pathophysiology

The prostate is the largest accessory sex gland of the male genital tract and plays an essential role in facilitating fertility. The main function of the prostate is to secrete the prostate fluid that contributes up to 30% semen volume. The prostate fluid contains a large number of factors essential for healthy reproduction by protecting sperm during its travel through the female genital tract [82]. The well-known functions of these prostatic factors include ejaculation controlling, semen clotting and liquefaction, sperm activation and capitation, and female genital tract remodeling for fertilization and implantation. The prostate is composed mainly of an epithelial duct and a small volume occupied by a stroma. Epithelial cells are responsible for fluid secretion, and stromal cells are essential for the maintenance of tissue homeostasis in both the physical structure and microenvironment, and they are necessary for the secretory function of epithelial cells. The prostate is the organ with the most prevalent diseases.

Three major pathogenic conditions affect the prostate, namely benign prostatic hyperplasia, prostate cancer, and prostatitis. All three morbidities are associated with the immune response in the prostate. Benign prostatic hyperplasia is extremely prevalent in males over 50 years old and almost 90% of men older than 80 years suffer from the disease [83]. Prostate cancer is the most common cancer in men older than 60 years and is the second most prevalent cause of cancer-related deaths in men, second only to lung cancer in men [84]. Moreover, prostatitis is the most common inflammation of the urogenital tract in men younger than 50 years, in which chronic noninfectious prostatitis without defined etiological factors encompasses more than 90% of cases [79]. All three pathological conditions in the prostate are associated to local inflammation.

4.2 Inflammation and the diseased prostate

Various leukocytes, including lymphocytes, macrophages, and mast cells, have been found in the prostatic stroma. These leukocytes are involved in prostatitis toward allo- and autoantigens [85]. A set of observations suggest the association between chronic prostatitis and benign prostatic hyperplasia, including: (1) leukocyte infiltration in benign prostatic hyperplasia; (2) positive correlation between a history of prostatitis and later development of benign prostatic hyperplasia; (3) non-steroidal anti-inflammatory drugs reduce the clinical symptoms; and (4) inflammation facilitates the progression of benign prostatic hyperplasia. In addition to immune cells, prostatic epithelial and stromal cells can be inducers of prostatitis, because these tissue-specific cells express several functional TLRs, including TLR4, TLR5, TLR7, and TLR9 [86]. The activation of TLRs in prostatic cells induces the expression of pro-inflammatory cytokines and chemokines, and these cytokines subsequently recruit and activate immune cells leading to inflammatory conditions.

The chronic inflammation in the prostate plays a role in cancer development. The association between TLR-initiated innate immune response and prostate cancer has been revealed [87]. The role of TLRs in prostate cancer seems complex and might be a "double-edged sword" in cancer progression. There is evidence showing that the activation of TLR3 in prostate cancer cells inhibits cancer cell growth by different mechanisms. TLR3 activation in the prostate cancer cells in vivo induces an innate immune response that increases infiltration of T lymphocytes and NK cells into the tumor, thereby suppressing cancer growth [88]. Moreover, TLR3 signaling induces apoptosis of prostate cancer cells in vitro [89]. By contrast, the TLR3 level is associated with the recurrence of prostate cancer in humans [90]. In addition, several other TLRs, including TLR2, TLR4, and TLR9 also facilitate the growth and invasion of prostate cancer cells. The mechanisms behind the TLR3 signaling inhibition of prostate cancer development and the facilitation of tumor progression by other TLRs remain to be clarified. In this regard, the specific effects of distinct TLR signaling pathways and TLR-mediated cytokine production on cancer cell growth and apoptosis should be focused. The clarification of TLR functions in regulating the prostate cancer development can aid in developing immunotherapy against the prostate tumorigenesis by manipulating the TLR signaling pathways. In fact, several drugs targeting TLR signaling have been used in the clinic for the treatment of cancer patients [91].

4.3 Adverse effect of prostatitis on male fertility

Prostatitis, the most common inflammation among urology outpatients younger than 50 years, is characterized by dysuria, nocturia, pelvic, and perineal pain, and ejaculatory disturbances [92]. Prostatitis comprises two subtypes of acute and chronic inflammation. Acute prostatitis is induced by an acute infection of bacteria, including mostly *E. coli* and sexually transmitted pathogens *N. gonorrhoeae* and *C. trachomatis*. Therefore, acute prostatitis is sensitive to antimicrobial treatment. By contrast, chronic prostatitis is a hallmark in nonbacterial inflammation, which represents more than 90% of all prostatitis cases. The etiology and mechanism underlying chronic prostatitis are poorly understood. Since prostatic secretions significantly contribute to the seminal plasma that plays a role in natural fertilization, inflammation in the prostate may alter the components of seminal plasma and thereby impairs fertility. While seminal plasma is not indispensable for successful reproduction using assisted technology, fertilization, and fetal development are compromised without exposure of spermatozoa and female genital tract to seminal plasma [82]. Acute prostatitis scarcely alters sperm parameters and impairs male fertility. However, chronic prostatitis has a negative effect on sperm parameters and male fertility. Persistently increased levels of inflammatory factors, including TNF-α, IL-6, IL-1β, IFN-γ, and reactive oxygen species, in the semen of chronic prostatitis patients are associated with abnormal sperm parameters and reduction male fertility [79]. Since both acute and chronic prostatitis rarely results in the obstruction of genital tracts that may lead to male infertility, the roles of the prostate on health male fertility have been neglected. However, the enthusiasm and interest in this area have been increasing in recent years. The prostate secretes numerous substances that regulate fertility through the protection of sperm function, modulation of immunity in the female tract, and preservation of proper embryo implantation. The consequences of prostatitis on fertility should be considered for the assessment of male infertility.

5. Innate immunity and inflammation in the seminal vesicles

The seminal vesicle is another major male accessory gland besides the prostate and produces approximately 70% of the seminal plasma volume. The production of prostaglandin and promotion of semen coagulation are two major functions of the seminal vesicles.

Prostaglandin affects sperm parameters and sperm-oocyte interaction [93]. Inflammation in the seminal vesicles can alter prostaglandin production and semen coagulation [94], thereby impairing sperm function and fertility. Microbial infections may lead to vesiculitis, which frequently occurs with prostatitis and is then termed prostate-vesiculitis [81]. The vesiculitis is considered as an etiological factor of hemospermia. The immunity in the seminal vesicle and the vesiculitis, as well as their effects on male fertility, is largely neglected and remain unknown. A recent study examined the expression and function of PRRs in seminal vesicle epithelial cells. TLR3, TLR4, and various viral DNA and RNA sensors are expressed in the epithelial cells. These PRRs can initiate the innate response in the seminal vesicle epithelial cells, thereby expressing pro-inflammatory cytokines and chemokines. PRR-initiated innate immune responses alter prostaglandin synthesis and semen coagulation. These observations provide insights into mechanisms underlying vesiculitis and its potential adverse effect on the functions of the seminal vesicle.

6. Conclusions

Infection and inflammation in the male reproductive system are major etiological factors for male infertility. The male reproductive system possesses a special immune environment to protect the organism from the sperm's antigens and prevent microbial infection. The innate immune responses in the male genital tracts are involved in the regulation of immune environment and their pathophysiology. While the immune regulation in the testis and epididymis has been intensively investigated, immunity in the accessory glands is less understood. Further research in the field will be helpful for understanding mechanisms underlying infectious and inflammatory male infertility or subfertility, which can aid in the development of preventive and therapeutic approaches for the inflammation in the reproductive system.

Acknowledgements

This work was supported by grants from CAMS Initiative for Innovative Medicine (Nos. 2017-I2M-B and R-06, 2017-I2M-3-007) and the State Key Research and Developmental Project of China (Nos. 2016YFA0101001 and 2018YFC1003902).

Conflict of interest

The authors declare no conflict of interest.

Author details

Fei Wang, Ran Chen and Daishu Han*
Institute of Basic Medical Sciences, Chinese Academy of Medical Sciences, School of Basic Medicine, Peking Union Medical College, Beijing, China

*Address all correspondence to: dshan@ibms.pumc.edu.cn

References

[1] O'Neill LA, Golenbock D, Bowie AG. The history of toll-like receptors - redefining innate immunity. Nature Reviews. Immunology. 2013;**13**:453-460. DOI: 10.1038/nri3446

[2] Iwasaki A, Medzhitov R. Control of adaptive immunity by the innate immune system. Nature Immunology. 2015;**16**:343-353. DOI: 10.1038/ni.3123

[3] Newton K, Dixit VM. Signaling in innate immunity and inflammation. Cold Spring Harbor Perspectives in Biology. 2012;**4**(3):a006049. DOI: 10.1101/cshperspect.a006049

[4] Haidl G, Haidl F, Allam JP, et al. Therapeutic options in male genital tract inflammation. Andrologia. 2019;**51**:e13207. DOI: 10.1111/and.13207

[5] Trojian TH, Lishnak TS, Heiman D. Epididymitis and orchitis: An overview. American Family Physician. 2009;**79**:583-587

[6] Fijak M, Pilatz A, Hedger MP, et al. Infectious, inflammatory and 'autoimmune' male factor infertility: How do rodent models inform clinical practice? Human Reproduction Update. 2018;**24**:416-441. DOI: 10.1093/humupd/dmy009

[7] Stein-Streilein J, Caspi RR. Immune privilege and the philosophy of immunology. Frontiers in Immunology. 2014;**5**:110. DOI: 10.3389/fimmu.2014.00110

[8] Tung KS, Teuscher C, Meng AL. Autoimmunity to spermatozoa and the testis. Immunological Reviews. 1981;**55**:217-255

[9] Yule TD, Montoya GD, Russell LD, et al. Autoantigenic germ cells exist outside the blood testis barrier. Journal of Immunology. 1988;**141**:1161-1167

[10] Fijak M, Schneider E, Klug J, et al. Testosterone replacement effectively inhibits the development of experimental autoimmune orchitis in rats: Evidence for a direct role of testosterone on regulatory T cell expansion. Journal of Immunology. 2011;**186**:5162-5172. DOI: 10.4049/jimmunol.1001958

[11] Meng J, Greenlee AR, Taub CJ, et al. Sertoli cell-specific deletion of the androgen receptor compromises testicular immune privilege in mice. Biology of Reproduction. 2011;**85**:254-260. DOI: 10.1095/biolreprod.110.090621

[12] Rothlin CV, Carrera-Silva EA, Bosurgi L, et al. TAM receptor signaling in immune homeostasis. Annual Review of Immunology. 2015;**33**:355-391. DOI: 10.1146/annurev-immunol-032414-112103

[13] Wang H, Chen Y, Ge Y, et al. Immunoexpression of tyro 3 family receptors—tyro 3, Axl, and Mer—and their ligand Gas6 in postnatal developing mouse testis. The Journal of Histochemistry and Cytochemistry. 2005;**53**:1355-1364. DOI: 10.1369/jhc.5A6637.2005

[14] Winnall WR, Muir JA, Hedger MP. Rat resident testicular macrophages have an alternatively activated phenotype and constitutively produce interleukin-10 in vitro. Journal of Leukocyte Biology. 2011;**90**:133-143. DOI: 10.1189/jlb.1010557

[15] Suarez-Pinzon W, Korbutt GS, Power R, et al. Testicular sertoli cells protect islet beta-cells from autoimmune destruction in NOD mice by a transforming growth factor-beta1-dependent mechanism. Diabetes. 2000;**49**:1810-1818. DOI: 10.2337/diabetes.49.11.1810

[16] Cheng X, Dai H, Wan N, et al. Interaction of programmed death-1

and programmed death-1 ligand-1 contributes to testicular immune privilege. Transplantation. 2009;**87**:1778-1786. DOI: 10.1097/TP.0b013e3181a75633

[17] Bellgrau D, Gold D, Selawry H, et al. A role for CD95 ligand in preventing graft rejection. Nature. 1995;**377**:630-632. DOI: 10.1038/377630a0

[18] D'Alessio A, Riccioli A, Lauretti P, et al. Testicular FasL is expressed by sperm cells. Proceedings of the National Academy of Sciences of the United States of America. 2001;**98**:3316-3321. DOI: 10.1073/pnas.051566098

[19] Zhao S, Zhu W, Xue S, et al. Testicular defense systems: Immune privilege and innate immunity. Cellular and Molecular Immunology. 2014;**11**:428-437. DOI: 10.1038/cmi.2014.38

[20] Kawai T, Akira S. The role of pattern-recognition receptors in innate immunity: Update on toll-like receptors. Nature Immunology. 2010;**11**:373-384. DOI: 10.1038/ni.1863

[21] Ori D, Murase M, Kawai T. Cytosolic nucleic acid sensors and innate immune regulation. International Reviews of Immunology. 2017;**36**:74-88. DOI: 10.1080/08830185.2017.1298749

[22] Loo YM, Gale M Jr. Immune signaling by RIG-I-like receptors. Immunity. 2011;**34**:680-692. DOI: 10.1016/j.immuni.2011.05.003

[23] Paludan SR, Bowie AG. Immune sensing of DNA. Immunity. 2013;**38**:870-880. DOI: 10.1016/j.immuni.2013.05.004

[24] Bhushan S, Meinhardt A. The macrophages in testis function. Journal of Reproductive Immunology. 2017;**119**:107-112. DOI: 10.1016/j.jri.2016.06.008

[25] Dejucq N, Chousterman S, Jegou B. The testicular antiviral defense system: Localization, expression, and regulation of 2′5′ oligoadenylate synthetase, double-stranded RNA-activated protein kinase, and mx proteins in the rat seminiferous tubule. The Journal of Cell Biology. 1997;**139**:865-873. DOI: 10.1083/jcb.139.4.865

[26] Dejucq N, Dugast I, Ruffault A, et al. Interferon-alpha and -gamma expression in the rat testis. Endocrinology. 1995;**136**:4925-4931. DOI: 10.1210/endo.136.11.7588226

[27] Le Tortorec A, Denis H, Satie AP, et al. Antiviral responses of human Leydig cells to mumps virus infection or poly I:C stimulation. Human Reproduction. 2008;**23**:2095-2103. DOI: 10.1093/humrep/den207

[28] Shang T, Zhang X, Wang T, et al. Toll-like receptor-initiated testicular innate immune responses in mouse Leydig cells. Endocrinology. 2011;**152**:2827-2836. DOI: 10.1210/en.2011-0031

[29] Xiong Y, Hales DB. The role of tumor necrosis factor-alpha in the regulation of mouse Leydig cell steroidogenesis. Endocrinology. 1993;**132**:2438-2444. DOI: 10.1210/endo.132.6.8504748

[30] Van der Hoek KH, Woodhouse CM, Brannstrom M, et al. Effects of interleukin (IL)-6 on luteinizing hormone- and IL-1beta-induced ovulation and steroidogenesis in the rat ovary. Biology of Reproduction. 1998;**58**:1266-1271

[31] Zhang Y, Li N, Chen Q, et al. Breakdown of immune homeostasis in the testis of mice lacking Tyro3, Axl and Mer receptor tyrosine kinases. Immunology and Cell Biology. 2013;**91**:416-426. DOI: 10.1038/icb.2013.22

[32] Xiong W, Chen Y, Wang H, et al. Gas6 and the tyro 3 receptor

tyrosine kinase subfamily regulate the phagocytic function of Sertoli cells. Reproduction. 2008;**135**:77-87. DOI: 10.1530/REP-07-0287

[33] Li N, Liu Z, Zhang Y, et al. Mice lacking Axl and Mer tyrosine kinase receptors are susceptible to experimental autoimmune orchitis induction. Immunology and Cell Biology. 2015;**93**:311-320. DOI: 10.1038/icb.2014.97

[34] Zhu W, Chen Q, Yan K, et al. RIG-I-like receptors mediate innate antiviral response in mouse testis. Molecular Endocrinology. 2013;**27**:1455-1467. DOI: 10.1210/me.2013-1075

[35] Sadler AJ, Williams BR. Interferon-inducible antiviral effectors. Nature Reviews. Immunology. 2008;**8**:559-568. DOI: 10.1038/nri2314

[36] Liu W, Han R, Wu H, et al. Viral threat to male fertility. Andrologia. 2018;**50**:e13140. DOI: 10.1111/and.13140

[37] Wu H, Shi L, Wang Q, et al. Mumps virus-induced innate immune responses in mouse Sertoli and Leydig cells. Scientific Reports. 2016;**6**:19507. DOI: 10.1038/srep19507

[38] Wu H, Zhao X, Wang F, et al. Mouse testicular cell type-specific antiviral response against mumps virus replication. Frontiers in Immunology. 2017;**8**:117. DOI: 10.3389/fimmu.2017.00117

[39] Jiang Q, Wang F, Shi L, et al. C-X-C motif chemokine ligand 10 produced by mouse Sertoli cells in response to mumps virus infection induces male germ cell apoptosis. Cell Death and Disease. 2017;**8**:e3146. DOI: 10.1038/cddis.2017.560

[40] Zhu W, Liu P, Yu L, et al. p204-initiated innate antiviral response in mouse Leydig cells. Biology of Reproduction. 2014;**91**:8. DOI: 10.1095/biolreprod.114.119396

[41] Riccioli A, Starace D, Galli R, et al. Sertoli cells initiate testicular innate immune responses through TLR activation. Journal of Immunology. 2006;**177**:7122-7130. DOI: 10.4049/jimmunol.177.10.7122

[42] Palladino MA, Johnson TA, Gupta R, et al. Members of the toll-like receptor family of innate immunity pattern-recognition receptors are abundant in the male rat reproductive tract. Biology of Reproduction. 2007;**76**:958-964. DOI: 10.1095/biolreprod.106.059410

[43] Wu H, Wang H, Xiong W, et al. Expression patterns and functions of toll-like receptors in mouse sertoli cells. Endocrinology. 2008;**149**:4402-4412. DOI: 10.1210/en.2007-1776

[44] Starace D, Galli R, Paone A, et al. Toll-like receptor 3 activation induces antiviral immune responses in mouse sertoli cells. Biology of Reproduction. 2008;**79**:766-775. DOI: 10.1095/biolreprod.108.068619

[45] Zhang X, Wang T, Deng T, et al. Damaged spermatogenic cells induce inflammatory gene expression in mouse Sertoli cells through the activation of toll-like receptors 2 and 4. Molecular and Cellular Endocrinology. 2013;**365**:162-173. DOI: 10.1016/j.mce.2012.10.016

[46] Nistal M, Riestra ML, Paniagua R. Focal orchitis in undescended testes: Discussion of pathogenetic mechanisms of tubular atrophy. Archives of Pathology and Laboratory Medicine. 2002;**126**:64-69. DOI: 10.1043/0003-9985(2002)126<0064:FOIUT>2.0.CO;2

[47] Goldacre MJ, Wotton CJ, Seagroatt V, et al. Immune-related disease before and after vasectomy: An epidemiological database study. Human Reproduction. 2007;**22**:1273-1278. DOI: 10.1093/humrep/dem010

[48] Fei W, Han D. Sertoli cell phagocytosis: An essential event for spermatogenesis. In: Mukherjee S, editor. Phagocytes. Rijeka, Croatia: IntechOpen; 2019

[49] Pelletier RM, Yoon SR, Akpovi CD, et al. Defects in the regulatory clearance mechanisms favor the breakdown of self-tolerance during spontaneous autoimmune orchitis. American Journal of Physiology. Regulatory, Integrative and Comparative Physiology. 2009;**296**:R743-R762. DOI: 10.1152/ajpregu.90751.2008

[50] Sun B, Qi N, Shang T, et al. Sertoli cell-initiated testicular innate immune response through toll-like receptor-3 activation is negatively regulated by Tyro3, Axl, and mer receptors. Endocrinology. 2010;**151**:2886-2897. DOI: 10.1210/en.2009-1498

[51] Liew FY, Xu D, Brint EK, et al. Negative regulation of toll-like receptor-mediated immune responses. Nature Reviews. Immunology. 2005;**5**:446-458. DOI: 10.1038/nri1630

[52] Dejucq N, Jegou B. Viruses in the mammalian male genital tract and their effects on the reproductive system. Microbiology and Molecular Biology Reviews. 2001;**65**:208-231. DOI: 10.1128/MMBR.65.2.208-231.2001

[53] Wang T, Zhang X, Chen Q, et al. Toll-like receptor 3-initiated antiviral responses in mouse male germ cells in vitro. Biology of Reproduction. 2012;**86**:106. DOI: 10.1095/biolreprod.111.096719

[54] Chen Q, Zhu W, Liu Z, et al. Toll-like receptor 11-initiated innate immune response in male mouse germ cells. Biology of Reproduction. 2014;**90**:38. DOI: 10.1095/biolreprod.113.114421

[55] Suzuki Y, Orellana MA, Schreiber RD, et al. Interferon-gamma: The major mediator of resistance against *Toxoplasma gondii*. Science. 1988;**240**:516-518

[56] Zhang D, Zhang G, Hayden MS, et al. A toll-like receptor that prevents infection by uropathogenic bacteria. Science. 2004;**303**:1522-1526. DOI: 10.1126/science.1094351

[57] Choi AM, Ryter SW, Levine B. Autophagy in human health and disease. The New England Journal of Medicine. 2013;**368**:651-662. DOI: 10.1056/NEJMra1205406

[58] Deretic V, Saitoh T, Akira S. Autophagy in infection, inflammation and immunity. Nature Reviews. Immunology. 2013;**13**:722-737. DOI: 10.1038/nri3532

[59] Wileman T. Autophagy as a defence against intracellular pathogens. Essays in Biochemistry. 2013;**55**:153-163. DOI: 10.1042/bse0550153

[60] Musso D, Roche C, Robin E, et al. Potential sexual transmission of Zika virus. Emerging Infectious Diseases. 2015;**21**:359-361. DOI: 10.3201/eid2102.141363

[61] Meinhardt A. Infection: A new threat on the horizon - Zika virus and male fertility. Nature Reviews. Urology. 2017;**14**:135-136. DOI: 10.1038/nrurol.2016.265

[62] Matusali G, Houzet L, Satie AP, et al. Zika virus infects human testicular tissue and germ cells. The Journal of Clinical Investigation. 2018;**128**: 4697-4710. DOI: 10.1172/JCI121735

[63] Robinson CL, Chong ACN, Ashbrook AW, et al. Male germ cells support long-term propagation of Zika virus. Nature Communications. 2018;**9**:2090. DOI: 10.1038/s41467-018-04444-w

[64] Joguet G, Mansuy JM, Matusali G, et al. Effect of acute Zika virus infection

on sperm and virus clearance in body fluids: A prospective observational study. The Lancet Infectious Diseases. 2017;**17**:1200-1208. DOI: 10.1016/S1473-3099(17)30444-9

[65] Huang JM, Huang TH, Qiu HY, et al. Studies on the integration of hepatitis B virus DNA sequence in human sperm chromosomes. Asian Journal of Andrology. 2002;**4**:209-212

[66] Muciaccia B, Corallini S, Vicini E, et al. HIV-1 viral DNA is present in ejaculated abnormal spermatozoa of seropositive subjects. Human Reproduction. 2007;**22**:2868-2878. DOI: 10.1093/humrep/dem288

[67] Hedger MP. Immunophysiology and pathology of inflammation in the testis and epididymis. Journal of Andrology. 2011;**32**:625-640. DOI: 10.2164/jandrol.111.012989

[68] Shum WW, Smith TB, Cortez-Retamozo V, et al. Epithelial basal cells are distinct from dendritic cells and macrophages in the mouse epididymis. Biology of Reproduction. 2014;**90**:90. DOI: 10.1095/biolreprod.113.116681

[69] Da Silva N, Barton CR. Macrophages and dendritic cells in the post-testicular environment. Cell and Tissue Research. 2016;**363**:97-104. DOI: 10.1007/s00441-015-2270-0

[70] Flickinger CJ, Bush LA, Howards SS, et al. Distribution of leukocytes in the epithelium and interstitium of four regions of the Lewis rat epididymis. The Anatomical Record. 1997;**248**:380-390. DOI: 10.1002/(SICI)1097-0185(199707)248:3<380:AID-AR11>3.0.CO;2-L

[71] Dorin JR, Barratt CL. Importance of beta-defensins in sperm function. Molecular Human Reproduction. 2014;**20**:821-826. DOI: 10.1093/molehr/gau050

[72] Zhao YT, Guo JH, Wu ZL, et al. Innate immune responses of epididymal epithelial cells to *Staphylococcus aureus* infection. Immunology Letters. 2008;**119**:84-90. DOI: 10.1016/j.imlet.2008.05.002

[73] Cheng L, Chen Q, Zhu W, et al. Toll-like receptors 4 and 5 cooperatively initiate the innate immune responses to uropathogenic *Escherichia coli* infection in mouse epididymal epithelial cells. Biology of Reproduction. 2016;**94**:58. DOI: 10.1095/biolreprod.115.136580

[74] Zhu W, Zhao S, Liu Z, et al. Pattern recognition receptor-initiated innate antiviral responses in mouse epididymal epithelial cells. Journal of Immunology. 2015;**194**:4825-4835. DOI: 10.4049/jimmunol.1402706

[75] Ludwig M, Johannes S, Bergmann M, et al. Experimental *Escherichia coli* epididymitis in rats: A model to assess the outcome of antibiotic treatment. BJU International. 2002;**90**:933-938

[76] Haidl G, Allam JP, Schuppe HC. Chronic epididymitis: Impact on semen parameters and therapeutic options. Andrologia. 2008;**40**:92-96. DOI: 10.1111/j.1439-0272.2007.00819.x

[77] Wang F, Liu W, Jiang Q, et al. Lipopolysaccharide-induced testicular dysfunction and epididymitis in mice: A critical role of tumor necrosis factor alphadagger. Biology of Reproduction. 2019;**100**:849-861. DOI: 10.1093/biolre/ioy235

[78] Schuppe HC, Pilatz A, Hossain H, et al. Urogenital Infection as a risk factor for male infertility. Deutsches Ärzteblatt International. 2017;**114**:339-346. DOI: 10.3238/arztebl.2017.0339

[79] Motrich RD, Salazar FC, Breser ML, et al. Implications of prostate inflammation on male fertility.

Andrologia. 2018;**50**:e13093. DOI: 10.1111/and.13093

[80] Motrich RD, Maccioni M, Molina R, et al. Reduced semen quality in chronic prostatitis patients that have cellular autoimmune response to prostate antigens. Human Reproduction. 2005;**20**:2567-2572. DOI: 10.1093/humrep/dei073

[81] La Vignera S, Vicari E, Condorelli RA, et al. Male accessory gland infection and sperm parameters (review). International Journal of Andrology. 2011;**34**:e330-e347. DOI: 10.1111/j.1365-2605.2011.01200.x

[82] Robertson SA, Sharkey DJ. Seminal fluid and fertility in women. Fertility and Sterility. 2016;**106**:511-519. DOI: 10.1016/j.fertnstert.2016.07.1101

[83] Fibbi B, Penna G, Morelli A, et al. Chronic inflammation in the pathogenesis of benign prostatic hyperplasia. International Journal of Andrology. 2010;**33**:475-488. DOI: 10.1111/j.1365-2605.2009.00972.x

[84] Siegel R, Ma J, Zou Z, et al. Cancer statistics, 2014. CA: A Cancer Journal for Clinicians. 2014;**64**:9-29. DOI: 10.3322/caac.21208

[85] Vykhovanets EV, Resnick MI, Marengo SR. The healthy rat prostate contains high levels of natural killer-like cells and unique subsets of CD4+ helper-inducer T cells: Implications for prostatitis. The Journal of Urology. 2005;**173**:1004-1010. DOI: 10.1097/01.ju.0000149130.06055.f2

[86] Konig JE, Senge T, Allhoff EP, et al. Analysis of the inflammatory network in benign prostate hyperplasia and prostate cancer. The Prostate. 2004;**58**:121-129. DOI: 10.1002/pros.10317

[87] Zhao S, Zhang Y, Zhang Q, et al. Toll-like receptors and prostate cancer. Frontiers in Immunology. 2014;**5**:352. DOI: 10.3389/fimmu.2014.00352

[88] Chin AI, Miyahira AK, Covarrubias A, et al. Toll-like receptor 3-mediated suppression of TRAMP prostate cancer shows the critical role of type I interferons in tumor immune surveillance. Cancer Research. 2010;**70**:2595-2603. DOI: 10.1158/0008-5472.CAN-09-1162

[89] Harashima N, Inao T, Imamura R, et al. Roles of the PI3K/Akt pathway and autophagy in TLR3 signaling-induced apoptosis and growth arrest of human prostate cancer cells. Cancer Immunology, Immunotherapy. 2012;**61**:667-676. DOI: 10.1007/s00262-011-1132-1

[90] Gonzalez-Reyes S, Fernandez JM, Gonzalez LO, et al. Study of TLR3, TLR4, and TLR9 in prostate carcinomas and their association with biochemical recurrence. Cancer Immunology, Immunotherapy. 2011;**60**:217-226. DOI: 10.1007/s00262-010-0931-0

[91] Vacchelli E, Galluzzi L, Eggermont A, et al. Trial watch: FDA-approved toll-like receptor agonists for cancer therapy. Oncoimmunology. 2012;**1**:894-907. DOI: 10.4161/onci.20931

[92] Khan FU, Ihsan AU, Khan HU, et al. Comprehensive overview of prostatitis. Biomedicine and Pharmacotherapy. 2017;**94**:1064-1076. DOI: 10.1016/j.biopha.2017.08.016

[93] Lee TC, Ho HC. Effects of prostaglandin E2 and vascular endothelial growth factor on sperm might lead to endometriosis-associated infertility. Fertility and Sterility. 2011;**95**:360-362. DOI: 10.1016/j.fertnstert.2010.08.040

[94] Liu B, Song Z, Su S, et al. Abnormal expression of Sg I is closely related to seminal vesiculitis. Urology. 2016;**88**:227 e229-227 e214. DOI: 10.1016/j.urology.2015.08.031

Section 3

Innate Immunity and Regenerative Medicine

Chapter 10

Stem Cell Therapy and Regenerative Medicine in Autoimmune Diseases

Bhuvaneshwari Sampath, Priyadarshan Kathirvelu and Kavitha Sankaranarayanan

Abstract

The role of immune system in our body is to defense against the foreign bodies. However, if the immune system fails to recognize self and non-self-cells in our body leads to autoimmune diseases. Widespread autoimmune diseases are rheumatoid arthritis, systemic lupus erythematosus, multiple sclerosis, type 1 diabetes, and more yet to be added to the list. This chapter discusses about how stem cell-based therapies and advancement of regenerative medicine endow with novel treatment for autoimmune diseases. Furthermore, in detail, specific types of stem cells and their therapeutic approach for each autoimmune condition along with their efficiency to obtain desired results are discussed. Ultimately, this chapter describes the recent trends in treating autoimmune diseases effectively using advanced stem cell research.

Keywords: autoimmune disease, stem cell therapy, regenerative medicine, hematopoietic stem cell lines, mesenchymal stem cells, rheumatoid arthritis

1. Introduction

Autoimmune disease [AID also called as autoimmune disorder] is a result of immunological imbalance and intolerance. In such a condition, an immune response is produced against the healthy tissues or substances present in our own body [1]. Though, there were roughly 67 autoimmune diseases known as per the reporting of American Autoimmune Related Disease Association [AARDA] in 1992 with 40 in suspicion, the number has grown to be in a range of 150 AID in 2016. With such a rapid incidence, AID has an impact on the social status and economy of the country too. With a mean age of onset at 65, AID targets the age group of 20–29 [2]. Gender-based studies continue to be a conundrum due to biased data reports and sexual dimorphism [3]. A series of events trigger AID, but the trigger that causes such a holocaust still remains unknown. Environmental factors, misregulation of immune system, and heredities are few common factors that influence AID out of the humungous list. Smoking, alcohol, industrial pollution, oral contraceptives, birth weight, breastfeeding, protein intake, geography, and socioeconomic status are some of the possible environmental triggers associated with AID. In case of misregulation of genes, the association of human leucocyte antigen (HLA) class II encoded HLA-DRB1-DQA1-DQB1 haplotype has been detected with several AIDs, including type 1 diabetes, Graves' disease, and rheumatoid arthritis.

2. Classification and types

AID is broadly classified into two types; systemic and organ specific. In systemic autoimmune disease, autoimmune response targets self-antigens that are distributed in various organs resulting in widespread tissue damage. Most affected areas include joints, skin, kidneys, heart, lungs, and red blood cells. On the other hand, antigens present over a particular organ are targeted in organ-specific autoimmune response [4]. Body parts that are affected by AID are represented in **Figure 1**. Prominent examples under both categories are considered and discussed in the following sections.

Systemic	Local disorders
Rheumatoid arthritis	Endocrinology: diabetes mellitus
Systemic lupus erythematosus (SLE)	Gastrointestinal: Crohn's disease
Scleroderma	Neurological: multiple sclerosis

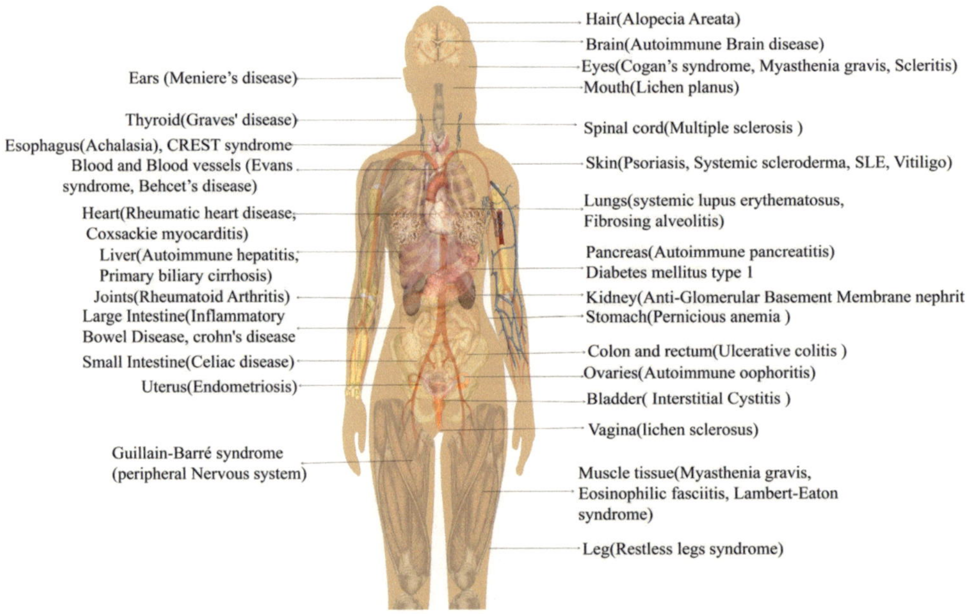

Figure 1.
Body parts affected by autoimmune disease condition.

Current therapies lead to symptoms alleviation concomitant with side effects. Hence, stem cells can be a potential answer to the proposed conundrum by bridging the gap between the problem and the solution. The following section condenses on the therapeutic potential of stem cells.

3. Stem cells

With a lead role to play, stem cells have been superheroes in regenerative medicine. Possessing the ability to self-renew, multilineage differentiation, mobility and homing, stem cells can treat many diseases by turning out to be any kind of cell through differentiation [5–8]. Stem cells can be classified into two major groups, which include embryonic stem cells (ESC) and adult stem cells (ASC). Other type of stem cell is induced pluripotent stem cells [iPSCs] that can be produced in the laboratory by reprogramming adult cells to express embryonic stem cell characteristics.

3.1 Embryonic stem cells (ESCs)

ESCs are derived from an embryo approximately 3 days after fertilization with totipotency to give rise to more than 220 cell types in the future. ESCs face a critical reception due to complications such as rejection, directed differentiation, and source of ESCs. Method of obtaining ESCs involves the destruction of blastocysts and those who believe that life begins at the blastocyst is a human life, and to destroy that is unacceptable and immoral, this procedure is equally risk to female donors being consented [8].

3.2 Adult stem cells (ASCs)

ASCs are derived from the adult human body, precursor cells that can efficiently mend the body to help promoting homeostasis. Major accessible sources of ASCs include bone marrow, adipose tissues, and blood. Hematopoietic stem cells (HSCs) can differentiate into immune stem cells (multipotent) and all kinds of blood cells including white blood cells, red blood cells, and platelets [9].

3.3 Induced pluripotent stem cells (iPSCs)

iPSCs can be generated in the laboratory by introducing reprogramming factor in the somatic cells that express the defining properties of embryonic stem cells. This provides an opportunity to generate pluripotent patient-specific cell lines, to generate model for human disease, and also used as a tool for drug development. Moreover, the tissues derived from iPSCs are almost identical with the donor cells; thus, it avoids the major rejection issue arose by the immune system [10].

3.4 Stem cells and their timeline

Once the Second World War came to an end, the world turned its concentration from research in destructive and nuclear weapons toward welfare of the human race after NATO. When such a scenario occurred and scientists started working on the betterment of human race, a path was paved for research on stem cells. The following table is a timeline containing significant events pertaining to stem cells [11, 12].

Year	Significant endeavors
1963	Self-renewing property of transplanted mouse bone marrow cells were first documented by Canadian researchers Ernest A McCulloch and James E Till
1968	First bone transplant was performed for leukemia
1978	Discovery of stem cell in human cord blood
1981	Embryonic stem cells were isolated from mice
1988	Embryonic stem cell lines generated from a hamster
1995	First embryonic stem cell line was derived from a primate
1996	First British and European stem cell company (ReNeuron) emerged
1997	Cloned lamb from stem cells

Year	Significant endeavors
1997	Leukemia origin was found as hematopoietic stem cell, thus indicated a possible proof of cancer stem cells
1998	James Alexander Thomson and his team cultivated the first human embryonic stem cells in a laboratory dish
1999–2000	Scientists discovered that manipulating adult mouse tissues could produce different cell types indicates bone marrow cells could produce different type of other cells
2001	For the first-time Christine Mummery and her team used stem cells to create beating heart cells outside the body
2002	Chunhui Xu and team found that heart muscle cells can be made from human embryonic stem cells
2003	Antonio Beltrami described a small population of stem cells in the heart that help in repair itself after damage
2004	Valérie Planat-Bénard and colleagues found that heart-like cells could be heart-like cells could be cultivated from adipose tissue
2004	First UK Stem Cell Bank (UKSCB) accredited
2006	Shinya Yamanaka of Japan reprogrammed adult cells and formed "induced pluripotent stem cells"
2007	Anthony Atala claimed that a new type of stem cell had been isolated in amniotic fluid
2007	Direct transformation of Human skin cells to iPS cells found by Shinya Yamanaka has become the revolutionary breakthrough in stem-cell biology
2009	Cardio 3 bioscience company performed lineage guided Stem cell transplant for heart failure and heart attack
2010	Geron company conducted first clinical trial for spinal cord injury and two successful human embryonic stem cell trials were conducted by Advanced Cell Technology for macular degeneration
2011	Geron terminated hESC trials and first hESC cell lines were generated
2012	Human embryonic stem cells show promising treatment for blindness
2013	-Advanced cell technology and Cardio 3 bioscience published clinical results -Brainstorm demonstrates positive results in phase 1 and phase 2 clinical trials -Aastrom-terminated phase 3 clinical study conducted for critical limb ischemia -Patient-specific human embryonic stem cells were produced
2013	Shoukhrat Mitalipov produce hESCs from fetal cells
2014	Masayo Takahashi performed the world's first trial for iPSC derived transplant to treat a form of age-related blindness
2014	Charles Vacanti Haruko Obokata at the Riken Center for Developmental Biology announced a breakthrough discovery of the concept pre-embryonic state. Dieter Egli of the New York Stem Cell Foundation and Young Gie Chung from CHA University independently produce hESCs from adult cells, using therapeutic cloning
2016	Jo Mountford and the University of Glasgow, culturing red blood cells from stem cells to make a limitless supply of clean blood for transfusion
2019	Xuefei Gao et al. for the first time derived the Expanded Potential Stem Cell lines (EPSCs) of pig and human cells which has the important implications for developmental biology, regenerative medicine, organ transplantation, disease modeling, and screening for drugs

4. Action of stem cells on AID

Hematopoietic stem cells (HSCs) were first employed to serve as a solution to leukemia and lymphoma. Eventually, after conducting trials over animal model experiments, hematopoietic stem cells found its application in destruction of self-reactive memory cells and in regeneration of self-tolerant immune cells; hence constructing a new functional immune system from hematopoietic precursors. HSC transplantation has been used to treat several types of AID over the past 15 years with a 30% decline in the progress of the AID [13–15].

Mesenchymal stem cells (MSCs) were another kind of multipotent stromal cells that were almost omnipotent, discovered by Frienden Stein et al. Being immunosuppressive, MSCs were found to suppress inflammation and downregulate pathogenic immune response triggered in Graft versus Host Disease (GVHD) and in AID such as multiple sclerosis, autoimmune diabetes, and rheumatoid arthritis [16, 17].

4.1 Hematopoietic stem cell transplantation

HSC transplantation to treat AID has been in progress since 1970s. HSCs are typically obtained from bone marrow, peripheral blood, or umbilical cord blood. Source of HSC may be autologous or allogeneic. Differentiation of hematopoietic stem cells into all kinds of blood and immune stem cells is represented in **Figure 2.**

4.1.1 Isolation of HSCs

Stem cells are mobilized from the bone marrow to peripheral blood, which will facilitate the collection of HSCs without general anesthesia or bone marrow harvest. The method of mobilization can be done by a variety of protocols. Majorly protocols make use of granulocyte colony stimulating factor or cyclophosphamide. Cyclophosphamide leads to rebound mobilization of stem cells as it is both immunosuppressive and myelosuppressive. Leukapheresis is used to collect the HSCs and purification is done by identifying stem cell containing $CD34^+$ markers or lineage-specific surface markers by negative selection of T or B cells. Negative selection process leads to T cell depletion in the auto graft [18, 19].

While isolation was first performed in murine model, efforts have been made to develop human models. Fluorescence-activated cell sorting (FACS) has been employed

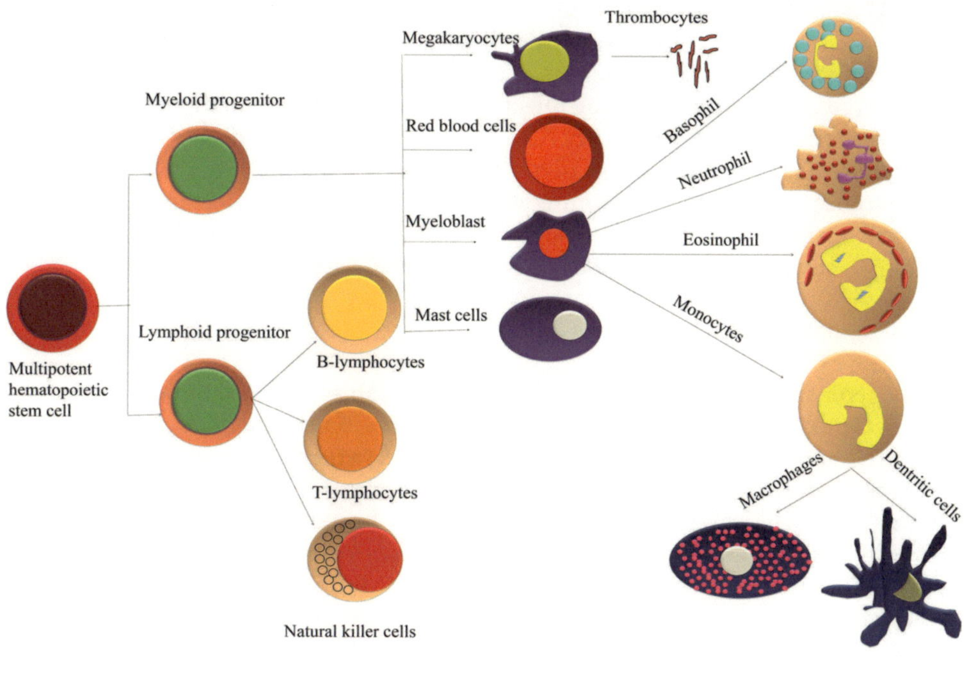

Figure 2.
Differentiation of multipotent hematopoietic stem cells.

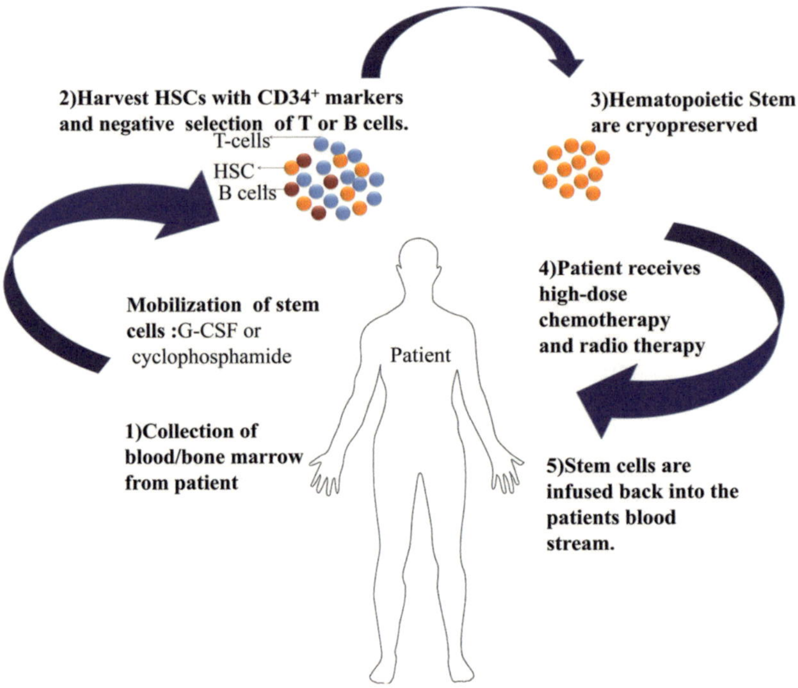

Figure 3.
Autologous stem cell transplantation.

for isolation, recognition, and quantification of smaller number of cells in a huge population. The technique is so accurate that there is a cent percent probability in purity [20].

4.1.2 Autologous HSC transplantation

Autologous HSC transplantation has been used to treat several AIDs such as rheumatoid arthritis, multiple sclerosis, Sjögren's syndrome, and systemic lupus erythematosus. As the name suggests, autologous HSC transplantation uses the subject's own stem cell avoiding tissue rejection. This kind of transplantation eradicates autoreactive immunologic memory through conditioning with highly active cytotoxic agents. Then, B and T cells can be introduced to auto antigens and undergo self-tolerance; in contrast, environmental factors that trigger autoimmunity might not occur again throughout the subject's lifetime [21–24]. The step wise protocol for autologous stem cell transplantation is depicted in **Figure 3.**

4.1.3 Allogenic HSC transplantation

Allogenic transplantation is done between two subjects whose human leucocyte antigen (HLA) match. Even though the HLA gene matches, the recipient will undergo immunosuppressive medications to tone down graft-versus-host disease. In this type, the donor may be closely related, syngeneic (identical twin of the patient) or may be unrelated but with HLA match. In allogeneic HSCTs process, the healthy stem cells are transferred to the recipient's bloodstream to reform a healthy immune system and this method appears to improve chances for cure and in long-term remission. However, there were also cases of rheumatoid arthritis subjects affected by drug-induced aplastic anemia, where relapse occurred since all the immune competent cells were from the donor [25–28]. Autoimmunity is treated using HSC transplantation and the following are the action of HSCs:

1. Developing tolerance by T regulatory cells.

2. Developing tolerance of autoreactive and alloreactive B cells.

3. Deleting alloreactive and autoreactive T cells in thymus.

4. Deleting peripheral autoreactive and alloreactive T cells.

5. Destruction of autoreactive B cells and T cells mediated by the immune system.

6. Immunosuppressive conditioning and autogenic HSC transplantation leading to immunomodulation.

The step wise protocol in allogeneic stem cell transplantation is represented in **Figure 4.**

4.2 Mesenchymal stem cell transplantation

MSCs possess the ability to differentiate both *in-vivo* and *in-vitro* into different lineages, which include adipose, bone, cartilage, muscle, and myelosupportive stroma (**Figure 5**) [29]**.** Isolation of MSCs can be done from bone marrow, skeletal muscle, adipose tissue synovial membranes, connective tissues in adults, cord blood, and products of placenta; each of which is defined by using phenotypic markers and their functional properties [12, 30–32]. Allogeneic MSCs can be transplanted into a patient without preconditioning and still have positive clinical effects on the subject without acute toxicity [33, 34]. MSCs possess the following abilities that make them a clinical success [35].

1. Homes to inflammation site when delivered intravenously following tissue injury.

2. Differentiates into a variety of cells.

3. Secretes multiple bioactive molecules that facilitate recovery of injured cells and inhibition of inflammation in return.

4. Lacks immunogenicity and possess immunomodulatory functions.

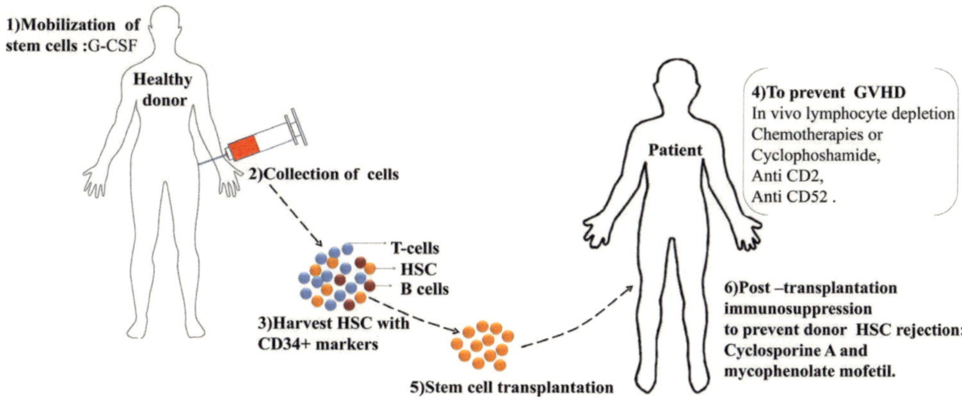

Figure 4.
Allogeneic stem cell transplantation.

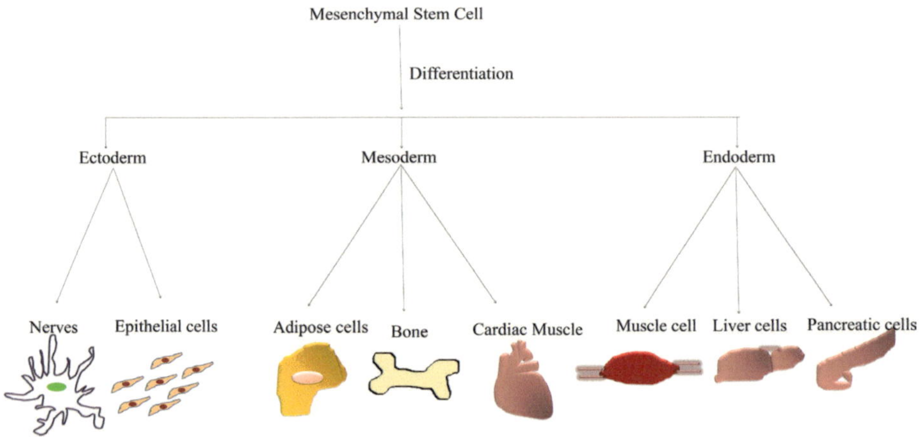

Figure 5.
Mesenchymal stem cell differentiation.

MSCs can also migrate and engraft at the inflammation site, when administered locally or systemically [36, 37]. Various cases of such an ability is discussed in the below paragraph.

Ortiz et al. found that murine MSCs could respond during a lung injury, ameliorating inflammation while adopting epithelium-like phenotype; when mice were injected with bleomycin [38]. Liu et al. found that MSCs could migrate to the site of injury in muscle tissues [39]. Yagi et al. found that the migration of MSCs was influenced by a variety of tyrosine kinase growth factor receptors such as platelet derived growth factor (PDGF) and IGF-1 and chemokines such as CCR2, CCR3, CCR4, or CCL5. Chemokines were found to lessen migration of MSCs in *in vitro* migration assays. MSCs are more privileged as they express low levels of HLA class I and HLA class II, CD40, CD80, and CD86 cannot be detected on the cell surface. Class I and class II molecules were found to increase when stimulated with interferon; thus, facilitating an increased efficiency of MSCs by being immunosuppressive and possessing immunological friendly phenotype [40, 41].
Animal models of experimental autoimmune encephalomyelitis were found to be successfully treated with MSCs, while the case was hazardous in collagen-induced arthritis (CIA). Autologous bone marrow-derived MSCs were found to be anti-proliferative toward stimulated T cells derived from normal subjects and AID subjects [42–44].

4.2.1 Isolation of MSC

Markers such as CD44, CD73, CD90, and CD105 for MSCs with the lack of CD14, CD31, CD33, CD34, and CD45 are used for isolating MSCs using cytofluorometric analysis. Due to their heterogeneity, the Mesenchymal and Tissue Stem Cell Committee of the International Society for Cellular Therapy (ICST) published three minimal criteria for MSCs [35]:

- MSCs should adhere to plastic when in standard culture conditions.
- Display of surface antigen expression pattern including $CD73^+$, $CD901^+$, $CD105^+$, $CD34^-$, $CD45^-$, $CD11b^-$, $CD14^-$, $CD19^-$, $CD79a^-$, HLA-DR^-.
- MSCs must be multipotent with the ability to differentiate into osteogenic, chondrogenic, and adipogenic lineage.

4.2.2 Autologous and allogeneic MSC therapy

MSC therapy has been widely applied and two different modes of therapy are discussed below. Isolation of cells is done based on the markers stated above and the cells are delivered using two approaches: intravenous and intra-arterial injection. In intravenous injection, the MSCs migrate to the affected site and can stay up to 13 months in the body. Moreover, route of administration is chosen based on the application used to resolve [45]. Autologous MSCT is obtained from self, while allogeneic MSCT is performed between individuals whose HLA expressions match. Both autologous and allogeneic MSC therapies are used to subside inflammation and hence are used to treat disease such as systemic lupus erythematosus, Crohn's disease, multiple system atrophy, multiple sclerosis, amyotrophic lateral sclerosis, and stroke [46].

4.2.3 MSC over HSC

HSCs have been associated with the risk of Graft versus Host Disease (GVHD). On the other hand, conducted preclinical studies prove that MSCs were successful. When MSCs were infused intravenously in leukemia subjects who were grafted with HSCs, the GVHD incidence was found to be loud (4). GVHD is suppressed by secretion of transforming growth factor β (TGF-β), prostaglandin E2, and indoleamine 2,3-dioxygenase, which in turn suppresses T cell proliferation and activation. Hence, MSCs were found to be compatible rather than HSCs.

5. Application of stem cells to treat AID

5.1 Systemic AID

Systemic autoimmune diseases are broad range of related diseases characterized by misregulation of immune system that gives rise to activation of immune cells and attack auto antigens, which result in multi-tissue/organ damages. In the following section, prominent examples of systemic AID and their therapy with stem cells are discussed briefly with proven animal model and clinical studies.

5.1.1 Rheumatoid arthritis (RA)

RA affects approximately 1.5% of the world population, and it is characterized by chronic joint inflammation, production of auto antibodies accompanied with various degrees of bone, and cartilage erosion triggered due to immunological self-intolerance [47–49]. It is a multifactorial disease, results from the combination of misregulation of genetic factor, immune system, and environmental exposure. However, the precise underlying mechanism of RA is not clear. In RA subjects, 80% of them comprise rheumatoid factor, an auto antibody specific to the Fc region of the IgG. There is no clear evidence about the source of inflammatory cytokines in specific stem cells. However, on contrast, RA can be transmitted as well as abolished by allogenic and autologous hematopoietic stem cell transplantation [50].

5.1.2 Current treatment methods

Current treatment methods include drugs, therapies, and surgeries. Drugs such as steroids and nonsteroidal anti-inflammatory drugs (NSAID); and disease-modifying antirheumatic drugs (DMARD) such as methotrexate, leflunomide, and plaquenil to slacken the progression of the disease and biological DMARD. Therapy

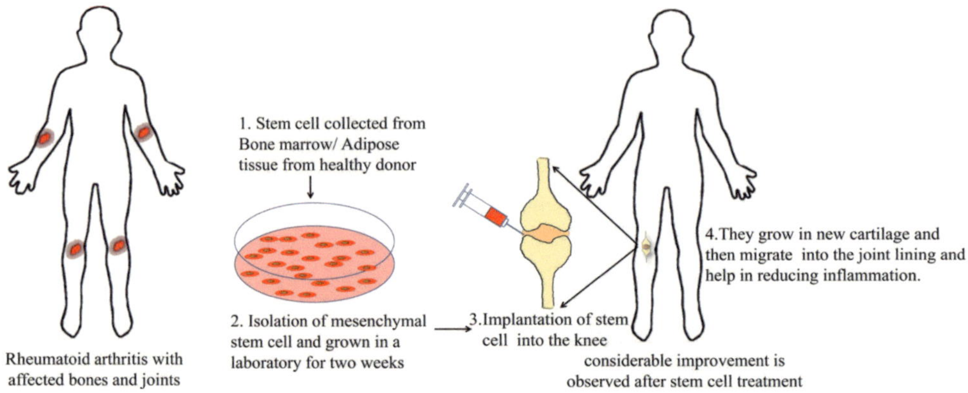

Figure 6.
Treatment method for rheumatoid arthritis using allogeneic mesenchymal stem cell transplantation.

includes cryotherapy, short wave or microwave diathermy, and physiotherapy. Heat and cold treatment can soothe pain. Surgery such as total knee replacement, tendon repair, or joint infusion is done according to the magnitude and site of damage.

Current therapy includes the use of hematopoietic stem cell transplant and mesenchymal stem cell transplant. However, recurrence was found to occur in majority of the cases irrespective of $CD34^+$ graft selection. Preclinical studies were performed and the trial consisted of three phases. The first two phases were found to establish potential characterization of the technique with high dose chemotherapy within safety limits. Improved response was noted with increased doses in the conditioning regimen and with post SCT therapy was studied but the graft manipulation was not taken into consideration. Phase three was attempted with widespread use of biological anti-rheumatic agents. Autologous SCT is considered in rare subjects, who resist both conventional and biological treatments [51, 52]. **Figure 6** represents the treatment method for rheumatoid arthritis using allogeneic mesenchymal stem cell transplantation.

5.1.2.1 Animal model study

Gonzalez et al. performed the study over a mice model. Injection of adipose-derived MSCs was followed by the decrease in inflammatory cytokines and chemokines with an expansion of Th1/Th17 cells and increase in IL-10. Together, it was found to induce peripheral tolerance by controlling self-antigen reactive T cells. Further, an increase in $CD4^+$, $CD25^+$, Fox $P3^+$, T-reg led to the suppression of self-reactive response. Other studies described the differences between *in vivo* and *in vitro* studies. In *in vitro* studies show that MSCs inhibit T cell proliferation by regulating IFN-γ levels while *in vivo* studies show that the transplantation did not have an effect on the progression of the disease. The real complication lies in the MSCs reaching the lymph and spleen nodes when injected [53]. Bouffi et al. demonstrated that the MSCs had an immunosuppressive effect involving a pathway regulated by prostaglandin 2. There were also evidences showing that T-reg cell induction was not influenced by the MSCs and hence choosing a different age group for the mice used in the study could have been a factor for the complications listed previously [43].

5.1.2.2 Intra-bone marrow-bone marrow transplantation (IBM-BMT)

SKG/Jcl mouse with T cell-mediated AD that mimics RA. BM cells of C57BL/6 J mice were transplanted to SKG/Jcl mice and there was no incidence of arthritis for

12 months with replacement of hematolymphoid cells with the donors' cells. IBM-BMT is a viable method and lends further credit to be tested in humans [54, 55].

5.1.2.3 Autologous stem cell transplantation (ASCT)

Elimination of mature autoreactive lymphocytes is done by manipulation of the graft by antithymocyte globulin (ATG) combined or with CD34$^+$ by itself. The method basically depends on G-CSF and throughout the process, immunosuppression is done to introduce the SCs into the subject [56].

5.1.2.4 Clinical trial studies

Moore et al. conducted a study over 33 subjects and the subjects were randomly incident to autologous transplantation of non-manipulated cells or selected CD34$^+$ cells. Non-manipulated stem cells were found to produce a better effect. Syngeneic transplantation of HSCs between twin brothers could control recurrence for at least 24 months. Healthy lymphocytes in a syngeneic transplantation could modify immunoregulation disorder. EBMT with 76 subjects out of which majority of the subjects responded well, but relapse rate was high [57, 58].

5.1.3 Systemic lupus erythematosus (SLE)

SLE is a rare chronic AD, which is characterized by upregulation of IFN-regulated gene transcripts. In SLE, antibodies are generated against nuclear and cytoplasmic antigen. Autoreactive plasma cells play a major role in inducing SLE and hence short-lived plasma blasts are found in positive subjects. Reduced stem cell proliferation, BM dysfunction, and decline in CD34$^+$ cells are associated with SLE. Hematopoietic system of SLE subjects had several defects due to the unbalanced expression of cytokines and growth factors. Transplantation of hematopoietic stem cells in MRL/lpr mice was found to reduce the occurrence. MSC transplantation was found to ameliorate the progression by inhibiting T lymphocytes and Th2 proliferation [59, 60].

5.1.3.1 Current treatment methods

Current treatment techniques include medication. The medication prescribed depends on the purpose served by the drug; disease modifying antirheumatic drugs, immunosuppressive drugs for immunomodulation, analgesics for analgesia, and intravenous immunoglobulins. In terminally ill cases, kidney transplant is an option while lesser silica, pesticides, and mercury levels can also have an impact on the subject. In recent years, hematopoietic stem cells and mesenchymal stem cells have been used for the drug-resistant SLE.

5.1.3.2 Animal model study

SLE animal model of W/BF1 mouse had a resultant decline in platelet count due to the production of anti-DNA antibodies and anti-platelet antibodies. These mice were found to possess lupus nephritis along with myocardial infarction, high WBC count, and hypertension. Transplantation of BM cells from normal mice was found to cure lupus nephritis, thrombocytopenia, and anti-phospholipid antibody syndrome. Normalization of platelet count was accompanied along with reduction in antiplatelet antibody levels and anti-phospholipid levels. BMT along with thymus transplantation in MRL/Lpr mouse could treat AID as the allogeneic T cells were naïve T cells that were resistant to apoptosis with lesser Fas expression [61, 62].

5.1.3.3 Clinical studies

The first autologous hematopoietic stem cell transplantation for SLE was performed by Marmont et al. [62]. In this approach, peripheral $CD34^+$ stem cell source was used after mobilization with CYC and granulocyte colony stimulating factor. According to European Group for Blood and Marrow Transplantation (EBMT), registry for HSCT in SLE patients showed around 80% of overall survival and 29% of disease-free survival and with a mortality rate of 15% suggests a goof efficacy and safety of autologous HSCT. In case of allogeneic HSCT, EBMT data showed that out of two patients, one patient died of infection and other had progressed disease after 3 years and thus clinical use of allogenic HSCT for SLE is limited. Mesenchymal stem cell transplantation was developed after 10 years of HSCT, and this approach showed a better efficacy with low cost compared to HSCT. Overall in the past 5 years, the advantage of stem cell therapies for SLE patients has increased tremendously with an initial development in high dose immunosuppression maintained by HSCT being followed by MSCT approach [63, 64].

5.1.4 Systemic scleroderma

Systemic sclerosis (SSc) is characterized by expansion of dysregulated fibroblast clones, which are uncontrollable and over expression of genes that constitute the extra cellular matrix, collagen type I in particular. SSc is characterized by high case-specific mortality as a result of internal organ disease and is also accompanied by other burdensome outcomes.

5.1.4.1 Classification

5.1.4.1.1 Limited (CREST syndrome)

Symptoms associated with limited syndrome are listed as follows.

- Raynaud's phenomenon (vasoconstriction with less blood supply in the hands resulting with a color transition from red, white to blue in cold conditions).
- Calcinosis (calcium deposition occurs in the nodules).
- Dysfunctional esophagus promoting difficulty in swallowing.
- Clerodactyly (fingers have thickened skin).
- Telangiectasias (features such as dilated capillaries in hand, face and mucous membrane).

5.1.4.1.2 Diffuse (systemic sclerosis)

Diffuse scleroderma shows up mostly in organs such as kidney, esophagus, heart, and lungs. It is either combinational or occurs separately. It is more fatal when it occurs in the lungs. Symptoms mostly include changes in the skin within a year, frictional rubs of tendons, lung, and GIT-associated complications, respectively [65, 66].

5.1.4.2 Current treatment methods

Current treatment method only focuses on treating some of the symptoms that softens the skin and to lessen the inflammation condition, but is not completely curative and still remains as a puzzle. Some patients may get benefit by exposure to heat but non-lethal manifestations such as fatigue, calcinosis, and anorectal dysfunction still remains as a challenge.

5.1.4.3 Clinical trials with autologous stem cell transplant for early diffuse SSc

In the first trail, American Scleroderma Stem Cell versus Immune Suppression Trial (ASSIST), 10 patients who received autologous HSCT compared with 9 patients who received 1.0 g/m^2 intravenous CYC. Adverse events were poorly documented in this study, but seven of nine were worsened. In the second trail, Autologous Stem Cell Transplantation International Scleroderma (ASTIS) 79 patients who received autologous HSCT compared with 77 patients who received 750 mg/m^2 IV CYC monthly for 12 months. This study results in higher mortality rate in the first year but had better long-term survival rate than those treated with CYC alone. The third trail scleroderma cyclophosphamide or transplantation (SCOT), 36 patients who received autologous HSCT compared with 39 patients given monthly IV CYC over 12 months. Despite the low numbers included in this study, the data demonstrated the efficacy of HSCT over CYC. Overall, the data provided by ASTIS and SCOT supports the HSCT over IV CYC [67, 68].

5.1.4.4 Clinical trial studies

A 16 year-old subject with localized scleroderma was selected. The subject possessed multiple plaque lesions along the trunk on 2008. By 2010, the subject had a progress in the condition involving the right half of the body further resulting in face asymmetry. Autologous HSCT was performed in 2011 by injecting fludarabine, cyclophosphamide and equine anti-thymocyte globulin along with GCSF and acyclovir till engraftment. Supplemental dose of co-trimoxazole was given for prophylaxis, irradiated single donor platelets, and red blood cells for supportive care, respectively. No post transplantation complication was found and the subject was found to be stable with no progress in lesions after 41 months. The subject had no progress in the disease condition and no immunosuppression was done once the transplantation came to an end [69].

6. Organ-specific AID

The following section discusses prominent examples in Organ specific AID and their association with stem cells.

6.1 Type 1 diabetes (T1D)

T1D is an autoimmune disease with a strong genetic component that tends to occur in childhood. As of 2014, an estimated 387 million people have diabetes worldwide, out of which T1D accounts for 5–10% worldwide. Characterized by the destruction of insulin-producing β cells by the auto antibody directed against it. Hence, introduction of insulin or islet replacement is necessary for homeostasis by regulating sugar levels [70].

Autoreactive $CD4^+$ and $CD8^+$ T cells target against islet cells. SCs could differentiate into insulin producing β cells. Sources of stem cells for diabetes therapy include embryonic stem cells (MSC), hematopoietic stem cells (HSC), and induced pluripotent stem cells (iPSCs). Following stem cells therapy, C-peptide secretion, HbA1c level, and insulin levels are monitored for better intervention of the efficiency of the technique. Monitoring response of T cells to HLA-A2-retsricted insulin B10, pre-pro-insulin, islet antigen, GAD65 and pre-pro islet amyloid polypeptide might hint the efficiency of SCT [71, 72].

6.1.1 Current treatment methods

Current treatment methods focus on producing insulin to regulate the blood sugar level. Insulin is injected into the subject and the blood sugar level is monitored at various time points. Islets cells are transplanted in certain cases along with immune suppression.

6.1.2 Animal model study

Soria et al. used mice model to study type 1 diabetes. In the study, mice derived ESCs were allowed to differentiate into insulin producing cells and were injected into a diabetic mouse. Secreted insulin could reverse glycaemia. Further SCT was also successful in streptozotocin induced diabetic mice. iPSCs obtained from mouse ESCs could synthesize insulin by cleaving pro-insulin into C-peptide and insulin. It is also evidential that ESC derived cells consisted of all β cell features except for production of insulin at high glucose levels [73].

Oh et al. chose BM-derived cells for their mice model. The experimenters found that when the medium was supplemented with DMSO and high glucose concentration, the cells transformed into insulin producing cells (IPC). Moreover, the cells could aggregate mimicking the islet cells. Blood sugar level regulation could be done up to 3 months successfully [74].

Xie et al. found that hBM-MSC were able to give rise to IPC with addition of Activin A. Differentiated cells could produce insulin in glucose-dependent manner and could regulate blood sugar level until a month in diabetes induced mice [75].

6.1.3 Clinical trial study

Hu et al. performed Type 1 diabetes over three human subjects during the year 2011. Selection criteria used for the study was that the onset period should be less than 60 days staying within a healthy BMI of 22. Two subjects were treated with BM-derived SCs delivered by liver puncture. Before the therapy, the subjects were positive for insulin cells Ab (ICA) and glutamic acid decarboxylase (GAD). After a 12-month follow up, the subject was found to be negative on ICA, GAD, and insulin antibody. Subject's serum also had increase in C peptide and a decline in blood glucose and HbA1C and glycosylated Hb.

Dr. Chen et al. studied the long-term effects of implanting Wharton's jelly-derived MSC (WJ-MSC) from umbilical cord. Twenty-nine subjects were used to participate in the study by dividing into two groups. Group 1 comprises of people injected with WJ-MSC and group 2 had people treated with insulin once in 3 months for a period of 21 months. The HbA1C and C-peptide levels were found to decline in group 1 subjects documenting the success of WJ-MSCs [76, 77].

6.2 Multiple sclerosis

Multiple sclerosis (MS) is a neurological disability in which the myelin sheath around the axons of the brain and the spinal cord are demyelinated. In MS, T and B cells initiate the inflammatory attack. Such damage cannot be reversed and hence causes many complications. No current therapy has found a solution to arrest the progress of the disease. SCT for MS subjects was found to replace neural precursors such as oligodendrocytes and myelin with attenuation in the autoimmune process locally [78, 79].

6.2.1 Current treatment methods

Current treatment methods include administering corticosteroids such as oral prednisone and intravenous methylprednisolone, oral treatments such as fingolimod, dimethyl fumarate, teriflunomide, siponimod and infusion treatments include ocrelizumab, natalizumab, alemtuzumab, and mitoxantrone. Beta interferons, glatiramer acetate, dimethyl fumarate, fingolimod, and teriflunomide are prescribed to reduce nerve inflammation. Plasma exchange (plasmapheresis) is another technique in which blood plasma is removed from blood cells and then mixed with albumin and put it back to the body. Other treatments such as physiotherapy, muscle relaxants (Zanaflex and Lioresal), and medications to reduce fatigue (amantadine and methylphenidate) are also applied when the magnitude of symptoms increase. Currently, stem cell therapy using MSCs or iPSCs shows great potential as treatment for MS. **Figure 7** shows iPSC-based therapeutics for multiple sclerosis.

6.2.2 Animal model study

Deng et al. performed a murine model study in mice affected by experimental autoimmune encephalomyelitis (EAE). EAE shared common features with human model of MS. During the study, it was evidential that the injection of MSCs could pacify myelin oligodendrocyte glycoprotein (MOG) induced EAE. There was a

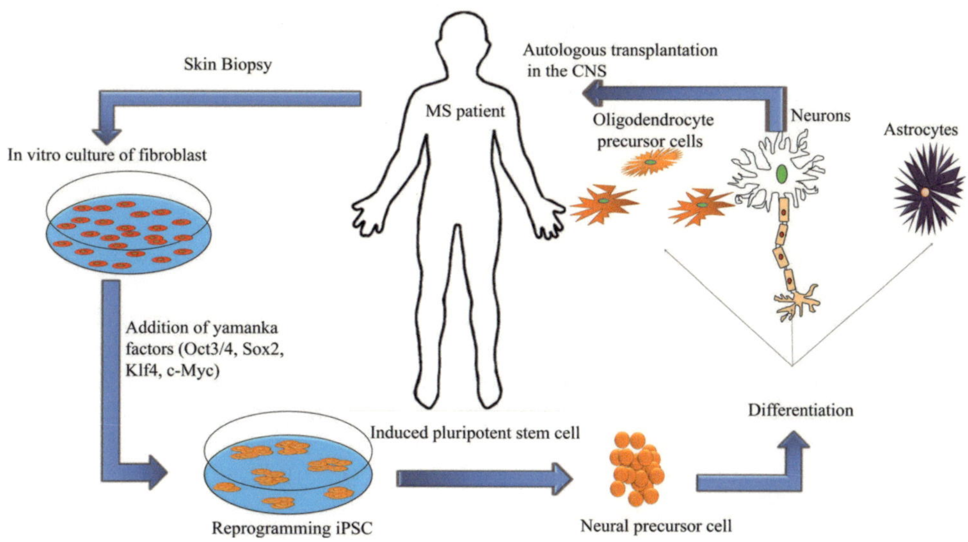

Figure 7.
Human iPSC-based therapeutics for multiple sclerosis.

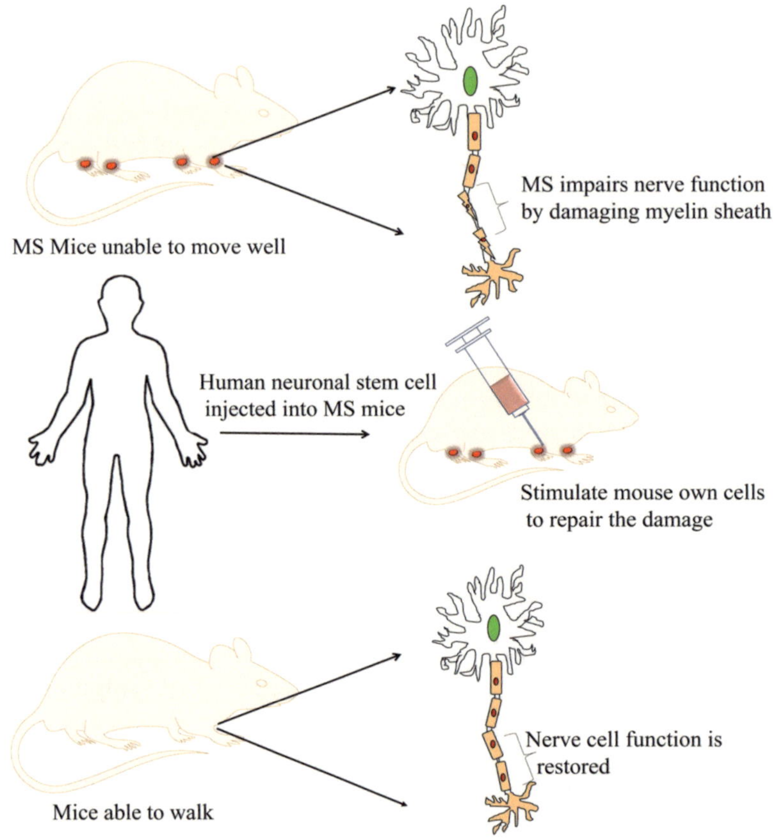

Figure 8.
Stem cell treatment for mice affected with multiple sclerosis.

decline in the infiltration of T cells, B cells, and macrophages into the brain and spinal cord. MSCs were found to migrate to the spleen and inflamed CNS to thereby have a neuroprotective effect on the CNS **Figure 8**. Further, such a therapy was also found to help in oligodendrogenesis and also increased the magnitude of symptoms. As a result, $CD8^+$ cells were more in the brain Deng et al. Carmen Marin-Bariasco et al. derived MCSs from SJL/JCrl mice. During the study, it was evidential that there was a modulation in the progress of the disease [80, 81].

6.2.3 Clinical trial study

The first reported AHSCT was performed on MS subjects in 1995 and re-myelinization occurred in the damaged sites. In 2002, HSCT was performed over 200 subjects affected with MS and there was reduction in inflammation at the CNS. However, complications such as infection and high T cell associated mortality and morbidity rates occurred during the study. Bonab et al. found that MSCs when injected intrathecally produced no adverse effects in 10 patients with non-responsive disease. Karusis et al. proceeded with phase one and two with 10–15 subjects. They found that there was no adverse effect produced during the follow up period of up to 28 months. Outcomes such as increment in $CD4^+$, $CD25^+$ regulatory T cells, decline in proliferative response of lymphocytes, and activation markers on dendritic cells. Connick et al. transplanted autologous MSCs in 10 subjects. The team found that such a therapy was found to reduce progress of general disability and improve visual function of subjects [82].

Riccardo et al. performed an autologous HSCT for severe progressive multiple sclerosis in a multicenter trial on 19 subjects with rapidly progressive MS with a score up to 7 on the scale for expanded disability status. After stem cell mobilization with CY and filgrastim, patients were conditioned with 1,3-bis[2-chloroethyl]-1-nitrosourea, etoposide, aracytin, melphalan, and followed by horse ATG. All patients showed clinical stabilization or improvement and three patients experienced deteriorate as a result of transplant related complication and 1 beyond the baseline. Among 19 patients, no death was reported after the change of conditioning to CY plus ATG. These studies indicate that HSCT is able to induce a prolonged clinical stabilization in severe progressive MS patients, resulting in both sustained treatment-free periods and improved quality of life [83].

6.3 Crohn's disease

Crohn's disease is characterized by recurring episodes of inflammation in the GI tract. The exact cause for Crohn's disease is unknown. Sources can be genetical, immunological, environmental, or even microbial. Hereditary transfer of Crohn's disease ends up with a higher probability and the incidence is 30 times higher in siblings. Crohn's disease is associated with the auto activation of T cells, especially Th1and Th17. Environmental factors such as dietary constituents and smoking also play a role in acquiring crohn's disease. High level of interleukins such as IL-21/IL-22, MMP 9 and fecal calprotectin are associated with the disease [84, 85].

6.3.1 Current treatment methods

On a major scale, drugs and techniques are employed to allay symptoms and consequences of the disease. Anti-inflammatory drugs such as corticosteroids are administered. Immune system suppressors are administered and drugs vary a wide range from cyclosporine to azathioprine. Antibiotics are and other medications such as anti-diarrheal, pain relievers, calcium and vitamin D supplements, vitamin B-12 shots and iron supplement. Surgery is done in certain cases and nutrition therapy is also recommended in certain cases.

6.3.2 Animal model study

Both canine and mice model were used to study Crohn's study. Cavazza et al. induced CD in 8-week year old mice by injecting Dextran sulfate sodium. The outcome obtained by the experimenters was a positive outcome, adipose-derived MSC could achieve the therapeutic effect and human cord derived blood platelets (hCBPL) were found to reduce colitis score [86].

Hoffman et al. used a dog affected with canine anal furunculosis as it had the same features as a human fistulizing Crohn's disease. Human ESCs derived MSCs were injected into six dogs used in the study. As a result, the interleukins associated with Crohn's disease (IL-2 and IL-6) were reduced after 2 months of post injection. After 3 months, the dogs were found completely free and after 6 months two dogs had relapse [87].

6.3.3 Clinical trial study

Molendijk et al. performed allogeneic bone marrow-derived mesenchymal stromal cells transplant to promote healing of refractory perianal fistulas in patients with Crohn's Disease. About 21 patients were randomly grouped and given injections of MSCs into the wall of curettage fistula in three different cell concentration and control (1×10^7 (n = 5), 3×10^7 (n = 5), and 9×10^7 (n = 5), and n = 6 of placebo cells). Fistula healing was observed and observed that there was nil association between local administration of allogenic MSCs and the adverse events in subjects affected by perianal fistulizing Crohn's disease [88]. Garcia-Olmo et al. performed clinical trials over 10 subjects (8 males and 2 females). In six subjects, there was complete cessation of suppuration of the fistula. Partial response was seen in three subjects with decline in suppuration. A year later there was a nil score on six cases with two cases where the incontinence score improved from the range 12-8 to 5. Nil adverse effects were observed throughout the experiment [89].

Ciccocioppo et al. studied the long-term effects of MSC therapy on eight subjects during a period of 7 years from 2007 to 2014. Disease remission was noted for up to 12 months with a gradual decline in between followed by remission again. The probability of a disease-free state declined from 88% in the first year to 37% in the last 4 years [90].

7. Trial results of stem cell therapy

Though stem cell therapy is found to be a promising solution, in the long run several complications follow. But there are also cases where the technique was found to be ineffective over certain subjects. Hence, impact of the technique as a whole is discussed in the following section.

A retrospective analysis of subjects affected by AID was taken into account. Nine hundred subjects were taken into consideration with a lead count of 345 on MS, followed by other AIDs such as the systemic sclerosis (175), SLE (85), RA (89), juvenile idiopathic arthritis (65), and idiopathic cytopenic purpura (37).

One third of the subjects responded completely while two third of the remaining had no response. There were 85% five-year survivals and 43% progress free survival during the whole period of study. On the better side, transplant related mortality was 1% for RA subjects while it came up to 11% for SLE and JIA. On the worse side, HSCT could cause acute toxicity and end up subjects with infection and bleeding. In Systematic Sclerosis subjects, HSCT was found to induce rapid fluid and electrolyte shifts. In juvenile idiopathic arthritis (JIA), fatal macrophage activation was found to occur as a result of immunosuppression. Though subjects were affected by fungal and other infections, there were also lethal outcomes due to second autoimmunity. Worst case scenario also included re-expression and relapse with a reduction in magnitude of the disease [91, 92].

8. Conclusion

Overall, this chapter highlighted the advances in the clinical use of stem cells in the treatment of an array of systemic and organ-specific autoimmune disorders. The treatment of many types of autoimmune diseases was conducted through the administration of autologous/allogenic HSCs and MSCs transplantation. Although the progress in clinical trials using stem cells in AID modification, immunomodulation, and regenerative purposes are certainly encouraging, still most of the treatment methods are still in the early stages. This is due to the clinical results reported are not clear about therapeutic efficacy and only a significant number of studies were conducted in humans, while most of them are conducted in animal models of immune-related diseases. This indicates the need for the conduction of randomized clinical trials in relevant immune-related diseases for the potential application of stem cell treatment. Due to the substantial variation among studies, comparing results from one stem cell-based medicinal product to another is very challenging. Therefore, uniform regulation of the clinical application of stem cells is highly indispensable for minimally invasive, customizable and individualized therapeutic method for a safe and successful treatment alternative. It is also important to determine the most appropriate source of stem cells that should be applied for the treatment of each autoimmune disease. In conclusion, despite the need for further studies, the treatment of immune-related diseases through the administration of stem cell is progressively ceasing due to many questions regarding the risks of stem cell application and its potential side effects that need better answers.

Acknowledgements

SB thankfully acknowledges the Department of Science and Technology (DST), New Delhi, India, for providing an INSPIRE Fellowship (Ref Number: IF130100).

Conflict of interest

The author declares no financial or commercial conflict of interest.

Author details

Bhuvaneshwari Sampath, Priyadarshan Kathirvelu and Kavitha Sankaranarayanan*
Ion Channel Biology Laboratory, AU-KBC Research, Madras Institute of Technology, Anna University, Chennai, India

*Address all correspondence to: skavitham@gmail.com

References

[1] Davidson A, Diamond B. Autoimmune diseases. The New England Journal of Medicine. 2001;**345**(5):340-350

[2] AARDA (American Autoimmune Related Disease Association). Available from: http://www.aarda.org/autoimmune-information/autoimmune-statistics/

[3] Hayter SM, Cook MC. Updated assessment of the prevalence, spectrum and case definition of autoimmune disease. Autoimmunity Reviews. 2012;**11**(10):754-765

[4] Fridkis-Hareli M. Immunogenetic mechanisms for the coexistence of organ-specific and systemic autoimmune diseases. Journal of Autoimmune Diseases. 2008;**5**(1)

[5] Kattman SJ, Witty AD, Gagliardi M, Dubois NC, Niapour M, Hotta A, et al. Stage-specific optimization of activin/nodal and BMP signaling promotes cardiac differentiation of mouse and human pluripotent stem cell lines. Cell Stem Cell. 2011;**8**(2):228-240

[6] Sui L, Geens M, Sermon K, Bouwens L, Mfopou JK. Role of BMP signaling in pancreatic progenitor differentiation from human embryonic stem cells. Stem Cell Reviews. 2013;**9**(5):569-577

[7] Konagaya S, Iwata H. Microencapsulationofdopamineneurons derived from human induced pluripotent stem cells. Biochimica et Biophysica Acta. 2015;**1850**(1):22-32

[8] Parekkadan B, Milwid JM. Mesenchymal stem cells as therapeutics. Annual Review of Biomedical Engineering. 2010;**12**:87-117

[9] van Laar JM, Tyndall A. Adult stem cells in the treatment of autoimmune diseases. Rheumatology. 2006;**45**(10):1187-1193

[10] Wertheim JA, Leventhal JR. Clinical implications of basic science discoveries: Induced pluripotent stem cell therapy in transplantation–A potential role for immunologic tolerance. American Journal of Transplantation. 2015;**15**(4):887-890

[11] Ratcliffe E, Glen KE, Naing MW, Williams DJ. Current status and perspectives on stem cell-based therapies undergoing clinical trials for regenerative medicine: Case studies. British Medical Bulletin. 2013;**108**:73-94

[12] Girlovanu M, Susman S, Soritau O, Rus-Ciuca D, Melincovici C, Constantin AM, et al. Stem cells-biological update and cell therapy progress. Clujul Medical. 2015;**88**(3):265-271

[13] Burt RK, Traynor AE. Hematopoietic stem cell transplantation: A new therapy for autoimmune disease. Stem Cells. 1999;**17**(6):366-372

[14] Gratwohl A, Baldomero H, Aljurf M, Pasquini MC, Bouzas LF, Yoshimi A, et al. Hematopoietic stem cell transplantation: A global perspective. Journal of the American Medical Association. 2010;**303**(16):1617-1624

[15] Copelan EA. Hematopoietic stem-cell transplantation. The New England Journal of Medicine. 2006;**354**(17):1813-1826

[16] English K. Mechanisms of mesenchymal stromal cell immunomodulation. Immunology and Cell Biology. 2013;**91**(1):19-26

[17] Hass R, Kasper C, Bohm S, Jacobs R. Different populations and

sources of human mesenchymal stem cells (MSC): A comparison of adult and neonatal tissue-derived MSC. Cell Communication and Signaling: CCS. 2011;**9**:12

[18] Gabus R, Borelli G, Ferrando M, Bodega E, Citrin E, Jimenez CO, et al. Mobilization of hematopoietic progenitor cells with granulocyte colony stimulating factors for autologous transplant in hematologic malignancies: A single center experience. Revista Brasileira de Hematologia e Hemoterapia. 2011;**33**(6):410-416

[19] Sheridan WP, Begley CG, Juttner CA, Szer J, To LB, Maher D, et al. Effect of peripheral-blood progenitor cells mobilised by filgrastim (G-CSF) on platelet recovery after high-dose chemotherapy. Lancet. 1992;**339**(8794):640-644

[20] Burt RK, Traynor AE, Craig R, Marmont AM. The promise of hematopoietic stem cell transplantation for autoimmune diseases. Bone Marrow Transplantation. 2003;**31**(7):521-524

[21] Bhatia S, Robison LL, Francisco L, Carter A, Liu Y, Grant M, et al. Late mortality in survivors of autologous hematopoietic-cell transplantation: Report from the bone marrow transplant survivor study. Blood. 2005;**105**(11):4215-4222

[22] Gooley TA, Chien JW, Pergam SA, Hingorani S, Sorror ML, Boeckh M, et al. Reduced mortality after allogeneic hematopoietic-cell transplantation. The New England Journal of Medicine. 2010;**363**(22):2091-2101

[23] Doubek M, Folber F, Koristek Z, Brychtova Y, Krejci M, Tomiska M, et al. Autologous hematopoietic stem cell transplantation in adult acute lymphoblastic leukemia: Still not out of fashion. Annals of Hematology. 2009;**88**(9):881-887

[24] Nakasone H, Kanda Y, Ueda T, Matsumoto K, Shimizu N, Minami J, et al. Retrospective comparison of mobilization methods for autologous stem cell transplantation in multiple myeloma. American Journal of Hematology. 2009;**84**(12):809-814

[25] Thomas ED, Blume KG. Historical markers in the development of allogeneic hematopoietic cell transplantation. Biology of Blood and Marrow Transplantation. 1999;**5**(6):341-346

[26] Porter DL. Allogeneic immunotherapy to optimize the graft-versus-tumor effect: Concepts and controversies. Hematology/The Education Program of the American Society of Hematology Education Program. 2011;**2011**:292-298

[27] Meshinchi S, Arceci RJ, Sanders JE, Smith FO, Woods WB, Radich JP, et al. Role of allogeneic stem cell transplantation in FLT3/ITD-positive AML. Blood. 2006;**108**(1):400

[28] Hinterberger W, Hinterberger-Fischer M, Marmont A. Clinically demonstrable anti-autoimmunity mediated by allogeneic immune cells favorably affects outcome after stem cell transplantation in human autoimmune diseases. Bone Marrow Transplantation. 2002;**30**(11):753-759

[29] Hoogduijn MJ, Crop MJ, Peeters AM, Van Osch GJ, Balk AH, Ijzermans JN, et al. Human heart, spleen, and perirenal fat-derived mesenchymal stem cells have immunomodulatory capacities. Stem Cells and Development. 2007;**16**(4):597-604

[30] Portmann-Lanz CB, Schoeberlein A, Huber A, Sager R, Malek A, Holzgreve W, et al. Placental mesenchymal stem cells as potential autologous graft for pre- and perinatal neuroregeneration. American Journal of Obstetrics and Gynecology. 2006;**194**(3):664-673

[31] Steigman SA, Fauza DO. Isolation of mesenchymal stem cells from amniotic fluid and placenta. Current Protocols in Stem Cell Biology. 2007

[32] Tirino V, Paino F, d'Aquino R, Desiderio V, De Rosa A, Papaccio G. Methods for the identification, characterization and banking of human DPSCs: Current strategies and perspectives. Stem Cell Reviews. 2011;7(3):608-615

[33] Rafei M, Birman E, Forner K, Galipeau J. Allogeneic mesenchymal stem cells for treatment of experimental autoimmune encephalomyelitis. Molecular Therapy: The Journal of the American Society of Gene Therapy. 2009;**17**(10):1799-1803

[34] Aggarwal S, Pittenger MF. Human mesenchymal stem cells modulate allogeneic immune cell responses. Blood. 2005;**105**(4):1815-1822

[35] Ullah I, Subbarao RB, Rho GJ. Human mesenchymal stem cells-current trends and future prospective. Bioscience Reports. 2015;**35**(2)

[36] Leibacher J, Henschler R. Biodistribution, migration and homing of systemically applied mesenchymal stem/stromal cells. Stem Cell Research & Therapy. 2016;**11**, 7

[37] Regmi S, Pathak S, Kim JO, Yong CS, Jeong JH. Mesenchymal stem cell therapy for the treatment of inflammatory diseases: Challenges, opportunities, and future perspectives. European Journal of Cell Biology. 2019;**14**(19)

[38] Ortiz LA, Gambelli F, McBride C, Gaupp D, Baddoo M, Kaminski N, et al. Mesenchymal stem cell engraftment in lung is enhanced in response to bleomycin exposure and ameliorates its fibrotic effects. Proceedings of the National Academy of Sciences of the United States of America. 2003;**100**(14):8407-8411

[39] Liu Y, Yan X, Sun Z, Chen B, Han Q, Li J, et al. Flk-1+ adipose-derived mesenchymal stem cells differentiate into skeletal muscle satellite cells and ameliorate muscular dystrophy in mdx mice. Stem Cells and Development. 2007;**16**(5):695-706

[40] Yagi H, Soto-Gutierrez A, Parekkadan B, Kitagawa Y, Tompkins RG, Kobayashi N, et al. Mesenchymal stem cells: Mechanisms of immunomodulation and homing. Cell Transplantation. 2010;**19**(6):667-679

[41] Bocelli-Tyndall C, Bracci L, Spagnoli G, Braccini A, Bouchenaki M, Ceredig R, et al. Bone marrow mesenchymal stromal cells (BM-MSCs) from healthy donors and auto-immune disease patients reduce the proliferation of autologous- and allogeneic-stimulated lymphocytes in vitro. Rheumatology. 2007;**46**(3):403-408

[42] Gerdoni E, Gallo B, Casazza S, Musio S, Bonanni I, Pedemonte E, et al. Mesenchymal stem cells effectively modulate pathogenic immune response in experimental autoimmune encephalomyelitis. Annals of Neurology. 2007;**61**(3):219-227

[43] Mao F, Xu WR, Qian H, Zhu W, Yan YM, Shao QX, et al. Immunosuppressive effects of mesenchymal stem cells in collagen-induced mouse arthritis. Inflammation Research: Official Journal of the European Histamine Research Society. 2010;**59**(3):219-225

[44] Schurgers E, Kelchtermans H, Mitera T, Geboes L, Matthys P. Discrepancy between the in vitro and in vivo effects of murine mesenchymal stem cells on T-cell proliferation and collagen-induced arthritis. Arthritis Research & Therapy. 2010;**12**(1):R31

[45] Meier RP, Muller YD, Morel P, Gonelle-Gispert C, Buhler LH.

Transplantation of mesenchymal stem cells for the treatment of liver diseases, is there enough evidence? Stem Cell Research. 2013;**11**(3):1348-1364

[46] Wei X, Yang X, Han ZP, Qu FF, Shao L, Shi YF. Mesenchymal stem cells: A new trend for cell therapy. Acta Pharmacologica Sinica. 2013;**34**(6):747-754

[47] McInnes IB, Schett G. The pathogenesis of rheumatoid arthritis. The New England Journal of Medicine. 2011;**365**:2205-2219

[48] Aletaha D, Landewe R, Karonitsch T, Bathon J, Boers M, Bombardier C, et al. Reporting disease activity in clinical trials of patients with rheumatoid arthritis: EULAR/ ACR collaborative recommendations. Arthritis and Rheumatism. 2008;**59**(10):1371-1377

[49] Nistala K, Wedderburn LR. Th17 and regulatory T cells: Rebalancing pro- and anti-inflammatory forces in autoimmune arthritis. Rheumatology. 2009;**48**(6):602-606

[50] Snowden JA, Passweg J, Moore JJ, Milliken S, Cannell P, Van Laar J, et al. Autologous hemopoietic stem cell transplantation in severe rheumatoid arthritis: A report from the EBMT and ABMTR. The Journal of Rheumatology. 2004;**31**(3):482-488

[51] Snowden JA, Kapoor S, Wilson AG. Stem cell transplantation in rheumatoid arthritis. Autoimmunity. 2008;**41**(8):625-631

[52] Alexander T, Bondanza A, Muraro PA, Greco R, Saccardi R, Daikeler T, et al. SCT for severe autoimmune diseases: Consensus guidelines of the European Society for Blood and Marrow Transplantation for immune monitoring and biobanking. Bone Marrow Transplantation. 2015;**50**(2):173-180

[53] Gonzalez MA, Gonzalez-Rey E, Rico L, Buscher D, Delgado M. Treatment of experimental arthritis by inducing immune tolerance with human adipose-derived mesenchymal stem cells. Arthritis and Rheumatism. 2009;**60**(4):1006-1019

[54] Azuma K, Osaki T, Wakuda T, Tsuka T, Imagawa T, Okamoto Y, et al. Suppressive effects of N-acetyl-D-glucosamine on rheumatoid arthritis mouse models. Inflammation. 2012;**35**(4):1462-1465

[55] Kushida T, Ueda Y, Umeda M, Oe K, Okamoto N, Iida H, et al. Allogeneic intra-bone marrow transplantation prevents rheumatoid arthritis in SKG/ Jcl mice. Journal of Autoimmunity. 2009;**32**(3-4):216-222

[56] Breban M, Dougados M, Picard F, Zompi S, Marolleau JP, Bocaccio C, et al. Intensified-dose (4 gm/m^2) cyclophosphamide and granulocyte colony-stimulating factor administration for hematopoietic stem cell mobilization in refractory rheumatoid arthritis. Arthritis and Rheumatism. 1999;**42**(11):2275-2280

[57] Moore J, Brooks P, Milliken S, Biggs J, Ma D, Handel M, et al. A pilot randomized trial comparing CD34-selected versus unmanipulated hemopoietic stem cell transplantation for severe, refractory rheumatoid arthritis. Arthritis and Rheumatism. 2002;**46**(9):2301-2309

[58] McColl G, Kohsaka H, Szer J, Wicks I. High-dose chemotherapy and syngeneic hemopoietic stem-cell transplantation for severe, seronegative rheumatoid arthritis. Annals of Internal Medicine. 1999;**131**(7):507-509

[59] Tsokos GC. Systemic lupus erythematosus. The New England Journal of Medicine. 2011;**365**(22):2110-2121

[60] Illei GG, Cervera R, Burt RK, Doria A, Hiepe F, Jayne D, et al. Current state and future directions of autologous hematopoietic stem cell transplantation in systemic lupus erythematosus. Annals of the Rheumatic Diseases. 2011;**70**(12):2071-2074

[61] Jacobi AM, Mei H, Hoyer BF, Mumtaz IM, Thiele K, Radbruch A, et al. HLA-DRhigh/CD27high plasmablasts indicate active disease in patients with systemic lupus erythematosus. Annals of the Rheumatic Diseases. 2010;**69**(1):305-308

[62] Marmont AM, van Lint MT, Gualandi F, Bacigalupo A. Autologous marrow stem cell transplantation for severe systemic lupus erythematosus of long duration. Lupus. 1997;**6**(6):545-548

[63] Jayne D, Passweg J, Marmont A, Farge D, Zhao X, Arnold R, et al. Autologous stem cell transplantation for systemic lupus erythematosus. Lupus. 2004;**13**(3):168-176

[64] Burt RK, Traynor A, Statkute L, Barr WG, Rosa R, Schroeder J, et al. Nonmyeloablative hematopoietic stem cell transplantation for systemic lupus erythematosus. Journal of the American Medical Association. 2006;**295**(5):527-535

[65] Nikpour M, Stevens WM, Herrick AL, Proudman SM. Epidemiology of systemic sclerosis. Best Practice & Research. Clinical Rheumatology. 2010;**24**(6):857-869

[66] Steen VD, Medsger TA Jr. Severe organ involvement in systemic sclerosis with diffuse scleroderma. Arthritis and Rheumatism. 2000;**43**(11):2437-2444

[67] van Laar JM, Farge D, Sont JK, Naraghi K, Marjanovic Z, Larghero J, et al. Autologous hematopoietic stem cell transplantation vs intravenous pulse cyclophosphamide in diffuse cutaneous systemic sclerosis: A randomized clinical trial. Journal of the American Medical Association. 2014;**311**(24):2490-2498

[68] Burt RK, Shah SJ, Dill K, Grant T, Gheorghiade M, Schroeder J, et al. Autologous non-myeloablative haemopoietic stem-cell transplantation compared with pulse cyclophosphamide once per month for systemic sclerosis (ASSIST): An open-label, randomised phase 2 trial. Lancet. 2011;**378**(9790):498-506

[69] Nair V, Sharma A, Sharma S, Das S, Bhakuni DS, Narayanan K, et al. Successful autologous hematopoietic stem cell transplantation for a patient with rapidly progressive localized scleroderma. International Journal of Rheumatic Diseases. 2015;**18**(3):366-371

[70] Yoon JW, Jun HS. Autoimmune destruction of pancreatic beta cells. American Journal of Therapeutics. 2005;**12**(6):580-591

[71] Mallone R, Roep BO. Biomarkers for immune intervention trials in type 1 diabetes. Clinical Immunology. 2013;**149**(3):286-296

[72] Velthuis JH, Unger WW, Abreu JR, Duinkerken G, Franken K, Peakman M, et al. Simultaneous detection of circulating autoreactive CD^{8+} T-cells specific for different islet cell-associated epitopes using combinatorial MHC multimers. Diabetes. 2010;**59**(7):1721-1730

[73] Soria B, Roche E, Berna G, Leon-Quinto T, Reig JA, Martin F. Insulin-secreting cells derived from embryonic stem cells normalize glycemia in streptozotocin-induced diabetic mice. Diabetes. 2000;**49**(2):157-162

[74] Oh SH, Muzzonigro TM, Bae SH, LaPlante JM, Hatch HM, Petersen BE. Adult bone marrow-derived cells trans-differentiating

into insulin-producing cells for the treatment of type I diabetes. Laboratory Investigation: A Journal of Technical Methods and Pathology. 2004;**84**(5):607-617

[75] Xie QP, Huang H, Xu B, Dong X, Gao SL, Zhang B, et al. Human bone marrow mesenchymal stem cells differentiate into insulin-producing cells upon microenvironmental manipulation in vitro. Differentiation; Research in Biological Diversity. 2009;**77**(5):483-491

[76] Mesples A, Majeed N, Zhang Y, Hu X. Early immunotherapy using autologous adult stem cells reversed the effect of anti-pancreatic islets in recently diagnosed type 1 diabetes mellitus: Preliminary results. Medical Science Monitor: International Medical Journal of Experimental and Clinical Research. 2013;**19**:852-857

[77] Hu J, Yu X, Wang Z, Wang F, Wang L, Gao H, et al. Long term effects of the implantation of Wharton's jelly-derived mesenchymal stem cells from the umbilical cord for newly-onset type 1 diabetes mellitus. Endocrine Journal. 2013;**60**(3):347-357

[78] Rodgers JM, Robinson AP, Miller SD. Strategies for protecting oligodendrocytes and enhancing remyelination in multiple sclerosis. Discovery Medicine. 2013;**16**(86):53-63

[79] Ryan M, Piascik P. Providing pharmaceutical care to the multiple sclerosis patient. Journal of the American Pharmaceutical Association. 2002;**42**(5):753-766

[80] Xiao J, Yang R, Biswas S, Qin X, Zhang M, Deng W. Mesenchymal stem cells and induced pluripotent stem cells as therapies for multiple sclerosis. International Journal of Molecular Sciences. 2015;**16**(5):9283-9302

[81] Marin-Banasco C, Suardiaz Garcia M, Hurtado Guerrero I, Maldonado Sanchez R, Estivill-Torrus G, Leyva Fernandez L, et al. Mesenchymal properties of SJL mice-stem cells and their efficacy as autologous therapy in a relapsing-remitting multiple sclerosis model. Stem Cell Research & Therapy. 2014;**5**(6):134

[82] Connick P, Kolappan M, Crawley C, Webber DJ, Patani R, Michell AW, et al. Autologous mesenchymal stem cells for the treatment of secondary progressive multiple sclerosis: An open-label phase 2a proof-of-concept study. The Lancet Neurology. 2012;**11**(2):150-156

[83] Saccardi R, Mancardi GL, Solari A, Bosi A, Bruzzi P, Di Bartolomeo P, et al. Autologous HSCT for severe progressive multiple sclerosis in a multicenter trial: Impact on disease activity and quality of life. Blood. 2005;**105**(6):2601-2607

[84] Jose FA, Heyman MB. Extraintestinal manifestations of inflammatory bowel disease. Journal of Pediatric Gastroenterology and Nutrition. 2008;**46**(2):124-133

[85] Ferrero-Miliani L, Nielsen OH, Andersen PS, Girardin SE. Chronic inflammation: Importance of NOD2 and NALP3 in interleukin-1beta generation. Clinical and Experimental Immunology. 2007;**147**(2):227-235

[86] Ferrer L, Kimbrel EA, Lam A, Falk EB, Zewe C, Juopperi T, et al. Treatment of perianal fistulas with human embryonic stem cell-derived mesenchymal stem cells: A canine model of human fistulizing Crohn's disease. Regenerative Medicine. 2016;**11**(1):33-43

[87] Forte D, Ciciarello M, Valerii MC, De Fazio L, Cavazza E, Giordano R, et al. Human cord blood-derived platelet lysate enhances the therapeutic activity of adipose-derived mesenchymal stromal cells isolated from Crohn's disease patients in a mouse model of colitis. Stem Cell Research & Therapy. 2015;**6**:170

[88] Molendijk I, Bonsing BA, Roelofs H, Peeters KC, Wasser MN, Dijkstra G, et al. Allogeneic bone marrow-derived mesenchymal stromal cells promote healing of refractory perianal fistulas in patients with Crohn's disease. Gastroenterology. 2015;**149**(4):918-927

[89] Garcia-Olmo D, Guadalajara H, Rubio-Perez I, Herreros MD, de-la-Quintana P, Garcia-Arranz M. Recurrent anal fistulae: Limited surgery supported by stem cells. World Journal of Gastroenterology. 2015;**21**(11):3330-3336

[90] Ciccocioppo R, Gallia A, Sgarella A, Kruzliak P, Gobbi PG, Corazza GR. Long-term follow-up of Crohn disease fistulas after local injections of bone marrow-derived mesenchymal stem cells. Mayo Clinic Proceedings. 2015;**90**(6):747-755

[91] Tyndall A. Successes and failures of stem cell transplantation in autoimmune diseases. Hematology. American Society of Hematology. Education Program. 2011;**2011**:280-284

[92] Műzes G, Sipos F. Issues and opportunities of stem cell therapy in autoimmune diseases. World Journal of Stem Cells. 2019;**11**(4):212-221

Chapter 11

Congenital and Acquired Interferonopathies: Differentiated Approaches to Interferon Therapy

Irina V. Nesterova, Svetlana V. Kovaleva, Valentina V. Malinovskaya, Galina A. Chudilova and Tatyana V. Rusinova

Abstract

This chapter reviews various interferon (IFN) system disturbances—interferonopathies. The authors describe clinical specifics of type I interferonopathy associated with overexpression of IFNα—which is a rare Mendelian genetic disease. Certain autoimmune diseases (systemic lupus erythematosus (SLE), vasculitis, immune dysregulation syndrome, etc.) are also characterized by overproduction of IFNα. Furthermore the most common interferonopathies are described—deficiencies of IFN, congenital or acquired IFNα/IFNβ and IFNγ deficiencies in children and adults. Deficiency of IFNα/IFNβ associated with severe recurrent viral infections and deficiency of IFNγ cause mycobacterial infection. Interferon-corrective therapy methods are described. The target therapy of type I interferonopathies (biologics) binds IFNα and normalizes the high level of IFNα. From the other side, patients with congenital IFNα deficiencies are needed in replacement IFN therapy. In case of acquired IFNα deficiency, the differentiated interferon-corrective therapy is performed. In both replacement and interferon-corrective therapies, recombinant human IFNα2b in complex with antioxidants (Viferon®) can be used, because their application is safe and has good clinical efficiency and no side effects.

Keywords: interferonopathies, interferon deficiency, interferon overexpression, IFN-corrective therapy

1. Introduction

Type I interferonopathies are congenital genetic disorder of the interferon (IFN) system, characterized by certain clinical symptoms resulting from the overproduction of IFNα [1–3]. In our opinion, the term interferonopathy means a general pathology of the interferon system, congenital or acquired, which includes the following types of disorders of the IFN system: *deficiency*; *paralysis* of the IFN system; inadequate response on viruses, bacteria, and mutated tumor cells; and *overproduction* of type I IFN. Interferons are the cornerstone of immune defense against viral infections and are also involved in the regulation of immune responses. Currently there are isolated type I, II, and III interferons in accordance with their ability to interact with the three types of receptors. Type I interferons include IFNα/IFNβ; type II interferons, IFNγ; and type III interferons, interferon-like cytokines (IL-29, IL-28A, IL-28B) [4].

2. Molecular mechanisms of the induction of type I interferon synthesis

The main role of type I interferons is to control viral infection. The synthesis and secretion of type I IFN is activated when our immune cells come in contact with viruses. Type I IFN is synthesized by epithelial cells, many cells of the immune system, including plasmacytoid dendritic cells (pDC) that recognize foreign or auto nucleic acids. Although both epithelial and pDC synthesize type I IFN simultaneously in different tissues, pDC-derived type I IFN actually exerts various immune responses through its cognate receptors on target cells. This results in modulation of diverse processes such as antigen presentation and activation of adaptive immunological process involving B and T cells [5]. For the synthesis of interferons in the body, cell activation is necessary. Toll-like receptors (TLRs); RIG-like receptors (RLRs), RIG-I; melanoma differentiation-associated protein 5 (MDA5); and cyclic GMP-AMP synthase (cGAS) participate in the recognition of foreign and host nucleic acid sites [6]. The main inducers of the synthesis of type I interferons are double-stranded and single-stranded RNA of viruses, as well as bacterial DNA [7]. RIG-like receptors recognize both single- and double-stranded viral RNAs, whereas cGAS (cyclic GMP-AMP synthase), in contrast, recognizes double-stranded DNA and RNA: DNA duplexes are formed during the replication of retroviruses and catalyze the synthesis of cGMP-AMP, which is the main agonist of the adapter protein—STING. After binding RNA, RIG-I and MDA5 bind the mitochondrial antiviral-signaling (MAVS) adapter protein. Both STING and MAVS stimulate downstream signaling cascades that include multiple kinases and finally lead to phosphorylation of IRF3 and induction of interferon synthesis [8]. Then type I IFN binds to the corresponding IFNAR

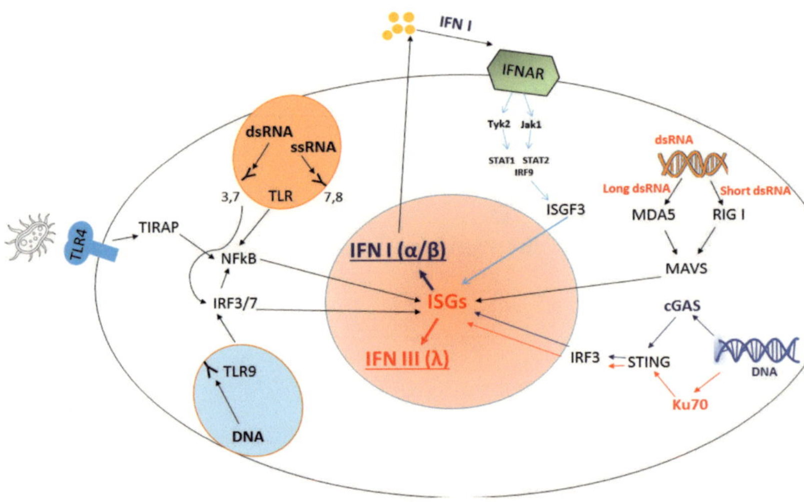

Figure 1.
Molecular mechanisms of the induction of type I and III interferon synthesis. PAMPs: dsRNA, double-stranded RNA; ssRNA, single-stranded RNA. Nucleic acid sensors: cGAS, cyclic GMP-AMP synthase; MDA5, *melanoma differentiation-associated protein 5; RIG-I, RIG-I-like receptor dsRNA helicase enzyme. Adaptor proteins: TIRAP, toll-interleukin 1 receptor (TIR) domain-containing adaptor protein; MAVS, mitochondrial antiviral-signaling protein; STAT, signal transducer and activator of transcription. Nuclear factors: IRF, IFN regulatory factor; NF-kB, nuclear factor kappa-light-chain-enhancer of activated B cells; IFNAR, IFNα receptor; ISGs, interferon-stimulated genes; Tyk, tyrosine kinase; Jak, Janus kinase.*

receptors located on the cell membrane, which leads to the activation of Tyk2 and Jak1 kinases, which undergo phosphorylation and activate signal transduction and transcription activation proteins (STAT1 and STAT2). As a result, a heterotrimeric complex is formed, known as IFN-stimulating gene factor-3 (ISGF3), which includes IFN regulatory factor 9 (IRF9). Janus kinase (Jak) activation is negatively regulated by IFNα-inducible proteins SOCS1 and SOCS3. The binding of ISGF3 promotes interferon-stimulated genes (ISGs), which leads to their transcriptional activation and the collective actions of hundreds of ISGs, resulting in the production of a large number of induced IFN, which inhibits both viral replication and the spread of viruses. Rapid type I IFN secretion and then rapid synthesis induce activity of congenital and adaptive immunity cells by activation of pro-inflammatory cytokines that activate cellular and humoral antiviral immune response [9] (**Figure 1**).

3. Interferonopathies classification

During acute viral infection, IFN level is significantly elevated, and more than 70% of cells acquire antiviral status, i.e., they are protected against virus penetration and are able to efficiently neutralize them. Type I IFN has several very important positive effects: direct and indirect antiviral effect, protective antiviral effect, antitumor effect, and immunomodulatory effect. At the same time, it was shown that increased production of IFN can lead to negative consequences similar to autoimmune reactions.

The information presented by several authors about interferon system pathologies is vast and diverse but is not well-systematized. All known defects of IFN system, including type I and II IFN, whether congenital or acquired, including various disorders (deficiency; inadequate response to contact with viruses, bacteria, and mutated or tumor cells; IFN system paralysis; IFN overexpression), are pathologies of IFN system. All those defects of IFN system are collectively known as interferonopathies. There is an urgent need to create a classification of congenital and acquired disorders of the IFN system. We believe that the classification of IFN pathology would help in determining the correct approaches to the differentiated choice of adequate treatment tactics.

Based on our own and others' experience, we have developed the interferonopathies classification as per the following table [1–3, 10–15] (**Table 1**).

3.1 Congenital type I interferonopathies associated with IFNα overexpression

Recently several studies have presented genetic and molecular disorders accompanying rare Mendelian diseases that are associated with type I IFN overexpression resulting from defects in intracellular nucleic acid metabolism, DNAse deficiency, or defects in sensor nucleic acid recognition. Genetic disorders—Mendelian diseases (Aicardi-Goutières syndrome, familial chilblain lupus, spondyenchondromatosis, proteasome-associated autoinflammatory syndrome, Singleton-Merten syndrome)—resulting in inadequately high type I IFN overexpression accompanied by a certain range of clinical disorders are called type I interferonopathies. Interferonopathies have rare pathology; their occurrence varies from 1:10,000 to 1:1,000,000 people. According to the literature, the most common syndrome is Aicardi-Goutières [16]. The frequency of some recently described genetic disorders (e.g., PRAAS2) cannot be counted [17]. Such disorders cause a great number of own nucleic acids in cell cytoplasm to appear. It results in active DNA recognition and pathological overexpression of type I IFN which launch autoimmunity hyperactivation, thus leading to autoimmune inflammation affecting the central and peripheral nervous system. It also results to skin and vessel damage (vasculitis), lung damage, etc. Therefore rapid

I. Congenital interferonopathies	II. Acquired—secondary interferonopathies
1. IFN deficiency	1. **IFN deficiency**
1.1 Interferon α deficiency (IFNα)	1.1 IFNα deficiency
1.2 Interferon β deficiency (IFNβ)	1.2 IFNβ deficiency
1.3 Interferon γ deficiency (IFNγ)	1.3 IFNγ deficiency
2. Interferon overexpression	**2. Interferon system paralysis**
2.1 IFNα overexpression	2.1 Blockage IFNα adequate response
2.1.1 Autoinflammatory syndromes and autoimmune diseases (systemic lupus erythematosus (SLE), systemic angiitis, dermatomyositis), Down syndrome	2.2 Blockage IFNβ adequate response 2.3 Blockage IFNγ adequate response **3. IFN overexpression**
2.1.2 Type I interferonopathies: Aicardi-Goutières syndrome (AGS), familial chilblain lupus (FCL), spondyenchondromatosis, proteasome-associated autoinflammatory syndrome (PRAAS), Singleton-Merten syndrome (SMS)	3.1 Cytokine storm

Table 1.
Classification of interferonopathies.

and efficient immune reaction to alien nucleic acids is positive when it causes type I IFN activation during pathogen invasion and antimicrobial protection. It becomes deleterious when it responds to own DNA which is due to the defect of own nucleic acid metabolism. Some neurological, vascular, and skin symptoms which are typical for type I interferonopathies are reviewed in such multifactorial diseases as exanthematous lupus erythematosus, widespread vasculitis, and autoimmune multiple myositis [6, 7, 18] (**Table 2**).

3.1.1 Target therapy by biologics in the treatment associated with type I IFN overexpression of type I IFN hyperproduction

Data available on genetic defects of intracellular nucleic acid metabolism greatly facilitate understanding of the mechanisms of insufficient immune activation, which can help in the development of new therapeutic approaches to the treatment of autoinflammatory and autoimmune diseases [1–3]. The progress in understanding immunopathogenesis mechanism makes it possible to set the exact targets for new therapeutic strategy development [1, 2]. The immune dysregulation syndrome is characterized by a high level of IFNα, a deficiency of regulatory T-lymphocytes, impaired functioning of B cells, and low content of low-density neutrophils. These neutrophils easily form neutrophilic extracellular traps (NET), and the resulting DNA complexes provoke an increase in IFNα synthesis, and then pDC recognizes autoDNA and produces IFNα [10, 11, 19]. These disorders are observed primarily in systemic lupus erythematosus. New approaches in treatment of SLE and other type I interferonopathies have been developed. Monoclonal antibody therapy in type I interferonopathies treatment with SLE is sifalimumab, rontalizumab against IFNα, and anifrolumab against IFNα receptor (IFNAR). Baricitinib (JAK1/JAK2 inhibitor) is currently at clinical studies (phases 2 and 3) in small cohort of patients with interferonopathies [20–22]. It is also known that treatment with baricitinib decreased disease signs and symptoms and allowed a significant reduction of corticosteroid treatment in patients with CANDLE and SAVI [23] (**Figure 2**).

3.2 Congenital interferonopathies associated with type I IFN deficiency

There are genetic defects in the synthesis of IFNα/IFNβ and IFNγ and defects in the receptors for IFNα and IFNγ (IFNAR and IFNGR) including genetic disorders associated with low IFN production according to recent studies. Those genetic

Syndrome	Responsible gene	Phenotypes
Aicardi-Goutières syndrome	TREX1, RNASEH2B, RNASEH2C, RNASEH2A, SANHD, ADAR, IFIH1	Encephalopathy, muscular dystonia, microcephaly, calcification of the basal ganglia in the substance of the brain, cramps, fever, increased acute phase blood markers, cytopenia, increased levels of interferon in the cerebrospinal fluid
Singleton-Merten syndrome	IFIH1 DDX58 RIG-I	Cardiovascular diseases with aortic calcification, osteoporotic manifestations, dental and skeletal abnormalities, psoriatic skin lesions
Proteasome-associated autoinflammatory syndromes Chronic atypical neutrophilic dermatosis with lipodystrophy and elevated temperature (CANDLE)	PSMB4 PSMB3 PSMB8 PSMB9 POMP	Erythematous skin lesions, panniculitis, lipodystrophy, arthritis with the development of joint contractures, myalgia, hepatomegaly, splenomegaly, calcification of the basal ganglia in the brain, fever, increased acute phase blood markers Recurrent fevers in the first months of life, along with characteristic skin lesions, lipodystrophy, violaceous swollen eyelids, arthralgias, extremity contractures, and delayed physical development as well as systemic inflammation markers have been identified as CANDLE-related clinical manifestations
STING-associated vasculopathy with onset in infancy (SAVI)	TMEM173	Vasculopathy with the formation of distal gangrene; necrosis; erythematous rash on the face, tip of the nose, and auricles; interstitial lung disease, arthralgia, fever
Spondyloenchondrodysplasia (SPENCD)	ACP5	Spondylometaphyseal dysplasia, stunting, calcification of the basal ganglia in the substance of the brain, arthropathy, thrombocytopenia, deficiency of cellular and humoral immunity
ISG15 deficiency	ISG15	Calcification of the basal ganglia in the substance of the brain, convulsions, mycobacterial infections
USP18 deficiency (pseudo-TORCH syndrome)	USP18	Cerebral hemorrhage and calcification, hepatomegaly, thrombocytopenia
Trichohepatoenteric syndrome 2	SKIV2L	Watery diarrhea, brittle and tangled hair, liver damage, mental retardation
Retinal vasculopathy with cerebral leukodystrophy (RVCL)	TREX1	The main characteristics of RVCL include the middle-age onset, the progressive visual loss due to retinal vasculopathy (telangiectasias, microaneurysms, and retinal capillary obliteration around the macula), and variable neurological manifestations such as dementia or migraine

Syndrome	Responsible gene	Phenotypes
Familial chilblain lupus	TREX1	Rare monogenic form of cutaneous lupus erythematosus; partly ulcerating acral lesions or painful bluish-red papules located in the fingers, toes, nose, and ears; arthralgias, affecting mainly large joints, without evidence of true arthritis; photosensitivity; or mouth ulcers
X-linked reticulate pigmentary disorder (XLPDR)	POLA1	Generalized hyperpigmentation intermingled with small hypomelanotic macules during early childhood, a distinctive face characterized by an upswept frontal hairline and arched eyebrows, as well as severe photophobia, recurrent respiratory infections, and severe gastrointestinal disorders

Table 2.
Genetic disorders associated with immune dysregulations and clinical characteristics of interferonopathies associated with type I IFN overexpression.

defects of IFNs are accompanied by clinical signs of severe recurrent viral and/or mycobacterial infection.

Congenital defects of type I IFN are associated with mutation of genes participating in synthesis of IFNα/IFNβ resulting to deficiency of various molecules (STAT1, UNC93B1, MCM4, TLR3, TRAF3, TRIF, TBK1) and decline level of IFNα/IFNβ. Deficiency of IFNγ, its receptor IFNGR (IFNγR1), and IL-12 plays an important role in IFNγ regulation [12, 24, 25]. Congenital defects of type I IFN have been globally systematized in 2015 by Bousfiha et al. [24]. It has been proven that it causes severe viral and bacterial intracellular infections which are the cause of deaths. Such patients are needed in replacement therapy with recombinant IFNα2b in complex

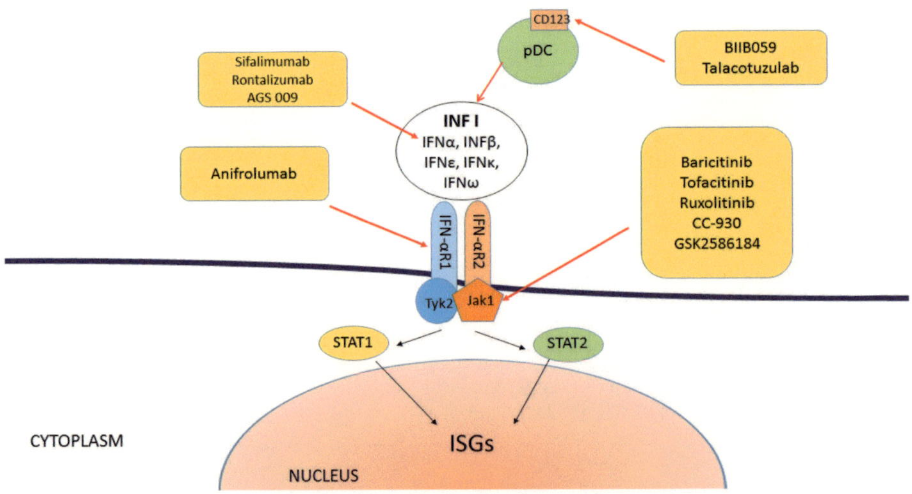

Figure 2.
Target therapies by biologics in the treatment of type I IFN overproduction. IFNAR, IFNα receptor; ISGs, interferon-stimulated genes; Tyk, tyrosine kinase; Jak, Janus kinase; pDC, plasmacytoid dendritic cell; STAT, signal transducer and activator of transcription.

with antioxidants. Congenital defects of IFNγR1 receptor are associated with severe intracellular mycobacterial infections. Combined genetic defects leading to deficiency of IFNα and IFNγ are associated with an autosomal recessive mutation in the STAT1 gene, which causes severe viral and mycobacterial infections [12, 24, 25] (**Table 3**).

3.3 Acquired: secondary interferonopathies

There are secondary acquired disorders in the IFN system, which cause a weakening of antiviral resistance in adults and children [12]. Different viruses can damage synthesis and production of IFN at various interferonogenesis stages. These secondary defects of the type I IFN lead to the occurrence of severe viral infections

Predominant susceptibility to viral infection		
Syndrome	**Responsible gene**	**Phenotypes**
Herpes simplex encephalitis (HSE)	AR (autosomal recessive inheritance): UNC 9381 TLR3 TRIF AD (autosomal dominant inheritance): TLR3 TRIF TRAF3 TBK1	Dominant clinical phenotype is HSE during primary infection with HSV1, usually between 3 months and 6 years of age Specific tests examining the TLR3 pathway marked decrease on the ability of patient's fibroblasts to produce IFNβ/IFNλ in response to TLR3 agonists and HSV1 infection
Warts, hypogammaglobulinemia, infection, myelokathexis (WHIM) syndrome	AD: CCXR4	Warts/human papilloma virus infection Neutropenia, reduced B cell numbers
Epidermodysplasia verruciformis	EVER1/TMC6, EVER2/TMC8	Human papilloma virus (group B1) infection and skin cancer
STAT1 deficiency STAT2 deficiency		Viral infections
CD16 deficiency		Severe viral infections
IRF7 deficiency		Severe influenza disease
Susceptibility to mycobacteria		
Syndrome	Responsible gene	Phenotypes
IRF8 deficiency	AR: IRF8	Susceptibility to mycobacteria, *Candida*, myeloproliferation
RORc deficiency	RORc	Susceptibility to mycobacteria, *Candida*
MSMD IL-12-IFNγ axis deficiency	AD: IFNGR1 Complete AR IFNGR1 and AR IFNGR2 Partial STAT1 LOF (AD), partial IFNGR1, partial IFNGR2, complete IL-12R1, complete IL-12B, complete ISG15, XL CYBB, IRF8, Tyk2, XL NEMO	Mycobacterial osteomyelitis Serious disseminated BCG and environmental mycobacteria infections (soft tissue, bone marrow, lungs, skin, bones, and lymph nodes), *Salmonella* spp., *Listeria monocytogenes*, and viruses Usually less severe

Table 3.
Genetic disorders and clinical characteristics of interferonopathies associated with type I IFN deficiency.

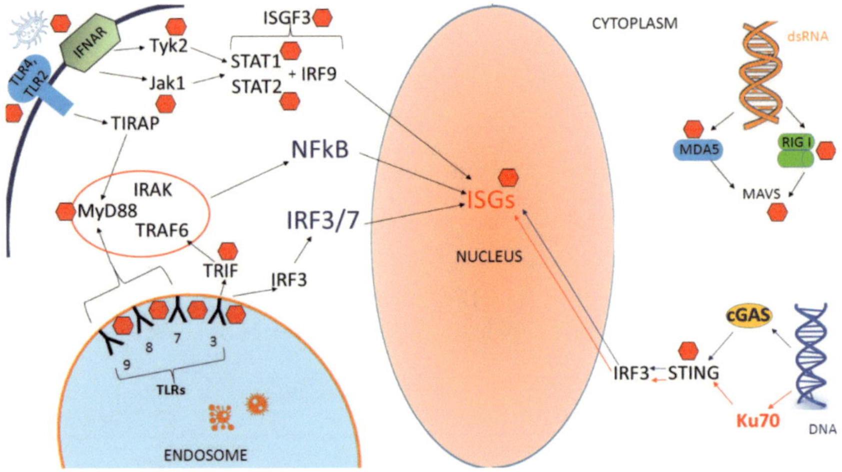

Figure 3.
Blockage of signaling pathways for the induction of interferon by viruses (red hexagons indicate the points of application of all herpesviruses, majority of respiratory viruses, chronic hepatitis B and C viruses, etc.). dsRNA, double-stranded RNA; IRF, IFN regulatory factor; IFNAR, IFNα receptor; ISGs, interferon-stimulated genes; Tyk, tyrosine kinase; Jak, Janus kinase; NF-kB, nuclear factor kappa-light-chain-enhancer of activated B cells; cGAS, cyclic GMP-AMP synthase; MAVS, mitochondrial antiviral-signaling protein; MDA5, melanoma differentiation-associated protein 5; STAT, signal transducer and activator of transcription; TRIF, TIR domain-containing adaptor inducing interferon-beta; Ku70, component of the nonhomologous end-joining pathway that repairs DNA double-stranded breaks.

(herpesviral encephalitis), recurrent acute respiratory viral infections (recARVI), chronic recurrent HSV1 infection, atypical chronic EBV infections, and other atypical cases of virus infection. It was shown that viruses can avoid the effects of IFN and inhibit the action and synthesis of IFN using various molecular mechanisms. Numerous studies demonstrated that a lot of viruses (all herpesviruses, majority of respiratory viruses, hepatitis B and C viruses, etc.) produce proteins capable of inhibiting synthesis and production of IFNα/IFNβ and IFNγ. Viruses can damage each stage of the expression of ISGs [9] (**Figure 3**).

Patients with recurrent acute respiratory viral infections and various chronic herpesvirus infections including recurrent herpes viral infections have secondary defects of IFN system. Immunocompromised children of various ages and adults may suffer from recARVI with the frequency of 10 to 16–24 and more times annually; almost in 100% of cases, it is associated with the presence of mono and mixed herpes viral infection. The frequency of recurrent chronic HSV1/HSV2 infection of facial and/or genital location in those patients may reach 16–24 times per year. Epstein-Barr virus may cause atypical virus infection associated with chronic fatigue syndrome [12].

3.3.1 Differentiated approaches to interferon therapy in patients with secondary interferonopathies

The problem of developing new approaches to the treatment of congenital and acquired defects of the IFN system is very acute [12, 26–28]. Acquired defects in the IFN system (93–96%) and impaired functioning of neutrophilic granulocytes (NG) are most often detected in patients with recurrent chronic herpes virus infections.

We conducted experiment in vitro to study the effect of recombinant IFNα2b (rIFNα2b) on NG in viral (cells from patients with HSV1/HSV2 infection) and bacterial (model infection by fMLP) infections. The study showed positive regulation of the negatively charged $IFN\alpha\beta R1^{+}IFN\gamma R^{+}TLR4^{+}NG$ phenotype in patients with various chronic herpesvirus infections under the influence of rIFNα2b in vitro.

It was noted that the number of NGs carrying IFNαβR1 and IFNγR and expression density of IFNαβR1 is increasing, wherein expression density of IFNγR and TLR4 is decreased [29]. rIFNα2b modulating effects on $CD16^{+}CD66b^{+}CD33^{+}CD11b^{+}$NG phenotype transformed by fMLP in experimental model of bacterial process in vitro, to promote remodeling of the pro-inflammatory NG phenotype into anti-inflammatory, have been shown [30]. Thus rIFNα2b has a protective effect on the NG phenotype according to experimental data.

In clinical practice, the use of parenteral IFN to correct disorders in the IFN system is very difficult due to the presence of numerous side effects. One should also bear in mind the inefficiency of short courses of IFN therapy for restoration of the normal IFN system functioning in recARVI, recurrent chronic herpes viral infection of facial or genital location, and papilloma virus infection of the skin and mucosa characterized by recurrent episodes when the frequency of recARVI and/or recurrent attacks of HSV1/HSV2 infection may reach 14–24 and more per year. For over 20 years, we have been developing interferon therapy programs using drugs in Russian production—rectal suppositories and gel of recombinant human IFNα2b (rIFNα2b+aox) in combination with antioxidants (vitamins E and C) (Viferon) [12–15, 26, 27]. During that period, we managed to demonstrate safety, antiviral, and immunomodulatory efficiency of this kind of IFN therapy, total absence of any side effects that are typical for parenteral IFN therapy, and total absence of antibodies against IFNα2b. Replacement therapy with rIFNα2b + aox is prescribed to patients with primary, genetically preconditioned, congenital or acquired IFN system disorders. In case of primary IFN system disorders, patients need a basic recovery therapy making it possible to eliminate viral antigens as much as possible; and then it is required to select dosage for permanent replacement therapy with rIFNα2b+aox. In case of acquired interferon deficiency, patients are prescribed with differentiated therapy with high, medium, and low doses of rIFNα2b+aox (**Figure 4**).

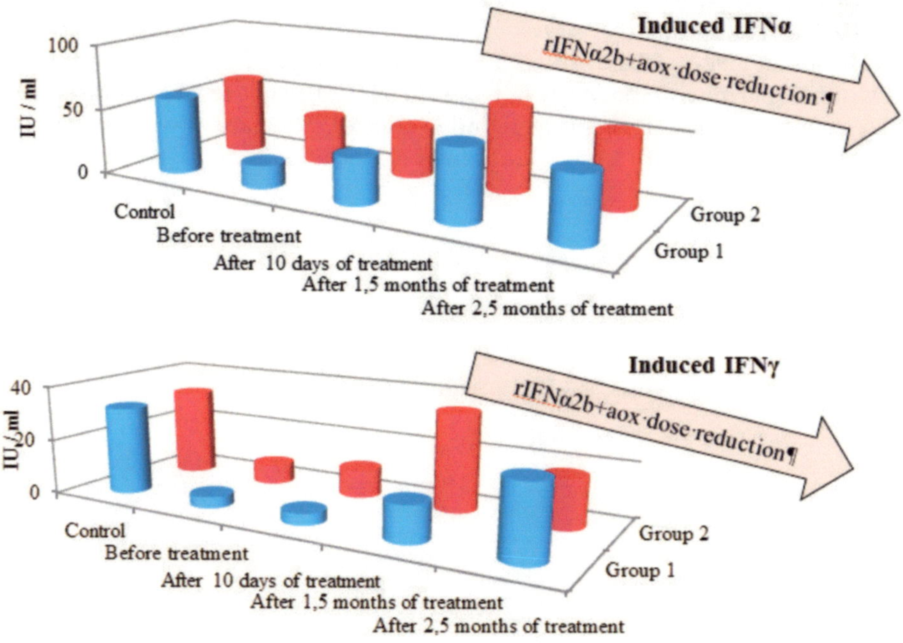

Figure 4.
Dynamics of changes in the system of IFN in immunocompromised children against the background of therapy with rIFNα2b+aox (Viferon).

At the same time, in case when we had treated the group of patients with combined immunodeficiency (defects of induced production of IFNα and IFNγ and dysfunctions of phagocytic and microbicidal activities of neutrophilic granulocytes) that was associated with recurrent acute respiratory viral infection and different chronic herpes viral coinfections, combined interferon and immunomodulatory therapy was used. The aim was to restore the levels of induced production of IFNα and IFNγ and to reconstruct dysfunctions of phagocytic and microbicidal activities of neutrophilic granulocytes and other deficient chains in antiviral immunity. One group of children, group 1, received an interferon therapy program (rIFNα2b+aox), and patients in group 2 received a program of combined interferon therapy (rIFNα2b+aox) and immunotherapy (glucosaminylmuramyldipeptide). The use of replacement and immunomodulatory mono-rIFNα2b+aox or in combination with immunotherapy (glucosaminylmuramyldipeptide) has helped us to receive very good clinical efficacies and has reached restoration of interferon status and normal functioning of neutrophilic granulocytes ($p < 0.05$) (**Figure 5**). At the same time, it is required to take into account both uneven viral infection syndrome manifestation and the rate of IFNα deficiency as well as peculiarities of immune system disorders in case of secondary immune deficiency [12–15, 27].

Here is an example illustrating the change in clinical, immune, and interferon status in immunocompromised children with recurrent acute respiratory viral infections under the influence of interferonotherapy.

Clinical case. Patient X, 3 years old. The child suffers from repeated acute respiratory viral infections 1–2 times per month (14–16 episodes per year); the duration of the acute period of respiratory viral infection is 7–10 days. The clinical symptoms of the disease were acute rhinitis, acute pharyngitis, acute laryngitis, acute tracheitis, febrile and subfebrile body temperature for 2–4 days, and severe symptoms of intoxication. The duration of the frequent incidence of acute respiratory viral infections is 2 years. The defects of the immune system are a decrease of $CD3^+CD4^+$ lymphocytes and $CD3^+CD8^+$ lymphocytes; a decrease of immunoregulatory index; neutropenia; a decrease of bacteria absorption and digestion processes by neutrophils; and a decrease of microbicidal activity of neutrophils. We tested spontaneous and Newcastle disease virus-induced IFN production during the incubation of peripheral blood (24 h, t 37°C in 5% CO_2). The level of induced IFNα in the patient was 4 IU/ml versus 58 IU/ml in control. The patient was prescribed rIFNα2b+aox therapy with a total duration of 2.5 months.

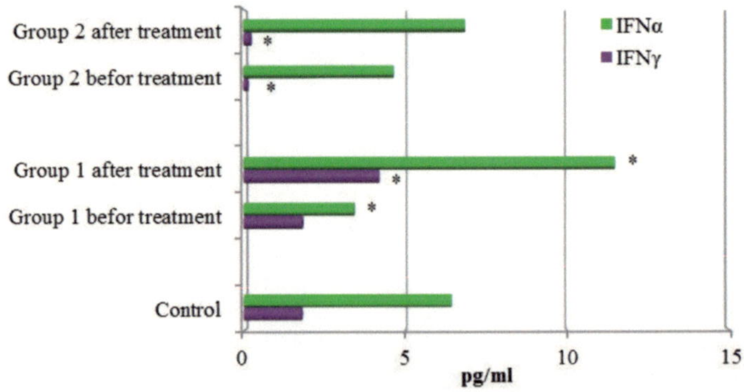

Figure 5.
*The state of the interferon system in immunocompromised children with recurrent respiratory infections on the background of differentiated programs interferon and immunotherapy. Note: group 1 received an interferon therapy program (rIFNα2b+aox); group 2 received a program of combined interferon therapy (rIFNα2b+aox) and immunotherapy (glucosaminylmuramyldipeptide); (*p < 0.05, reliability in relation to control).*

Treatment program:

- Local intranasal use of rIFNα2b+aox (Viferon gel, 36,000 IU/g), two to three times a day, 6 weeks.
- Systemic rectal application of rIFNα2b+aox suppositories according to a "step-by-step" scheme:

300,000 IU per day, 10 days.
300,000 IU per day three times a week, 2 weeks.
300,000 IU per day two times a week, 2 weeks.
150,000 IU per day two times a week, 2 weeks.
150,000 IU per day once a week, 2 weeks.

Conducted local and systemic interferon therapy led to a reduction in the frequency of acute respiratory viral infections to three episodes per year lasting 5–7 days, proceeding in a milder form. Rehabilitation of immunity parameters occurred after 2.5 months of interferonotherapy, and the level of induced IFNα was normalized to 64 IU/ml.

4. Conclusion

Summing up the above information, we may conclude that new biological drugs based on mAb are effective and safe, and they are able to neutralize IFNα overexpression in type I interferonopathies, both in Mendelian's diseases and in autoimmune disorders. At the same time, local and system use of rIFNα2b+aox (Viferon) in congenital and acquired IFN system defects associated with viral infection syndrome, where a differential dosage is selected individually taking into account the rate of deficiency and an adequate, extended course of therapy is optimal because it is associated with positive clinical and immunological effects without any negative and side effects. Our more than 20-year experience has shown that using recIFNα2b+aox in patients with congenital or acquired IFN system defects had demonstrated positive clinical effect and is safe [31]. IFN (rIFNα2b+aox) therapy can be used with very good clinical efficacy in cases of primary or secondary defects of induced production of IFNα and IFNγ. From the other side, it is very important that in patients with a genetic predisposition to the manifestation of autoimmune diseases, primarily vasculitis and systemic lupus erythematosus, we do not recommend to use IFN therapy.

Author details

Irina V. Nesterova[1,2]*, Svetlana V. Kovaleva[1], Valentina V. Malinovskaya[3], Galina A. Chudilova[1] and Tatyana V. Rusinova[1]

1 Kuban State Medical University, Krasnodar, Russia

2 RUDN University, Moscow, Russia

3 N.F. Gamaleya Federal Research Centre of Epidemiology and Microbiology, Moscow, Russia

*Address all correspondence to: inesterova1@yandex.ru

References

[1] Crow YJ. Type I interferonopathies: A novel set of in born errors of immunity. Annals of the New York Academy of Sciences. 2011;**1238**:91-98. DOI: 10.1111/j.1749-6632.2011.06220.x

[2] Günther C, Schmidt F, König N, Lee-Kirsch MA. Type I interferonopathies. Systemic inflammatory diseases triggered by type I interferons. Zeitschrift für Rheumatologie. 2016;**75**(2):134-140. DOI: 10.1007/s00393-015-0027-5

[3] Lee-Kirsch MA. The Type I interferonopathies. Annual Review of Medicine. 2017;**68**(14):297-315. DOI: 10.1146/annurev-med-050715-104506

[4] Ferreira VL, Borba HHL, Bonetti AF, Leonart LP, Pontarolo R. Cytokines and Interferons: Types and Functions, Autoantibodies and Cytokines. Wahid Ali Khan: IntechOpen; 2018. DOI: 10.5772/intechopen.74550

[5] Flessa CM, Argiriou E, Mavragani CP. Type I interferonopathies: From pathophysiology to clinical expression. In: Efthimiou P, editor. Auto-Inflammatory Syndromes. Cham: Springer; 2019. DOI: 10.1007/978-3-319-96929-9_10

[6] Kato H, Oh SW, Fujita T. RIG-I-like receptors and type I interferonopathies. Journal of Interferon & Cytokine Research. 2017;**37**(5):207-213. DOI: 10.1089/jir.2016.0095

[7] Ng CS, Kato H, Fujita T. Fueling type I interferonopathies: Regulation and function of type I interferon antiviral responses. Journal of Interferon & Cytokine Research. 2019;**39**(7):383-392. DOI: 10.1089/jir.2019.0037

[8] Zevini A, Olagnier D, Hiscott J. Cross-talk between the cytoplasmic RIG-I and STING sensing pathways. Trends in Immunology. 2017;**38**(3):194-205

[9] Taylor KE, Mossman KL. Recent advances in understanding viral evasion of type I interferon. Immunology. 2013;**138**(3):190-197

[10] Crow YJ. Human type I interferonopathies. Cytokine. 2015;**76**(1):58

[11] Rodero M, Crow YJ. Type I interferon-mediated monogenic autoinflammation: The type I interferonopathies, a conceptual overview. The Journal of Experimental Medicine. 2016;**213**(12):2527-2538

[12] Nesterova IV, Malinovskaya VV, Tarakanov VA, Kovaleva SV. Interferon and Immune Therapy in the Treatment of Children and Adults Suffering from Long and Often Recovering Symptoms of a Respiratory Infections. USA, UK, Russia: Capricorn Publishing, Inc.; 2004. 158 p

[13] Nesterova IV, Kovaleva S, Chudiliva G, Lomtatidze L. Defects in functioning of interferon and immune systems and their correction in chronic active Epstein-Barr infection. In: Allergy, Asthma & Immunophysiology: From Basic Science to Clinical Application. Medimond International Proceedings. 2012. pp. 187-191

[14] Nesterova IV, Kovaleva SV, Kleschenko EI, Chudilova GA, Lomtatidze LV, Shinkareva ON, et al. IFN- and immune-therapy approach optimization in rehabilitation of immunocompromised children with recurrent respiratory and herpetic viral infections. Pediatrics. 2014;**93**(3):66-72

[15] Nesterova IV. IFNs in clinician's practice: Best friends or dangerous enemies. Allergology and Immunology. 2016;**17**(3):189-191

[16] Yu JE, Orange JS, Demirdag YY. New primary immunodeficiency diseases:

Context and future. Current Opinion in Pediatrics. 2018;**30**(6):806-820. DOI: 10.1097/MOP.0000000000000699

[17] Poli MC, Ebstein F, Nicholas SK, de Guzman MM, Forbes LR, Chinn IK, et al. Heterozygous truncating variants in POMP escape nonsense-mediated decay and cause a unique immune dysregulatory syndrome. American Journal of Human Genetics. 2018;**102**:1126-1142

[18] Bousfiha A, Jeddane L, Picard C, Ailal F, Gaspar HB, Al-Herz W, et al. The 2017 IUIS phenotypic classification for primary immunodeficiencies. Journal of Clinical Immunology. 2018;**38**(1):129-143. DOI: 10.1007/s10875-017-0465-8

[19] Caielli S, Athale S, Domic B, et al. Oxidized mitochondrial nucleoids released by neutrophils drive type I interferon production in human lupus. The Journal of Experimental Medicine. 2016;**213**(5):697-713

[20] Khamashta M, Merrill J, Werth V, et al. Sifalimumab, an anti-interferon-α monoclonal antibody, in moderate to severe systemic lupus erythematosus: A randomised, double-blind, placebo-controlled study. Annals of the Rheumatic Diseases. 2016;**75**(11):1909-1916

[21] Kalunian K, Merrill J, Maciuca R, et al. A phase II study of the efficacy and safety of Rontalizumab (rhuMAb interferon-α) in patients with systemic lupus erythematosus (ROSE). Annals of the Rheumatic Diseases. 2016;**75**(1):196-202

[22] Furie R, Khamashta M, Merrill J, et al. Anifrolumab, an anti-interferon-α receptor monoclonal antibody, in moderate-to-severe systemic lupus erythematosus. Arthritis & Rhematology. 2017;**69**(2):376-386

[23] Montealegre Sanchez GA, Macias WL, Goldbach-Mansky R, et al. JAK1/2 inhibition with baricitinib in the treatment of autoinflammatory interferonopathies. The Journal of Clinical Investigation. 2018;**128**(7):3041-3052. DOI: 10.1172/JCI98814

[24] Bousfiha A, Jeddane L, Al-Herz W, et al. The 2015 IUIS phenotypic classification for primary Immunodeficiencies. Journal of Clinical Immunology. 2015;**35**(8):727-738. DOI: 10.1007/s10875-015-0198-5

[25] Andersen L, Mørk N, Reinert L, et al. Functional IRF3 deficiency in a patient with herpes simplex encephalitis. JEM. 2015;**212**(9):1371-1379

[26] Nesterova IV, Sepiashvily RI. Immunologic medications and contemporary immune therapy in clinical immunology and medicine. Allergology and Immunology. 2000;**3**:18-28

[27] Nesterova IV. Targeted immunomodulatory therapy: Current reality and prospects. Allergology and Immunology. 2015;**1**:18-20

[28] Nesterova IV. Differentiated approaches to interferon-correcting therapy of congenital and acquired interferonopathies. In: Simbirtsev AS, Lavrenov GV, editors. Immunotherapy in the Practice of an ENT Doctor and Therapist. Dialog; 2018. pp. 153-166

[29] Nesterova IV, Nguyen LTD, Malinovskaya VV, Chudilova GA, Lomtatidze LV, Kovaleva SV. The experimental reprogramming of phenotype of subpopulation of neutrophilic granulocytes IFNa/ßR+IFNγR+TLR4+ of healthy persons. Allergy. 2018;**73**(S105):585-586

[30] Pavlenko VN, Rusinova TV, Nesterova IV. The differentiated effects of recombinant IFNα2b on membrane expression of CD16, CD66b, CD33, CD11b of non-transformed and in vitro

transformed neutrophilic granulocytes in healthy children. In: Allergy, Asthma, COPD, Immunophysiology & Norehabilitology: Innovative Technologies. Bologna, Italy: Filodiritto International Proceedings; 2019. pp. 239-247

[31] Nesterova IV, Kovaleva SV, Chudilova GA, Lomtatidze LV, Shinkareva ON, Malinovskaya VV. Viferon® efficiency and security in mono- and combined immune therapy in children with recurrent respiratory infections. Pharmateca. 2017;**4**:48-56

Chapter 12

Immune Dysfunction and Antiretroviral Therapy Challenges in Children and Adolescents Living with Human Immunodeficiency Virus

Mirvat Said and Adam W. Bartlett

Abstract

Human immunodeficiency virus (HIV) infection results in progressive decline in immune function ultimately leading to acquired immunodeficiency syndrome (AIDS) characterised by increased susceptibility to opportunistic infections and malignancies. In addition, it causes immune dysfunction, which manifests as a persistent inflammatory state due to dysregulation of cytokine production. Antiretroviral therapy (ART) not only improves immune function but also mitigates systemic immune activation associated with disease progression. Early initiation of ART in children living with HIV has led to a growing cohort surviving into adolescence and beyond. As such, they will experience lifelong exposure to an array of physiologic processes associated with systemic infection, immune dysfunction and antiretroviral medications. This leaves them not only susceptible to a range of morbidities associated with chronic inflammation, immune dysregulation, and drug toxicity but also vulnerable to treatment fatigue leading to issues with treatment adherence and engagement in care. Children experience additional barriers to maintaining suppressive ART due to limited paediatric-friendly formulations that are palatable and contribute to regimen complexity. Tolerability and durability of long-term ART are integral in optimising outcomes for children and adolescents living with HIV and maximising viability of future ART regimens throughout adulthood.

Keywords: HIV, antiretroviral therapy, children, adolescents, paediatric formulations, adherence, HIV drug resistance, morbidity, immune dysfunction

1. Introduction

In 2018, an estimated 2.8 million children and adolescents aged between 0 and 19 years were living with human immunodeficiency virus (HIV) globally [1]. There were approximately 1.6 million adolescents between the ages 10 and 19 years, including an estimated 190,000 with newly infected HIV in 2018, majority of whom reside in sub-Saharan Africa [1, 2]. Although the number of new HIV infections among children less than 10 years of age has declined by 41% from an estimated 280,000 in 2010 to an estimated 160,000 in 2018, mostly due to successful strategies for prevention of mother to child transmission (pMTCT), we are far from meeting the global target goals of less than 20,000 new paediatric infections by 2020 [1, 3].

HIV targets CD4 T cells that play an important role in both humoral and cell-mediated immune responses to pathogens [4]. HIV infection causes immunodeficiency through depletion of CD4 T cells, defective function of CD4 T cells and macrophages, and dysregulation of cytokine production. This results in immune dysfunction, which manifests as increased susceptibility to opportunistic infections and a heightened immune activation state [5]. HIV disease progression in children not on ART is rapid compared to adults, and is associated with a mortality of up to 52% by the age of 2 years in sub-Saharan Africa [6].

The Joint United Nations Programme on HIV/AIDS (UNAIDS) launched the 90–90-90 targets in 2014 with the aim to diagnose 90% of all people with HIV, provide ART to 90% for those diagnosed, and attain viral suppression in 90% of those treated by 2020 to end the AIDS epidemic by 2030 [7, 8]. In support of the targets, the World Health Organisation (WHO) implemented the "treat all" policy in 2015 that recommends initiating anti-retroviral therapy (ART) as soon as practical to all people with HIV infection regardless of age or disease stage with the aim to improve quality of life, maximise immune preservation and potential for immune reconstitution, and reduce risk of transmission [9]. Despite this, only 56% of children under the age of 15 years were accessing treatment in 2018 with considerable geographic variation in ART coverage (**Table 1**) [10]. Coordinated efforts to speed up access and availability of HIV treatment for children by stakeholders and development of optimal paediatric formulations still lags behind that of adults.

Early initiation of ART in children living with HIV has led to a growing cohort surviving into adolescence and beyond, transforming the paradigm of HIV infection from a terminal disease into a chronic condition [11, 12]. This has created additional management challenges related to long-term ART-associated morbidity and treatment fatigue. Of growing concern, despite treatment scale up, HIV mortality is increasing among older adolescents (15–19 years) whilst mortality in other age groups is declining [13]. This reflects gaps in adolescent HIV care to address complex management challenges faced by this vulnerable group including lack of engagement in care and poor treatment adherence [13].

To achieve the global targets, HIV programs need to address management challenges in children and adolescents including scaling up access to paediatric-friendly

Region	Living with HIV N	Receiving antiretroviral therapy N (%)
Asia and the Pacific	110,000	87,908 (80)
Caribbean	11,000	4982 (45)
East and Southern Africa	1,100,000	679,921 (62)
Latin America	31,000	15,861 (51)
Middle east and North Africa	9900	3666 (37)
West and Central Africa	450,000	132,216 (29)
Global	1,700,000	947,243 (56)

Complete data not reported for Eastern Europe, Central Asia, Western and Central Europe, and North America.

Table 1.
Estimates of children (0–14 years of age) living with HIV and receiving antiretroviral therapy in 2018 (UNAIDS) [10].

ART formulations, developing strategies to maximise engagement in care and ART adherence, and improve capacity to recognise and manage treatment failure to optimise ART durability and tolerability [8].

This chapter gives an overview of HIV-related immune dysfunction and discusses management challenges for children and adolescents living with HIV. It further outlines the multifaceted approaches to address these challenges to optimise outcomes for this vulnerable population.

2. Immune dysfunction in children and adolescents with human immunodeficiency virus

HIV binds to receptors on CD4 T cells, internalises into the cell and replicates itself [14]. Through this process, the virus progressively destroys the infected CD4 T cells resulting in depletion of the cells and immunodeficiency, thereby increasing susceptibility to opportunistic infections (**Figure 1**). Destruction of CD4 T cells can be countered by the generation of new CD4 T cells in the setting of immune activation, however this process may not restore all functionally important CD4 T cells and is not sustainable in the long-term [15]. Furthermore, HIV-related immune system activation and chronic inflammation has been associated with neurodevelopmental impairment, cardiovascular disease, and clinical HIV disease progression regardless of the CD4 count [16, 17].

ART arrests the HIV life cycle at various stages, thereby inhibiting replication of the virus. This restores cellular immunity resulting in a decline in incidence of opportunistic infections and improves survival. Following ART initiation, the incidence of the majority of opportunistic infections decreases to less than 2.5% [19]. ART has also been shown to reduce systemic immune activation that reduces inflammatory-mediated disease progression [5].

Early initiation of ART in children is critical for immune reconstitution and long-term immune preservation. Mathematical modelling using data from large

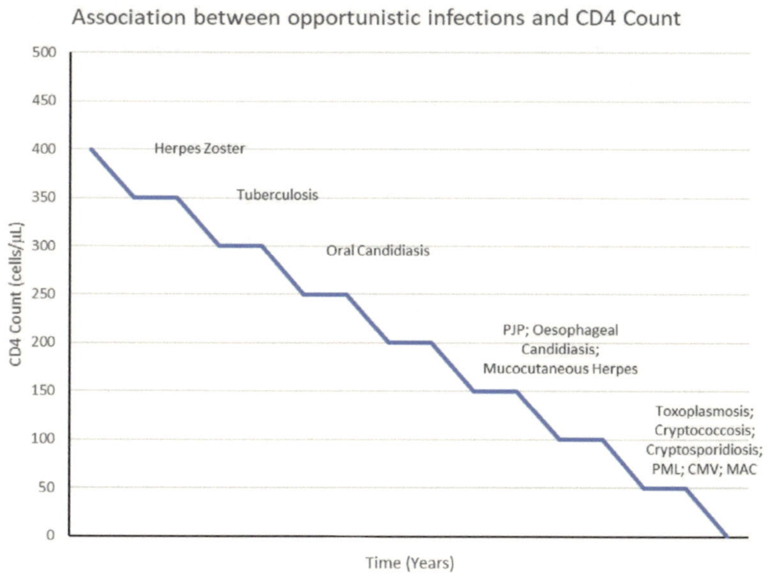

Figure 1.
Association between opportunistic infections and CD4 count [18]. CMV: Cytomegalovirus; MAC: Mycobacterium avium *complex; PJP:* Pneumocystis jirovecii *pneumonia; PML: Progressive multifocal leukoencephalopathy.*

European and African cohorts indicate that both age and CD4 count at ART initiation are important determinants of CD4 cell recovery [20, 21]. Children initiating ART under the age of 5 years have better potential for CD4 count recovery, with the potential for long-term CD4 count recovery diminishing every year after the age of 5 years that ART is initiated.

3. Antiretroviral therapy challenges in children and adolescents

There are limitations to paediatric-friendly ART formulations that pose important obstacles to maintaining good adherence in children. Most available paediatric ART formulations have poor palatability and constitute complex regimens [9]. Children often have to take a combination of pills and liquid formulations with short dosing intervals making it challenging for caregivers to administer. Fixed dose combinations (FDCs), which combine two or three antiretroviral medications have been shown to improve adherence [9]. Nonetheless, there are few available FDCs in non-pill form, limiting administration to older children who can swallow large pills. Complex storage and transportation is another challenge particularly with the liquid formulations that require cold chain storage [9].

ART drug development for children still lags behind that of adults despite major achievements in improving access to ART worldwide. There are a number of barriers slowing drug development for children. These include, lack of economic incentives for manufacturers contributed by the small paediatric market, with children making up less than 10% of people living with HIV and declining numbers of new paediatric HIV infections. The technical and pharmacokinetic complexities related to development of formulations that are safe, palatable and allow dosing across various ages and weight is challenging, requiring extensive research. Furthermore, outdated procurement practices and gaps in supply chains, delay in regulatory approvals at an individual country level, and stagnant government policies can result in further delay in uptake of new ART into treatment programs [11, 12].

The WHO ART recommendations for children are evolving with development of new antiretroviral medications. Most recently in 2018, WHO updated its treatment guidelines and now recommends the new integrase strand transfer inhibitor (INSTI)-based ART regimens including dolutegravir (DTG) for children older than 4 weeks of age and raltegravir (RAL) for neonates as a first line option (**Table 2**) [3]. INSTIs are comparatively efficacious, have a high barrier to resistance, and are better tolerated than protease inhibitors (PIs). However, the use of DTG is restricted by the only available DTG formulation (50 mg tablet) being approved for use in children who are at least 20 kg, with DTG dosing guidance for children less than

	Preferred	Alternative	Special circumstances***
Neonates	AZT + 3TC + RAL	AZT + 3TC + NVP	AZT + 3TC + LPV/r
Children	ABC + 3TC + DTG*	ABC + 3TC + LPV/r ABC + 3TC + RAL	ABC + 3TC + EFV** AZT + 3TC + EFV** AZT + 3TC + LPV/r AZT + 3TC + NVP AZT + 3TC + RAL ABC + 3TC + RAL

**DTG approved in children > = 20 kg.*
***From 3 years of age.*
****Where no alternatives are available.*

Table 2.
WHO 2018 recommendations for first line paediatric antiretroviral regimens [23].

20 kg still under development [3]. This along with the limited manufacturing capacity and cost limit accessibility of these regimens limit the use of INSTIs in children.

In the interim, lopinavir/ritonavir (LPV/r)-based regimens remain the only available first line ART for infants and young children as recommended by WHO. They are superior and more effective than the non-nucleoside reverse transcriptase inhibitors (NNRTIs) such as nevirapine (NVP) and efavirenz (EFV) [11]. There is also comparatively high resistance to NNRTIs, particularly in children whose mothers received NNRTIs to prevent mother to child transmission, which further limits the use of this drug class as first line [22]. **Table 2** illustrates the WHO recommended first line paediatric ART regimens for neonates and children.

LPV/r is available as an oral solution, heat stable tablets, oral pellets and oral granules. The LPV/r liquid formulation was developed for ease of administration in younger children and infants. However, the liquid formulations have issues with palatability and require caregivers to measure out the precise amount of liquid. Additionally, liquid formulations need cold chain storage, which can be an obstacle in resource limited settings [9]. In 2015, the USA food and drug formulation tentatively approved LPV/r oral pellets, and subsequently approved LPV/r oral granules in August 2018 [24]. The pellet and granule formulations offer advantages over the liquid form as they are easy to administer across dosing ranges, easy to store and transport, and palatable. However, there has been a slow uptake of these formulations in low and middle-income countries due to limited manufacturing capacity, making policy makers reluctant to transition to LPV/r based ART regimens [24].

Alternative PIs such as darunavir/ritonavir (DRV/r) may be an option for children above 3 years of age who have failed first line LPV/r-based treatment. Studies have shown darunavir is not only effective in ART paediatric-experienced patients but also has low rates of resistance among children with prolonged PI exposure [25]. However, DRV/r is not recommended for children under the age of 3 years due to its toxicity profile. Although DRV/r is available as an oral solution, there is a paucity of data regarding its tolerability in young children.

With evolving optimal, new antiretroviral treatment options, children continue to be exposed to several classes of ART throughout their lives. Appropriate sequencing of age- and weight-based ART regimens, along with long-term treatment-related morbidities remains unclear and needs ongoing evaluation.

The innovations for paediatric-friendly formulations should strive towards safe, effective and palatable ART for children [9]. Several strategies have been put in place to improve development and access to paediatric-friendly formulations including:

- The Global Accelerator for Paediatric Formulations (GAP-f), a new mechanism working to support and formalise collaboration across sectors, and accelerate both upstream (strict drug regulatory authority filing and approvals, formulation development by innovators and generics, and generic manufacturing) and downstream processes (national treatment policy, management of supply chains and market uptake) to ensure that children are able to access the new optimal paediatric antiretroviral medications [26].

- Improving incentives for paediatric formulation development. Pharmaceutical companies Cipla and Mylan that developed LPV/r pellets and granules respectively, have committed to increase supply to meet the growing demand of these formulations until DTG generic formulation and dosing for children less than 20 kg becomes available. It will be essential for the Antiretroviral Procurement-Working Group (APWG) to regularly keep track of demands to ensure supply is not stripped [3].

- Investment in development of palatable, safe and effective antiretroviral medications and simplified regimens for ease of administration. FDC, ideally once a day, formulations would be beneficial in particular LPV/r-based regimens for children who cannot swallow tablets and potentially serve as an alternative for children who do not tolerate DTG (**Table 3**) [3]. Development of a FDC in the pipeline is ABC/3TC/DTG, a WHO recommended regimen, in a single formulation which will simplify treatment. However, there remains uncertainty as to the timing of its availability on the market and dosing guidance for children under 20 kg [3].

- Accelerating availability of DTG dosing for infants and young children through provision of incentives. Recently Unitaid provided financial incentives to the pharmaceutical companies Mylan and Macleods working to develop a generic DTG 10 mg dispersible tablet, which would be suitable for children <20 kg [3]. ViiV also committed to filing their 5 mg dispersible DTG with the US Food and Drug Administration (FDA) (**Table 3**).

- National programs should rapidly take up new formulations as they become available and timely provide necessary training to health care workers responsible for prescribing these medications.

Although there has been some progress with development of paediatric-friendly formulations discussed above, accelerating these processes remains a priority in order to achieve treatment success. Ongoing research and regular surveillance on access and uptake of the new ART formulations at national and global level is vital to recognise and address challenges that may arise. Additionally, longitudinal studies are needed to examine the long-term effects of exposure to several classes of ART in the paediatric population and optimal sequencing strategies of ART to limit treatment failure.

3.1 Non-adherence to antiretroviral treatment among children and adolescents

ART leads to improved survival and reduced HIV transmission rates, however high levels of adherence are required for sustained effects. Adherence to antiretroviral therapy is critical to achieve virological suppression, immune reconstitution and ultimately improved clinical outcomes among children and adolescents living with HIV [27]. Sub-optimal adherence includes missed doses, treatment interruptions or discontinuation, and sub-therapeutic dosing [28]. Non-adherence to ART is not only a barrier to achieve treatment success but is also a driver of resistance limiting future treatment options, and potentially increasing the risk of secondary transmission of drug resistant virus.

Adherence is a complex, dynamic process that can vary throughout the course of treatment and needs to be assessed continuously [29]. Adherence behaviour is influenced by multiple factors that can interplay with each other at different stages. These include patient factors, family and carer factors, patient-provider relationship, socio-cultural and medication related factors [28, 29]. It is therefore important to work closely with family, caregivers, and children and adolescents to best understand the barriers to adherence specific to the individual in order to provide tailored support.

Assessment of adherence should be routine with every clinic visit. A systematic review on ART adherence in adolescents with HIV found almost 40% are non-adherent to treatment [30]. In resource-limited settings, the range of ART adherence has been reported from 49–100% [29]. A contributing factor to this wide

Formulation	Q3 2019	Q4 2019	Q1 2020	Q2 2020	Q3 2020	Q4 2020
LPV/r			Cipla pellet and Mylan granule capacity expansion			
ABC/3TC/ LPV/r	Cipla FDA filing		Mylan FDA filing	Cipla FDA approved (t)	Mylan FDA approved (t)	
DTG		ViiV DTG 5 mg DT FDA filing	Mylan and Macleods DTG 10 mg DT scored FDA filing		ViiV DTG 5 mg DT FDA approval	Mylan and Macleods DTG 10 mg scored DT FDA approval (t)

3TC: Lamivudine; ABC: Abacavir; DT: Dispersible tablet; FDA: Food drug administration; LPV: Lopinavir; Q: Quarter; r: Ritonavir; (t): tentative.

Table 3.
Estimated timeline for development of key paediatric antiretroviral therapies [3].

variation is the lack of standardised methods for measuring adherence. In most instances, adherence is measured subjectively by self or caregiver reporting, and hence is subject to self-enhancement and recall bias. Objective measures of adherence are costly and include HIV viral load, pill counting, electronic dose monitoring and drug detection in biological samples [13, 28, 29]. Discrepancies between pill count and viral load results of 40% have been reported [29].

3.1.1 Barriers to antiretroviral therapy among adolescents living with HIV

Adolescence is a period of dramatic neurocognitive and physiological change and often includes experimentation with sexual behaviour, alcohol and recreational drug use. Furthermore, there is a decrease in the engagement of health care services that may influence adherence [12, 30]. The rising mortality for older adolescents (15–19 years) living with HIV has been attributed to complex challenges with ART adherence and poor retention in care [30].

Several factors contribute to adherence behaviour in this age group. These include [27]:

- Lifestyle barriers such as forgetting to take medications, worrying about disclosure of HIV status, and varied schedules (e.g., schooling)

- Physical factors such as feeling well that may result in complacency and neglecting to take antiretroviral medications.

- Medication related barriers including: (i) treatment fatigue, an important factor particularly in adolescents with perinatal HIV infection who have experienced multiple ART regimens over a prolonged period; (ii) complexity of ART regimens (e.g., pill burden and frequent dosing); (iii) toxicities and adverse effects associated with ART that lead to a reluctance in taking medications.

- Lack of health literacy, poor treatment knowledge and/or understanding of the importance of treatment adherence.

- Disclosure status, including non-disclosure of HIV status to adolescents with perinatal HIV infection by caregivers.
- Structural factors such as lack of transport to travel to the clinics to obtain medications [27].

3.1.2 Barriers to antiretroviral therapy among children with HIV

The obstacles to adherence in children include limited paediatric-friendly formulations (as previously discussed), barriers associated with caregivers that includes forgetting doses, incorrectly measuring liquid formulations, and changes in routine resulting in delays in administration. Some caregivers place responsibility for managing medications to older children before they are developmentally prepared to undertake such a task. Socio-cultural factors that influence adherence across the age groups include poverty, violence, substance abuse, poor mental health and lack of social supports.

3.1.3 Interventions to improve adherence to antiretroviral therapy in children and adolescents

Interventions to improve adherence should be tailored to the individual's needs. Strategies to improve adherence can be grouped as medication-related, patient/family-related, and health care provider-related strategies.

3.1.3.1 Medication-related strategies

- Efforts should be made to simplify regimens with regards to the number of pills or volume of liquid required to be administered and reduce dosing frequency. When feasible, once daily FDC antiretroviral regimens should be prescribed to lessen pill burden and hence improve adherence particularly for older children and adolescents [28].
- Pill swallowing training [28, 31].
- Adherence support through medication education, blister packs and refill reminders.
- Wider access to paediatric-friendly formulations that are palatable.
- Minimising drug toxicities through regular monitoring for adverse effects and potential drug–drug interactions, and a proactive approach in trialling different dosing strategies or switches in ART regimens as required (if feasible).

3.1.3.2 Patient/family-related strategies

Health care providers should evaluate potential barriers to adherence, discuss goals of therapy, importance of optimising adherence, and strategies to support adherence prior initiating ART [28]. Demonstration of drug administration equipment such as use of syringes and medication cups, and ensuring supply of these is important. Provision of information and adherence tools such as written and visual aids may be useful, however this should take into consideration literacy levels of the caregiver.

The use of behaviour modification techniques such as positive reinforcements and provision of incentives to encourage medication compliance can be effective. Trained community outreach workers can provide directly observed therapy and closer adherence support to children with poor adherence particularly those living in regional or remote settings [28]. This can be resource intensive, and therefore not a sustainable long-term solution in low- and middle-income countries. Early recognition and treatment of mental health disorders such as depression, which may impact adherence, should be addressed. This will depend on availability of appropriate mental health services.

In instances where the child has not been informed of their HIV status, timing of HIV disclosure should be discussed with the caregivers. A systematic review evaluating ART adherence and disclosure demonstrated mixed results, with some studies showing improved adherence while other studies finding worse adherence [28, 32]. The decision to disclose HIV status should take into consideration the needs of the child and family, cognitive capacity of the child and psychosocial situation.

For adolescents living with HIV, it is paramount to engage them in management decisions. Use of electronic devices to support adherence such as mobile applications that serve as reminders to take medications and sending text message reminders may be useful. A systematic review found the two most effective interventions were a phone-based counselling approach with adherence monitors and weekly individual and family counselling [13].

3.1.3.3 Health care provider related strategies

Health care providers can improve adherence by establishing a rapport with the patient and family, fostering a trusting relationship and encouraging open communication.

Creating child and adolescent centred multidisciplinary health care settings has been shown to improve treatment outcomes. Adolescent-friendly clinics providing peer counselling, peer navigators and psychosocial supports at clinics and school have demonstrated substantial improvement in retention of adolescents and young adults living with HIV [13]. Such a service also presents a supportive environment to discuss adherence barriers, provide reproductive health education, mental health and disclosure supports, and social activities to promote not only adherence but also retention.

Factors that influence adherence are complex and dynamic, and need to be continuously assessed [29]. A multidisciplinary approach to address adherence challenges is necessary. There remains a paucity of evidence supporting interventions that improve adherence particularly in adolescents. As such, there is a need for evidence-based innovative interventions that are feasible, sustainable and importantly tailored to the individual patient. Ultimately, this will improve treatment outcomes, reduce resistance to ART and be an important step forward towards achieving global targets [13].

3.2 Antiretroviral treatment-related morbidity

The benefits of early initiation and improved access to ART are well recognised, resulting in significant reduction in HIV related morbidity and mortality. Nonetheless, children and adolescents with early exposure to ART experience an array of multisystem morbidities including metabolic complications, increased risk of cardiovascular disease, and neuropsychological challenges. In this section, we discuss morbidities associated with ART faced by children and adolescents.

3.2.1 Metabolic complications

Lipodystrophy syndrome involves redistribution of body fat, which can manifest as lipoatrophy (decrease subcutaneous fat in the face and limbs) with or without central adiposity (lipohypertrophy) [33]. The prevalence of lipodystrophy among children living with HIV can range from 1–57% [34–36]. Studies from sub-Saharan Africa reported a lipodystrophy prevalence of 27–30% among children aged 1 to 18 years, with older children and use of stavudine (d4T) being major risk factors [33].

The pathogenesis of ART-related lipodystrophy is not well understood and felt to be multifactorial including direct effects on lipid metabolism, mitochondrial toxicity and genetic predisposition. Lipodystrophy is a complication of the NRTIs including d4T and zidovudine (AZT). PIs have also been implicated but to a lesser extent. Lipodystrophy has been described to most likely develop during puberty, and such body changes can result in stigmatisation potentially leading to poor adherence and treatment failure.

The diagnosis of lipodystrophy is usually clinical, particularly in resource-limited settings. Anthropometric measurements may be used, which are inexpensive but require experience and standardisation. The use of dual energy X-ray absorptiometry to assess fat distribution is restricted by cost in resource-limited settings [33]. Active clinical surveillance for fat maldistribution particularly in children receiving antiretroviral drugs associated with lipodystrophy and monitoring of lipid profile is necessary.

Insulin resistance and dyslipidemias are commonly linked to lipodystrophy, potentially increasing the lifetime risk of cardiovascular disease. PIs have been associated with elevated triglycerides, low-density lipoprotein cholesterol and total cholesterol. Children receiving LPV/r have been shown to have higher low-density lipoprotein cholesterol and triglyceride levels compared to children receiving NVP [37]. Although the long-term risk of cardiovascular disease in children on ART remains uncertain, the observed elevation in cholesterol levels at a young age is a predictor of long-term risk of premature atherosclerotic disease. Lipid profiles should be obtained from children and adolescents prior initiation of ART and ideally monitored every 6–12 months.

Insulin resistance is less common in children compared to adults. Impaired glucose homeostasis has been reported in 8–35% of children with HIV on ART, which includes impaired glucose tolerance, impaired fasting glucose and type 2 diabetes mellitus [33]. Prolonged exposure to high insulin levels may increase the risk of type 2 diabetes mellitus, a risk factor for cardiovascular disease. Management of insulin resistance includes lifestyle modifications (e.g., diet and exercise), as well as switching to a PI-sparing regimen.

3.2.2 Cardiovascular disease

Cardiovascular complications of HIV infection was recognised early in the epidemic particularly in adults. Evidence suggests children and adolescents with perinatal HIV infection may be at risk of cardiovascular disease due to long term viral effects and exposure to certain classes of ART, especially NRTIs and PIs [33].

Potential cardiovascular risk factors for children with perinatal HIV Infection include dyslipidaemia associated with PIs, as well as heightened vascular inflammation and endothelial dysfunction that may predispose to future atherosclerosis, however supporting data remains limited [33]. HIV related cardiomyopathy has been reported as a potential cardiovascular complication, with a suggested pathogenesis involving mitochondrial toxicity associated with NRTIs (e.g., zalcitabine, didanosine (ddI), d4T and AZT), viral cytopathic effects on cardiac myocytes, and increased cytokine production within the myocardium [38].

Children with perinatal HIV infection remain at risk of long-term cardiovascular disease and thus warrant close surveillance [38]. This may be a challenge in resource-limited settings where diagnostic screening for cardiovascular disease may not be readily available. Therefore, an emphasis should be placed on prevention strategies such as lifestyle modifications, whilst more cost-effective cardiovascular monitoring needs to be evaluated [33].

3.2.3 Lactic acidosis

Hyperlactatemia is a well-known complication of ART. It can vary in severity from asymptomatic to life threatening. In children, the estimated prevalence of mild to moderate hyperlactatemia is 35–50%, with severe forms being rare [33].

Lactic acidosis has been associated with NRTI-induced mitochondrial toxicity. D4T and ddI have the greatest effect, with AZT, 3TC, tenofovir disoproxil fumarate (TDF) and abacavir (ABC) having a lesser effect on the mitochondria. Most children with hyperlactatemia are asymptomatic [33]. The clinical presentation of lactic acidosis is non-specific and can include malaise, abdominal pain, vomiting, muscle weakness and dyspnoea. Supportive laboratory findings include elevated transaminases, lactate dehydrogenase deficiency, amylase, lipase, increased anion gap on venous blood gas and a raised lactate level. Diagnosis requires a high index of clinical suspicion and confirmed with raised venous lactate level [33].

Management of lactic acidosis involves ceasing the offending antiretroviral agent and switching to an agent that is less likely to cause mitochondrial toxicity. In severe forms, NRTI–sparing regimens are advisable [33].

3.2.4 Bone disease

Children and adolescents with perinatal HIV infection are considered at increased risk for lower bone mineral density (BMD) due to the effects of a chronic viral infection and exposure to ART (particularly TDF), though the evidence is mixed [38]. Some studies illustrate significant BMD loss among children treated with TDF-containing salvage regimens [39, 40], whilst other studies demonstrate no association [38]. Other risk factors associated with low BMD include advanced HIV stage and a high viral load [33]. Furthermore, HIV infection is an established cause of pubertal delay which may influence bone mass and subsequent risk of osteoporosis and fractures. Long-term use of certain contraceptives such as depot medroxyprogesterone acetate may contribute to loss of BMD in adolescent females [41].

Longitudinal data are required to further evaluate bone density changes through puberty while on ART to guide treatment regimens and identify bone disease among children and adolescents with perinatal HIV infection [38]. In addition, further exploration to identify interventions to minimise the long-term risk of osteoporosis are needed.

3.2.5 Psychological complications

Children and adolescents with perinatal HIV infection are at increased risk of mental and behavioural disorders. This is influenced by several factors including long-term chronic disease management, psychosocial stressors, stigma and the neurocognitive impact of HIV infection. The most common mental health disorders reported include anxiety, depression, behavioural disorders, learning difficulties and attention deficit hyperactivity disorder [38]. HIV health care providers should be trained to integrate screening of mental health and behavioural disorders into routine care of these children and refer to appropriate services where available. This is an integral component of the holistic long-term management of HIV that will ultimately serve to improve ART adherence, engagement in care, neurocognitive development and social relationships [41].

3.2.6 Reproductive health complications

Clinicians managing adolescents of childbearing potential should assess their fertility intentions and review the potential drug–drug interactions between ART and contraception options to avoid adverse outcomes. There are potential interactions between NRTIs and some PIs with oral contraceptives that reduces their efficacy; whereas RAL does not interact with oestrogen-based contraceptives. There is preliminary data to suggest DTG may increase the risk of neural tube defects [41], and this should be a consideration when discussing ART regimen options during pregnancy. As part of comprehensive HIV care, reproductive health education should be provided to adolescents including risks of sexual transmission of HIV (and other infections) and perinatal HIV transmission, contraception, and access to family planning services.

3.3 Resistance to antiretroviral therapy

Development of HIV drug resistance resulting in treatment failure is a growing concern. Resistance to ART limits alternative treatment options, fuels progression of HIV disease and threatens the success of treatment programs [42]. There are three classes of HIV drug resistance. Acquired drug resistance (ADR) develops when HIV mutations emerge while on antiretroviral medications. Transmitted drug resistance (TDR) occurs through the transmission of resistant HIV. Pre-treatment drug resistance (PDR) is detected in anti-retroviral naïve patients as a result of TDR or following exposure to ART through pMTCT strategies [42]. PDR is a strong predictor of treatment failure and should inform recommended first line ART regimens. The WHO advises a national PDR prevalence of greater than 10% to an antiretroviral drug or drug class as an indication to switch to a different empiric first line ART regimen [42].

Studies across sub-Saharan Africa have shown virological failure ranging between 13 and 64% [43–47], with the proportion of antiretroviral drug resistance around 90% among those with virological failure [45, 46]. National surveys on HIV drug resistance in newly diagnosed HIV infection in infants and children less than 18 months of age across sub-Saharan African countries found an overall prevalence of HIV drug resistance to one or more antiretroviral drugs of 54.1% [22]. NNRTI resistance was present in 53%, predominantly in pMTCT-exposed children; and NRTI resistance present in 8.9%, which was largely driven by d4T and lamivudine (3TC)/emtricitabine (FTC) resistance reflecting the d4T/3TC backbone used at the time in these countries. A systematic literature review on PDR from 13 sub-Saharan African countries found a PDR prevalence of 42.7% in pMTCT-exposed children compared to 12.7% in pMTCT-unexposed children [48]. This study also demonstrated an increase in PDR in pMTCT–unexposed children from 0% in 2004 to 26.8% in 2013, which likely reflects NNRTI TDR from pregnant and/or breastfeeding women to their children. These findings are supported by a systematic literature review on PDR in children starting ART in low-and middle-income countries, which found a median prevalence of NNRTI resistance of 49.3% and more than 50% of pMTCT-exposed children with NNRTI resistance [49].

The high prevalence of PDR to NNRTIs in children supports recommendations to commence children on PI- or DTG-based regimens as the preferred first line due to lower levels of PDR and higher barriers to resistance. Longitudinal observational data from an Asian cohort of children and adolescents receiving second line PI-based regimens showed acquired PI resistance of less than 10% [50]. PI PDR prevalence rates in infants and young children in sub-Saharan African countries are reported as less than 3% [42]. This is likely due to the low rate of maternal PI-based regimens and the higher barrier to resistance for boosted PI regimens. However, widespread use of PI-based regimens in resource-limited settings has been hindered by the lack of access to appropriate paediatric-friendly formulations, prohibitive costs and procurement issues [42].

It is paramount to sustain high levels of viral suppression among children and adolescents with HIV in order to minimise development of HIV drug resistance [42]. To help achieve this, scaling up of viral load monitoring and HIV genotyping at initiation of treatment and throughout treatment is necessary to better understand the prevalence of antiretroviral drug resistance in children and adolescents. This will allow early recognition of treatment failure and guide treatment adjustments to subsequent suppressive ART regimens. Continued efforts to improve treatment adherence, prioritising first line therapies with high genetic barriers to resistance and ensuring availability of third line therapy is critical to reaching the UNAIDS 90-90-90 target goals [46].

4. Conclusion

Successful efforts of pMTCT programs has resulted in declining numbers of HIV infection in children worldwide. However, a considerable number of children and adolescents are living with HIV who require lifelong ART. HIV causes progressive CD4 T cell-related immunodeficiency and chronic immune system activation that results in an array of infectious and non-infectious morbidities and mortality. Early ART initiation is integral in achieving the treatment goals of maximising sustained viral suppression, optimising immunologic status, reducing HIV-related morbidity, and increasing survival. The global scale-up of ART has transformed HIV into a manageable chronic disease, however children and adolescents living with HIV continue to face unique management challenges with respect to ART and supportive care.

ART adherence and engagement in care are key to achieving the goals of therapy. Current challenges to ART adherence include the limited availability of paediatric-friendly formulations, lack of simplified regimens, and psychosocial complexities of managing children and adolescents through periods of marked biopsychosocial development. To improve ART adherence in children there is a need to accelerate development of paediatric-friendly formulations that are palatable and safe for children, with simplified regimens that are easy to administer and able to be transported and stored with minimal resources. Investment and collaboration across public and private sectors are integral to promote access to such paediatric-friendly ART globally. A holistic, multidisciplinary approach to managing children and adolescents living with HIV through the provision of comprehensive child and adolescent HIV health services that provide psychosocial support, surveillance and management of disease- and treatment-related morbidity, mental health screening, and reproductive health counselling is necessary to optimise engagement in care and treatment outcomes. Ongoing efforts to implement effective strategies to identify and manage treatment failure, through upscaling of HIV viral load testing, enhanced adherence support, and access to alternate effective ART regimens are required to maximise ART durability, minimise the development of antiretroviral resistance, and preserve future ART options.

Children and adolescents living with HIV have been and will continue to be exposed to various ART regimens throughout their lives as new antiretroviral agents become available and novel ART regimens introduced. The long-term impact of lifelong exposure to multiple antiretroviral agents with regards to treatment response and morbidity is uncertain and needs ongoing evaluation. Longitudinal studies are essential to provide data on long-term treatment outcomes and antiretroviral drug toxicities to inform optimal sequencing of ART regimens.

A coordinated approach incorporating all stakeholders involved in addressing HIV in children and adolescents throughout their life course is necessary to navigate the challenges in reaching the successive targets set for children and adolescents in overcoming the epidemic. This will require sustained financial, research, and political commitment to best inform HIV models of care for this vulnerable group.

Conflict of interest

The authors have no conflicts of interests to declare.

Author details

Mirvat Said[1] and Adam W. Bartlett[2,3]*

1 Department of Paediatrics and Child Health, Sydney Children's Hospital, Sydney, Australia

2 Department of Immunology and Infectious Diseases, Sydney Children's Hospital, Sydney, Australia

3 School of Women's and Children's Health, UNSW Sydney, Australia

*Address all correspondence to: adam.bartlett@health.nsw.gov.au

References

[1] United Nations International Children's Emergency Fund. HIV and AIDS. 2019. Available from: https://www.unicef.org/hiv

[2] United Nations International Children's Emergency Plan. Adolescent HIV prevention. 2019. Available from: https://data.unicef.org/topic/hivaids/adolescents-young-people/

[3] Initiative CHA. HiV market report - the state of HiV treatment, testing, and prevention in low-and middle-income countries; 2019

[4] Akase IE, Musa BOP, Obiako RO, Ahmad Elfulatiy A, Mohammed AA. Immune dysfunction in HIV: A possible role for pro- and anti-inflammatory cytokines in HIV staging. Journal of Immunology Research. 2017;**2017**:4128398

[5] Alvarez P, Mwamzuka M, Marshed F, Kravietz A, Ilmet T, Ahmed A, et al. Immune activation despite preserved CD4 T cells in perinatally HIV-infected children and adolescents. PLoS One. 2017;**12**(12):e0190332

[6] Newell ML, Coovadia H, Cortina-Borja M, Rollins N, Gaillard P, Dabis F. Mortality of infected and uninfected infants born to HIV-infected mothers in Africa: A pooled analysis. Lancet. 2004;**364**(9441):1236-1243

[7] Kisesa A, Chamla D. Getting to 90-90-90 targets for children and adolescents HIV in low and concentrated epidemics: Bottlenecks, opportunities, and solutions. Current Opinion in HIV and AIDS. 2016;**11**(Suppl 1):S1-S5

[8] Bain LE, Nkoke C, Noubiap JJN. UNAIDS 90-90-90 targets to end the AIDS epidemic by 2020 are not realistic: Comment on "can the UNAIDS 90-90-90 target be achieved? A systematic analysis of national HIV treatment cascades". BMJ Global Health. 2017;**2**(2):e000227

[9] Schlatter AF, Deathe AR, Vreeman RC. The need for pediatric formulations to treat children with HIV. AIDS Research and Treatment. 2016;**2016**:1654938

[10] HIV/AIDS JUNPo. Joint United Nations Programme on HIV/AIDS. 2018. Available from: https://aidsinfo.unaids.org

[11] Nebot Giralt A, Nostlinger C, Lee J, Salami O, Lallemant M, Ouma O, et al. Understanding the acceptability and adherence to paediatric antiretroviral treatment in the new formulation of pellets (LPV/r): The protocol of a realist evaluation. BMJ Open. 2017;**7**(3):e014528

[12] Bartlett AW, Truong KH, Songtaweesin WN, Chokephaibulkit K, Hansudewechakul R, Ly PS, et al. Characteristics, mortality and outcomes at transition for adolescents with perinatal HIV infection in Asia. AIDS. 2018;**32**(12):1689-1697

[13] Enane LA, Vreeman RC, Foster C. Retention and adherence: Global challenges for the long-term care of adolescents and young adults living with HIV. Current Opinion in HIV and AIDS. 2018;**13**(3):212-219

[14] Abul K, Abbas AHL. Pillai Shiv. Basic Immunology Functions and Disorders of the Immune System. Canada: Elsevier; 2016

[15] Okoye AA, Picker LJ. CD4(+) T-cell depletion in HIV infection: Mechanisms of immunological failure. Immunological Reviews. 2013;**254**(1):54-64

[16] Warriner AH, Burkholder GA, Overton ET. HIV-related metabolic comorbidities in the current ART era.

Infectious Disease Clinics of North America. 2014;**28**(3):457-476

[17] Hearps AC, Martin GE, Rajasuriar R, Crowe SM. Inflammatory co-morbidities in HIV+ individuals: Learning lessons from healthy ageing. Current HIV/AIDS Reports. 2014;**11**(1):20-34

[18] Weisner AM, Johnson AP, Lamagni TL, Arnold E, Warner M, Heath PT, et al. Characterization of group B streptococci recovered from infants with invasive disease in England and Wales. Clinical Infectious Diseases. 2004;**38**(9):1203-1208

[19] MR, Drouin O, Bartlett G, Nguyen Q, Low A, Gavriilidis G, et al. Incidence and prevalence of opportunistic and other infections and the impact of antiretroviral therapy among HIV-infected children in Low- and middle-income countries: A systematic review and meta-analysis. Clinical Infectious Diseases. 2016;**62**(12):1586-1594

[20] Picat MQ, Lewis J, Musiime V, Prendergast A, Nathoo K, Kekitiinwa A, et al. Predicting patterns of long-term CD4 reconstitution in HIV-infected children starting antiretroviral therapy in sub-Saharan Africa: A cohort-based modelling study. PLoS Medicine. 2013;**10**(10):e1001542

[21] Lewis J, Walker AS, Castro H, De Rossi A, Gibb DM, Giaquinto C, et al. Age and CD4 count at initiation of antiretroviral therapy in HIV-infected children: Effects on long-term T-cell reconstitution. The Journal of Infectious Diseases. 2012;**205**(4):548-556

[22] Jordan MR, Penazzato M, Cournil A, Vubil A, Jani I, Hunt G, et al. Human immunodeficiency virus (HIV) drug resistance in African infants and young children newly diagnosed with HIV: A multicountry analysis. Clinical Infectious Diseases. 2017;**65**(12):2018-2025

[23] Organization WH. World Health Organization. Policy Brief. The 2018 Optimal formulary and limited-use list for Paediatric ARVS. WHO. 2018. Available from: https://apps.who.int/iris/bitstream/handle/10665/273153/WHO-CDS-HIV-18.15-eng.pdf?ua=1.

[24] Malati CY, Golin R, O'Brien L, Sugandhi N, Srivastava M, Larson C, et al. Pursuing use of optimal formulations for paediatric HIV epidemic control - a look at the use of LPV/r oral pellets and oral granules. Journal of the International AIDS Society. 2019;**22**(4):e25267

[25] Boerma RS, Boender TS, van Hensbroek MB, Rinke de Wit TF, Sigaloff KC. Sequencing paediatric antiretroviral therapy in the context of a public health approach. Journal of the International AIDS Society. 2015;**18**(Suppl 6):20265

[26] Penazzato M, Lewis L, Watkins M, Prabhu V, Pascual F, Auton M, et al. Shortening the decade-long gap between adult and paediatric drug formulations: A new framework based on the HIV experience in low- and middle-income countries. Journal of the International AIDS Society. 2018;**21**(Suppl 1):e25049

[27] Agwu AL, Fairlie L. Antiretroviral treatment, management challenges and outcomes in perinatally HIV-infected adolescents. Journal of the International AIDS Society. 2013;**16**:18579

[28] Panel on Antiretroviral Therapy and Medical Management of Children Living with HIV. Guidelines for the Use of Antiretroviral Agents in Pediatric HIV Infection. 2019. Available from: http://aidsinfo.nih.gov/contentfiles/lvguidelines/pediatricguidelines.pdf

[29] Martelli G, Antonucci R, Mukurasi A, Zepherine H, Nostlinger C. Adherence to antiretroviral treatment among children and adolescents in

Tanzania: Comparison between pill count and viral load outcomes in a rural context of Mwanza region. PLoS One. 2019;**14**(3):e0214014

[30] Kim SH, Gerver SM, Fidler S, Ward H. Adherence to antiretroviral therapy in adolescents living with HIV: Systematic review and meta-analysis. AIDS. 2014;**28**(13):1945-1956

[31] Garvie PA, Lensing S, Rai SN. Efficacy of a pill-swallowing training intervention to improve antiretroviral medication adherence in pediatric patients with HIV/AIDS. Pediatrics. 2007;**119**(4):e893-e899

[32] Nichols J, Steinmetz A, Paintsil E. Impact of HIV-status disclosure on adherence to antiretroviral therapy among HIV-infected children in resource-limited settings: A systematic review. AIDS and Behavior. 2017;**21**(1):59-69

[33] Barlow-Mosha L, Eckard AR, McComsey GA, Musoke PM. Metabolic complications and treatment of perinatally HIV-infected children and adolescents. Journal of the International AIDS Society. 2013;**16**:18600

[34] European Paediatric Lipodystrophy Group. Antiretroviral therapy, fat redistribution and hyperlipidaemia in HIV-infected children in Europe. AIDS. 2004;**18**(10):1443-1451

[35] Alam N, Cortina-Borja M, Goetghebuer T, Marczynska M, Vigano A, Thorne C. Body fat abnormality in HIV-infected children and adolescents living in Europe: Prevalence and risk factors. Journal of Acquired Immune Deficiency Syndromes. 2012;**59**(3):314-324

[36] Kinabo GD, Sprengers M, Msuya LJ, Shayo AM, van Asten H, Dolmans WM, et al. Prevalence of lipodystrophy in HIV-infected children in Tanzania on highly active antiretroviral therapy. The Pediatric Infectious Disease Journal. 2013;**32**(1):39-44

[37] Arpadi S, Shiau S, Strehlau R, Martens L, Patel F, Coovadia A, et al. Metabolic abnormalities and body composition of HIV-infected children on Lopinavir or Nevirapine-based antiretroviral therapy. Archives of Disease in Childhood. 2013;**98**(4):258-264

[38] Vreeman RC, Scanlon ML, McHenry MS, Nyandiko WM. The physical and psychological effects of HIV infection and its treatment on perinatally HIV-infected children. Journal of the International AIDS Society. 2015;**18**(Suppl 6):20258

[39] Gafni RI, Hazra R, Reynolds JC, Maldarelli F, Tullio AN, DeCarlo E, et al. Tenofovir disoproxil fumarate and an optimized background regimen of antiretroviral agents as salvage therapy: Impact on bone mineral density in HIV-infected children. Pediatrics. 2006;**118**(3):e711-e718

[40] Purdy JB, Gafni RI, Reynolds JC, Zeichner S, Hazra R. Decreased bone mineral density with off-label use of tenofovir in children and adolescents infected with human immunodeficiency virus. The Journal of Pediatrics. 2008;**152**(4):582-584

[41] Panel on Antiretroviral Therapy and Medical Management of Children Living with HIV. Specific Issues in Antiretroviral Therapy for Adolescents Living with HIV Infection. 2019. Available from: https://aidsinfo.nih.gov/guidelines/html/2/pediatric-arv/78/specific-issues-in-antiretroviral-therapy-for-adolescents-living-with-hiv-infection

[42] Nuttall J, Pillay V. Antiretroviral resistance patterns in children with HIV infection. Current Infectious Disease Reports. 2019;**21**(2):7

[43] Kebe K, Thiam M, Diagne Gueye NR, Diop H, Dia A, Signate Sy H, et al. High rate of antiretroviral

drug resistance mutations in HIV type 1-infected Senegalese children in virological failure on first-line treatment according to the World Health Organization guidelines. AIDS Research and Human Retroviruses. 2013;**29**(2):242-249

[44] Salou M, Dagnra AY, Butel C, Vidal N, Serrano L, Takassi E, et al. High rates of virological failure and drug resistance in perinatally HIV-1-infected children and adolescents receiving lifelong antiretroviral therapy in routine clinics in Togo. Journal of the International AIDS Society. 2016;**19**(1):20683

[45] Muri L, Gamell A, Ntamatungiro AJ, Glass TR, Luwanda LB, Battegay M, et al. Development of HIV drug resistance and therapeutic failure in children and adolescents in rural Tanzania: An emerging public health concern. AIDS. 2017;**31**(1):61-70

[46] Cisse AM, Laborde-Balen G, Kebe-Fall K, Drame A, Diop H, Diop K, et al. High level of treatment failure and drug resistance to first-line antiretroviral therapies among HIV-infected children receiving decentralized care in Senegal. BMC Pediatrics. 2019;**19**(1):47

[47] Makadzange AT, Higgins-Biddle M, Chimukangara B, Birri R, Gordon M, Mahlanza T, et al. Clinical, Virologic, immunologic outcomes and emerging HIV drug resistance patterns in children and adolescents in public ART Care in Zimbabwe. PLoS One. 2015;**10**(12):e0144057

[48] Boerma RS, Sigaloff KC, Akanmu AS, Inzaule S. Boele van Hensbroek M, Rinke de Wit TF, et al. alarming increase in pretreatment HIV drug resistance in children living in sub-Saharan Africa: A systematic review and meta-analysis. The Journal of Antimicrobial Chemotherapy. 2017;**72**(2):365-371

[49] World Health Organization. HIV drug resistance report. Global perspective on all forms of HIV drug resistance in both adults and children focussing on systematic literature reviews of nationally representative surveys in LMICs; 2017

[50] Prasitsuebsai W, Teeraananchai S, Singtoroj T, Truong KH, Ananworanich J, Do VC, et al. Treatment outcomes and resistance patterns of children and adolescents on second-line antiretroviral therapy in Asia. Journal of Acquired Immune Deficiency Syndromes. 2016;**72**(4):380-386

Section 4

Innate Immunity and Autoimmunity

Chapter 13

Innate Immunity and Autoimmune Diseases

Marcela Catalina Fandiño Vargas

Abstract

The innate immune response is responsible for the initial defense against invading pathogens and signs of damage; in turn, it activates the adaptive immune response to result in highly specific and lasting immunity, mediated by the clonal expansion of antigen-specific B and T lymphocytes. Inflammation is the acute response to infection and tissue damage to limit aggression to the body. It is a complex reaction of vascularized tissues to infection, toxin exposure or cell injury that includes extravasation of plasma proteins and leukocytes. Paradoxically, uncontrolled and prolonged inflammation can result in secondary damage and the development of immune pathology in the host. The components of the innate immune system have recently been studied as responsible mechanisms in various chronic diseases such as diabetes mellitus, atherosclerosis, asthma and allergies, among others. Autoimmune disease is an attack on auto tissues by the adaptation of the immune system. In general, such diseases are characterized by autoantibodies and/or autoreactive lymphocytes directed at antigens against themselves. The innate immune system is often considered an effector of self-reactive lymphocytes, but also provides protection. Studies in mice with specific gene-directed mutations show that defects in innate immune system proteins may predispose to the development of a systemic lupus erythematosus-like syndrome (lupus) characterized by autoantibodies against double-stranded DNA (ds DNA) or nuclear components. This seems to be due to a failure in the removal of apoptotic cells or nuclear waste. These observations imply that the innate immune system has a general protective role against autoimmune disease. For example, in systemic diseases such as lupus, innate immunity is important in the elimination of nuclear antigens and, therefore, in the improvement of tolerance to B lymphocytes. Alternatively, in specific organ disorders such as type diabetes 1 o Crohn's disease, the innate immune system can be protective by eliminating pathogens that trigger or exacerbate the disease or regulate the presentation of antigens for T lymphocytes. Discuss various disease models in which the innate immune system could provide a protective role, deficiencies in the regulation of B lymphocyte signaling through the antigen/receptor or in the clearance of lupus antigens, (dsDNA and nuclear proteins), can lead to a disease similar to lupus. The repertoire of B cells seems to be very biased toward self-activity, as, possibly, that of the T-cell. This tendency toward self-activity is not surprising because B and T cells are positively selected against highly conserved autoantigens.

Keywords: toll, antigens, dendritic cells, lymphocytes, lupus, PAMP, DAMP tolerance, autoimmunity

1. Introduction

The human immune system has two major divisions: innate and acquired. We will talk about innate immunity. Innate immunity can be defined as the first line of defense against pathogens, which represents a great machinery to create an adequate and definitive systemic response to prevent infections and maintain homeostasis of the organism. The elements of innate immunity include external physical barriers, humoral and cellular effector mechanisms. This type of immunity recognizes pathogens such as bacteria and viruses. This works thanks to the phagocytosis of the pathogens with the consequent induction of inflammatory reactions. It also has a critical role in the activation and regulation of adaptive immunity. This immunity has the ability to develop an induced response during primoinfection. This response is specific due to the expression of cell surface pattern recognition (PRR) receptors, which are capable of recognizing complex polysaccharides, glycolipids, lipoproteins, and nucleic acids. We know that pathogens contain in their structure various components that act as substances strange (antigens) and this in turn will induce an innate immune response that will subsequently activate the adaptive response. It is imperative to recognize that the important exploration of these innate mechanisms is essential for the understanding of the complex events involved in human innate immunity and is also crucial for the discovery of new antimicrobials, antitumor drugs, and immunomodulators with therapeutic applications [1]. Innate immunity, which is considered a simple immune system, is essential for the onset of acquired immunity and has been found to play an important role in the pathogenesis of the disease age [2]. Among them, it recognizes nucleic acids derived from pathogens. The innate immune pattern recognition (PRR) receptor recognizes self-derived nucleic acids. Innate pattern recognition receptors regulate antigens for the presentation and subsequent responses of B cells and T cells, for example, physiological management of autoantigens, induction of immature dendritic cells to detect tolerant signals to T cells. The activation of toll-like receptors (TLR), NOD type receptors (NLR) or Helicases similar to RIG (RLH) by molecular agents associated with pathogens where the patterns will induce dendritic cell maturation, costimulation.

T cell activation and production of antibodies by B cells. Therefore, recognition of innate patterns is now being considered as a central element of immunity modulation. There are at least 80 different autoimmune diseases discovered so far, which in the US alone, affect 20 million people [3]. These pathologies are established systemically or in a specific organ, but require for their expression certain conditions that are the result of multifactorial processes that involve a deregulation of the innate immune system and therefore adaptive that lead the body to erroneous responses with the subsequent attack itself of their own tissues. The innate immune system as discussed above is the first line of immediate defense against invading microorganisms that links to the adaptive response. Specific cells of the innate immune system, which are dendritic cells (DC) (antigen presenting), which are cells with an important and critical role in promoting the responses of B and T cells. This type of immunity is critical to maintain homeostasis and prevent microbial invasion, eliminating a wide variety of pathogens and contributing to the activation of the adaptive immune response.

2. The entrance door: PAMP/DAMP

It is the control point. A dendritic type receptor that bears the title of "access gate" for innate cellular immunity: this basically consists of a type of toll-like receptor. It has been found that it plays a fundamental role as a sensor in the recognition of pathogens in the innate immune system [4].

3. Let's talk about PAMP

This pattern recognition receptor acts on bacteria and viruses (PAMP) [5]. The innate immune response in immunological terms controls the infection and prevents its spread. And more recently it is known that to induce this series of reactions against pathogens, in addition to the existence of antigens, another series of molecules in the pathogens is required. These molecules are known as pathogen-associated molecular patterns (PAMPs). PAMPs play and interact with a series of receptors that are mainly present in phagocytic cells (macrophages), and these "gate" receptors have been called recognition patterns to pathogen-associated molecular patterns (PRRs). These receptors contain other subfamilies where we can find toll type receptors (TLRs), NOD type receptors (NLRs), RIG-1 type receptors (RLRs), and lectin C type (CLRs). This molecular pattern related to the associated damage known as DAMP comes to behave as a type of alert that recognizes signals and most importantly this does not involve pathogen detection. The main molecular recognition patterns (PRRs) include TLR and NLR receptors, also known as nucleotide binding oligomerization domains. TLR is the homologous receptor that has already been identified in the Drosophila genetic code, and that to date some TLRs have been found in humans mainly in the cell surface, membrane, and lipids [6]. Types 1, 2, 4, 5, and 6 are those that recognize proteins, nucleic acids located in the endoplasmic reticulum and those that are found in the endosomal membranes. 3, 7, 8, and 9 detect lipopolysaccharides in the outer membrane of gram-negative bacteria (endotoxins). The TLR4 type, which transmits inflammatory signals, is the best known in general and the most studied of the TLR. This receptor responds to MyD88, which becomes a station at the central point of the inflammation signal, and corresponds to the first phase of activation of the transcription factor NF-κB pathway (nuclear factor-kappa B), which a In turn, production begins and a kind of "chain reaction" of inflammatory cytokines to eliminate pathogens [7]. Meanwhile, these TLR receptors are incorporated into PAMPs, which by recognizing nucleic acids act as an inflammatory cytokine. Receptors that mediate innate immune responses, such as toll-like receptors (TLR) and specific C-type lectin receptors (CLR) that recognize associated molecular patterns (PAMP), have been implicated in autoimmune disease mechanisms, both directly through self-recognition ligands and indirectly through the regulation of immune homeostasis [8, 9].

4. DAMPs: (the antigenic gift of cells)

In intracellular infections, in addition to antigens and PAMPs, the participation of another series of molecules that participate in the activation of the immune response is necessary. Recently, some studies have shown that cells can die from a type of immunogenic "apoptosis" and thus expose their nuclear or cytoplasmic molecules to their membrane. These have a way of stimulating the immune response, thanks to their activity. They are also released during the process of necrosis and have been given the name of molecular patterns associated with damage or warning signs, the famous DAMPs. The NLR receptor is present in the cytoplasm. It has the particularity of recognizing not only PAMP but also several DAMP among them [uric acid, cholesterol, sterols crystals, extracellular ATP (adenosine triphosphate), silica] or even recognizing exogenous DAMP such as asbestos, origin of aseptic inflammation, such as gout, arteriosclerosis, and silicosis [10]. It is clear that it is a cause and attracts attention. The abnormalities in the immune system that are the basis and fertilizer for autoimmune diseases are mainly caused by an abnormal acquired immunity [11]. In recent years, in contrast to the concept that autoimmune or auto-inflammatory diseases are mainly due to abnormal innate immunity, it is attracting more attention.

5. Innate immunity cells: "soldiers of the first line of defense"

Dendritic cells, macrophages, and other myeloid cells also play an important role in the innate immune response, both as antigen presenting cells as effector cells that mediate the tissue damage [12–14]. Therefore, they are fundamental and will be as in conflicts, "the first line of defense" in the face of a bacterial or other stimulus. We will also take them into account in relation to autoimmune diseases, because of their responsiveness and because they are important mediators of innate immunity, an interest has arisen in this potential to contribute to the pathology of these diseases. Proinflammatory cytokines: mainly TNFa (tumor necrosis factor alpha), induce the activation of endothelial cells, resulting in an increase in the expression of different adhesion molecules (CD62E, CD62P, ICAM-1, and VCAM-1). This causes the leukocytes to roll over them, and during this bearing, they are activated by the intracellular signals that are generated through their adhesion molecules and different chemokine receptors, which interact with the ligands found on the surface of the cells endothelial. Subsequently, these activated leukocytes adhere firmly to the endothelium, change their morphology (cell polarization) and carry out their transendothelial migration, and then migrate to the inflammatory focus, guided by the gradient of chemotactic substances that are released. Macrophages are multifunctional antigen presenting cells, with an important role in innate immunity and, therefore, in the inflammation process [15]. Macrophages are found in almost all organs, and recent studies have demonstrated their multifunctionality and heterogeneous capacities established by their numerous subpopulations, adaptation in specific tissue microenvironments and different stages of maturation. For example, during a bacterial infection, classically activated macrophages show inflammatory functions (type 1 or M1 macrophages), while with alternative activation (by Th2 type cytokines, such as IL-4 or IL-13), macrophages acquire anti-inflammatory functions (type 2 macrophages or M2). In addition to depletion or inhibition of macrophage function, reprogramming of M2 has also been explored. Recently, it has been shown that paracoccin, a protein contained in a fungal human pathogen, induces the repolarization of M1 macrophages through interaction with toll as a receptor (TLR) 4, being a new possible immunotherapeutic agent for pathologies related to M2 macrophages. Macrophage-related therapies have been proposed for various autoimmune and inflammatory pathologies. In the case of PPARγ and PPARδ, which are nuclear receptors that control different genes associated with M2 macrophages, and their agonists have been proposed as a therapy directed at macrophages to induce M2 pathways. In addition, the demonstration that TLR9 receptor signaling can reverse the aberrant M2 macrophage phenotype.

Dendritic cells (DC) are professional antigen presenting cells (APC), often referred to as "orchestra directors of the innate immune response" due to their ability to capture, process, and present antigens to T cells. Depending on the nature of the antigen may exhibit an immunogenic or tolerogenic effect, which will be defined by cytokine secretion. They are often considered tolerogenic, because they have autoantigens in the absence of costimulation and, together with anti-inflammatory stimuli, (TGF-β), can promote the induction of regulatory T cells and/or induce anergy of T cells [16]. After activation by proinflammatory stimuli, they mature and generate an expression of costimulatory molecules and the major histocompatibility complex (HCM) class II, which causes a potent response of specific T cells to the antigens. Therefore, they play a fundamental role in maintaining self-tolerance, and on the other hand, they initiate the response against foreign antigens for their subsequent elimination by effector immune cells. In a state of aberrant hyperreactivity, they could contribute to perpetuating immune responses, backed by evidence of a high frequency of immunogenic infiltration [17]. Due to their ability to modulate the cellular response, they have been considered a powerful target for immune modulation. Strategies such as pharmacological modulation to affect their maturation status and genetic engineering to improve their tolerance or immunogenic properties for the treatment of autoimmune diseases have been studied. In several murine models, they were transduced to express IL-4 and were able to prevent disease in 12-week NOD mice. In a murine model of collagen-induced arthritis (CIA), it was shown that the injection of dendritic cells with tolerogenic activity improves the clinical and the outcome of the disease. Although the treatment was found to be safe and feasible, other studies are needed to evaluate the efficacy of cellular treatment in autoimmunity.

6. Innate lymphoid cells (ILC): "the element of surprise"

They are a growing family of immune cells that reflect the phenotypes and functions of T cells. Natural killer cells (NK) can be considered innate homologs of cytotoxic CD8 + T cells, while ILC1, ILC2, and ILC3 correspond to innate homologs of T cells CD4 + (TH1), TH2, and TH17. However, in contrast to T cells, they do not express antigen receptors or undergo clonal selection and expansion when stimulated [4]. The ILCs react and respond to the signs of tissue damage and produce a series of cytokines, which direct the immune response and this adapts to contain the lesion. Therefore, these cells can control or unleash the immune response. As with B cells and T cells, these also originate from the common lymphoid lineage but the specific transcription factors of these suppress and modify their development until the generation of the different types of ILC. The precursors of these can migrate from their primary production site in infected and injured tissues, where they complete their maturation, in a process very similar to the differentiation of virgin T cells into TH effectors. The cytokines produced by local cells, as well as some trauma and stress response ligands as well as bacterial and dietary compounds regulate the maturation and activation of ILC in effectors that play an important role in early immune responses to pathogens in particular has been found relationship with symbionts, helminths, and allergens. The cytokines they produce induce innate responses in stromal, epithelial, and myeloid cells that in turn will regulate the activity of dendritic cells and will also play a central role in the transfer of information between ILC and T cells. ILCs by activating DC found in tissues to migrate to the lymph nodes, where they cause specific T-type cellular responses. ILCs also regulate T cells directly through the presentation of peptide antigens through CMH type II. However, ILCs are also involved in autoimmunity, because their cytokine production can exacerbate and exaggerate the inflammatory process.

7. The eye of the hurricane: autoimmunity and innate immune system. How can an autoimmune disease use machinery of innate immunity?

Recent research has revealed new knowledge about the respective roles of these cells in relation to cellular and humoral immunity as well as the extension to adaptive immunity [18]. There is talk of a recent study in which a genetically modified mouse prototype model was developed with an autoimmune disease similar to lupus that does not require to express the adaptive immune system machinery, but is triggered directly by the innate immune response [19]. For many autoimmune diseases, we largely know the roles that key cells (T cells and B cells) play and for example are evident in the success of existing therapies (anti-CD3 and anti-CD20). Then knowing this, each of the functions of myeloid cells, and in general of the innate immune response cells, can "autoimmune" disease occur in the absence of adaptive immunity and these cells act as effectors in disease progression? The answer to this could be yes [20]. The most recent example is the study of mice eaten by moths that have been genetically modified to have deficiencies in hematopoietic cells, and to express an autoimmune disease characterized by alopecia (giving a "peeled or eaten by moths") and edema in their legs. These were also accompanied by high antibody titers, with renal and pulmonary functions being compromised due to immune complex deposits [21, 22]. However, in another study, mice with deficiency in hematopoietic cell phosphatase were crossed with mice that lack the recombinase-1 activator gene (RAG-1) that caused a subsequent deficiency in the production of T and B cells and found that the disease autoimmune had progressed normally in the absence of an adaptive immune response [22, 23] even though these mice lacked high antibody titers and immune complex deposits, and they exhibited all other symptoms of the disease. Subsequently, although the onset and progression of the disease could not be defined, it was concluded that the autoimmune disease of this type of mice was mediated by an aggressive response of macrophages and other myeloid cells. Now, a study with murine models is also described, with mice with a genetic alteration associated with the deficiency in the enzyme α-mannosidase type II (αM-II) where there is premature aging with the clinical expression and the characteristic symptoms of SLE and Lupus nephritis (high titers of anti-DNA antibodies, glomerulonephritis, and renal compromise due to deposition of immunoglobulins in the kidney) that seems to be driven by a mechanism that also seems to involve the innate immune system [12, 24, 25]. In the case of the murine model, evidence was provided that the abnormal presence of hybrid glycoprotein structures acts as a trigger for the induction of an innate immune response mediated by members of the C-type lectin family that is specific for mannose. Serum mannose-binding lectins (MBL-A and MBL-B) are soluble lectins that mediate innate immunity to pathogenic bacteria and fungi that express glucans (mannose). It is also believed that the macrophage of the mannose receptor cell surface (MMR) participates in innate immune responses, and its expression has been documented in mesangial renal cells [26, 27]. In mice with αM-II deficiency, MBL lectins are deposited in renal glomeruli which, when they express high levels of mannose glucans in mesangial cells, also express higher levels of MMR, which can bind mannose ligands in the serum. Monocyte chemoattractant protein 1 (MCP-1) levels, produced by activated mesangial cells, represent the entry of activated macrophages. By aberrantly expressing mannose-containing glucans in mice with αMII deficiency, they act as triggers for an innate immune response mediated by mannose-specific C-type lectins programmed to recognize mannose glucans as PAMP.

The second point in importance is the role of antibodies in stimulating the innate immune response. How can this be to the production of autoantibodies in autoimmune diseases, such as our old friend, lupus? Systemic lupus erythematosus (SLE) is an autoimmune disease that translates inflammation and exaggerated immune responses and thus with a large generalized associated tissue damage. We are clear that innate immunity plays a great role in its development and sequentially its clinical expression, and it has been shown that defects found in any of the immune recognition pathways will promote autoimmunity. First, dendritic cells and macrophages activated by TLR receptors can regulate the differentiation of self-reactive B cells through the expression of CD40 and the action of IL-6. Second, by nucleic acids. These can activate and in a powerful and disorderly way certain TLR and RLH receptors; therefore, these are normally protected from immune recognition by multiple mechanisms (epigenetic modifications, nuclear compartmentalization, and the rapid elimination of cells that have entered apoptosis and extracellular compartments by a type of DNase and RNAse enzymes). These immune complexes containing chromatin or circulating RNA particles can avoid being "digested" by these enzymes in the extracellular space and facilitate the uptake of the complex in intracellular compartments through Fc receptor-mediated endocytosis (FcR) in dendritic cell-mediated uptake or by B cell receptor (BCR) in B cells. And it has also been confirmed by studies with lupus-prone mice deficient in TLR receptors and their respective signaling molecules. As an exception, mice with TLR-9 deficiency with a predisposition to lupus produce more autoantibodies against it, indicating that TLR-9s have additional functions in the regulation of systemic autoimmunity. Innate pattern recognition (PRR) receptors regulate the production of autoantibodies associated with lupus and self-reactive T cells by modulating the presentation of autoantigens and also contribute directly to the end result that is tissue or organ injury secondary to autoimmunity. In general, it is believed that this "injury" or tissue damage is generated from the deposition of the immune complex, complement activation, and subsequent release of cytokines and chemokines to trigger local inflammation. This concept has been redefined. For example in glomerulonephritis, in the glomerular immune complex, deposits are not always associated with innate and adaptive immune responses. These are traditionally seen as separated from each other, but emerging evidence suggests that they overlap and interact with each other. Recently discovered cell types, particularly innate lymphoid cells and myeloid cell-derived suppressors that are gaining increasing attention. It is a rapidly evolving field with molecular pathways and new types of discovered cells and multiple constantly changing paradigms. In general, it is believed that many autoimmune diseases are triggered by aggressive responses of adaptive immunity by an automatic antigen system, resulting in tissue damage and pathological sequelae.

The third point is undoubtedly the role of infectious agents, which have the potential to trigger an exaggerated immune response, through molecular imitation, polyclonal activation or antigen release. For example, there are certain diseases that respond to certain infectious autoantigen peptides. This is the case of multiple sclerosis, where T cells are activated by Epstein-Barr virus peptides, type A flu, and human papilloma and that react with the myelin autoantigen peptide [28]. In this case, the viral infection could cause the activation of the lymphocytes, and the autoantigen could maintain this activation, even after the eradication of the infectious agent. Microbial infection can also cause polyclonal activation of lymphocytes, and this is the underlying mechanism in increasing the incidence of autoimmunity in murine models exposed to microbial pathogens [29]. Microbes (viruses or bacteria) that destroy cells also cause an inflammatory response and also the release of antigens that have been previously captured and this could also result in autoimmunity. There is another important point. Inflammation, even in the absence of infection, can trigger polyclonal activation and self-activity. This is that through the activation of annergic cells, by inflammatory mediators or the activation of new self-reactive cells in an inflammatory environment for example in the context of ischemia of any tissue, tissue autoreactivity could be caused and because not at a systematic level [3]. Within non-infectious detonators, we have those of the hormonal type that in many autoimmune diseases are more common in women than in men. Drugs can also alter the immune repertoire. One of the most common and studied procainamide induces antinuclear antibodies and sometimes induces a lupus-like syndrome. And even some substances produced by the same cells can act as haptens and make autoantigens immunogenic, for example, CD1 T cells, with receptors (gamma/delta), CD4+, CD25+, and cytokine-producing agents that monitor activity, reduce, and control self-reactive cells, and they can become pathogenic. As some must complete their maturation in the thymus, and others the activation of autoantigens in the periphery, in these processes alterations in the number and function of regulatory cells that can contribute to autoimmunization can be generated.

8. "War": mechanisms of tissue damage of innate immunity

Upon contact with the stimulus, whether microbial or of any substance, the recruitment and activation of macrophages will begin. The macrophages will serve as the primary effector cells that cause tissue damage and loss. And it has been concluded that the vast majority of autoimmune diseases could be explained by an aberrant adaptation as an immune response to the antigens themselves. On the other hand, autoimmunity as a disease contrasts with innate immunity. The first in which the term autoinflammatory was used was the periodic fever syndrome related to the TNF receptor (tumor necrosis factor), whose causative gene is TRAPS 4 and which was directly related to the presence of genetic abnormalities associated with innate immunity autoinflammatory diseases that are generally considered as a group of diseases where we can find an active responsibility for aberrant innate immunity and in which T cells are not detected and include TRAPS, cryopyrine-associated syndrome (secondary to mutations in the NLRP3 gene in children) (CAPS), Familial Mediterranean Fever (FMF), Bechet's disease, Still's disease in adults, Crohn's disease, Gout, Type 2 diabetes, and various metabolic disorders [30]. The mechanism of its many initiation is still unclear, but the symptoms and diseases themselves are caused by the collapse of immune tolerance. Thymus autoreactivity and subsequent and completely abnormal inactivation of receptive and regulatory (Th) T cells suppress the reaction to the foreign antigen. The other part of the aberrant response of the innate immune response is carried out in the recipients of recognition of autoimmune patterns and diseases recognized by nucleic acids (PRR). This recognition is transmembrane due to its location in the cell and is divided into two general and cytoplasmic phases. This receptor is found in the endoplasmic reticulum or endosome and is directly related to autoimmune diseases (SLE = TLR7/9). When comparing the sequence of own nucleic acids and pathogen derivatives by means of the TLR7/9TLR9 receptors, it is noted that it contains unmethylated CpG sequences, and these are derived from pathogens that in turn recognize a type of single stranded DNA. TLR7 on the other hand recognizes single stranded RNA derived from viruses and other types, as well as messenger RNA (mRNA). From this, TLR7/9 is self-sufficient, and this receptor can strictly distinguish between conventional nucleic acids and pathogen derivatives. Stimulates an immune response in response to auto-nucleic acid. In other words, viruses and infected cells are captured by endosomes, and these nucleic acids are recognized by TLR7/9. In the case of SLE, the TLR7/9 receptor, due to the genetic modification secondary to the aberrant response to the own nucleic acids that were released and transferred to the endosome and therefore increases the genetic expression of the type I IFN and is known as the "IFN signature." This signature of IFN is directly related to SLE, rheumatoid arthritis (RA), and systemic sclerosis (SSc) and its effects, suggesting the importance of type I IFN in autoimmune disease [31]. It also activates and stimulates plasma cells that in turn produce large amounts of type I IFN.

The TLR7/9 receptor also mediates the response of plasmacytoid cells and is considered an IFN type I producing cell, which through the TLR7/9 Fc receptor activates the signal to induce the production of non-protein IFN type I histone in the core (HMGB1), to subsequently activate DAMP. The balance between TLR7 and TLR9 is also considered important for inflammation and immune response. The other transmembrane PRR receptors TLR3 and TLR8 also recognize double stranded and single stranded RNA. On the other hand, the cytoplasmic PRR receptor, type RIG-I, and MDA-5 normally identifies a specific structure of single stranded RNA. By recognizing double-stranded RNA, the specific proteins of these DAI, IFI16, and DDX41 receptors induce the production of IFN and inflamasome and, in turn, the production of IL-1β and IL-18.

Conflict of interest

"The authors declare no conflict of interest."

Author details

Marcela Catalina Fandiño Vargas
Unidad de Reumatologia y Osteoporosis, Hospital Central Ignacio Morones Prieto, San Luis Potosi, Mexico

*Address all correspondence to: doctorcita.markitty@hotmail.com

References

[1] Davidson A, Diamond B. Autoimmune diseases. New England Journal of Medicine. 2001;**345**:5. DOI: 10.1056/NEJM200108023450506

[2] Banchereau J, Briere F, Caux C, Davoust J, Lebecque S, Liu YJ, et al. Immunobiology of dendritic cells. Annual Review of Immunology. 2000;**18**:767-811

[3] Hackstein H, Thomson AW. Dendritic cells: Emerging pharmacological targets of immunosuppressive drugs. Nature Reviews. Immunology. 2004;**4**(1): 24-34. DOI: 10.1038/nri1256

[4] Waldner H. The role of innate immune responses in autoimmune disease development. Autoimmunity Reviews. 2009;**8**:400-404

[5] Gasteiger G, D'Osualdo A, Schubert DA, Weber A, Bruscia EM, Hartl D. Cellular innate immunity: An old game with new players. Journal of Innate Immunity. 2017;**9**:111-125. DOI: 10.1159/000453397

[6] Gasteiger G, D'Osualdo A, Schubert DA, Weber A, Bruscia EM, Hartl D, et al. Cellular innate immunity: An old game with new players. Dendritic-cell control of pathogen-driven T-cell polarization. Nature Reviews. Immunology. 2003;**3**(12):984-993

[7] Olive E. Pattern recognition receptors: Sentinels in innate immunity and targets of new vaccine adjuvants. Expert Review of Vaccines. 2012;**11**(2):237-256. DOI: 10.1586/ ERV.11.189

[8] Kisielow P, Bluthmann H, Staerz UD, Steinmetz M, von Boehmer H. Tolerance in T-cell-receptor transgenic mice involves deletion of nonmature CD4 + 8+ thymocytes. Nature. 1988;**333**(6175): 742-746

[9] Abbas, Lichtman, Pillai. Mecanismos de la autoinmunidad. Inmunología celular y molecular: 9 edition. Philadelphia, PA: Elsevier; 2018

[10] Medzhitov R, Preston-Hurlburt P, Janeway CA Jr. A human homologue of the drosophila toll protein signals activation of adaptive immunity. Nature. 1997;**388**(6640):394-397

[11] Romo R, Pérez-Martíınez D, Ferrer CC. Innate immunity in vertebrates: An overview. In: Immunology. Vol. 148. Hoboken, NJ: John Wiley & Sons Ltd; 2016. pp. 125-139

[12] Bettelli E, Oukka M, Kuchroo VK. TH-17 cells in the circle of immunity and autoimmunity. Nature Immunology [Internet]. Springer Science and Business Media LLC; 2007 Apr;**8**(4):345-350. DOI: http://dx.doi.org/10.1038/ni0407-345

[13] Geijtenbeek TBH, van Vliet SJ, Engering A, 't Hart BA, van Kooyk Y. Self- and Nonself-Recognition by C-Type Lectins on Dendritic Cells. Annual Review of Immunology [Internet]. Annual Reviews; 2004 Apr;**22**(1):33-54. DOI: http://dx.doi.org/10.1146/annurev.immunol.22.012703.104558

[14] Green RS, Stone EL, Tenno M, Lehtonen E, Farquhar MG, Marth JD. Mammalian N-Glycan Branching Protects against Innate Immune Self-Recognition and Inflammation in Autoimmune Disease Pathogenesis. Immunity [Internet]. Elsevier BV; 2007 Aug;**27**(2):308-320. DOI: http://dx.doi.org/10.1016/j.immuni.2007.06.008

[15] Manzo A, Vitolo B, Humby F, Caporali R, Jarrossay D, Dell'accio F, et al. Mature antigen-experienced T helper cells synthesize and secrete the B cell chemoattractant CXCL13 in the inflammatory environment of

the rheumatoid joint. Arthritis and Rheumatism. 2008;**58**:3377-3387

[16] Capolunghi F, Rosado MM, Cascioli S, Girolami E, Bordasco S, Vivarelli M, et al. Pharmacological inhibition of TLR9 activation blocks autoantibody production in human B cells from SLE patients. Rheumatology (Oxford, England). 2010;**49**:2281-2289

[17] Kumar H, Kawai T, Akira S. Pathogen recognition by the innate immune system. International Reviews of Immunology. 2011;**30**(1):16-34. DOI: 10.3109/08830185.2010.529976

[18] Cañas C. Autoimmunity and autoinflammation. Acta Médica Colombiana. 2011;**36**(2):78-84. Source: http://www.actamedicacolombiana.com/anexo/articulos/v36n2a05.htm

[19] Akira S, Takeda K. Toll-like receptor signalling. Nature Reviews. Immunology. 2004;**4**(7):499-511

[20] Passineau MJ, Fahrenholz T, Machen L, Zourelias L, Nega K, Paul R, et al. Alpha-galactosidase a expressed in the salivary glands partially corrects organ biochemical deficits in the fabry mouse through endocrine trafficking. Human Gene Therapy. 2011;22:293-301

[21] Linehan SA, Martínez-Pomares L, Stahl PD, Gordon S. Mannose Receptor and Its Putative Ligands in Normal Murine Lymphoid and Nonlymphoid Organs: In Situ Expression of Mannose Receptor by Selected Macrophages, Endothelial Cells, Perivascular Microglia, and Mesangial Cells, but not Dendritic Cells. Journal of Experimental Medicine [Internet]. Rockefeller University Press; 1999 Jun 21;**189**(12):1961-1972. DOI: http://dx.doi.org/10.1084/jem.189.12.1961

[22] Marshak-Rothstein A, Rifkin IR. Immunologically Active Autoantigens: The Role of Toll-Like Receptors in the Development of Chronic Inflammatory Disease. Annual Review of Immunology [Internet]. Annual Reviews; 2007 Apr;**25**(1):419-441. DOI: http://dx.doi.org/10.1146/annurev.immunol.22.012703.104514

[23] Mensah-Brown E, Shahin A, Parekh K, Hakim AA, Shamisi MA, Hsu D, et al. Functional Capacity of Macrophages Determines the Induction of Type 1 Diabetes. Annals of the New York Academy of Sciences [Internet]. Wiley; 2006 Nov 1;**1084**(1):49-57. DOI: http://dx.doi.org/10.1196/annals.1372.014

[24] Ohtsubo K, Marth JD. Glycosylation in Cellular Mechanisms of Health and Disease. Cell [Internet]. Elsevier BV; 2006 Sep;**126**(5):855-867. DOI: http://dx.doi.org/10.1016/j.cell.2006.08.019

[25] Robinson MJ, Sancho D, Slack EC, LeibundGut-Landmann S, Reis e Sousa C. Nature Immunology. 2006;**7**:1258-1265

[26] Ronnblom L, Alm GV. A pivotal role for the natural interferon alpha producing cells (plasmacytoid dendritic cells) in the pathogenesis of lupus. The Journal of Experimental Medicine. 2001;**194**(12):F59-F63

[27] Lastra MD. Facultad de quimica UNAM. Notas de Autoinmunidad. 2004

[28] Dong L, Ito S, Ishii KJ, Klinman DM. Suppressive oligodeoxynucleotides delay the onset of glomerulonephritis and prolong survival in lupus prone NZB×NZW mice. Arthritis and Rheumatism. 2005;**52**(2):651-658

[29] Pisetsky DS. The role of innate immunity in the induction of autoimmunity. Autoimmunity Reviews. 2008;**8**(1):69-72

[30] Eberl G, Colonna M, Di Santo JP, McKenzie ANJ. Innate lymphoid cells: A new paradigm in immunology. Science [Internet]. American Association for the

Advancement of Science (AAAS); 2015 May 21;**348**(6237):879-887. DOI: http://dx.doi.org/10.1126/science.aaa6566

[31] Kramer JM, Klimatcheva E, Rothstein TL. CXCL13 is elevated in Sjogren's syndrome in mice and humans and is implicated in disease pathogenesis. Journal of Leukocyte Biology. 2013

Chapter 14

Precision Medicine of Autoimmune Diseases

Ayodeji Ajayi, Oluwadunsin Adebayo and Emmanuel Adebayo

Abstract

Genomic-based information is an essential key to precise therapy referred to as personalized medicine. Its application in autoimmune disease treatment will bring the required breakthrough in medicine. Autoimmune diseases are the disease conditions where the body's immune system recognizes and generate an immune response against self-antigens. There exist different approaches of which precision medicine data can be utilized in the clinical management of autoimmune diseases; this includes diagnosis, prognosis, stratification and treatment response prediction. Different markers exist to guide clinical decision while several others are still being identified and proposed. This chapter highlights data and databases in precision medicine of autoimmune diseases and the pathway for data sharing. The precision medicine of selected autoimmune diseases was discussed, and the different biomarkers utilized in the diagnosis, prognosis, stratification and response monitoring of such condition were considered.

Keywords: autoimmune diseases, databases, genomic data, personalized medicine, precision medicine

1. Introduction

The functional responsibility of the immune system (humoral and cell-mediated alike) is to protect against infection by destroying various infectious agents when such agents attack the body or are introduced through vaccination [1]. The functioning of the immune system is coordinated and maintained by a sequence of highly regulated and physiological mechanisms which aids the identification and recognition of both body cells and foreign cells [2].

The body's immune units usually coexist with other cells of the body that carries a self-marker molecule. Immune reactions are only triggered when an antigen which could be a microbe, part of a microbe or a molecule is presented to the surface of the cell and perceived by the body defenses [3].

The immune system of humans is made up of two divisions which are innate and acquired immunity. The innate immunity forms the first line of defense immediately after infectious agents are recognized by the body while acquired immunity functions in the removal of pathogens at the later phase of infection [3].

When the immune system is stimulated, it targets and destroys foreign units. Still, in some abnormal situation, the immune system might be insensitive to antigens, hypersensitive to antigens or recognize the cells with self-marker as foreign cells [2].

There are disease conditions that affect the immune system, which leads to different degree and types of conditions known as the Immune diseases. Diseases of the immune system include inherited and acquired immunodeficiency and immune-proliferative disorders which includes malignancies of the immune system (multiple myeloma, lymphoma, and leukemia), autoimmune diseases (rheumatoid arthritis), and immune hypersensitivities (allergies) [4]. Inherited immunodeficiency, also is known as primary immunodeficiency, refers to a large number of immune disorders which alters either or both development and function of the immune system. Primary immunodeficiency implies conditions resulting from loss of function, a gain of function or loss of expression due to monogenic germline mutations [5]. External and environmental factors can induce an adverse effect on the immune system, and this is regarded as secondary or acquired immunodeficiency, which is encountered commonly in clinical practice and could arise from quite a number of conditions [6].

The evolvement of medical practices especially diagnosis and treatment from the usual "one size fits all" approach to a more genetic and detailed patient stratification in a bit to acquire more information about the disease condition and the patient is known as personalized medicine [7].

The complexity of the body defense system and the ability of the cells associated with it to shift between different activation states under physiological and pathological conditions are some of the reasons for diversity in the treatment approach. The immune diseases at times are diverse, and this result in variations in response to therapy. The difference in the disease course also create reasons why there should be the identification of personalized marker for diagnosis of immune disease. Therefore, the use of genetic assessment to determine the best possible therapeutic approach from the numerous available options with different mechanisms, risks, and efficacy are essential [7, 8].

The Precision medicine data types, genomic data in precision medicine, genomic and personalized medicine databases, data sharing, access and use are discussed in this chapter. Also, the use of genomic methods and data in the understanding, diagnosis of diseases using specific biomarkers, monitoring of prognosis using prognosis biomarkers, personalized treatment of immune disorders, monitoring of response to treatment using response biomarkers are also described in this chapter.

2. Precision medicine of specific autoimmune diseases

2.1 What is immunity?

Immunity is the ability of the body to prevent infection by resisting the invasion of such a body by harmful microorganism knows as infectious agents. Immunity can be categorized broadly into two types which are:

i. Innate or Natural Immunity and

ii. Acquired Immunity

2.1.1 Innate or natural immunity

The initial host protection against diseases- causing agents is the innate immunity which is mediated by phagocytes. Through germline-encoded pattern-recognition receptors (PRRs), the innate immunity of the human body recognizes microorganisms invading its body. For the immune cells to be activated, different classes of the PRRs, which include Toll-like receptors and cytoplasm receptors recognize distinct and important microbial component of invading microorganisms thereby activating immune cells [3, 9].

2.1.1.1 Mechanism of action of innate or natural immunity

Immediately after the detection of non-self-agents by PRRs which could be exhibited on the outer membrane of the cell, in intracellular parts, or released in the bloodstream and fluids of the body tissues, the PRRs then perform the function of opsonization, stimulation of complement and coagulation outflow, phagocytosis, initiation of pro-inflammatory signaling pathways, and inception of apoptosis. These cascades of intracellular signaling induce the expression of overlapping and unique genes which are involved in the inflammatory immune responses and essential in precision medicine. The reaction by the innate immune system is carried out by phagocytes (neutrophils, monocytes, and macrophages), inflammatory mediators releasing cells (basophils, mast cells, and eosinophils), and natural killer (NK) cells [3, 10].

2.1.2 Acquired immunity

Acquired immunity is the immunity that is developed against an infectious agent by the body after the previous encounter with a pathogen or a type of immunity developed by a child by the exchange of protective materials from mother to child before and after birth or by the injection of such substances. The mediation of adaptive immunity is the function of clonally distributed T and B Lymphocytes whose characteristics are the possession of specificity and memory. Many at times, activation of the innate immune response can trigger acquired immunity. The generation of Helper T cells subsets and the production of cytokines influence adaptive immunity [11, 12].

2.1.2.1 Mechanism of action of acquired immunity

When naïve T-helpers cells are stimulated by Antigen-presenting Cells otherwise known as APCs, they differentiate into two subsets of T helper (T_H) cells such as T_H1 and T_H2. Interferon-γ (IFN-γ) is produced by T_H1 cells that solely promote cellular immunity. T_H2 cells, on the other hand, produce interleukin 4, 5, 10 and 13 (IL-4, IL-5, IL-10, and IL-13). Whereas, IL-12 is the propelling source of T_H1 separation while IL-4 stimulates T_H2 distinction. T_H2 is majorly involved in the promotion of humoral immunity [12].

2.1.3 Pathogenesis of immune diseases

The occurrence of the immunological disease is consequent to the dysregulation of numerous and different part of the human immune system. Fundamentally, the response of the immune system recognizes and eliminates antigens but tolerates its tissues. However, predominant immunopathology lesion is the basis on which the characterization of immune-mediated diseases is based. Immune-mediated disorders can be grouped into immediate hypersensitivity, autoimmunity, immune-complex disease, and delayed-type hypersensitivity. Autoimmunity can be further classified into those mediated by adaptive immunity and those mediated by innate immunity. Most of the disorders lie between the two, which will be best described

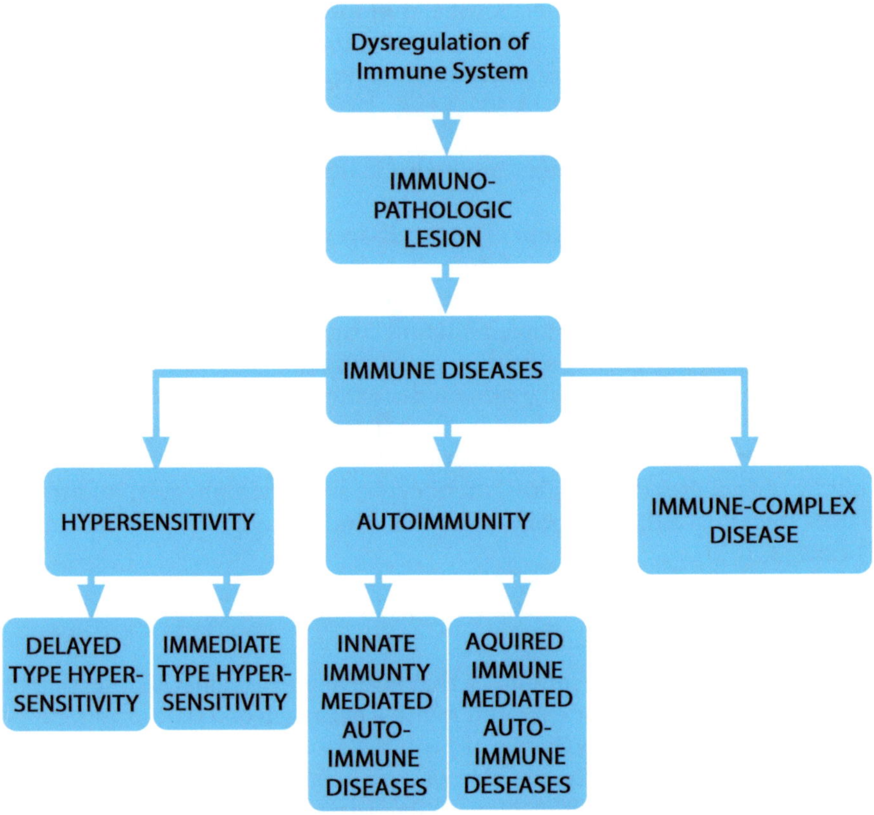

Figure 1.
Schematic representation of the pathogenesis of immune diseases.

as positive pathological feedback between innate and adaptive immune mechanisms [13]. **Figure 1** below represents the pathogenesis of immune diseases.

2.2 Personalized medicine

Personalized medicine is the process of tailoring the diagnostic procedures, treatment, and preventive measures towards the characteristics of individual patients to get an optimal outcome for each patient while emphasizing easy accessibility and cost-effectiveness [14]. In the practice of personalized medicine, the characteristics of an individual, including the uniqueness of its genetic profile guide the clinical decision in the treatment. Prognostic, diagnostic and predictive biomarkers are always being searched to guide these clinical decisions, at the same time, ensure that the best treatment is offered to the right patient at the best time [15]. The division of personalized medicine is illustrated in **Figure 2**.

While the method of application of precision medicine is given in **Figure 3.**

2.2.1 Personalized medicine and genomic data

Generally, personalized medicine compose of a vast collection of genetic data. The development of power systems has helped to increase the effective use of big data in personalized or precisions medicine over time. Also, the evolution of genomics data offers limitless possibilities in the design of clinical procedures, diagnostic, prevention, addressing and prediction of most favorable therapeutics for many diseases that are related to different regions and lineage [16].

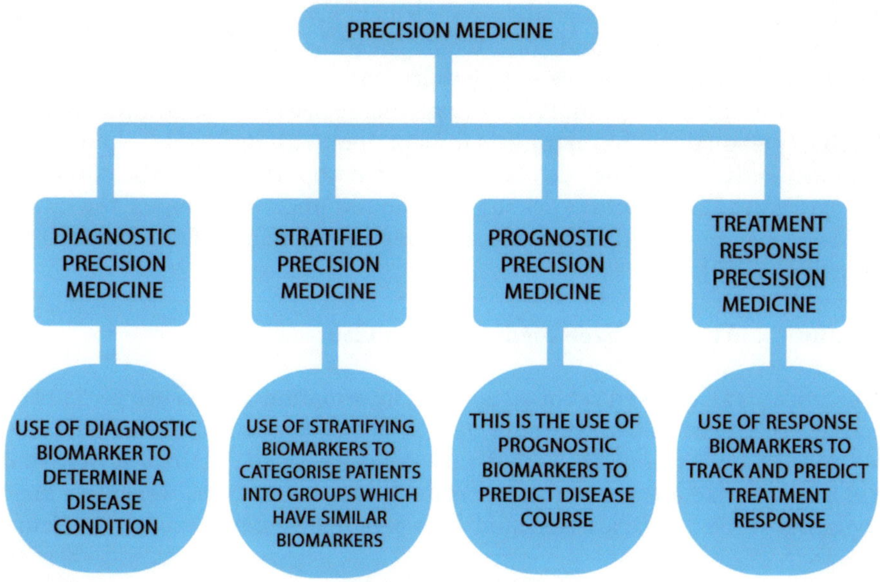

Figure 2.
Diagram showing the different division of precision medicine.

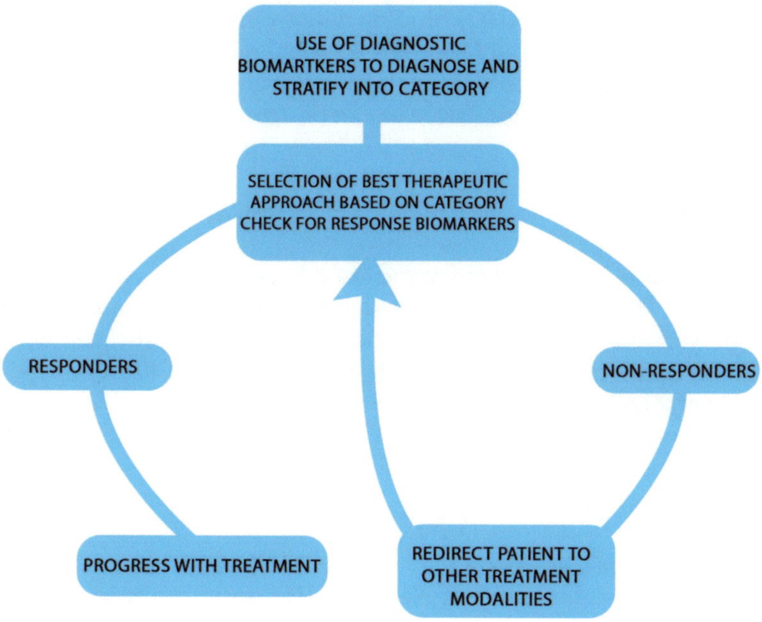

Figure 3.
A flow chart representing application method of precision medicine.

2.2.2 Precision medicine data types

The systematic collection of patient information is now accumulating and gaining complexity, as seen in the case of neuroimaging, which is currently producing above ten petabytes of data every year [17]. Studies in the field of precision medicine research make use of relevant data types such as Imaging data (CT, PET, UltraSound and MRI), bio-sample data (serum, plasma and urine value), molecular data, genomics data (nucleotide sequences), proteomic profiling data (mass spectrometry), digital pathology data, biomedical instrument data (blood pressure, heart rate and insulin level) and clinical data (death/survival data, demographics and medical-based questionnaire) and others [18].

Some of the achievement in Precision medicine has led to solutions, such as the birth of personalized brain models for a patient with intractable epilepsy [19] and the success in epigenetics mechanism of hematopoiesis [20]. The combination and integration of these data types require a sound understanding of the different fields of informatics (data science, data management and data curation) and bioinformatics [18].

2.3 Genomic and personalized medicine database

A database is an ordered set of structured information or data usually controlled by the database management system (DBMS) in an electronic computer. The data, DBMS and the applications associated with them are called database system or database in short. Each database contains certain types of data; here, we will be introducing some of the database associated with personalized or precision medicine.

2.3.1 Immune epitope database (IEDB)

The IEDB is a free to use database that is very useful in vaccine and drug development, this database catalogs data such as experimental data on antibodies, Major histocompatibility complex (MHC) binding data from different antigenic sources, Helper T lymphocyte (HTL) and Cytotoxic T Lymphocytes (CTL) epitopes for human and other animal species. This database also aids in prediction and analysis of varieties of epitopes [21]. This database can be accessed through https://www.iedb.org/.

2.3.2 Prostate cancer related lifestyle database (PCaLiStDB)

Lifestyle medicine is the study of association between lifestyle, chronic and immune diseases. PCaLiStDB is a lifestyle database that is channeled towards precision in the prevention of prostate cancer and other diseases associated with lifestyle. The data found in this database are lifestyle associated genes, lifestyle type biomarkers and personalized lifestyle-disease associated predictors [22]. The database link is http://www.sysbio.org.cn/pcalistdb/.

2.3.3 Clinical genome resources (ClinGen)

ClinGen database provides data that are of clinical importance, this database is funded by the National Institute of Health (NIH), and it is aimed at collecting necessary data for use in precision medicine and research. Data such as clinically relevant gene and variants are retrieved from this database in making precise diagnosis and treatment [23]. This database is accessed via https://clinicalgenome.org/.

2.3.4 Personal genome project (PGP)

One of the breakthroughs of medical informatics is the personal genome project database. This is an open-access database that is channeled towards the development of a tool for personalized medicine and advancing research. The database provides a wide range of data for different regions (PGP-UK, PGP-AUSTRIA, PGP-CHINA, PGP-CANADA and PGP-UNITED STATE, etc.). Data such as Genome,

Methylome, transcriptome and phenotype data are retrieved from this database for use in the procedure of precise medicine [24]. The database can be linked through https://www.personalgenomes.org/.

2.3.5 Online mendelian inheritance in man (OMIM)

This database was initiated in the early 1960s, and the online version was created in 1985. OMIM is an open-access database that is mainly built for professionals concerned with genetic disorders, a genetics researcher and advance students in medicine. Data such as human gene, genetic disorders, clinical features, phenotype and genes are available [25, 26]. This database address is https://www.omim.org/.

2.3.6 Human gene mutation database (HGMD)

This is a variant-related database that collates already known gene lesion that is responsible for human inherited diseases. The database includes precision medicine data such as gene symbol, genomics coordinates, splicing, different disease, phenotype and mutations in the human genome [27, 28]. This database is accessible via http://www.hgmd.cf.ac.uk/ac/index.php.

2.3.7 Clinical genome database (CGD)

Clinical Genomic Database fills the critical niche in the field of clinical and genomic medicine; it also encompasses medically significant genetic data with available interventions. For each entry in the database, the CGD gives out data such as allelic conditions, gene symbol, clinical categorization (both manifestation and interventions), affected age groups mode of inheritance and pathogenic mutation for all diseases so far captured [27]. This database can be accessed via https://research.nhgri.nih.gov/CGD/.

2.3.8 Other database related to precision/personalized medicine

There are other ongoing database projects to improve the existing ones, an example of this is The Human Variome Project [29]. Also, there are many websites and databases linked to precision medicine that this chapter cannot introduce all. **Table 1** below provides more of the database related to precision medicine in general and their links [30].

2.4 Genomic and personalized medicine data utilization

Data sharing is the potential inherent in the exchange of the same data resource with many applications or users; it encompasses the transferring of copies, accessing and enabling the reuse of data. Data can be open access (publicly available) or controlled (restricted), also, sharing data encompasses both sharing of primary (in case of nucleotide sequences) and secondary data (already used or analyzed data) [31].

Figure 4 above illustrates that precision medicine data encompasses both hospital data (information), GIS and PGHD. Sharing of the Precision medicine information (clinical data) can be accessed openly or otherwise restricted, whereby authorization will be needed by an authorized person to access and use the specified data for therapeutic, diagnostic and research purpose.

Database	Link
Pathway Interaction Database	http://pid.nci.nih.gov/
VirusMINT (interaction between viral protein and human)	http://mint.bio.uniroma2.it/virusmint/Welcome.do
AutDB (animal model resources)	http://autism.mindspec.org/autdb/AMHome.do
Pathogen Interaction Gateway ((host and pathogen interaction)	http://molvis.vbi.vt.edu/ or http://pathogenportal.net/pig/
NetPath (signal transduction)	http://www.netpath.org/
Entrez – (encompasses sub-Databases)	http://www.ncbi.nlm.nih.gov/sites/gquery
GeneCards	http://www.genecards.org/
Human Genome Resources	http://www.ncbi.nlm.nih.gov/projects/genome/guide/human/
Ensembl Human Genome Browser	http://www.ensembl.org/Homo_sapiens/Info/
Online Mendelian Inheritance in Man (OMIM)	http://www.ncbi.nlm.nih.gov/omim/
GeneCards	http://www.genecards.org/
Entrez Gene	http://www.ncbi.nlm.nih.gov/sites/entrez?db=gene
National Institute of Neurological Disorders and Stroke (NINDS): Clinical and Translational Resources	http://www.ninds.nih.gov/research/scientific_resources/clinical/
Database of Genotypes and Phenotypes (dbGaP)	http://www.ncbi.nlm.nih.gov/sites/entrez?db=gap
NIH Chemical Genomics Center	http://www.ncgc.nih.gov/]
Gene Expression Omnibus	http://www.ncbi.nlm.nih.gov/geo/
ENCODE Project: ENCyclopedia of DNA Elements, NHGRI	http://www.genome.gov/ENCODE/
PubChem	http://pubchem.ncbi.nlm.nih.gov/
PhenX Toolkit	https://www.plienxtoolkit.org/
Human Genome Project, NHGRI	http://www.genome.gov/10001772
NCBI BioSystems	http://www.ncbi.nlm.nih.gov/biosystems/
National Human Genome Research Institute (NHGRI)	http://genome.gov
Kyoto Encyclopedia of Genes and Genomes	http://www.genome.jp/kegg/
HUPO Brain Proteome Project	http://www.hbpp.org/5602.html
ExPASy Proteomics Server	http://expasy.org/
HUPO: Human Proteome Organization	http://www.hupo.org/
European Proteomics Association (EuPA)	http://www.eupa.org/

Table 1.
Database linked to precision medicine in general and their links [30].

2.5 Precision medicine of specific autoimmune diseases

Autoimmune diseases are disease conditions where the immune system respond to self-antigens as a result of damage or dysfunction or disorder in the tissues. It is controlled by a whole lot of factors of which host gene and environment play a vital role. It could affect the entire body, selected systems or selected organs and an

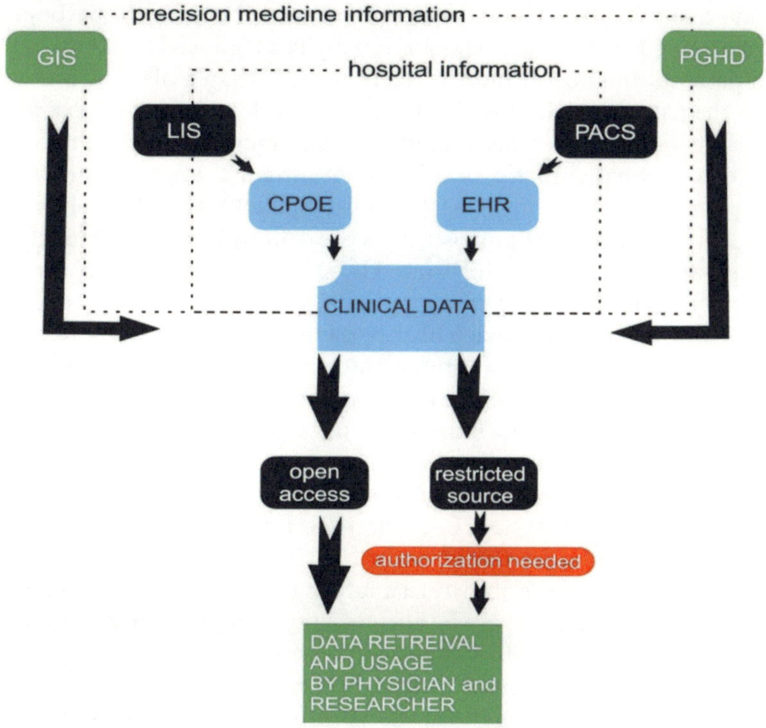

Figure 4.
*Semantic diagram of genomic and personalized medicine data sharing (**LIS**: Laboratory Information System; **GIS**: Genome Information System; **EHR**: Electronic Health Record; **PGHD**: Person Generated Health Data; **PACS**: Picture Archives and Communication System; **CPOE**: Computerized Physician Order Entry).*

interplay between genetic makeup with environmental factors and the self-antigen presented for recognition controls which organ or system of the body that will become the target of the immune system [32, 33].

The precision medicines of the following autoimmune diseases are discussed below:

1. Multiple Sclerosis
2. Myasthenia Gravis
3. Pernicious Anemia
4. Rheumatoid Arthritis
5. Sjogren Syndrome
6. Lupus Erythematosus
7. Type 1 Diabetes

2.5.1 Genomic assessment of multiple sclerosis

Multiple sclerosis is an inflammatory/autoimmune disorder that selects the myelin in the central nervous system which is capable of affecting patients of all age and causing neurologic disability when not adequately managed [34, 35]. More than

200 loci have been identified as an independent contributor to the pathogenesis of multiple sclerosis [36]. Multiple sclerosis is usually diagnosed between age 30 and 50 in most patients and occurs more often in females than male. The best way to understand the pathogenesis of multiple sclerosis is to address it from a multifactorial perspective with a model that proposes the interaction among genetic, epigenetic, infectious, dietary, climatic, or other environmental effects, together with sunlight exposure, and smoking. These interacting factors leads to self-intolerance and depreciation of immune homeostasis in the central Nervous system [34]. The brain and spinal cord tissues get infiltrated by stimulating peripheral mononuclear cells, and this leads to the loss of myelin, gliosis, which often leads to neurological dysfunction. Two primary approach of treatment has been given to the patient with multiple sclerosis due to the autoimmune model of the pathogenesis of such disease [34]. The former treatment is the use of global immunosuppressive agents which are aggressive. At the same time, the latter is the use of more specific agents to target specific elements of the immune system.

The contribution of common variants to multiple sclerosis has been probed, and different HLA alleles variants have been modeled for their contribution to multiple sclerosis and were found to be almost as common in control as it is in the sample as it was observed that OR of the statistical analysis tends towards 1 with an increase in sample size [37]. Biomarkers are important in the genetic assessment of Multiple Sclerosis as they possess the ability to express diverse aspects of multiple sclerosis heterogeneity. They also help in the diagnosis, stratification, and disease course prediction, identification of beneficial therapies and development of a precise treatment based on the predicted treatment response. As of 2016, MRI has turned to the most appropriate tool in the diagnosis of MS. The recommendation for brain MRI is the use of 1.5 T field strength, but 3.0 T is deemed preferable. However, using 7 T field strength has been supported by recent evidence to detect central vein in brain lesion of MS patients, but this can also be depicted using T2-weighted sequences at 3 T which help in the differentiation from microangiopathic lesions. The use of MRI for the diagnosis of MS seems simplified but its complexity sets in the differentiation of MS from other disease conditions like neuromyelitis optical spectrum disorders (NMOSD) which also has short spinal cord lesion at the onset. T2-weighted and contrast-enhanced T1-weighted brain MRI are recommended for the monitoring of disease progression while MRI of the spinal cord is not encouraged. Other than the MRI biomarkers there exist a few body fluid biomarker which could mark different stages of MS disease and differentiate each step from other similar disease conditions [34].

Body fluid biomarkers can be divided into three main groups, including those marking the early phase of MS, those associated with disease course and those associated with treatment response. Low vitamin D level in Cerebrospinal fluid is a marker of the initial stage of MS. Astrocyte-derived chitinase 3-like 1 (CHI3L1) in the CSF is also a prognostic marker of which an increased level of CHI3L1 in the CSF is a significant independent risk factor connected with the progression of disability in multivariate Cox regression models. Utilizing a proteomic approach and verification of result with ELISA confirmed that CHI3L1 would be the best predictors of the conversion to MS in CIS patients. CSF CHI3L1 level with MRI and age were the best predictors of MS risk in a multivariable analysis. Neurofilaments (NF-L) has also been implicated as a biomarker in the early phase of MS [36, 37].

Transcriptional regulator high-mobility group box protein 1 help differentiates patients with relapse-onset MS from patients from primary progressive MS. Proteomic studies show that two isoforms of vitamin D-binding protein and apolipoprotein E permit discrimination between MS patients with aggressive and benign disease courses [36]. During the disease course, calcium-binding protein secretogranin-1 is decreased in the CSF when compared with the early phase of MS. Stable MS patients, when compared with relapsing patients, possess an increase in B cell activating factors in their plasma samples. Solute carrier family 9, subfamily A (SLC9A9) is a biomarker associated with the non-response to IFN beta. Upregulation of the NLR family, pyrin domain containing 3 (NLRP3) inflammasome is also a biomarker for non-responsive IFN beta treatment. Biomarkers of glatiramer acetate response are feedback gene to complement 32 (RGC-32), FasL, and IL-21. Up-regulated mRNA expression levels of RGC-32 and FasL and reduced expression of IL-21 seen in peripheral blood cells from responders in contrast to non-responders forms the basis for the use of these biomarkers [34, 36, 37].

2.5.2 Genomic assessment of myasthenia gravis

Myasthenia gravis (MG) is an autoimmune disease treated with chronic immunosuppression due to the actions of autoantibodies against the diverse structure of the neuromuscular intersection [38]. The variation of the patient's response to treatment and the variation in side effects to such treatment is the justifying reason for the recognition of the biological markers to predict the effectiveness of each treatment in each patient. Presence of anti-AChR antibodies is a beneficial biomarker in the diagnosis of MG. Still, it cannot judge disease severity as no specific correlation was found between MG severity and anti-AChR antibodies level [39]. MiR-323b-3p, −409-3p, −485-3p, −181d-5p, and − 340-3p has been predicted and suggested as response biomarker to project immunosuppressive drug sensitivity in MG patients.

The miRNAs can be tested in the blood, which would make it a potent response biomarker for treatment response, and any patient detected not to respond as expected will be addressed to other treatments thereby increasing cost-effectiveness. MiR-323b-3p, −409-3p and − 485-3p were downregulated in Non-responding patients while miRNA-181d-5p, and − 340-3p were upregulated in the Non-responding patients [39, 40]. A significant association has been identified between patient's response to azathioprine and two haplotypes, the TPMT3E haplotype in the thiopurine S-methyltransferase and a haplotype in the ATP-binding cassette sub-family C member 6 transporter. The glucocorticoid therapy non-responsive MG patients were found to possess a genetic variant in the secreted phosphoprotein 1 (SPP1) gene encoding osteopontin, which associates it with the non-responsive group [40].

2.5.3 Genomic assessment of pernicious anemia

Pernicious anemia (PA), is an autoimmune disease which results from a long-standing infection by *Helicobacter pylori* and the end-stage of atrophic body gastritis (ABG). The condition which is still active gradually phased out by an autoimmunity reaction that depletes the gastric mucosa irreversibly. The deficiency of vitamin B12 has been implicated in the etiology also. Therefore the goal of a clinician in treating pernicious anemia may be to avert the signs and symptoms of anemia itself, manage its complications such as damage to the nerve and heart tissues, and identifying the specific cause where precision medicine comes in [41]. The National Heart, Lung, and Blood Institute (NHLBI) are currently carrying out basic and clinical researches that could incorporate precision medicine and improve the treatment of the condition.

2.5.4 Genomic assessment of rheumatoid arthritis

Rheumatoid arthritis (RA) is a heterogeneous disease which can range from mild, self-limiting arthritis to fast progressive joint damage. It is triggered by a complex interaction between the human genetic makeup and the environment. Still, both environmental influence and genetics cannot exhaustively account for the heterogenic clinical features of the disease condition. It is also characterized by synovial hyperplasia and joint destruction, which can lead to joint deformity or [42].

Currently, the treatment of RA is based on the control of inflammation with which an effective therapy that comes early ensures a drastic reduction in the risk of joint damage, mortality and disability. As of 2017, major researches has focused on the identification of biomarkers that can predict patient's response to only Methotrexate (MTX) which is the first non-biologic therapeutic agent administered. Also, TNF inhibitors (TNFi) has been established to be ineffective in about 30% of patients but remains the first choice of available biologic therapeutic agents. Solute carrier family 19 member 1 (SLC19A1) gene possess the most consistent and relevant evidence. It is one of the many transport carriers that allow the transport of MTX into the cell [43].

Anti- CCP antibodies a genomic marker associated with poor prognosis as it relates to the severity of disease and the extent of damage caused on the joint, HLA-DRB1 alleles coding for shared epitope is another marker for severity in RA [44].

2.5.5 Genomic assessment of sjogren syndrome

Sjögren's syndrome (SS) is a form of B cell hypersensitivity which is manifested in the formation of excess autoantibodies and a strong propensity for NHL of B cell emergence [45]. About 5% of patients of primary SS are at risk of lymphoma development. However, it is vital to have a specific biomarker to identify such patient early to be able to monitor and detect early and select appropriate therapy. The diagnostic biomarkers will guide in the diagnosis, and the predictive biomarkers are meant to show another aspect of clinical decision. Cytopenias is an established prognostic biomarker for the development of lymphoma [46]. A lot of proposed biomarkers in the assessment of SS are yet to be confirmed in more extensive studies before adoption into clinical use [47].

2.5.6 Genomic assessment of systemic lupus erythematosus

The systemic lupus erythematosus (SLE) has a broad spectrum of signs and symptoms which varies among patients and involves numerous organs with skin, joints, kidneys, lungs and CNS included. It is a chronic inflammatory autoimmune disease [48]. An association has been established between SLE and human leukocyte antigen (HLA) haplotypes (HLA-DR3; DR9; DR15; DQA1*0101 especially). The extensive association has also been found between vitamin D matching up with serum concentrations and vitamin D-receptor genomic binding domains [49].

2.5.7 Genomic assessment of type 1 diabetes

The type 1 diabetes (T1D) takes place as a result of autoimmune beta-cell destruction, which leads to insufficient production of insulin and results in hyperglycemia [50]. Although the role of precision medicine in type 1 diabetes is not well defined, patient with T1D severity varies with difference in their pancreatic autoantibodies profile and the rate at which their beta cells destroy [51].

In genetic studies (an important feature of precision medicine), identification of over 50 genetic signals in notably HLA region has been found to influence T1D predisposition [52]. The diagnostic biomarkers (serum biomarkers) use in the diagnosis of T1D includes the combination of glucose, C-peptide, glycated molecules and autoantibodies established for T1D. Still, these molecules often mark the late stage of the disease [53].

So far, advance in genomic research introduces the administration of islet autoantigens or peptides into a recipient with the risk of T1D; these studies suggest promising changes in immune regulation of islet autoimmunity. The challenges remain dosing frequency, dosage, route of administration, and adjuvants use.

2.6 Future perspectives

i. A systemic follow up of variant genes like the TNFRSF1A that is connected with multiple sclerosis risk should be closely investigated by researchers. This gene could give an essential perception of the etiology of multiple sclerosis and new treatment strategies.

ii. Myasthenia gravis-related loci may display their involvement in the pathogenesis of immune disease by increasing immune response, repression of the mechanism involved in immune suppression, alteration of procedure that differentiates between autologous and heterologous molecular configuration through immune tolerance, therefore investigations into Single nucleotide polymorphisms (SNPs) in the general population that is associated with Myasthenia gravis will improve diagnosis, therapy and its outcome.

iii. Genome editing technologies have been used with a degree of success in the treatment of sickle cell disease and β-thalassemia, this could be introduced into the precise treatment of pernicious anemia with proper study of the gene encoding for mitochondrial transport of vitamin B12.

iv. Rheumatoid arthritis research should focus on discovering more associated genes and their resultant effects. Transcriptomic and epigenomic strategies should also be used in discovering biomarkers of response to treatments and pathways that are related to therapies. Integration of genetic, clinical and environmental data are also crucial in achieving the aim of precision medicine in the treatment of rheumatoid arthritis.

v. Selection of novel treatments could be achieved for sjogren syndrome by identification of genetic risk factors like that of profound interferon signaling pathway by IRF5 and STAT 4 genes.

vi. Prevention of systemic lupus erythematosus by assaying genetic profile, developing new biomarkers of immune activation and alteration is the precise future treatment of this condition.

vii. Investigations into genes and pathways of type 1 diabetes may reveal on time the pathogenic role of the destruction of β-cell and production of clinical disease by the innate and adaptive immune system. Type 1 Diabetes Genetics Consortium (T1DGC) international have resources that could help in diagnosis, interventions, and monitoring outcomes of treatment of type 1 diabetes.

As the era of 'Big Health Data' continues, it is essential for the diagnosis, prognosis and treatment monitoring efforts on autoimmune diseases to take advantage of the data and different machine learning and deep learning algorithms to establish patterns and clusters within the disease groups. This will help in the identification of more relevant biomarkers and also help in the easy transition of biomarker researches to the bedside.

Indeed, the application of precision medicine in autoimmune diseases depends on the progress of next-generation sequencing program, which at the same time will strive to provide not only a whole-exome, or transcriptome, but at an exact process that is cost-efficient.

3. Conclusions

The information provided by the Genomics data is an indispensable component of precision medicine as it holds the key to the explanation in individual variability and evolution [54]. But, the clinical use of genomic data still needs to be improved on to overcome challenges stated by Kim et al. [55] like:

a. The incongruity between the form of genomic and clinical information: as a result of extensive (several tens of gigabytes of sequence) data in the genomic data, clinical data cannot be processed in the clinical practice without additional processing [55, 56].

b. The difference in the properties of genomic data and observational data used in the clinical settings: given that the genomic workflows hold a large number of data, data obtained from this workflows is undoubtedly different from systems parallel to the clinical plan [57].

c. Difficulty in mapping the genomic and clinical data for medical interpretation: as seen in the case of targeted sequencing, where most data are processed before medical analysis [58].

Also, there is no international validation for biomarkers in use; there is a need for international collaboration to validate biomarkers presently in existence.

Overcoming these challenges will open up more opportunities for the use of genomic data in clinical practices.

Conflict of interest

The authors declare no conflict of interest.

Author details

Ayodeji Ajayi*, Oluwadunsin Adebayo and Emmanuel Adebayo
Department of Physiology, Ladoke Akintola University of Technology, Ogbomoso, Nigeria

*Address all correspondence to: aajayi22@lautech.edu.ng

References

[1] Musa M. Immune mechanism: A "double-edged sword." Malaysian J Med Sci. 2013;20(3):61-67.

[2] Corra L. CHILDREN AND NOISE - Children ' s Health and the Environment. World Heal Organ. Published online 2009:1-67.

[3] Akira S, Uematsu S, Takeuchi O. Pathogen recognition and innate immunity. Cell. 2006;124(4):783-801. doi:10.1016/j.cell.2006.02.015

[4] Shurin MR, Smolkin YS. Immune-mediated diseases II congress: Summary. J Immunotoxicol. 2008;5(2):159-162. doi:10.1080/15476910802129604

[5] Tangye SG, Al-Herz W, Bousfiha A, et al. Human Inborn Errors of Immunity: 2019 Update on the Classification from the International Union of Immunological Societies Expert Committee. J Clin Immunol. 2020;40(1):24-64. doi:10.1007/s10875-019-00737-x

[6] Chinen J, Shearer WT. Secondary immunodeficiencies, including HIV infection. J Allergy Clin Immunol. 2010;125(2 SUPPL. 2):S195-S203. doi:10.1016/j.jaci.2009.08.040

[7] Delhalle S, Bode SFN, Balling R, Ollert M, He FQ. A roadmap towards personalized immunology. npj Syst Biol Appl. 2018;4(1). doi:10.1038/s41540-017-0045-9

[8] Gafson A, Craner MJ, Matthews PM. Personalised medicine for multiple sclerosis care. Mult Scler. 2017;23(3):362-369. doi:10.1177/1352458516672017

[9] Janeway CA, Medzhitov R. Innate immune recognition. Annu Rev Immunol. 2002;20(2):197-216. doi:10.1146/annurev.immunol.20.083001.084359

[10] Tangye SG, Al-Herz W, Bousfiha A, et al. Molecules, Cells, and Tissues of Immunity. Annu Rev Immunol. 2004;20(3):1-67. doi:10.1016/j.cell.2006.02.015

[11] Akira S, Takeda K, Kaisho T. Toll-like receptors : critical proteins. 2001;2(8).

[12] Doolan DL, Dobaño C, Baird JK. Acquired immunity to Malaria. Clin Microbiol Rev. 2009;22(1):13-36. doi:10.1128/CMR.00025-08

[13] Elias K, Siegel R, O'Shea JJ. Molecular and cellular basis of immunity and immunological diseases. Prim Rheum Dis Thirteen Ed. Published online 2008:94-107. doi:10.1007/978-0-387-68566-3_4

[14] Galli SJ. Toward precision medicine and health: Opportunities and challenges in allergic diseases. J Allergy Clin Immunol. 2016;137(5):1289-1300. doi:10.1016/j.jaci.2016.03.006

[15] Jackson, Sarah E, and John D Chester. "Personalised cancer medicine." International journal of cancer vol. 137,2 (2015): 262-6. doi:10.1002/ijc.28940

[16] Cirillo D, Valencia A. Big data analytics for personalized medicine. Curr Opin Biotechnol. 2019 Aug;58:161-167. doi: 10.1016/j.copbio.2019.03.004. Epub 2019 Apr 6. PMID: 30965188.

[17] Dinov ID. Volume and Value of Big Healthcare Data. J Med Stat Inform. 2016;4:3. doi: 10.7243/2053-7662-4-3. PMID: 26998309; PMCID: PMC4795481.

[18] Hulsen T, Jamuar SS, Moody AR, Karnes JH, Varga O, Hedensted S, Spreafico R, Hafler DA, McKinney EF. From Big Data to Precision Medicine. Front Med (Lausanne). 2019 Mar 1;6:34. doi: 10.3389/fmed.2019.00034.

[19] Timothée Proix, Fabrice Bartolomei, Maxime Guye, Viktor K. Jirsa, Individual brain structure and modelling predict seizure propagation, Brain, Volume 140, Issue 3, March 2017, Pages 641-654, doi: 10.1093/brain/awx004.

[20] José María Fernández, Victor de la Torre, David Richardson, Romina Royo, Montserrat Puiggròs, Valentí Moncunill, Stamatina Fragkogianni, Laura Clarke, Paul Flicek, Daniel Rico, David Torrents, Enrique Carrillo de Santa Pau, Alfonso Valencia. The BLUEPRINT Data Analysis Portal, Cell Systems, 2016, ISSN 2405-4712, doi:10.1016/j.cels.2016.10.021.

[21] Yan Q. Immunoinformatics and Systems Biology Methods for Personalized Medicine. In: Yan Q. (eds) Systems Biology in Drug Discovery and Development. Methods in Molecular Biology (Methods and Protocols), 2010, vol 662. Humana Press, Totowa, NJ. doi:10.1007/978-1-60761-800-3_10

[22] Yalan Chen, Xingyun Liu, Yijun Yu, Chunjiang Yu, Lan Yang, Yuxin Lin, Ting Xi, Ziyun Ye, Zhe Feng, Bairong Shen, PCaLiStDB: a lifestyle database for precision prevention of prostate cancer, Database, Volume 2020, 2020, baz154, doi:10.1093/database/baz154

[23] Rehm HL, Berg JS, Brooks LD, Bustamante CD, Evans JP, Landrum MJ, Ledbetter DH, Maglott DR, Martin CL, Nussbaum RL, Plon SE. ClinGen—the clinical genome resource. New England Journal of Medicine. 2015 Jun 4;372(23):2235-42. DOI: 10.1056/NEJMsr1406261

[24] Miriam S. Reuter, Susan Walker, Bhooma Thiruvahindrapuram, Joe Whitney, Iris Cohn, Neal Sondheimer, Ryan K.C. Yuen, Bret.t Trost, Tara A. Paton, Sergio L. Pereira, Jo-Anne Herbrick, Richard F. Wintle, Daniele Merico, Jennifer Howe, Jeffrey R. MacDonald, Chao Lu, Thomas Nalpathamkalam, Wilson W.L. Sung, Zhuozhi Wang, Rohan V. Patel, Giovanna Pellecchia, John Wei, Lisa J. Strug, Sherilyn Bell, Barbara Kellam, Melanie M. Mahtani, Anne S. Bassett, Yvonne Bombard, Rosanna Weksberg, Cheryl Shuman, Ronald D. Cohn, Dimitri J. Stavropoulos, Sarah Bowdin, Matthew R. Hildebrandt, Wei Wei, Asli Romm, Peter Pasceri, James Ellis, Peter Ray, M. Stephen Meyn, Nasim Monfared, S. Mohsen Hosseini, Ann M. Joseph-George, Fred W. Keeley, Ryan A. Cook, Marc Fiume, Hin C. Lee, Christian R. Marshall, Jill Davies, Allison Hazell, Janet A. Buchanan, Michael J. Szego, Stephen W. Scherer. The Personal Genome Project Canada: findings from whole genome sequences of the inaugural 56 participants. CMAJ 2018, 190 (5) E126-E136; doi: 10.1503/cmaj.171151

[25] Madeleine P. Ball, Joseph V. Thakuria, Alexander Wait Zaranek, TomClegg, AbrahamM. Rosenbaum, Xiaodi Wu, Misha Angrist, Jong Bhak, Jason Bobe, MatthewJ. Callow, Carlos Cano, MichaelF. Chou, Wendy K.Chung, Shawn M. Douglas, Preston W. Estep, Athurva Gore, Peter Hulick, Alberto Labarga, Je-Hyuk Lee, Jeantine E. Lunshof, Byung Chul Kim, Jong-Il Kim, Zhe Li, Michael F. Murray, Geoffrey B. Nilsen, Brock A. Peters, Anugraha M. Raman, Hugh Y. Rienhoff, Kimberly Robasky, Matthew T.Wheeler, Ward Vandewege, Daniel B. Vorhaus, Joyce L. Yang, LuhanYang, John Aach, Euan A. Ashley, Radoje Drmanac, Seong-Jin Kim, Jin Billy Li, Leonid Peshkin, Christine E. Seidman, Jeong-Sun Seo, Kun Zhang, Heidi L. Rehm, George M. Church. A public resource for clinical use of genomesSciences. 2012, 109 (30) 11920-11927; doi:10.1073/pnas.1201904109

[26] (OMIM) Online Mendelian Inheritance in Man An Online Catalog of Human Genes and Genetic Disorders Updated October 14, 2020. (Retrieved from https://omim.org/ on 15th of October 2020)

[27] Benjamin D. Solomon, Anh-Dao Nguyen, Kelly A. Bear, and Tyra G. Wolfsberg PNAS June 11, 2013;110(24):9851-9855; doi:10.1073/pnas.1302575110

[28] (HGMD) The Human Gene Mutation Database, at the Institute of Medical Genetics in Cardiff retrieved on 15/10/2020 from http://www.hgmd.cf.ac.uk/ac/index.php

[29] (HVP) Human Variome Project retrieved on 15/10/2020 from https://www.humanvariomeproject.org/

[30] Stimson, Nancy F. "Personalized medicine: selected web resources." Dialogues in clinical neuroscience vol. 2009;11(4):464-9.

[31] Alessandro Blasimme, Marta Fadda, Manuel Schneider, and Effy Vayena. Data Sharing For Precision Medicine: Policy Lessons And Future Directions, Health Affairs, 2018;37(5):702-709,

[32] Marrack P, Kappler J, Kotzin BL. Autoimmune disease: why and where it occurs. Nat Med. 2001;7(8):899-905. doi: 10.1038/90935.

[33] Ngo, S. T., Steyn, F. J., & McCombe, P. A. Gender differences in autoimmune disease. Frontiers in Neuroendocrinology, 2014;35(3):347-369. doi:10.1016/j.yfrne.2014.04.004

[34] Bose G, Freedman MS. Precision medicine in the multiple sclerosis clinic: Selecting the right patient for the right treatment. Mult Scler J. 2020;26(5):540-547. doi:10.1177/1352458519887324

[35] Baranzini SE, Oksenberg JR. The Genetics of Multiple Sclerosis: From 0 to 200 in 50 Years. Trends Genet. 2017;33(12):960-970. doi:10.1016/j.tig.2017.09.004

[36] Chitnis T, Prat A. A roadmap to precision medicine for multiple sclerosis. Mult Scler J. 2020;26(5):522-532. doi:10.1177/1352458519881558

[37] Comabella M, Sastre-Garriga J, Montalban X. Precision medicine in multiple sclerosis: Biomarkers for diagnosis, prognosis, and treatment response. Curr Opin Neurol. 2016;29(3):254-262. doi:10.1097/WCO.0000000000000336

[38] Meriggioli MN, Sanders DB. Autoimmune myasthenia gravis: emerging clinical and biological heterogeneity. Lancet Neurol. 2009;8(5):475-490. doi:10.1016/S1474-4422(09)70063-8

[39] Cavalcante P, Mizrachi T, Barzago C, et al. MicroRNA signature associated with treatment response in myasthenia gravis: A further step towards precision medicine. Pharmacol Res. 2019;148(June). doi:10.1016/j.phrs.2019.104388

[40] Mantegazza R, Cavalcante P. Diagnosis and treatment of myasthenia gravis. Curr Opin Rheumatol. 2019;31(6):623-633. doi:10.1097/BOR.0000000000000647

[41] Annibale B, Lahner E, Fave GD. Diagnosis and management of pernicious anemia. Curr Gastroenterol Rep. 2011;13(6):518-524. doi:10.1007/s11894-011-0225-5

[42] Guo S, Xu L, Chang C, Zhang R, Jin Y, He D. Epigenetic Regulation Mediated by Methylation in the Pathogenesis and Precision Medicine of Rheumatoid Arthritis. Front Genet. 2020;11(August):1-9. doi:10.3389/fgene.2020.00811

[43] Bluett J, Barton A. Precision Medicine in Rheumatoid Arthritis. Rheum Dis Clin North Am. 2017;43(3):377-387. doi:10.1016/j.rdc.2017.04.008

[44] Kaneko Y, Takeuchi T. Targeted antibody therapy and relevant novel biomarkers for precision medicine for

rheumatoid arthritis. Int Immunol. 2017;29(11):511-517. doi:10.1093/intimm/dxx055

[45] Goules A V., Tzioufas AG. Lymphomagenesis in Sjögren's syndrome: Predictive biomarkers towards precision medicine. Autoimmun Rev. 2019;18(2):137-143. doi:10.1016/j.autrev.2018.08.007

[46] Retamozo S, Brito-Zerón P, Ramos-Casals M. Prognostic markers of lymphoma development in primary Sjögren syndrome. Lupus. 2019;28(8):923-936. doi:10.1177/0961203319857132

[47] Baldini C, Ferro F, Elefante E, Bombardieri S. Biomarkers for Sj ogren ' s syndrome. Published online 2018.

[48] Chen L, Wang YF, Liu L, Bielowka A, Ahmed R, Zhang H, Tombleson P, Roberts AL, Odhams CA, Cunninghame Graham DS, Zhang X, Yang W, Vyse TJ, Morris DL. Genome-wide assessment of genetic risk for systemic lupus erythematosus and disease severity. Hum Mol Genet. 2020 Jun 27;29(10):1745-1756. doi: 10.1093/hmg/ddaa030. PMID: 32077931; PMCID: PMC7322569.

[49] Lever E, Alves MR, Isenberg DA. Towards precision medicine in systemic lupus erythematosus. Pharmgenomics Pers Med. 2020;13:39-49. doi:10.2147/PGPM.S205079

[50] Kalra, S., Das, A.K., Bajaj, S. et al. Utility of Precision Medicine in the Management of Diabetes: Expert Opinion from an International Panel. Diabetes Ther 2020;11:411-422 . doi:10.1007/s13300-019-00753-5

[51] Mohan V, Unnikrishnan R. Precision diabetes: Where do we stand today?. Indian J Med Res. 2018;148(5):472-475. doi:10.4103/ijmr.IJMR_1628_18

[52] Miriam S Udler, Mark I McCarthy, Jose C Florez, Anubha Mahajan, Genetic Risk Scores for Diabetes Diagnosis and Precision Medicine, Endocrine Reviews, 40 (6) 2019, Pages 1500-1520, doi:10.1210/er.2019-00088

[53] Yi L, Swensen AC, Qian WJ. Serum biomarkers for diagnosis and prediction of type 1 diabetes. Transl Res. 2018 Nov;201:13-25. doi: 10.1016/j.trsl.2018.07.009.

[54] Collins, F. S. & Varmus, H. A new initiative on precision medicine. New England Journal of Medicine 372, 793-795 (2015). DOI: 10.1056/NEJMp1500523

[55] Kim, H.J., Kim, H.J., Park, Y. et al. Clinical Genome Data Model (cGDM) provides Interactive Clinical Decision Support for Precision Medicine. Sci Rep 2020;10:1414. doi:10.1038/s41598-020-58088-2

[56] Lubin Ira M., Nazneen Aziz, Lawrence J. Babb, Dennis Ballinger, Himani Bisht, Deanna M. Church, Shaun Cordes, Karen Eilbeck, Fiona Hyland, Lisa Kalman, Melissa Landrum, Edward R. Lockhart, Donna Maglott, Gabor Marth, John D. Pfeifer, Heidi L. Rehm, Somak Roy, Zivana Tezak, Rebecca Truty, Mollie Ullman-Cullere, Karl V. Voelkerding, Elizabeth A. Worthey, Alexander W. Zaranek, Justin M. Zook. Principles and Recommendations for Standardizing the Use of the Next-Generation Sequencing Variant File in Clinical Settings. 2017, ISSN 1525-1578, doi:10.1016/j.jmoldx.2016.12.001.

[57] Kho, A., Rasmussen, L., Connolly, J. et al. Practical challenges in integrating genomic data into the electronic health record. Genet Med 2013;15, 772-778. doi:10.1038/gim.2013.131

[58] Roukos, D. H. Next-generation, genome sequencing-based biomarkers: concerns and challenges for medical practice. Biomarkers in medicine 2010;4:583-586. doi:10.2217/bmm.10.70.

Chapter 15

Multiplex Technology for Biomarker Immunoassays

Haseeb Ahsan and Rizwan Ahmad

Abstract

The simultaneous measurement of different substances from a single sample is an emerging area for achieving efficient and high-throughput detection in several applications. Although immunoanalytical techniques are established and advantageous over alternative screening analytical platforms, one of the challenges for immunoassays is multiplexing. While ELISA is still commonly used to characterise a single analyte, laboratories and organisations are moving towards multiplex immunoassays. The validation of novel biomarkers and their amalgamation into multiplex immunoassays confers the prospects of simultaneous measurement of multiple analytes in a single sample, thereby minimising cost, time and sample. Therefore, the technological advancement in clinical sciences is helpful in the identification of analytes or biomarkers in test samples. However, the analytical bioanalysers are expensive and capable of detecting only a small amount or type of analytes. The simultaneous measurement of different substances from a single sample called multiplexing has become increasingly important for the quantification of pathological or toxicological samples. Although multiplex assays have many advantages over conventional assays, there are also problems that may cause apprehension among clinicians and researchers. Hence, many challenges still remain for these multiplexing systems which are at early stages of development.

Keywords: biomarkers, multiplex assays, immunoassays, autoantibodies, ELISA, analytes

1. Introduction

An early and accurate diagnosis of a specific disease plays an important role in its effective treatment, especially in an emergency where an immediate decision needs to be made (such as in stroke or sepsis) for the treatment of patient, and the rapid and precise identification of the pathological condition is vital. However, in many instances, the clinical evidence based on a single analyte or biomarker is not adequate for an appropriate diagnosis of a disease or monitoring of its treatment. The biomarkers have a pathophysiological significance and clinical application which may have a profound impact on the diagnosis and treatment of the patient. While contemporary singleplex techniques such as enzyme-linked immunosorbent assay (ELISA) and biomarker kits are able to accurately analyse a single analyte, the monitoring of more complex, multifactorial diseases such as cancer and autoimmune and neurodegenerative diseases requires the analysis of multiple biomarkers in order to implement optimised therapeutic regimen [1]. In addition, it is advantageous to screen various analytes simultaneously for a rapid, cost-effective and reliable quantification [2, 3]. The development of technologies for the analysis of whole genome (genomic) and total proteins (proteomic) has ushered in a new era of biomarker discovery, which has yielded numerous new biomarkers. In the future, they will have a significant impact on clinical diagnostics and therapeutics [2]. Since the advent of the proteomic era, multiplex immunoassays now constitute valuable tools for efficiently harnessing information available to expedite observation, monitoring and treatment of diseases. While the availability and implementation of commercial multiplex immunoassays for research applications is expanding rapidly, incorporation of the technology within routine clinical diagnostics is in infancy due to operational and quality control issues such as robust automation, time constraint, operational cost, etc. [1]. The multiplex assays are now replacing conventional ELISAs to save time, material and labour costs and allow efficient handling of a large number of samples to enhance the overall throughput. With increasing running costs, a major focus of immunoassays has shifted towards cost-effectiveness and convenience of handling a large number of samples rather than results and reliability of the assay.

Most of the diagnostic methods rely on immunoassays or enzymatic reactions and strongly depend on the sample (e.g. size of sample, patient-to-patient variation) and assay conditions (e.g. temperature, humidity, etc.) [3]. An ideal device for emergency testing should offer high performance, sensitivity, multiplexing capability, short turnaround time, low system complexity, low-cost fabrication and minimised user intervention [3]. Therefore, the multiplexed diagnostic devices capable of high-throughput analysis of several parameters have recently become important in the last couple of decades which are able to analyse different markers simultaneously, e.g. RNAs, metabolites, proteins, cells, etc. [3]. Multiplex immunoassays confer several advantages over widely used singleplex assays including increased efficiency, greater output per sample volume and higher throughput prediction leading to detailed diagnosis facilitating personalised medicine. Nonetheless, relatively few protein multiplex immunoassays have been validated for in vitro diagnostics in clinical settings [1]. The emerging need and demand for novel biomarkers (e.g. aptamers) or targets (e.g. circulating RNAs and DNA, tumour cells, miRNAs, etc.) and their applications for diagnostic, prognostic and therapeutic implications, including therapeutic drug monitoring, will shape the future of multiplex systems [3]. The validation of novel biomarkers and their incorporation into multiplex immunoassays confers the prospect of simultaneous measurement of multiple analytes in a single patient sample, thereby minimising assay costs, time and sample volume. Moreover, the advent of multiplex technology complements the progressive shift towards personalised medicine with holistic, molecular analyses of diseases through the identification and characterisation of biomarkers to accommodate greater diagnostic resolution between closely related disease phenotypes. The multiplex immunoassays will continue to garner popularity and secure a mainstream role in research and eventually clinical settings [1].

The singleplex or conventional ELISA immunoassays have assumed a 'workhorse' role in the highly sensitive qualitative and quantitative detection of analytes within heterogeneous samples for over half a century. Moreover, the advent of hybridoma technology as a means of generating monoclonal antibodies (MAbs) has facilitated the generation of highly robust, antibody-based assays for standardisation and automation [4]. Both the singleplex and multiplex ELISAs adopt a common 'sandwich' format (capture antibody-analyte-detection antibody). The multiplex ELISA adopts chemiluminescent/fluorescent reporter systems as enzymatic reporters are chemically incompatible for simultaneous analysis of multiple localised targets. The concept of immunoassay for diagnostics was conceived in 1963 by Joseph G. Feinberg and A.W. Wheeler when they developed a 'microspot' technique as a means of detecting autoimmune antibody and tissue antigens, whereby thyroglobulin immobilised in a microspot on cellulose acetate strips were incubated with serum from autoimmune thyroiditis patients [5]. The microspot assay has the ability to detect low levels of both autoantibody and antigen; has the advantage of being simple, sensitive, objective and quick; and requires minute quantities of serum and antigen. In 1989, Ekins postulated the miniaturisation of immunoassays (i.e. reduction of the capture antibody concentration) and outlined the fundamental microarray multiplex technology principles and envisioned their potential application in research and clinical diagnostics [6]. The large-scale screening in the postgenomic era has encompassed applications, ranging from functional analysis of unknown genes to identification of disease-related gene products, screening in drug discovery and clinical diagnostics [1].

2. Principle of multiplex assays

The multiplex assays require complex technology such as PCR, ELISA, microarrays, gel electrophoresis, capillary electrophoresis, Sanger sequencing, etc. The fluorescence spectroscopy measurements are important due to its compatibility with biochemical assays, small size of sample, ease of conjugation to potential molecules, affordability, stability, robustness and detection with less expensive optical instruments [7]. Mass spectrometry (MS) can identify molecules without separation. For example, matrix-assisted laser desorption/ionisation mass spectrometry (MALDI-MS) is employed in hospitals to characterise antigens [1]. However, these bioanalysers are bulky, expensive and capable of detecting only a small amount or type of analyte [1]. Thus, there are still many challenges of these systems due to their cost or their detection capability [1, 2].

Multiplexing as a new technology emerged a few decades ago for the detection and quantification of first nucleic acids and then proteins. Using this technology, a range of biomarker molecules can be identified and quantified for health and disease. Multiplexing is a significant technology for the analysis of thousands of analytes with high reproducibility, miniaturised protocols and increased output. In spite of the advantages of ELISA and PCR, the multiplex immunoassay allows for a large number of analyses in a short amount of time through simultaneous measurement of expression of genes in a single sample by reducing time, labour and cost. The omics profiling using whole genome, epigenome, transcriptome, proteome and metabolome may also offer detailed information and aid personalised medicine [8]. Precision medicine is based on advanced omic technologies, such as next generation sequencing, protein and gene microarray, laser capture microdissection in the correlation of genomics, epigenomics, proteomics and metabolomics with the clinical phenotypes of the individual patient. The development of multiplex genotyping technologies and high-throughput genomic profiling allows the analysis of the patient genome from peripheral blood or biopsy samples [9].

Assaying for soluble antigens and antibodies as biomarkers of various diseases has always been an invaluable diagnostic and research tool. Currently, ELISA has potentially replaced agglutination, complement fixation, precipitation and immunodiffusion in diagnosis and research. The possibility of automation of the test procedure is one of the main reasons in transition from the classical serological tests to ELISA. Despite its advantages, ELISA is capable of measuring one analyte or a few analytes at a time, which may be a limiting factor where multiple markers need to be evaluated. Although the original idea of multiplexed assays involves antigen-antibody interaction, a vast knowledge comes from planar DNA microarrays. The chemical synthesis of oligonucleotides, DNA sequencing and PCR are major breakthrough technologies which greatly accelerated the accumulation of genomic and transcriptomic data. These key technologies also enabled the construction of DNA microarrays with the aid of bioinformatics, optics and microfluidics [10]. Systems biology research demands a complete snapshot of measurable parameters possible at a time in order to analyse and describe biological systems. The DNA microarray fulfils the requirement in the detection of cellular changes such as genetic polymorphisms, mutations, methylation patterns and transcript abundancies. However, these features usually give an indirect assessment of the cellular physiology. A more concrete and detailed picture of cellular machinery may be depicted by means of high-throughput analysis of proteins and metabolites. The DNA microarray involves surface immobilisation of DNA probes with sequences deduced from either nucleic acid or protein database. It has always been an intricate task to fabricate high-throughput multiplexed protein arrays compared to DNA arrays since proteins are physicochemically more diverse, complicated and often fragile in nature. Today, 60-plex cytokine measurement panels are available, in contrast to thousands of DNA probes in one planar microarray [10–14]. The real benefits of multiplexing assays may be achieved by miniaturisation of immunoassays. The real potential of microspot mediated analyte detection was discovered by Roger Ekins [6], in which the number of capture molecules can be attached onto a large macrospot surface far exceeding a tiny microspot surface. Consequently, macrospots consume analyte molecules in the sample to yield higher total signal intensity; even after all analyte molecules have been consumed, the available binding sites still remain on the assay surface. In contrast, decreasing spot size enhances the occupancy of capture molecules with analytes. Therefore, capture molecule concentration is directly related to assay surface area and total signal intensity and inversely related to signal density [15].

Multiplexed in situ tagging (MIST) is a reliable strategy which makes use of convertible DNA antibody barcoded arrays. It assays hundreds of molecular targets in a single cell, with high throughput and sensitivity. MIST technique was created to overcome the limitations of prevailing microfluidic-based methods [16]. One of the common limitations of conventional microfluidic-based single-cell protein assays was low multiplexity, which is often linked to fluorescence spectrum overlap, due to the phenotypically similar cell populations exhibiting a large degree of intrinsic heterogeneity [17]. Recently, technological advancements in single-cell proteomics have allowed highly multiplexed measurement of multiple parameters simultaneously by using barcoded microfluidic ELISAs and mass cytometry techniques [18]. It was achieved by integrating a multicolor, multicycle molecular profiling technique with barcoded microbead antibody arrays and a DNA-encoded antibody library [15, 16].

3. Planar and suspension multiplex immunoassays

The multiplex assays use immunoassay principles in which high-affinity capture ligands are immobilised in parallel assays. The systems use either antibodies or proteins/peptides as binding molecules to capture circulating proteins or autoantibodies. The multiplex immunoassays are divided into two classes: planar assays and suspension microsphere assays. In planar assay, the capture ligands are immobilised on a two-dimensional support and the fluorescent or chemiluminescent signals identified. In suspension assay, the capture ligands are immobilised on colour- or size-coded microspheres, and flow cytometry is used to detect assay-specific fluorescent signals (**Figure 1**) [19]. Planar arrays can be produced in two formats, either slide based or microtitre based. The slide-based format supports various layouts where repeated or individual assays composed of specific sets of antibodies are printed robotically upon the activated slide surface. The microtitre-based immunoassays harbour antibodies within the wells of a standard protein-binding plate as in conventional ELISA. The suspension immunoassay employs thousands of micrometre-sized plastic microbeads infused with a chemiluminescent/fluorescent dye and a functionally activated surface, prior to linking with a specific capture antibody. The detection antibodies are individually labelled with a single chemiluminescent/fluorescent reporter added upon completion of incubation and washing stages. Each bead accommodates a 'sandwich' consisting of the captured target analyte and the cognate reporter-conjugated detection antibody. The bead analyte reporter constructs are subjected to analysis in a flow chamber bead separation in which the lasers excite the chemiluminescent/fluorescent reporters and emitted light is collected by a series of detectors for quantitative analysis (**Figure 2**) [1].

In the planar microarray format, the capture molecules are spotted on the modified glass surface to form a grid, and micrometric-sized beads may also be used as assay surfaces. A mixture of beads with a library of capture molecules is called 'bead array' or 'liquid array'. A suspension of the bead library is allowed to interact with the sample and reporter molecules [20]. Flow cytometry is used to measure the assay signals in the beads. Multiplex bead array assays (MBAA) offer multiple analyses at a time, but there may be discrepancies in measurements of certain cytokines by using ELISA or MBAA. In many experimental setups, it was not possible to test the same pair of capture and reporter antibodies in both tests which may originate from multiplexing itself. Possible cross-reactivity between the target analytes and other interfering substances present in the sample may cause matrix effect. Therefore, precaution is required in multiplexing assays that work perfectly in monoplex systems [15].

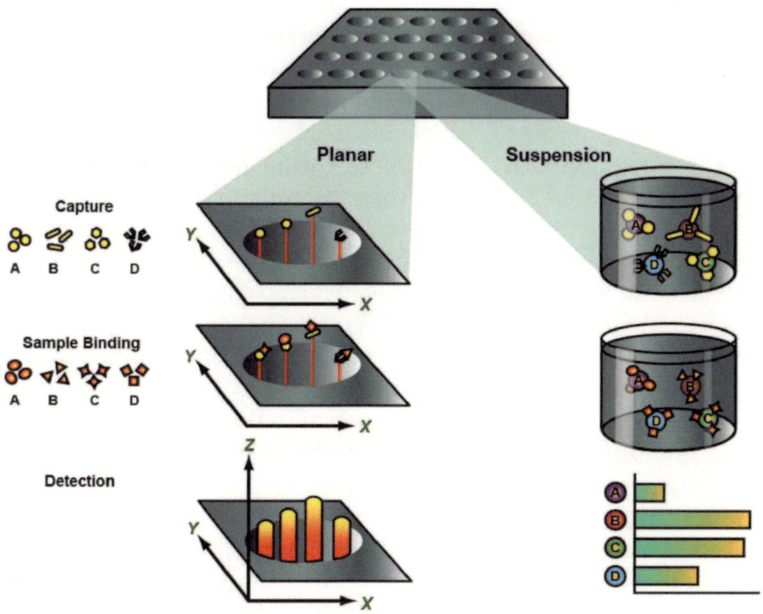

Figure 1.
Planar and suspension multiplexed immunoassay formats. In planar assays, capture ligands are immobilised on a rigid two-dimensional support and probed with sample. In suspension assays, capture ligands are immobilised on colour- or size-coded microspheres. Assays are distinguished by coding attributes, and flow cytometry is used to detect assay-specific fluorescent signal (adapted and reproduced from [19], with permission).

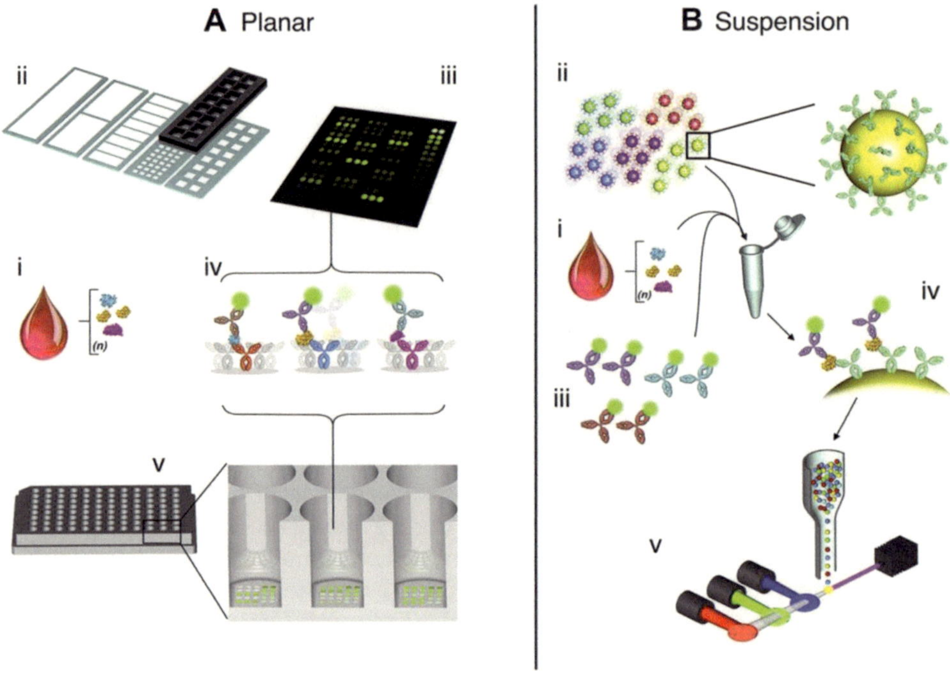

Figure 2.
Multiplex assays include planar-based assays and suspension-based assays. (A) Planar arrays (B) Suspension assay. Both assays us the serum sample extracted from blood as the starting point. (adapted and reproduced from [1], with permission from WILEY-VCH Verlag GmbH & Co).

4. Multiplex immunoassays for autoimmune diseases (AD)

A substantial percentage of the population carries detectable levels of circulating autoantibodies without developing clinical symptoms. Autoantibodies are also present in the sera of patients with systemic autoimmune diseases, such as systemic lupus erythematosus (SLE), rheumatoid arthritis (RA), etc. many years before clinical disease onset. The detection of antinuclear antibodies (ANA) has long been an important tool in the early diagnosis of autoimmune connective tissue diseases [21–25]. Antinuclear antibodies are detected in a substantial population, yet few individuals are diagnosed with the autoimmune disease, although some ANA-positive healthy individuals eventually develop clinical autoimmunity [26]. The correct use and interpretation of serologic testing for diagnosing autoimmune diseases present a challenge to clinicians for several reasons: (a) the sensitivity and specificity of laboratory tests for autoimmune disease and (b) detection of autoantibodies using different techniques such as indirect immunofluorescence (IF) or MBAA give different results. Multiplex microarray-based ELISA assays provide consistent results when compared with ELISA-based tests with the added advantage of reduced labour and the complete autoantibody profile in a single test. Autoantibody biomarkers assist in diagnosis and monitoring of disease activity, predicting disease onset, classifying disease subsets and defining prognosis. Despite various methods being used for autoantibody profiling, new techniques continue to be developed to facilitate diagnosis and therapeutic monitoring in connective tissue diseases [26, 27]. It is critical to evaluate new methods along with those being used in laboratories in order to assess their performance as well as to identify deficiencies of methods that are in current use such as methodology based on ELISA, western blot immunoassays, etc. [27].

Immunoassays are generally not considered as multiplex assays even though double immunodiffusion or immunoprecipitation can detect many specific autoantibodies in a single run [28]. Multiplex technology is considered to be beneficial for the simultaneous detection of different autoantibodies related to autoimmune diseases. The autoantibody profiling of patients may be useful for determining the concentration of specific autoantibodies, which may display different trends over time, both for diagnosis and prognosis, e.g. celiac disease, anti-phospholipid syndrome (APS), etc. which are characterised by the presence of autoantibodies of different isotypes. Nowadays, multiplex technology has achieved high analytical accuracy and provides results comparable or superior to the manual and automated monoplex technology [29]. Multiplex autoantibody assays can detect many specific autoantibodies in a single run, whereas the traditional ELISA uses a single antigen to detect only a single autoantibody. Thus, in a multiplex assay, a combination of native antigens or antigenic peptides is used to detect many autoantibodies. The multiplex assays include line immunoassay (LIA), MBAA, and solid-phase antigen microarray (protein microarray). LIA is similar to the dot blot or western blot (immunoblot) in which a diluted serum is incubated with a strip that has several specific antigens in different areas on a strip. In MBAA, beads of different sizes and fluorochromes with different colours or intensities are coated with specific antigens for the detection of specific autoantibody. In antigen microarrays, different specific antigens are coated on a slide/membrane, and the strips, mixture of beads or slide/membrane with multiple antigens are incubated with the diluted serum, and many specific autoantibodies can be determined simultaneously. While new multiplex immunoassays have certain advantages over conventional assays, using them without complete understanding or validation against classic or standard assays may lead to concerns, confusions and conflicts in autoantibody immunoassays in clinical settings [28].

5. Conclusion

Although immunoanalytical techniques are established and advantageous over alternative screening analytical platforms, one of the challenges for immunoassays is multiplexing. The simultaneous on-site measurement of different substances from a single sample called multiplex testing has become increasingly important for in vitro quantification of pathological or toxicological samples. The multiplex assays have recently gained importance for clinical diagnostics, with emerging applications in the developing world. The multiplex assays have several advantages such as performing many reactions on the sample and the ability to provide more information from the sample in a fast and efficient manner. Hence, the technological advancements in clinical sciences are helpful in the identification of analytes or biomolecules in pathological samples. While ELISA is still commonly used, many laboratories and organisations are moving towards multiplex immunoassays. The ELISA formats are able to accurately diagnose and characterise a single analyte; however, their amalgamation into multiplex immunoassays confers the prospects of simultaneous measurement of multiple analytes in a single sample, thereby minimising cost, time and volume.

Acknowledgements

RA is thankful to Professor Ali Ibrahim Al-Sultan, Dean, College of Medicine, IAU, Dammam, for all the help and support during the preparation of this manuscript.

Author details

Haseeb Ahsan[1] and Rizwan Ahmad[2*]

1 Faculty of Dentistry, Department of Biochemistry, Jamia Millia Islamia, New Delhi, India

2 College of Medicine, Imam Abdulrahman bin Faisal University, Dammam, Saudi Arabia

*Address all correspondence to: ahmadriz.biochem@gmail.com

References

[1] Tighe PJ, Ryder RR, Todd I, Fairclough LC. ELISA in the multiplex era: Potentials and pitfalls. Proteomics. Clinical Applications. 2015;**9**(3-4):406-422

[2] Ahsan H. The biomolecules of beauty: Biochemical pharmacology and immunotoxicology of cosmeceuticals. Journal of Immunoassay and Immunochemistry. 2019;**40**(1):91-108

[3] Dincer C, Bruch R, Kling A, Dittrich PS, Urban GA. Multiplexed point-of-care testing - xPOCT. Trends in Biotechnology. 2017;**35**(8):728-742

[4] Kohler G, Milstein C. Continuous cultures of fused cells secreting antibody of predefined specificity. Nature. 1975;**256**:495-497

[5] Feinberg JG, Wheeler AW. Detection of auto-immune antibody and tissue antigens by the 'microspot' technique. Journal of Clinical Pathology. 1963;**16**:282-284

[6] Ekins RP. Multi-analyte immunoassay. Journal of Pharmaceutical and Biomedical Analysis. 1989;**7**:155-168

[7] Rajagopal A, Scherer A, Homyk A, Kartalov E. Supercolor coding methods for large-scale multiplexing of biochemical assays. Analytical Chemistry. 2013;**85**(16):7629-7636

[8] Tanase C, Albulescu R, Neagu M. Updates in immune-based multiplex assays. Journal of Immunoassay and Immunochemistry. 2019;**40**(1):1-2

[9] Popa M-L, Albulescu R, Neagu M, Hinescu ME, Tanase C. Multiplex assay for multiomics advances in personalized precision medicine. Journal of Immunoassay and Immunochemistry. 2019;**40**:3-25

[10] Bumgarner R. DNA microarrays: Types, applications and their future. In: Ausubel FM, editor. Current Protocols in Molecular Biology. John Wiley & Sons, Inc.; 2013. p. 22, Unit–22. 1

[11] Elshal MF, McCoy JP. Multiplex bead array assays: Performance evaluation and comparison of sensitivity to ELISA. Methods (San Diego Calif.). 2006;**38**:317-323

[12] Hartmann M, Roeraade J, Stoll D, Templin MF, Joos TO. Protein microarrays for diagnostic assays. Analytical and Bioanalytical Chemistry. 2009;**393**:1407-1416

[13] Tanase CP, Albulescu R, Neagu M. Application of 3D hydrogel microarrays in molecular diagnostics: Advantages and limitations. Expert Review of Molecular Diagnostics. 2011;**11**:461-464

[14] Wu D, Dinh TL, Bausk BP, Walt DR. Long-term measurements of human inflammatory cytokines reveal complex baseline variations between individuals. The American Journal of Pathology. 2017;**187**:2620-2626

[15] Engin ED. The use of multiplexing technology in the immunodiagnosis of infectious agents. Journal of Immunoassay & Immunochemistry. 2019;**40**:109-122

[16] Zhao P, Bhowmick S, Yu J, Wang J. Highly multiplexed single-cell protein profiling with large-scale convertible DNA-antibody barcoded arrays. Advanced Science (Weinheim, Baden-Wurttemberg Germany). 2018;**5**:1800672

[17] Lu Y, Xue Q, Eisele MR, Sulistijo ES, Brower K, Han L, et al. Highly multiplexed profiling of single-cell effector functions reveals deep

functional heterogeneity in response to pathogenic ligands. Proceedings of the National Academy of Sciences of the United States of America. 2015;**112**:E607-E615

[18] Huang L, Michael SA, Chen Y, Wu H. Current advances in highly multiplexed antibody-based single-cell proteomic measurements. Chemistry - An Asian Journal. 2017;**12**:1680-1681

[19] Ellington AA, Kullo IJ, Bailey KR, Klee GG. Antibody-based protein multiplex platforms: Technical and operational challenges. Clinical Chemistry. 2010;**56**(2):186-193

[20] Doran TM, Kodadek T. A liquid array platform for the multiplexed analysis of synthetic molecule-protein interactions. ACS Chemical Biology. 2014;**9**:339-346

[21] Ahmad R, Ahsan H. Role of peroxynitrite-modified biomolecules in the etiopathogenesis of systemic lupus erythematosus. Clinical and Experimental Medicine. 2014;**14**(1):1-11

[22] Ahmad R, Ahsan H. Singlet oxygen species and systemic lupus erythematosus: A brief review. Journal of Immunoassay and Immunochemistry. 2019;**40**(4):343-349

[23] Ahmad R, Hussain A, Ahsan H. Peroxynitrite: Cellular pathology and implications in autoimmunity. Journal of Immunoassay and Immunochemistry. 2019;**40**(2):123-138

[24] Ahsan H. 3-nitrotyrosine: A biomarker of nitrogen free radical species modified proteins in systemic autoimmunogenic conditions. Human Immunology. 2013;**74**(10):1392-1399

[25] Ahsan H. Selfie: Autoimmunity, boon or bane. Journal of Immunoassay and Immunochemistry. 2017;**38**(3):235-246

[26] Slight-Webb S, Lu R, Ritterhouse LL, Munroe ME, Maecker HT, Fathman CG, et al. Autoantibody-positive healthy individuals display unique immune profiles that may regulate autoimmunity. Arthritis & Rheumatology. 2016 Oct;**68**(10):2492-2502

[27] Pérez AC, Kumble S, Kumble KD, Cañizal MCA, Jiménez LM, Diez LA, et al. Evaluation of a multiplex ELISA for autoantibody profiling in patients with autoimmune connective tissue diseases. Autoimmune Diseases. 2014;**2014**:896787

[28] Satoh M, Tanaka S, Chan EK. The uses and misuses of multiplex autoantibody assays in systemic autoimmune rheumatic diseases. Frontiers in Immunology. 2015;**6**:181

[29] Tozzoli R, D'Aurizio F, Villalta D, Bizzaro N. Automation, consolidation, and integration in autoimmune diagnostics. Autoimmunity Highlights 2015;6(1-2):1-6

Chapter 16

Neutropenia in Primary Immunodeficiency Diseases

Neslihan Edeer Karaca

Abstract

Phagocytes including neutrophil granulocytes and macrophages are important cells of the innate immune system whose primary function is to ingest and destroy microorganisms. Neutrophils help their host fight infections by phagocytosis, degranulation, and neutrophil extracellular traps. Neutrophils are the most common type of circulating white blood cells and the principal cell type in acute inflammatory reactions. A total absence of neutrophils or a significant decrease in their number leads to severe immunodeficiency that renders patients vulnerable to recurrent infections by *Staphylococcus aureus* and Gram-negative bacteria being the most life-threatening. Neutropenia may be classified as mild, moderate or severe in terms of numbers in the peripheral blood, and intermittent, cyclic, or chronic in terms of duration. Besides well-known classic severe congenital neutropenia, chronic neutropenia appears to be associated with an increasing number of primary immunodeficiency diseases (PIDs), including those of myeloid and lymphoid lineage. A comprehensive overview of the diverse clinical presenting symptoms, classification, aetiological and genetic etiologies of chronic isolated and syndromic neutropenia is aimed to be reviewed.

Keywords: immune system, neutropenia

1. Introduction

Inborn errors of immunity, traditionally called primary immunodeficiency diseases (PIDs) are a group of genetic defects that interfere with a component of the human immune system. Over the past decade, substantial knowledge has been gained regarding the genetic abnormalities involved in the pathogenesis of PIDs. More than 400 distinct disorders with 430 gene defects have been reported in the 2019 International Union of Immunological Societies (IUIS) phenotypical classification of human inborn errors of immunity [1]. Despite developmental changes in normal values for white blood cell counts during childhood and discrepancies in the mean value of neutrophil counts observed in people from different ethnicities, an absolute neutrophil count of less than 1500/μL is accepted as neutropenia. Absolute neutrophil count (ANC) is determined by multiplying the total leukocyte count by the percentage of segmented neutrophils and bands in the peripheral blood. Neutropenia may be defined as mild neutropenia, with an ANC of 1000–1500/μL; moderate neutropenia, with an ANC of 500–1000/μL; severe neutropenia, with an ANC <500/μL or agranulocytosis (ANC <200/μL). Neutropenia is defined to be chronic if it lasts longer than 3 months. Neutropenia may be chronic, intermittent, or cyclic. Peripheral neutrophil granulocyte counts show sinusoidal variation with 21 days in cyclic neutropenia.

Neutropenia is a common hematological manifestation of several PIDs with diverse genetic defects varying from congenital defects of phagocytes, to combined immunodeficiencies, and is often discovered in the course of an evaluation for acute infection. Congenital neutropenias associated with primary immunodeficiency diseases range from isolated severe congenital neutropenia to complex inherited disorders that comprise intellectual disabilities, organ abnormalities, facial dysmorphism or skin hypopigmentation. In IUIS classification, congenital defects of phagocytes are listed in two main groups; I. Defects of phagocyte number (neutropenia), and II. Functional defects of phagocytes [1]. In addition to the IUIS classification, chronic or intermittent neutropenia can be observed in other inborn errors of immune system, such as X-linked agammaglobulinemia, CD40L deficiency, reticular dysgenesis, WHIM syndrome, or in diseases of immune dysregulation. Defective myeloid cell differentiation, defective release of granulocytes from the bone marrow, enhanced apoptosis, or increased destruction of peripheral blood granulocytes are the main pathophysiological mechanisms underlying chronic severe or intermittent neutropenia in PID patients [1–3]. Primary immunodeficiency disorders associated with chronic or intermittent neutropenia are listed in **Table 1**.

Neutropenia increases host susceptibility to bacterial and fungal infections, primarily from their endogenous flora in the gut, mouth and skin as well as from nosocomial organisms, and usually presents with infections of mucous membranes, gingiva, and skin. *Staphylococcus aureus*, Gram-negative bacteria, and fungi are the most common pathogens. The most common presenting features of neutropenia are fever, aphthous stomatitis, and gingivitis. Recurrent gingivitis with multiple dental caries may lead to teeth loss. The spectrum of infections varies from localized cellulitis, furunculosis, perirectal inflammation, sinusitis, and otitis media to more severe infections such as pneumonia, colitis, intestinal perforation with peritonitis, deep tissue abscess, and sepsis. Patients with severe congenital neutropenia develop severe bacterial infections in the first year of life. In some cases, inherited neutropenia may predispose to acute myeloid leukemia (AML) or myelodysplastic syndrome (MDS).

1.1 Primary genetic defects of severe congenital neutropenia

Severe congenital neutropenia (SCN) comprise multiple hereditary syndromes, with or without extrahaematopoietic manifestations. It is characterized by an arrest in myeloid maturation at the promyelocyte-myelocyte stage and an absence of mature neutrophils in the bone marrow. Regardless of the molecular etiology, congenital neutropenia is rare with an estimated prevalence of <1/100,000 [4–6]. Patients are prone to recurrent infections such as otitis, sinusitis, gingivitis, stomatitis, skin infections, pneumonia, deep abscesses, and septicemia beginning in their first months of life. Furthermore, SCN patients have an increased risk of malignant transformation, AML, or MDS.

Mutations in numerous genes have been identified in SCN [1, 4, 5]. The prevalence of some genetic subtypes of SCN is dependent on ethnicity. Autosomal dominant heterozygous mutations of *ELANE* or *ELA2*, encoding neutrophil elastase, a serine protease that is stored in the azurophilic granules, are the most prevalent cause of SCN in Caucasians [7–9]. *ELANE* mutations lead to accelerated apoptosis as a result of abnormal protein folding in the endoplasmic reticulum and altered function through mutant neutrophil elastase mislocalization. ELANE deficiency is also responsible for cyclic neutropenia, which is characterized by regular

	Inheritance	Gene	Pathogenic mechanism	Clinical and laboratory features in addition to neutropenia
Severe Congenital Neutropenia				
Neutrophil elastase deficiency	AD, sporadic	ELANE / ELA2	Activation of the unfolded protein response (UPR), apoptosis of myeloid progenitor cells	Leukemia and myelodysplastic syndrome predisposition
HAX1 deficiency	AR	HAX1	Destabilization of mitochondrial membrane potential, abgrogated G-CSFR signaling, enhanced apoptosis of myeloid and neuronal cells	Leukemia and myelodysplastic syndrome predisposition, mental retardation, seizures
Glucose-6-phosphatase deficiency	AR	G6PC3	Impaired intracellular glucose homeostasis, dysglycosylation and UPR lead to enhanced apoptosis of myeloid cells	Thrombocytopenia, visible superficial veins, congenital heart defects, uropathy, cryptorchidism
X-linked neutropenia	XL	WASP	Disturbed actin polymerization, altered cytoskeletal responses, defective mitosis and cytokinesis	Lymphopenia, leukemia predisposition
Jagunal homolog 1 deficiency	AR	JAGN1	Aberrant N-glycosylation of multiple proteins, elevated apoptosis	CSF3 hypo/un-responsiveness
GFI1 deficiency	AD	GFI1	Impaired neutrophil differentiation, lymphoid immunodeficiency	Monocytosis, lymphopenia
SEC61A1 deficiency	AD	*SEC61A1*	Disturbed protein translocation, and dysregulation of the UPR	Recurrent sinopulmonary infections, skin abscess, oral aphthosis and enteritis
Bi-allelic CSF3R deficiency	AR	CSF3R	Transmembran GCSF receptor/intracellular signaling	CSF3 unresponsiveness
Somatic mutation of CSF3R	No genetic inheritance	CSF3R		
Disorders of molecular trafficking				
Chediak-Higashi syndrome	AR	LYST	Defective biogenesis of lysosomes, cytotoxic granules and melanosomes	Partial oculocutaneous albinism, recurrent infections, fever, hepatosplenomegaly, bleeding tendency, neurological dysfuntions, giant lysosomes (leukocytes), hair shaft anomaly

	Inheritance	Gene	Pathogenic mechanism	Clinical and laboratory features in addition to neutropenia
Griscelli syndrome type IIb	AR	RAB27a	Defective priming of cytotoxic granules and melanosomes	Recurrent infections, fever, hepatosplenomegaly, specific hair shaft anomaly
Cohen syndrome	AR	COH1, VPS13B	Altered vesicle sorting and transport	Psychomotor retardation, microcephaly, facial dysmorphism, hypotonia, joint laxity, obesity, retinochoroidal dsytrophy
Hermansky-Pudlak syndrome	AR	AP3P1	Defective endosome formation and lysosomal protein sorting,in immune cells	Recurrent infections, pulmonary fibrosis
VPS45 deficiency	AR	VPS45	Defective endosomal trafficking leads to impaired differentiation and motility and increased apoptosis of myeloid and mesenchymal cells	Myelofibrosis, nepromegaly, hepatomegaly, mental retardation
P14 deficiency	AR	LAMTOR2	Aberrant distribution of late endosomes, defective MAPK and ERK signaling, diminished phagocytosis	Growth delay, short stature, oculocutaneous hypopigmentation, partial albinism, coarse facial features
Disorders of molecular processing				
Shwachman-Diamond syndrome	AR	SBDS	Mitotic spindle destabilization, genomic instability, enhanced apoptosis	Exocrine pancreas deficiency, metaphyseal dysplasia, mental retardation, cardiomyopathy
Dyskeratosis congenita	XL	DKC1	Dysfunctional telomere maintenance	Skin pigmentation, nail dysplasia, oral leucoplakia, pulmonary fibrosis, stenosis of the oesophagus, liver diseas
	AD	TERC		
	AR	TERT		
Metabolic diseases				
Glycogen storage disease type 1b	AR	SLC37A4	Impaired intracellular glucose homeostasis	Hypoglycemia, fasting hyerlacticacidemia, hepatomegaly
Barth syndrome	XL	TAZ1	Mitochondrial dysfunction, destabilization of mitochondrial respiratory chain complexes, increased apoptosis in myeloid cells	Cardiomyopathy, endomyocardial fibrosis

	Inheritance	Gene	Pathogenic mechanism	Clinical and laboratory features in addition to neutropenia
Pearson syndrome	Mitochondrial	Deletion of mtDNA	Variably sized mtDNA deletion, variable heteroplasmy, and mosaicism lead to metabolic disorder/energy failure and apoptosis in affected tissues	Bone marrow failure, vacuoles in erythroid precursors, exocrine pancreas insufficiency, hepatopathy, nephropathy, endocrinopathy, neuromuscular degeneration
Other PID diseases				
X-linked agammaglobulinemia	XL	BTK	unclear	Recurrent bacterial infections, hypogammaglobulinemia, absent B cells
Hyper IgM syndrome	AR	CD40	Abrogated CD40LG:CD40-signalling, autoimmunity	Class-switch recombination deficiency, combined immunodeficiency oppurtunistic infections, biliary tract and liver disease, Cryptosporidium infections, intermittent neutropenia
	XL	CD40L		
WHIM syndrome	AD	CXCR4	Constitutively activated CXCR4 impairs chemokinesis of neutrophils, dendritic cells and B cells from the bone marrow	Warts, hypogammaglobulinemia, immunodeficiency, myelokathexis
Reticular dysgenesis	AR	AK2	Defective mitochondrial adenine nucleotide homeostasis, enhanced apoptosis	Lymphopenia (T-B-NK- SCID), deafness
GATA2 deficiency	AD	GATA2	Complex ontogenic dysregulation of hematopoiesis and vascularization, reduced numbers of hematopoietic stem cells	Sensoryneural deafness, lymphoedema, pulmonary alveolar proteinosis
STK4/MST1 deficiency	AR	STK4	Disturbed mitochondrial membrane potential, enhanced apoptosis	Intermittent neutropenia, bacterial, viral (HPV, EBV, molluscum), candidal infections, lymphoproliferation, combined immunodeficiency, congenital heart defects
Cartilage-Hair hypoplasia	AR	RMRP	Defective ribosome assembly, aberrant cell cycle control and telomere function	Short-limbed dwarfism, metaphyseal dysostosis, sparse hair, autoimmunity, lymhopenia, bone marrow failure, lymphoma predisposition

Table 1.
Inheritance patterns, pathogenic mechanisms and important hematological or extrahematopoietic features of primary immunodeficiency diseases associted with neutropenia.

oscillations with the ANC ranging from normal to <200/μL with a periodicity of the 21 days (±4 days) cycle [9]. Recessive disorders, such as *HAX1, G6PC3, JAGN1* are usually diagnosed in consanguineous families. Mutations in *TAZ* (Barth syndrome) and *WAS* have X-linked inheritance [4, 5, 10, 11]. Digenic or multigenic mutations have been also reported in SCN patients [12]. Some mutations are linked to the geographic origin [10, 13–15]. Autosomal recessive (AR) *HAX1* mutations account for about 15% of SCN patients, mostly from consanguineous Kurdish patients from the Middle East. AR *G6PC3* mutation has a high prevalence (25%) in Israel among Arameans [13, 16].

Several genetic defects have been identified as being responsible for SCN and there is currently no clear genotype–phenotype correlation for this syndrome. Patients with *ELANE*, germline *CSF3R* mutations and *WAS*/X-linked severe congenital neutropenia usually present without extrahaematopoietic manifestations [10, 13, 14].

Homozygous mutations in the antiapoptotic gene *HAX1*, encoding ubiquitously expressed HCLS1-associated protein X-1 protein, are the defects identified in classic Kostmann syndrome [10]. HAX1 is critical for the maintenance of inner mitochondrial membrane potential and is an important regulator of myeloid homeostasis. There are 2 HAX-1 isoforms. HAX1 mutations affecting both isoforms (mainly p.Q190X and p.R86X) cause SCN frequently accompanied by neurological involvement (mental retardation, developmental delay, and seizures). Mutations affecting only one isoform (mainly p.W44X in Turkish patients) lead to SCN without neurological symptoms [14].

GFI1 (Growth factor independent 1) is a zinc finger transcription factor important in myeloid and lymphoid differentiation. Dominant-negative GFI1 mutations cause a severe maturation arrest of myeloid cells [4, 13].

Inactivating, X-linked mutations in *WAS,* the Wiscott–Aldrich syndrome gene, are responsible for the classical immune deficiency, microthrombocytopenia, autoimmunity, bleeding diathesis, and predisposition to lymphoma. Apart from classic Wiscott–Aldrich syndrome, XLN is a rare familial form of SCN caused by autosomal dominant gain-of-function mutations in *WAS*. WAS protein participates in the dynamic regulation of actin polymerization. The gain-of-function mutations cause an overactive protein, leading to elevated actin polymerization, defective cytokinesis, increased apoptosis, and neutropenia [13, 15].

Mutations in *G6PC3* (glucose-6-phosphatase catalytic unit 3) were found to be a cause of SCN in 2009 [16]. G6PC3 is involved in the final step of the gluconeogenic and glycogenolytic pathway. Neutrophils of the patients have an increased sensitivity to apoptosis. Associated findings include congenital heart defects, urogenital abnormalities, inner ear hearing loss, and venous angiectasia [16, 17].

Homozygous mutations in protein jagunal homolog 1 (*JAGN1*) has been described as one of the causes of SCN in 2014 by Boztug *et al* [18]. JAGN1 deficient neutrophils show ultrastructural defects in the endoplasmic reticulum, absence of granules, defective N-glycosylation of multiple proteins, increased endoplasmic reticulum stress, and intracellular calcium activation leading to accelerated apoptosis. The phenotypic spectrum of JAGN1 deficiency includes short stature, scoliosis, hip dysplasia, amelogenesis imperfecta, facial dysmorphism, pyloric stenosis, urogenital and cardiac abnormalities. Besides neutropenia, hypogammaglobulinemia, low class-switched memory B cells, and CD4+ T cell lymphopenia are reported in JAGN1-deficient patients. This form of SCN does not respond to Colony Stimulating Factor 3 (CSF3), formerly called granulocyte colony-stimulating factor (GCSF) treatment [18, 19].

Colony stimulating factor 3 (CSF3), the main growth factor that controls both the proliferation and differentiation of myeloid progenitor cells into neutrophils, is the primary ligand for granulocyte colony-stimulating factor receptor

(G-CSFR). G-CSFR is encoded by the colony-stimulating factor 3 receptor gene (*CSF3R*). Somatic *CSF3R* gene mutations occur on a background of inherited mutations affecting genes such as *ELANE*, *HAX1*, and *G6PC3*. Acquired point mutations are localized within the intracellular domain of the receptor, and give rise to the truncated form of the receptor. This type of receptor introduces a premature stop codon and hampers its ability to transduce signals required for neutrophil differentiation. Patients who do not respond to CSF3 should be checked for *CSF3R* mutations [3–5, 7, 20]. Despite acquired *CSF3R* mutations, congenital forms of *CSF3R* mutations are localized within the extracellular or, rarely, the transmembrane domain of the receptor [20, 21].

Recently, an autosomal dominant mutation in *SEC61A1* was reported in a patient with SCN who was born to nonconsanguineous Belgian parents [22]. SEC61A1, encoding the α-subunit of the Sec61 complex controls the endoplasmic reticulum protein transport and passive calcium leakage. The mutation resulted in diminished protein expression, disturbed protein translocation, an increase in calcium leakage from the endoplasmic reticulum, and dysregulation of the unfolded protein response. The index patient presented with recurrent sinopulmonary infections, skin abscess, oral aphthous lesions, and enteritis, and responded well to CSF3 treatment.

Compensatory monocytosis, hypereosinophilia, and polyclonal hypergammaglobulinaemia appeared to be frequently associated with neutropenia and inversely proportional to its severity in SCN patients [13, 23].

Treatment of severe chronic neutropenia should focus on the prevention of infections. It includes antimicrobial prophylaxis, generally with trimethoprim-sulfamethoxazole, and also Colony Stimulating Factor 3 (CSF3). Prior to the era of filgastrim/CSF3 therapy, most patients died of infectious complications within the first 1–2 years of life despite antibiotic prophylaxis. More than 95% of SCN patients respond to CSF3 treatment with an increase in the ANC, a decrease in infections, and a great improvement in life expectancy [24, 25]. The dose and frequency of injection of CSF3 vary widely. For most patients, 5–8 micrograms (mcg) per kilogram (kg) of body weight of CSF3 given as a daily subcutaneous injection is usually sufficient. SCN is a premalignant condition. Studies showed the cumulative incidence of malignant transformation towards AML/MDS as about 22% after 8–15 years of CSF3 treatment [13, 25–27]. Patients who do not respond to filgrastim or who require high doses (>8–10 mcg/kg/day) and patients who develop AML or MDS should be considered for hematopoietic stem cell transplantation (HSCT). The strongly increased AML/MDS risk is a feature shared between *ELANE, HAX1,* and XLN SCN patients. A major risk factor for leukemogenesis in patients with severe congenital neutropenia is the expansion of hematopoietic clones with somatic (acquired) mutations in the gene encoding the G-CSF receptor (*CSF3R*). Due to the risk of developing AML or MDS, regular monitoring with blood counts, and yearly bone marrow aspiration and biopsy, including karyotyping, cytogenetic analysis, and fluorescence *in situ* hybridization should be performed. The most common cytogenetic feature is monosomy 7, which is detectable in approximately two-thirds of malignancies, but other recurrent cytogenetic abnormalities are also observed, such as trisomy 21 or trisomy 18 [23, 27].

1.2 Disorders of molecular processing

Shwachmann-Diamond syndrome and dyskeratosis congenita are in the group of diseases due to defective ribosomal biogenesis and RNA processing.

1.2.1 Shwachmann-Diamond syndrome

Shwachmann-Diamond syndrome is an autosomal recessive bone marrow failure syndrome characterized by neutropenia, exocrine pancreatic insufficiency, hepatic dysfunction, short stature and a wide spectrum of skeletal abnormalities. In addition to neutropenia, some children with SDS have defects in neutrophil chemotaxis or in the number and function of T, B and natural killer cells [28]. Bone marrow examination revealing condensed chromatin and hyposegmented neutrophils are in favor of Shwachman-Diamond syndrome.

1.2.2 Dyskeratosis congenita

Dyskeratosis congenita is a disorder of telomerase activity, usually presenting with neutropenia or pancytopenia due to bone marrow failure, cutaneous findings such as nail dystrophy, leukoplakia, malformed teeth, palmar hyperkeratosis, and hyperpigmentation of the skin [28, 29].

1.3 Disorders of metabolism

1.3.1 Glycogen storage disease type Ib

Glycogen storage disease type Ib is caused by mutations in the *SLC37A4* gene, encoding glucose-6-phosphate translocase (G6PT). It is characterized by hypoglycemia, excessive glycogen accumulation in the liver and kidney, neutropenia, and susceptibility to bacterial infections [4, 30].

1.3.2 Barth syndrome

Barth syndrome is a rare X-linked genetic disease characterized by cardiomyopathy, skeletal myopathy, growth delay, neutropenia, and increased urinary excretion of 3-methylglutaconic acid. Neutropenia can be constant, intermittent, or cyclic. Disabling mutations or deletions of *TAZ* gene, encoding tafazzin (a mitochondrial acyltransferase) cause the disorder by reducing remodeling of cardiolipin, a principal phospholipid of the inner mitochondrial membrane [31]. Survival is poor, largely depending on the severity of heart failure and the availability of a heart transplant.

1.3.3 Pearson syndrome

Pearson syndrome is an extremely rare mitochondrial disorder presenting with early-onset transfusion-dependent macrocytic sideroblastic anemia, neutropenia, and thrombocytopenia [32]. Additional clinical findings are failure to thrive, exocrine pancreatic insufficiency, and liver dysfunction. Bone marrow analyses show characteristic vacuolization of erythroid and myeloid precursor cells and ringed sideroblasts.

1.4 Vesicular trafficking disorders

Autosomal recessive vesicular trafficking disorders are caused by defects in the biogenesis or intracellular trafficking of lysosomes and related endosomal organelles [33]. Neutropenia, low natural killer and cytotoxic T lymphocyte activities and abnormal platelet functions can be observed in the patients.

1.4.1 Chediak-Higashi syndrome

Chediak-Higashi syndrome (CHS) is a rare autosomal recessive lysosomal disorder characterized by frequent infections, oculocutaneous albinism, bleeding diathesis, progressive neurologic deterioration and a high risk of developing hemophagocytic lymphohistiocytosis characterized by pancytopenia, high fever, and lymphohistiocytic infiltration of liver, spleen, and lymph nodes [33, 34]. Treatment of accelerated phase is difficult with poor prognosis. Observation of giant cytoplasmic granulations helps dicrimination of CHS from other PIDs with partial albinism and neutropenia.

1.4.2 Griscelli syndrome type 2

Griscelli syndrome type 2 (GS2) is a rare, autosomal recessive immunodeficiency caused by mutations in *RAB27A*, clinically characterized by pigmentary dilution of the skin and the hair and predisposition to uncontrolled T-lymphocyte and macrophage activation syndrome (known as hemophagocytic syndrome), leading to death in the absence of bone-marrow transplantation. Most patients also develop periods of lymphocyte proliferation and activation, leading to their infiltration in many organs, such as the nervous system, causing secondary neurological damage [34–36].

1.4.3 Hermansky-Pudlac syndrome type 2

Hermansky-Pudlac syndrome type 2 (HPS-2) is caused by mutations in the *AP3B1* gene, have prominent facial features, a tendency toward bleeding, neutropenia, oculocutaneous albinism and high risk for rapidly fibrosing lung disease during early childhood [37].

Examination of the hair shaft of patients with partial albinism can be helpful diagnostically, as irregular large melanin granules can be seen in Griscelli syndrome type 2, poorly distributed regular melanin granules in CHS, and small pigment clumps in Hermansky-Pudlac syndrome type 2 [34].

1.4.4 P14 deficiency

A ubiquitously expressed endosomal protein MAPBPIP or p14, encoded by the *LAMTOR2* (Late Endosomal/Lysosomal Adaptor, MAPK and MTOR Activator 2) gene, is crucial for the function of neutrophils, B cells, cytotoxic T cells and melanocytes. Adaptor molecule p14 defects cause an immunodeficiency syndrome associated with growth delay, short stature, oculocutaneous hypopigmentation, partial albinism, coarse facial features, lymphoid deficiency, neutropenia, and recurrent bronchopulmonary infections [38].

1.4.5 Cohen syndrome

Cohen syndrome, associated with an arrest of myeloid differentiation is caused by an AR mutation of the vacuolar protein sorting 13 homolog B (*VPS13B*, also referred to as *COH1*) gene on chromosome 8q22.2. It has diverse clinical manifestations including failure to thrive, hypotonia, microcephaly, craniofacial and limb anomalies, short stature, obesity, hypermobile joints, mental retardation, and neutropenia [39, 40].

1.4.6 VPS45 deficiency

Vacuolar sorting protein 45 (VPS45) is a peripheral membrane protein that controls membrane fusion through the endosomal/lysosomal trafficking and the release of inflammatory mediators. Autosomal recessive inherited *VPS45* deficiency is a severe primary immune deficiency characterized by neutropenia, myelodysplasia, progressive bone marrow fibrosis, impaired migration, endocytosis, and degranulation of neutrophils, megathrombocytopenia, increased cell apoptosis leading to overwhelming bacterial infections, and early death. Organomegaly, nephromegaly, neuromotor developmental delay, and osteosclerosis are also observed in VPS45 deficient patients [41–43]. Recombinant CSF3 therapy is not sufficient to achieve improvement in ANC counts. An early diagnosis of the condition is important as therapeutic options are currently limited to early hematopoietic stem cell transplantation.

1.5 Well-known primary immunodeficiency diseases associated with neutropenia

Primary immunodeficiency diseases are characterized by recurrent or chronic infections, autoimmunity, inflammation, allergy, or malignancy as a consequence of genetic alterations affecting the immune system. These disorders were initially considered to be rare, but many patients with PIDs have been recognized over the 3 decades with the increase in awareness and availability of better diagnostic facilities. The prevalence and distribution of the ten groups of inborn errors of immunity vary worldwide. Additionally, patients with the same disease may present a different clinical profile and outcome. Due to the limited number of registries, inconsistency in diagnostic criteria, different clinical phenotypes, and lack of molecular diagnosis, the global perspective of these diseases remains unclear. Reports from several PID registries in different countries show a prevalence of 1:8500 to 1:100000 for symptomatic patients [44–46]. Predominantly B-cell deficiencies encompass the main category of PIDs. Although the exact data about the frequency is lacking, a great number of immune deficiencies are known to be associated with mild or severe neutropenia as a result of close interactions both in their ontogeny and during their functional life of myeloid and lymphoid cells. Most of such cases of neutropenia are observed at diagnosis and may recover once appropriate therapy is administered, such as parenteral immunoglobulin replacement in B cell deficiency.

1.5.1 Bruton's disease

X-linked agammaglobulinemia (XLA) is a rare primary immune deficiency characterized by the absence of circulating B cells with a severe reduction in all serum immunoglobulin levels due to mutations in the *BTK* gene. B cells show a developmental arrest in the bone marrow at the pro-B to the pre-B stage in the presence of mutations in *BTK*. Most XLA patients present with recurrent bacterial infections such as otitis, sinusitis, and sinopulmonary infections, developing after 7 to 9 months of age when transplacental maternal immunoglobulin G (IgG) levels decrease below protective levels. *Streptococcus pneumoniae* and *Haemophilus influenzae* are the most common responsible encapsulated pathogens. Patients are specifically susceptible to Enterovirus family, and mostly to poliovirus, coxsackie virus (hand, foot, and mouth disease), and Echoviruses. These may cause severe central nervous system conditions as chronic encephalitis, meningitis, and death. Prevalence is approximately 1 per 10,000 [47, 48]. Almost 30% of XLA patients are reported to have profound neutropenia at the time of diagnosis and the resolving of neutropenia after initiation of regular IVIG replacement therapy [49–51]. The direct involvement of BTK in neutrophil development is not clear.

1.5.2 CD40LG deficiency (Hyper IgM syndrome type I)

The Hyper IgM (HIGM) syndromes are a group of rare genetic disorders leading to loss of T cell-driven immunoglobulin class switch recombination (CSR) and/or defective somatic hypermutation (SHM) with elevated or normal serum IgM and decreased IgG, IgA, and IgE. The most common causes are mutations in the CD40 Ligand (CD154) (*CD40LG*) gene leading to X-linked HIGM (XHIGM) in males. Interaction between CD40L expressed by the T_{helper} subset and its receptor CD40 on B cells induces B cell proliferation, CSR, and SHM. Patients with HIGM are highly susceptible to recurrent sinopulmonary infections, *Pneumocystis jiroveci* pneumonia, and chronic diarrhea due to *Cryptosporidium* infection that may lead to sclerosing cholangitis, hepatitis, and liver cirrhosis [52–55]. About 50% of XHIGM patients have chronic, cyclic or intermittent neutropenia, as a consequence of chronic infection or autoimmunity. Studies also revealed multiple functions of the CD40/CD40L interactions on stromal cells by enhancing the expression of granulopoiesis growth factors [56]. Decreased interaction between T cells and bone marrow stromal cells, resulting in reduced production of G-CSF is one of the mechanisms of neutropenia in XHIGM patients.

1.5.3 Severe combined immunodeficiency

Severe combined immunodeficiency (SCID) syndromes are characterized by a block in T lymphocyte differentiation that is variably associated with abnormal development of other lymphocyte lineages (B and/or natural killer [NK] cells), leading to death early in life unless treated urgently by hematopoietic stem cell transplantation. The overall frequency is estimated to 1 in 75 000–100 000 births [44, 57]. Reticular dysgenesis, caused by a mutation in the *adenylate kinase 2 (AK2)* gene is an autosomal recessive disease with granulocytopenia as well as pancytopenia, lack of innate and adaptive immune responses, and sensorineural deafness [1, 57]. Mitochondrial *adenylate kinase (AK)* regulates levels of adenosine diphosphate. AK2 deficiency results in increased apoptosis of myeloid and lymphoid precursors. This form is one of the rarest and most severe types of SCID. Severe infections occur earlier than in other forms of SCID due to profound neutropenia, in addition to markedly decreased T and NK cells.

1.5.4 Wiskott Aldrich syndrome

Wiskott Aldrich syndrome (WAS) results from a loss of function mutation in Wiskott-Aldrich syndrome protein (*WASP*) and presents with recurrent infections, eczema and microthrombocytopaenia [58]. In its classical form, significant combined immune deficiency, autoimmune complications, and risk of hematological malignancy necessitate early correction with stem cell transplantation or gene therapy. In Wiskott-Aldrich syndrome, neutropenia usually accompanies frequent autoimmune disorders. It is different from the milder form, X-linked thrombocytopenia (XLT) that is caused by the activating *WASP* mutations.

1.5.5 WHIM syndrome

WHIM syndrome (WHIM) is an autosomal dominant congenital immune deficiency with susceptibility to human papillomavirus infection-induced warts, B cell lymphopenia, hypogammaglobulinema, bone marrow myelokathexis (increase in the granulocyte pool, with hyper mature dystrophic neutrophils), and neutropenia [59]. Gain-of-function mutations in the G protein-coupled chemokine receptor *CXCR4* are causal in this disease. Mutations in this protein lead to increased responsiveness to its chemokine ligand CXCL12 and retention of neutrophils in the bone marrow. Intravenous immunoglobulin (IVIg) and CSF3 have been documented to prevent infections in patients with hypogammaglobulinemia and neutropenia, respectively. Granulocyte colony-stimulating factor can increase neutrophil counts but does not affect cytopenias. CXCR4 antagonist plerixafor has been used to increase absolute lymphocyte, monocyte, and neutrophil counts in the peripheral blood in a dose-dependent manner, correct neutropenia, and other cytopenias in WHIM syndrome [60, 61].

1.5.6 Cartilage-hair hypoplasia

Cartilage-hair hypoplasia is a rare form of skeletal dysplasia, but also a syndromic primary immunodeficiency disorder due to a mutation in the RNase MRP RNA gene (RMRP), a non-coding RNA gene. The main clinical features are chondrodysplasia, short-limbed short stature, sparse and fine hair, Hirschsprung disease, macrocytic anemia, defective T cell-mediated immunity and predisposition to severe infections and cancer [62].

1.5.7 STK4/MST1 deficiency

Biallelic mutations in *STK4*, encoding MST1 have been identified in patients with CD4 lymphopenia accompanying multiple bacterial and viral infections, EBV-related lymphoproliferative disorder and mucocutaneous candidiasis [63–65]. MST1 deficiency has overlapping features with other PIDs involving defects in actin cytoskeletal reorganization, such as DOCK8 deficiency and Wiskott-Aldrich Syndrome. Hypergammaglobulinemia, progressive loss of naive T cells, reduced in vitro T-cell proliferation and defective in LFA-1-mediated adhesion and chemotaxis are the immunological disturbances identified in these patients. Clinically, these disorders and MST1 deficiency may behave very similarly. A thorough diagnostic workup including molecular genetic testing is advised to inform decision-making around stem cell transplantation, which will often be required.

1.5.8 GATA2 deficiency

GATA2 is a transcription factor required for stem cell homeostasis. Clinical presentation of GATA2 deficiency varies from typical Emberger syndrome (deafness and lymphoedema), MonoMac syndrome (susceptibility to mycobacteria, myelodysplasia, cytogenetic abnormalities, myeloid leukemias, pulmonary alveolar proteinosis) [66]. A significant proportion of patients have monocytopenia and macrocytosis in addition to mild neutropenia.

2. Diagnostic work-up in chronic neutropenia

Children with a history of recurrent or unusual infections present a diagnostic challenge. A high index of suspicion could lead to an early diagnosis and treatment of underlying immune deficiency disease. Several points should be taken into consideration in the examination of the patient. These are;

DOI: http://dx.doi.org/10.5772/intechopen.97297

A. Age at the first detection of neutropenia;

B. The indication that required performing a complete blood cell count (CBC) (mild infection/fever, severe infection, fungal infection, aphthous, gingivitis stomatitis, diarrhea, developmental delay);

C. A family history of neutropenia, consanguinity, pregnancy losses, or infectious deaths, and geographic origin;

D. The presence of any severe infections, bacterial or fungal;

E. A physical examination that focuses on the oral cavity (ulceration, gingivitis or stomatitis), skin, lungs, and perirectal area for infection is important. Lymphadenopathy and hepatosplenomegaly must be determined.

F. The presence of any congenital malformation and/or any organ dysfunction;

G. The complete blood count with differential, performed at the time of diagnosis (including the ANC, absolute eosinophil count, absolute monocyte count, absolute lymphocyte count, hemoglobin levels, and platelet levels).

H. Some specific cytological abnormalities observed on the blood, such as large granular lymphocytes, suggestive of Chediak-Higashi syndrome.

The initial workup may also reveal a particular etiology, such as viral infections. After this screening evaluation, bone marrow aspiration, immunological tests (e.g., immunoglobulin G, A, M, E levels, T and B immunophenotype), pancreas markers (serum trypsinogen, fecal elastase), and auto-antibodies against neutrophils may help to determine the diagnosis. The diagnoses according to the system involvement are depicted in **Figure 1**.

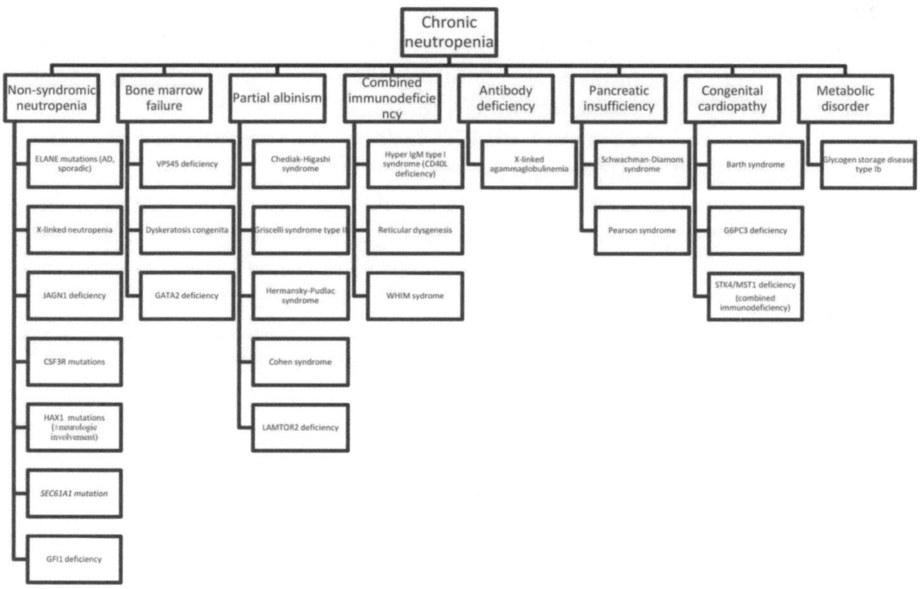

Figure 1.
Differential diagnosis of chronic neutropenia according to system involvements.

Targeted next-generation sequencing panels on the initial genetic investigations, followed-by whole-exome sequencing appears to be the most efficient strategy to identify the molecular etiology. In addition, the search for pathogenic copy-number variants or for regions of homozygosity in the case of consanguineous individuals should be considered. Mutations in some genes such as CSF3R and GATA2 can be either germline or somatic. As hematopoietic cells may acquire somatic mutations, non-hematopoietic tissue may be tested to distinguish germline versus somatic mutations. Buccal swabs or saliva samples may be contaminated by hematopoietic cells. Therefore, the germline status of a mutation should therefore be confirmed by analyzing DNA extracted from non-hematopoietic tissue, such as nails, hair follicles, or fibroblasts.

3. Treatment and follow-up

Treatment of severe chronic neutropenia in PIDs should focus on the prevention of infections, the management of associated organ dysfunction, and the prevention of leukemic transformation. The management of neutropenia will require a flexible, empiric, and patient-centered approach based on the use of cytokines and HSCT with consideration of antibiotic prophylaxis. Although many different genetic mutations may cause neutropenia, the clinical picture is similar. Most SCN patients find great benefit from subcutaneous CSF3 administration, which causes a significant decrease in the frequency of severe bacterial infections and increases the quality of life. The starting dose is 5 mcg/kg with dose modification according to the patient's absolute neutrophil count and the rate of infections. It should be kept in mind that neutropenia in JAGN1 and VPS45 deficiencies do not respond to CSF3. Patients who do not respond to CSF3 or who require high doses (>8–10 mcg/kg/day) and patients who develop AML or MDS should be treated by HSCT.

The treatment of neutropenia should be decided on a patient basis for the other disease groups. For example, patients with Shwachmann-Diamond syndrome require transfusions, pancreatic enzymes, antibiotics, and CSF3. The only definitive therapy for marrow failure is HSCT. Neutropenia, which is frequently detected at the time of diagnosis in XLA (Bruton agammaglobulinemia) patients, improves with regular IVIG replacement. XHIGM (CD40 Ligand deficiency) patients can be cured by HSCT. Future treatment strategies including gene therapy or novel genome editing technologies using CRISPR/Cas9 or TALEN systems will permit the correction of monogenic neutropenia disorders.

The rate of hematological malignancy in many of the inherited neutropenia disorders, regardless of genetic subtype, is far higher than that observed in the general population. The rate of transformation is not precisely documented, but the leukemic transformation has been reported in patients with *WAS, HAX1, G6PC3, SLC37A4* or acquired *CSF3R* gene mutations, whereas no transformation has been observed in patients with *VPS13B or CXCR4* mutations so far [25, 27]. Leukaemogenesis in CN is a multi-step process. In addition to germline mutations, several genetic mutations may occur in myeloid cells. Annual bone marrow examination should be performed to rule out malignant haemopathies, and determine cellularity, assess myeloid maturation, and detect some features that are typical of a precise etiology in the case of chronic neutropenia.

Blood neutrophils and monocytes are the cells both produced in the bone marrow, circulate in the blood, and are recruited to sites of inflammation. Compensatory monocytosis help SCN patients overcome infections. Although both are actively phagocytic, they differ in significant ways. The neutrophil response is more rapid and the lifespan of these cells is short, whereas monocytes become macrophages in the tissues, can live for long periods, and maintain tissue integrity by eliminating/repairing damaged cells. Over the recent years, an increasing amount of knowledge has been gained in the field of phagocytic cell subpopulations [67, 68]. In addition to their protective role against invading pathogens, the field has highlighted roles for inflammatory conditions including sterile injury, tumor development, atherosclerosis, and autoimmunity. With regard to their high plasticity, neutrophils and macrophages are shown to acquire an anti-tumorigenic N1/M1 or a pro-tumorigenic N2/M2 phenotype, respectively. The impact of M1 macrophages which have overlapping features with N1 subsets of neutrophils need further investigation in PIDs.

4. Conclusion

Neutropenia is a common hematologic manifestation of a wide range of diseases. Paying careful attention to associated features of a patient provides valuable clues leading to a narrow spectrum of differential diagnosis. Genetic investigation may be helpful in making a definitive diagnosis. This is of utmost importance since timely diagnosis helps the patient benefit from available therapeutic modalities such as HSCT and CSF3 administration.

Author details

Neslihan Edeer Karaca
Department of Pediatrics, Faculty of Medicine, Ege University, Izmir, Turkey

*Address all correspondence to: neslihanedeer@gmail.com

References

[1] Bousfiha A, Jeddane L, Picard C, Al-Herz W, Ailal F, Chatila T, Cunningham-Rundles C, Etzioni A, Franco JL, Holland SM, Klein C, Morio T, Ochs HD, Oksenhendler E, Puck J, Torgerson TR, Casanova JL, Sullivan KE, Tangye SG. Human Inborn Errors of Immunity: 2019 Update of the IUIS Phenotypical Classification. J Clin Immunol. 2020;40:66-81.

[2] Cunningham-Rundles C. Hematologic complications of primary immune deficiencies. Blood Rev 2002;16:61-4.

[3] Rezaei N ,Moazzami K, Aghamohammadi A, Klein C. Neutropenia and primary immunodeficiency diseases. Int Rev Immunol 2009;28:335-66.

[4] Donadieu J ,Beaupain B , Fenneteau O, Bellanné-Chantelot C. Congenital neutropenia in the era of genomics: classification, diagnosis, and natural history. Br J Haematol 2017;179:557-574.

[5] Welte K, Zeidler C, Dale DC. Severe congenital neutropenia. Semin Hematol. 2006;43:189-95.

[6] Ancliff PJ. Congenital neutropenia. Blood Rev. 2003;17:209-16.

[7] Germeshausen M, Deerberg S, Peter Y, Reimer C, Kratz CP, Ballmaier M. The spectrum of ELANE mutations and their implications in severe congenital and cyclic neutropenia. *Hum Mutat.* 2013;34:905-914.

[8] Horwitz MS, Corey SJ, Grimes HL, Tidwell T. ELANE mutations in cyclic and severe congenital neutropenia: genetics and pathophysiology. *Hematol Oncol Clin North Am.* 2013;27:19-41.

[9] Nanua S, Murakami M, Xia J, et al. Activation of the unfolded protein response is associated with impaired granulopoiesis in transgenic mice expressing mutant Elane. *Blood.* 2011;117:3539-3547.

[10] Klein C, Grudzien M, Appaswamy G, et al. HAX1 deficiency causes autosomal recessive severe congenital neutropenia (Kostmann disease). *Nat Genet.* 2007;39:86-92.

[11] Devriendt K, Kim AS, Mathijs G, et al. Constitutively activating mutation in WASP causes X-linked severe congenital neutropenia. *Nat Genet.* 2001;27:313-317.

[12] Germeshausen M, Zeidler C, Stuhrmann M, Lanciotti M, Ballmaier M, Welte K. Digenic mutations in severe congenital neutropenia. *Haematologica.* 2010;95:1207-1210.

[13] Touw IP. Game of clones: the genomic evolution of severe congenital neutropenia. Hematology Am Soc Hematol Educ Program 2015;2015:1-7.

[14] Yılmaz Karapınar D, Patıroğlu T, Metin A, Çalışkan Ü, Celkan T, Yılmaz B, Karakaş Z, Karapınar TH, Akıncı B, Özkınay F, Onay H, Yeşilipek MA, Akar HH, Tüysüz G, Tokgöz H, Özdemir GN, Aslan Kıykım A, Karaman S, Kılınç Y, Oymak Y, Küpesiz A, Olcay L, Keskin Yıldırım Z, Aydoğan G, Gökçe M, İleri T, Aral YZ, Bay A, Atabay B, Kaya Z, Söker M, Özdemir Karadaş N, Özbek U, Özsait Selçuk B, Özdemir HH, Uygun V, Tezcan Karasu G, Yılmaz Ş. Homozygous c.130-131 ins A (pW44X) mutation in the HAX1 gene as the most common cause of congenital neutropenia in Turkey: Report from the Turkish Severe Congenital Neutropenia Registry. Pediatr Blood Cancer. 2019;66:e27923.

[15] Keszei M, Record J, Kritikou JS, Wurzer H, Geyer C, Thiemann M, Drescher P, Brauner H, Köcher L, James J, He M, Baptista MA, Dahlberg CI, Biswas A, Lain S, Lane DP, Song W, Pütsep K, Vandenberghe P, Snapper SB, Westerberg LS. Constitutive activation of WASp in X-linked neutropenia renders neutrophils hyperactive. J Clin Invest. 2018 ;128:4115-4131.

[16] Boztug K, Appaswamy G, Ashikov A, Schaffer AA, Salzer U, Diestelhorst J, Germeshausen M, Brandes G, Lee-Gossler J, Noyan F, Gatzke AK, Minkov M, Greil J, Kratz C, Petropoulou T, Pellier I, Bellanne-Chantelot C, Rezaei N, Monkemoller K, Irani-Hakimeh N, Bakker H, Gerardy-Schahn R, Zeidler C, Grimbacher B, Welte K, Klein C. A syndrome with congenital neutropenia and mutations in G6PC3. New England Journal of Medicine, 2009; 360, 32-43.

[17] Kiykim A, Baris S, Karakoc-Aydiner E, Ozen AO, Ogulur I, Bozkurt S, Ataizi CC, Boztug K, Barlan IB G6PC3 Deficiency: Primary Immune Deficiency Beyond Just Neutropenia. J Pediatr Hematol Oncol. 2015;37:616-22.

[18] Boztug K, Jarvinen PM, Salzer E, et al. JAGN1 deficiency causes aberrant myeloid cell homeostasis and congenital neutropenia. *Nat Genet.* 2014;46:1021-1027.

[19] Baris S, Karakoc-Aydiner E, Ozen A, Delil K, Kiykim A, Ogulur I, Baris I, Barlan IB. JAGN1 Deficient Severe Congenital Neutropenia: Two Cases from the Same Family. J Clin Immunol. 2015. PMID: 25851723

[20] Skokowa J, Steinemann D, Katsman-Kuipers JE, et al. Cooperativity of RUNX1 and CSF3R mutations in severe congenital neutropenia: a unique pathway in myeloid leukemogenesis. *Blood.* 2014;123:2229- 2237.

[21] Triot A, Järvinen PM, Arostegui JI, et al. Inherited biallelic CSF3R mutations in severe congenital neutropenia. Blood. 2014;123:3811-3817.

[22] Van Nieuwenhove E, Barber JS, Neumann J, Smeets E, Willemsen M, Pasciuto E, Prezzemolo T, Lagou V, Seldeslachts L, Malengier-Devlies B, Metzemaekers M, Haßdenteufel S, Kerstens A, van der Kant R, Rousseau F, Schymkowitz J, Di Marino D, Lang S, Zimmermann R, Schlenner S, Munck S, Proost P, Matthys P, Devalck C, Boeckx N, Claessens F, Wouters C, Humblet-Baron S, Meyts I, Liston A. Defective Sec61alpha1 underlies a novel cause of autosomal dominant severe congenital neutropenia. J Allergy Clin Immunol. 2020;146:1180-1193.

[23] Jean Donadieu, Odile Fenneteau, Blandine Beaupain, Nizar Mahlaoui, Christine Bellanné Chantelot. Congenital neutropenia: diagnosis, molecular bases and patient management. Orphanet J Rare Dis 2011;6:26

[24] Dale DC, Bonilla MA, Davis MW, et al. A randomized controlled phase III trial of recombinant human granulocyte colony-stimulating factor (filgrastim) for treatment of severe chronic neutropenia. *Blood.* 1993; 81:2496-2502.

[25] Freedman MH, Bonilla MA, Fier C, et al. Myelodysplasia syndrome and acute myeloid leukemia in patients with congenital neutropenia receiving G-CSF therapy. *Blood.* 2000;96:429-436.

[26] Rosenberg PS, Zeidler C, Bolyard AA, et al. Stable long-term risk of leukaemia in patients with severe congenital neutropenia maintained on G-CSF therapy. *Br J Haematol.* 2010;150:196-199.

[27] Donadieu J, Leblanc T, Bader Meunier B, Barkaoui M, Fenneteau O, Bertrand Y, Maier-Redelsperger M,

Micheau M, Stephan JL, Phillipe N, Bordigoni P, Babin-Boilletot A, Bensaid P, Manel AM, Vilmer E, Thuret I, Blanche S, Gluckman E, Fischer A, Mechinaud F, Joly B, Lamy T, Hermine O, Cassinat B, Bellanné-Chantelot C, Chomienne C; French Severe Chronic Neutropenia Study Group. Analysis of risk factors for myelodysplasias, leukemias and death from infection among patients with congenital neutropenia. Experience of the French Severe Chronic Neutropenia Study Group. Haematologica. 2005;90:45-53.

[28] Nelson AS, Myers KC Diagnosis, Treatment, and Molecular Pathology of Shwachman-Diamond Syndrome. Hematol Oncol Clin North Am. 2018;32:687-700.

[29] Wegman-Ostrosky T, Savage SA. The genomics of inherited bone marrow failure: from mechanism to the clinic. Br J Haematol. 2017;177:526-542.

[30] Sarajlija A, Djordjevic M, Kecman B, Skakic A, Pavlovic S, Pasic S, Stojiljkovic M. Impact of genotype on neutropenia in a large cohort of Serbian patients with glycogen storage disease type Ib. Eur J Med Genet. 2020;63:103767.

[31] Barth PG, Valianpour F, Bowen VM, Lam J, Duran M, Vaz FM, Wanders RJ. X-linked cardioskeletal myopathy and neutropenia (Barth syndrome): an update. Am J Med Genet A. 2004;126A:349-54.

[32] Muraki K, Nishimura S, Goto Y, Nonaka I, Sakura N, Ueda K. The association between haematological manifestation and mtDNA deletions in Pearson syndrome. J Inherit Metab Dis. 1997;20:697-703.

[33] Dotta L, Parolini S, Prandini A, Tabellini G, Antolini M, Kingsmore SF, Badolato R. Clinical, laboratory and molecular signs of immunodeficiency in patients with partial oculo-cutaneous albinism. Orphanet J Rare Dis. 2013 Oct 17;8:168.

[34] Valente NY, Machado MC, Boggio P, Alves AC, Bergonse FN, Casella E, Vasconcelos DM, Grumach AS, de Oliveira ZN. Polarized light microscopy of hair shafts aids in the differential diagnosis of Chediak-Higashi and Griscelli-Prunieras syndromes. Clinics (Sao Paulo). 2006 ;61:327-32.

[35] Meeths M, Bryceson YT, Rudd E, Zheng C, Wood SM, Ramme K, Beutel K, Hasle H, Heilmann C, Hultenby K, Ljunggren HG, Fadeel B, Nordenskjöld M, Henter JI. Clinical presentation of Griscelli syndrome type 2 and spectrum of RAB27A mutations. Pediatr Blood Cancer. 2010 ;54:563-72.

[36] Ohishi Y, Ammann S, Ziaee V, Strege K, Groß M, Amos CV, Shahrooei M, Ashournia P, Razaghian A, Griffiths GM, Ehl S, Fukuda M, Parvaneh N. Griscelli Syndrome Type 2 Sine Albinism: Unraveling Differential RAB27A Effector Engagement. Front Immunol. 2020;11:612977.

[37] Bowman SL, Bi-Karchin J, Le L, Marks MS. The road to lysosome-related organelles: Insights from Hermansky-Pudlak syndrome and other rare diseases. Traffic. 2019 ;20:404-435.

[38] Bohn G, Allroth A, Brandes G, Thiel J, Glocker E, Schäffer AA, Rathinam C, Taub N, Teis D, Zeidler C, Dewey RA, Geffers R, Buer J, Huber LA, Welte K, Grimbacher B, Klein C. A novel human primary immunodeficiency syndrome caused by deficiency of the endosomal adaptor protein p14. Nat Med. 2007;13:38-45.

[39] Rodrigues JM, Fernandes HD, Caruthers C, Braddock SR, Knutsen AP. Cohen Syndrome: Review of the Literature. Cureus. 2018 ;10:e3330.

[40] Beck KD, Wong RW, Gibson JB, Harper CA. Nonleaking cystoid macular edema in Cohen syndrome. J AAPOS. 2019 ;23:38-39.e1.

[41] Stepensky P, Saada A, Cowan M, Tabib A, Fischer U, Berkun Y, Saleh H, Simanovsky N, Kogot-Levin A, Weintraub M, Ganaiem H, Shaag A, Zenvirt S, Borkhardt A, Elpeleg O, Bryant NJ, Mevorach D. The Thr224Asn mutation in the VPS45 gene is associated with the congenital neutropenia and primary myelofibrosis of infancy. Blood. 2013 ;121:5078-87.

[42] Shah RK, Munson M, Wierenga KJ, Pokala HR, Newburger PE, Crawford D.A novel homozygous VPS45 p.P468L mutation leading to severe congenital neutropenia with myelofibrosis. Pediatr Blood Cancer. 2017 ;64.

[43] Meerschaut I, Bordon V, Dhooge C, Delbeke P, Vanlander AV, Simon A, Klein C, Kooy RF, Somech R, Callewaert B. Severe congenital neutropenia with neurological impairment due to a homozygous VPS45 p.E238K mutation: A case report suggesting a genotype-phenotype correlation. Am J Med Genet A. 2015 ;167A:3214-8.

[44] GathmannB, GrimbacherB, BeauteJ, et al. The European internet-based patient and research database for primary immunodeficiencies: results 2006-2008. Clin Exp Immunol. 2009;157:3-11.

[45] AbolhassaniH, AghamohammadiA, FangM, et al. Clinical implications of systematic phenotyping and exome sequencing in patients with primary antibody deficiency. Genet Med. 2019;21:243-251.

[46] Condino-NetoA, SorensenRU, Gomez RaccioAC, et al. Current state and future perspectives of the Latin American Society for Immunodeficiencies (LASID). Allergol Immunopathol (Madr). 2015;43:493-497

[47] Stonebraker JS, Farrugia A, Gathmann B. ESID Registry Working Party, Orange JS. Modeling primary immunodeficiency disease epidemiology and its treatment to esti- mate latent therapeutic demand for immunoglobulin. J Clin Immunol. 2014;34: 233-244.

[48] Gathmann B, Goldacker S, Klima M, Belohradsky BH, Notheis G, Ehl S, Ritterbusch H, Baumann U, Meyer-Bahlburg A, Witte T, Schmidt R, Borte M, Borte S, Linde R, Schubert R, Bienemann K, Laws HJ, Dueckers G, Roesler J, Rothoeft T, Krüger R, Scharbatke EC, Masjosthusmann K, Wasmuth JC, Moser O, Kaiser P, Groß-Wieltsch U, Classen CF, Horneff G, Reiser V, Binder N, El-Helou SM, Klein C, Grimbacher B, Kindle G. The German national registry for primary immunodeficiencies (PID). Clin Exp Immunol. 2013;173:372-380.

[49] Kozlowski C, Evans DIK: Neutropenia associated with X-linked agammaglobulinemia. J Clin Pathol 44:388-390, 1991 16.

[50] Farrar JE, Rohrer J, Conley ME: Neutropenia in X-linked agammaglobulinemia. Clin Immunol Immunopathol 81:271-276, 1996

[51] Hirokazu Kanegane, Hiromichi Taneichi, Keiko Nomura, Takeshi Futatani, Toshio Miyawaki. Severe neutropenia in Japanese patients with x-linked agammaglobulinemia. J Clin Immunol 2005 ;25:491-5.

[52] Winkelstein JA, Marino MC, Ochs H, Fuleihan R, Scholl PR, Geha R, et al. The X-linked hyper-IgM syndrome: clinical and immunologic features of 79 patients. Medicine (Baltimore). 2003;82:373-84.

[53] Levy J, Espanol-Boren T, Thomas C, Fischer A, Tovo P, Bordigoni P, et al. Clinical spectrum of X-linked hyper-IgM syndrome. J Pediatr. 1997;131 (1 Pt 1):47-54.

[54] Leven EA, Maffucci P, Ochs HD, Scholl PR, Buckley RH, Fuleihan RL, Geha RS, Cunningham CK, Bonilla FA, Conley ME, Ferdman RM, Hernandez-Trujillo V, Puck JM, Sullivan K, Secord EA, Ramesh M, Cunningham-Rundles C. Hyper IgM Syndrome: a Report from the USIDNET Registry. J Clin Immunol. 2016 ;36:490-501.

[55] Al-Saud B, Al-Mousa H, Al-Ahmari A, Al-Ghonaium A, Ayas M, Alhissi S, Al-Muhsen S, Al-Seraihy A, Arnaout R, Al-Dhekri H, Hawwari A. Hematopoietic stem cell transplant for hyper-IgM syndrome due to CD40L defects: A single-center experience. Pediatr Transplant. 2015;19:634-9.

[56] Mavroudi I, Papadaki HA. The role of CD40/CD40 ligand interactions in bone marrow granulopoiesis. ScientificWorldJournal. 2011;11:2011-9.

[57] Chinn IK, Shearer WT. Severe combined immunodeficiency disorders. *Immunol Allergy Clin North Am* 2015;35:671-94.

[58] Candotti F. Clinical Manifestations and Pathophysiological Mechanisms of the Wiskott-Aldrich Syndrome. J Clin Immunol. 2018;38:13-27.

[59] Badolato R, Donadieu J; WHIM Research Group. How I treat warts, hypogammaglobulinemia, infections, and myelokathexis syndrome. Blood. 2017 ;130:2491-2498.

[60] McDermott DH, Liu Q, Velez D, Lopez L, Anaya-O'Brien S, Ulrick J, Kwatemaa N, Starling J, Fleisher TA, Priel DA, Merideth MA, Giuntoli RL, Evbuomwan MO, Littel P, Marquesen MM, Hilligoss D, DeCastro R, Grimes GJ, Hwang ST, Pittaluga S, Calvo KR, Stratton P, Cowen EW, Kuhns DB, Malech HL, Murphy PM. A phase 1 clinical trial of long-term, low-dose treatment of WHIM syndrome with the CXCR4 antagonist plerixafor.Blood. 2014;123:2308-16.

[61] McDermott DH, Pastrana DV, Calvo KR, Pittaluga S, Velez D, Cho E, Liu Q, Trout HH 3rd, Neves JF, Gardner PJ, Bianchi DA, Blair EA, Landon EM, Silva SL, Buck CB, Murphy PM. Plerixafor for the Treatment of WHIM Syndrome. N Engl J Med. 2019 ;380:163-170.

[62] Ammann RA, Duppenthaler A, Bux J, Aebi C. Granulocyte colony-stimulating factor-responsive chronic neutropenia in cartilage-hair hypoplasia. J Pediatr Hematol Oncol. 2004 ;26:379-81.

[63] Abdollahpour H, Appaswamy G, Kotlarz D, Diestelhorst J, Beier R, Schaffer AA, et al. The phenotype of human STK4 deficiency. Blood. 2012;119:3450-3457.

[64] Nehme NT, Pachlopnik Schmid J, Debeurme F, Andre-Schmutz I, Lim A, Nitschke P, et al. MST1 mutations in autosomal recessive primary immunodeficiency characterized by defective naive T-cell survival. Blood. 2011;119:3458-3468.

[65] Dang TS, Willet JD, Griffin HR, Morgan NV, O'Boyle G, Arkwright PD, Hughes SM, Abinun M, Tee LJ, Barge D, Engelhardt KR, Jackson M, Cant AJ, Maher ER, Koref MS, Reynard LN, Ali S, Hambleton S. Defective Leukocyte Adhesion and Chemotaxis Contributes to Combined Immunodeficiency in Humans with Autosomal Recessive MST1 Deficiency. J Clin Immunol. 2016;36:117-22.

[66] Hsu AP, McReynolds LJ, Holland SM. GATA2 deficiency. Curr

Opin Allergy Clin Immunol. 2015;15:104-9.

[67] Kely Campos Navegantes, Rafaelli de Souza Gomes, Priscilla Aparecida Tártari Pereira, Paula Giselle Czaikoski, Carolina Heitmann Mares Azevedo, Marta Chagas Monteiro. Immune modulation of some autoimmune diseases: the critical role of macrophages and neutrophils in the innate and adaptive immunity. J Transl Med. 2017; 15: 36.

[68] Ling Wu, Xiang H.-F. Zhang. Tumor Associated Neutrophils and Macrophages—Heterogenous but Not Chaotic. Front Immunol. 2020; 11: 553967.

Section 5

Innate Immunity and Cancer: Double Edge Aspect

Chapter 17

Role of Interferon in Cancer Metabolism

Vaishali Chandel and Dhruv Kumar

Abstract

Interferons (IFNs), a pleotropic cytokine that has long been regarded as an important effector molecule, are increasingly recognized due to their role in cancer and in antitumor immune response regulation. Interferons broadly alter cellular functions in response to viral and other infections. Dysregulation of interferon has been implicated in cancer, autoimmune disorders, and pathogenesis of chronic viral infections. However, the association between interferons and cancer cell metabolism is poorly understood. Emerging evidence suggests the importance of lipid, energy, and amino acid metabolic pathway in regulating interferon response against cancer. Additionally, viruses exploit and modulate the host cell and induce the major metabolic reprogramming causing cancer. In response, interferons upregulate the transcription of large number of interferon stimulating gene (ISG) whose products play a major role in the innate and adaptive immune response against viral infection. Immense research is being done on understanding the role of IFNs in cancer metabolism. Therefore, systematic evaluation of these associations between interferons and cancer metabolism may have important implications for the development of anticancer therapeutics targeting IFN, minimizing toxicity, and limiting off-target effects.

Keywords: interferons, cancer, cancer metabolism

1. Introduction

The interferons (IFNs) are a family of pleotropic cytokines, which play an important role in anticancer immune response. IFNs broadly modulate cellular functions in response to viral and other infections. These modulations include changes in membrane composition, proliferation, metabolism, protein synthesis, and the nutritional microenvironment [1]. Interferons (IFN) are classified as three major types distinguished by their nature, sequence identity, and distribution of cognate receptors [1]. The type I human IFN encodes a family of 17 distinct proteins (IFNα 13 subtypes, IFNβ, IFNε, IFNκ, and IFNω) consisting of IFNα/β receptor 1 (IFNAR1) and IFNα/β receptor 2 (IFNAR2) subunits that bind to their cognate receptor. The type 1 IFN is located on chromosome 9p. Engagement of receptor activates the receptor-associated protein tyrosine kinases Janus kinase 1 (JAK1) and tyrosine kinase 2 (TYK2), phosphorylating and activating signal transducer and activator of transcription 1 (STAT1) and STAT2 transcription factors [2]. IFNγ is the only single type II IFN, which binds to IFNγ receptor 1 (IFNGR1) and IFNγ receptor 2 IFNGR2 subunits. The type III IFNs consist of IFNλ1, IFNλ2, IFNλ3, and IFNλ4, which bind the IFNλ receptor 1 (IFNLR1) [3] (**Figure 1**). Pattern

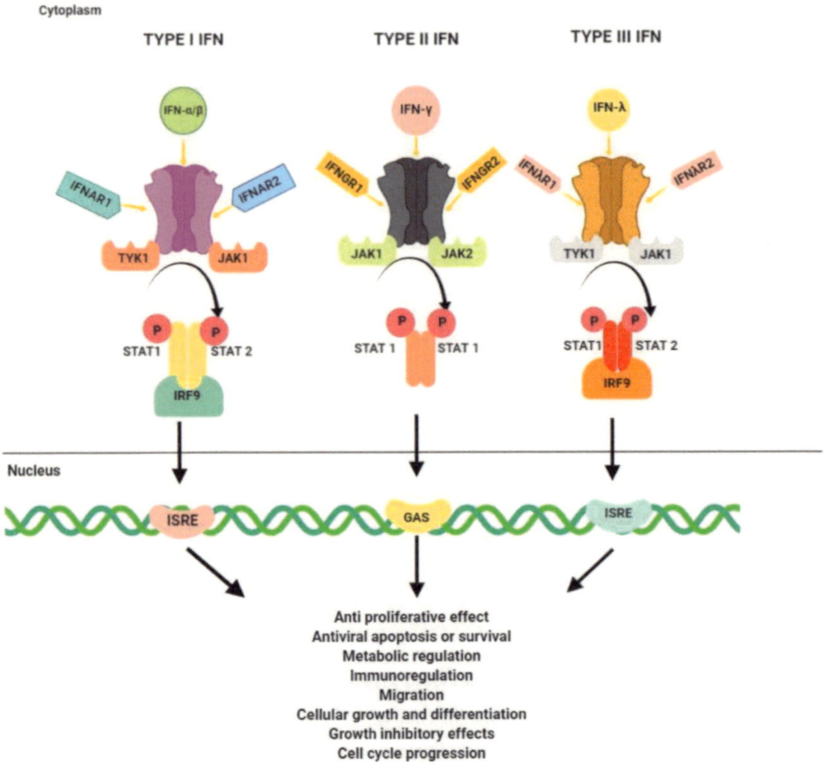

Figure 1.
Interferon signaling and role in cancer. Type I IFN encodes IFNα/β consisting of IFNα/β receptor 1 and 2 subunits that bind to their cognate receptor. Engagement of receptor activates JAK1 and TYK2, phosphorylating and activating STAT1 and STAT2 transcription factors. IFNγ binds to IFNγ receptor 1 and 2 subunits. The type III IFNs consist of IFNλ1, which bind the IFNλ receptor 1 and 4. Activation of the three types of interferons mediates downstream signaling pathway in cancer and leads to effector responses such as anti-proliferative, antiviral apoptosis, metabolic regulation, immunoregulation, migration, cellular growth and differentiation, growth inhibitory effects, and cell cycle progression. JAK1: Janus kinase 1; TY2: tyrosine kinase 2; STAT: signal transducer and activator of transcription.

recognition receptor (PRR) pathways activate the expression of type I and type III IFNs. However, cytokines expressed by natural killer (NK) cells and T cells, including IL12 and IL18, or mitogens induce type II IFN [4, 5]. Additionally, mammalian target of rapamycin (mTOR) also activates the expression of IFNs. Integration of mTOR complex 1 (mTORC1) with the major class of energy and nutrient sources [glucose, amino acids, adenosine triphosphate (ATP), and lipids] leads to the cellular activation and translation [6]. mTORC1 activation is important to induce and activate interferon regulatory transcription factor (IRF) such as IRF5 and IRF7, to initiate and maximize the production of type I IFN [7]. The transcription of majority of interferon stimulated genes (ISGs) is mediated by type I IFNs and IFNγ [8]. Three major families of ISGs play a major role in antiviral host immune response; RNA-activated protein kinase (PRK), Mx protein (Myxovirus resistance 1) and ribonuclease L (RNase L) [9]. They are responsible for inhibiting viral replication. PKR is induced by IFN and is a RNA-dependent kinase that phosphorylates eIF2α, the translation initiation factor 2α mediating inhibition of viral and cellular translation. Binding of dsRNA activates OAS and stimulates the activity of RNase L causing protein expression inhibition by cellular and viral ssRNA cleavage [10]. In addition, Mx proteins are GTPases, which trap and inhibit viral replication by sensing nucleocapsid-like viral structures [11]. The production of IFNs is important since they regulate tumorigenesis and mediate metabolic reprogramming by direct or indirect means [1, 12]. IFN plays a major role in cancer metabolism. Cellular metabolism is a complex and fundamental biological process involving catabolism to fuel cellular reactions by the breakdown of macromolecules to generate energy in the form of adenosine triphosphate (ATP) and anabolism that delivers nutrients such as amino acids, carbohydrates, and fatty acids for the synthesis of macromolecules [13]. As compared to the normal cells, the metabolic activities in cancer cells are altered, and these alterations facilitate and support the malignant properties of cancer cells. Therefore, metabolic reprogramming is one of the major hallmarks of cancer [14]. In order to meet biosynthetic and bioenergetic demands to facilitate rapid proliferation, cancer cells perform increased glycolysis even under anaerobic conditions (Warburg phenomenon) [15]. Thus, the conversion of glucose to lactic acid by glucose metabolism fulfills energy demands in cancer cells, as opposed to mitochondrial oxidative phosphorylation in normal cells [16]. Additionally, reliance on glycolysis by cancer cells is a useful adaptation in order to sustain in a hypoxic microenvironment. This glycolytic switch is mediated by various mechanisms [17]. For example, the best described canonical pathway mediating the regulation of tumor cell metabolism is the PI3K-Akt pathway [18]. PI3K-Akt pathway promotes the activity of glucose transporter (GLUT) and stimulates the glycolytic process and production of lactate through activating several glycolytic enzymes such as hexokinase (HK) and phosphofructokinase (PFK). Mechanistically, PI3K-AKT signaling activates mammalian target of rapamycin (mTOR), which activates the transcription factor in turn, hypoxia-inducible factor-1 (HIF-1). HIF-1 cooperation with other transcription factors such as p53, c-Myc, and Oct1 activates transcription of multiple genes involved in glycolytic metabolism, such as HK [19], GLUT-1 and GLUT-3 [20, 21], lactate dehydrogenase (LDH) [22], and phosphoglycerate kinase [23], as well as for pH regulation, such as carbonic anhydrase IX (CAIX) [24] and Na^+/H^+ exchanger 1 (NHE1) [25], and suppressors of TCA cycle, such as pyruvate dehydrogenase kinase (PDK) [26]. However, metabolic alteration in cancer cell is not only defined to glucose metabolism, but it is directly interconnected with various other metabolic pathways such as amino acid metabolism through the intermediate 3-phosphoglycerate, pentose phosphate pathway (PPP) by the glucose-6-phosphate intermediate, and metabolism of fatty acids (FA) by pyruvate into Krebs cycle [27].

Therefore, it is important to understand the role of interferons in cancer cell metabolism for the development of novel interventions to treat cancer.

2. Interferons and cancer metabolism

2.1 Type I IFN signaling and cancer metabolism

The correlation between the type I IFN and cancer metabolism in cancer is shown in several studies [1, 7, 12, 28, 29]. However, the mechanism underlying this altered metabolism is poorly understood and not widely studied because of the complexity in regulation by various cellular extrinsic and intrinsic signals [30]. Signaling pathway, including JAK/STAT, ERK/MAP, p38, and PI3/AKT, regulate the metabolic process [28]. Additionally, it has been shown that IRF also plays a major role in regulating metabolism in cancer [31]. The JAK/STAT signaling pathway plays an important role in regulating development, immune function, and apoptosis [32]. It regulates the expression of early response genes [33]. STAT1 and STAT3 alter the gene expression in glucose metabolism, gluconeogenesis, Krebs cycle, and mitochondrial oxidative phosphorylation (OXPHOS). Apart from this metabolic pathway, STAT 1 and STAT3 play a key role in modulating lipid metabolism in cancer [32, 34, 35]. Also, they have been shown to alter the cellular respiration process and mitochondrial function. The function of mitochondria is decreased due to PPARG coactivator-1 α (PGC-1 α) repression, a master regulator in mitochondrial biogenesis [36]. Alternatively, STAT3 localizes in mitochondria and interacts with complex I and II of the electron transport chain (ETC), thereby increasing the oxidation process [37]. Most importantly, while these modifications in metabolic pathways are needed to mount functional immune responses, changes associated with STAT activation may lead to the pathogenic processes during activation of IFN. Specifically, signaling mediated by STAT1 has been shown to mediate tumorigenesis and resistance to chemotherapy and ionizing radiation by upregulating the expression of genes involved in glucose metabolism, Krebs cycle, and OXPHOS [38]. Alternatively, alterations driven by STAT3 in mitochondrial metabolism lead to drug resistance in cancer patients by controlling the mitochondrial transition pore opening [39]. However, further study is needed to understand how these STAT mediated mechanisms facilitate to functional and nonfunctional type I IFN responses. Apart from the STAT signaling pathway, AKT/mTOR signaling has been shown to play an important role in type I IFN effector function regulation. The two complexes of mTOR (mTORC1 and mTORC2) [40, 41] have differential effects on type I IFN responses. mTORC1 plays a key role in ISGs translation [42], whereas mTORC2 performs transcription of IFN-dependent gene via interferon-stimulated response elements [43]. Additionally, mTOR in response to hormonal and environmental signals coordinates metabolism centrally [44]. Also, it has been associated with lipogenesis, adipogenesis, ribosomal biogenesis, and pyrimidine synthesis [45–48]. Previous studies have identified the correlation between mTOR signaling, OXPHOS, fatty acid oxidation (FAO), and glycolysis with type I interferons [49]. A major important regulator of interferon responses is IRFs, which centrally regulate the development of immune cell and effector function [29, 31, 50, 51]. The best described IRFs, IRF4, regulate the expression of the major molecules, which are important for aerobic glycolysis [50] and for suppressing the expression of lipogenic gene involved in lipogenesis and lipolysis activation [52]. In a similar manner, IRF5 upregulates the glycolytic process via activation of AKT and glycolytic gene induction in inflammatory macrophages [29]. Many studies have reported abnormalities in expression of IRF and their role in metabolic diseases such as cancer with poor prognosis, insulin resistance, atherosclerosis, and hepatic steatosis [53–55].

2.2 Type I IFN and altered bioenergetics

Metabolic reprogramming in cancer cells is closely linked to effector function and cellular activation [56]. Bioenergetic pathways include glucose metabolism, tricarboxylic acid cycle (TCA), FAO, OXPHOS, electron transport chain (ETC), and pentose phosphate pathway (PPP) [56]. Since metabolic reprogramming is needed to meet the biosynthetic and bioenergetic demands of the cells, recent studies suggest that metabolites (succinate and citrate) and enzyme pyruvate kinase M2 may play a key role and act as transcription factor and signaling molecule to mediate the immune function and inflammatory processes [27, 57, 58]. An important characteristic of type I IFN in cancer metabolism is upregulated glucose metabolism [59]. The metabolic shift is important to quickly generate ATP to meet energy demands of the cell. In fibroblasts, PI3/AKT signaling is important for type I IFN-associated shift and leads to increased uptake of glucose in the cell [60]. Alternatively, STAT1 mediates aerobic glycolysis in human squamous cell carcinoma [61]. Also, upregulated expression of 6-phosphofructo-2-kinase/fructose-2,6-biphosphatase 3 (PFKFB3) has been shown in variety of tumors [62] (**Table 1**). Furthermore, the metabolic shift from OXPHOS to glycolysis contributes to Warburg phenomenon,

Gene	Role in metabolism	Cancer type	Reference
PFKFB3	Regulator of glycolysis. Associated with many aspects of cancer, including metabolism, carcinogenesis, cancer cell proliferation, vessel aggressiveness, drug resistance, and tumor microenvironment	Liver, breast, head, and neck	[62]
SC4MOL	Protection against virus attack and important contributor in sterol metabolism	Breast, nonsmall cell lung cancer	[63]
SCAP	IFN-driven regulation of lipid metabolism	Brain cancer	[64, 65]
SREBP1/2	IFN-driven regulation of lipid metabolism	Colon, lung, pancreatic	[66]
CH25H	Regulate cellular functions and influence various physiological processes such as cholesterol metabolism, membrane fluidity regulation, and intracellular signaling pathways in cancer	Breast cancer	[67]
CYP27A1	Affects estrogen receptor function by the antagonism of estrogen action and also by the direct modulation of the receptor function modulating metabolism	Breast cancer	[68]
IDO1	Prevents viral proliferation and regulates lipid metabolism and inflammation	Breast, lung, pancreatic, leukemia	[69]
NOS2	Cytostatic and cytotoxic effects against tumor cells	Glioblastoma, melanoma, breast	[70]

Table 1.
Type I IFN immunometabolic gene response in the progression of various cancers.

tumor metastasis, and growth [71]. In cancer cells, decreased rate of mitochondrial OXPHOS is accompanied with the glycolytic shift in immune cells [72]. Consistent with these findings, mouse L929 cell triggered with type I IFN showed signs of reduced OXPHOS and production of ATP [73]. Also, CD4+ T cells isolated from multiple sclerosis patients treated with IFN-β underwent OXPHOS impairment in a dose-dependent manner as compared to healthy individuals [74]. A single nucleotide polymorphism (SNP) in PGC-1α, a gene involved in the mitochondrial biogenesis, was shown to be associated with reduced intracellular ATP production levels and altered therapeutic response to IFN-β in patients [74]. However, other studies suggest that bioenergetic reprogramming in cancer driven by IFN may be context and cell type dependent [75]. Mouse plasmacytoid DCs (pDCs) stimulated by IFN-α are linked with upregulated glycolytic genes in turn increased glycolysis, OXPHOS, and FAO to meet the energy demand of the cells [76]. mTOR activation mediates upregulation of OXPHOS and FAO and is important in mounting an immune response. In T cell, stimulation of CD8+ memory T cells by IFN-α is associated with upregulated OXPHOS, whereas effector T cell stimulation has not been shown to alter the activity of OXPHOS [77]. Additionally, in the reverse Warburg phenomenon, cancer cells induce aerobic glycolysis in cancer-associated fibroblasts (CAFs), present in the tumor stroma. CAFs generate pyruvate, lactate, and ketone bodies that enter the TCA cycle in cancer cells for mitochondrial OXPHOS. In fact, these tumor-associated stromal cells, for example, tumor-associated macrophages (TAMs), already vary from their original cells and have epigenetic and genetic changes, which result in altered metabolic profiles. Therefore, cancer cells influence each other not only in terms of growth factor or cytokines, such as IFN, but also on dependency on metabolic pathways. TAMs, for example, derive their ATP from OXPHOS rather than aerobic glycolysis.

2.3 IFN response and lipid metabolism

A wide variety of studies have recognized the role of type I IFNs in modulating lipid metabolism in cancer [7]. Lipids are the major constituent in plasma membrane and various other cellular compartments such as the endoplasmic reticulum, nuclear membrane, Golgi apparatus, lysosomes, and endosomes [7]. Alongside, lipids function as signaling molecules to regulate the majority of cellular processes, including inflammatory, metabolic, and innate immune responses [78]. A number of viruses causing cancer, such as Epstein-Barr Virus (EBV), Hepatitis B Virus (HBV), Hepatitis C Virus (HCV), Human Herpesvirus 8 (HHV8), Human Papillomavirus (HPV), Human T-cell Lymphotropic Virus 1 (HTLV), and Merkel Cell Polyomavirus hijack cholesterol and fatty acid (FA) biosynthesis of host to support replication and survival of virus [79]. To counteract this process, de novo cholesterol and lipid synthesis is decreased, and cholesterol and FA import is mediated by type I IFNs. After 30 min of exposure to IFN, STAT2-driven reprogramming occurs and is independent of ISG expression [64]. Decreased de novo cholesterol and lipid synthesis is a complex mechanism and needs further research to be done upon. Several studies have reported the role of sterol regulatory element-binding protein 2 (SREBP2)/SREBP cleavage-activating protein (SCAP) pathway in IFN-driven regulation of lipid metabolism [64, 65] (**Table 1**). SPREBP1 and SREBP2 are recruited by SCAP, a chaperone protein to the nucleus. In the nucleus, SPREBP1 and SREBP2 transcription factors regulate cholesterol and lipid metabolism, respectively. Knock out of SREBP2 or SCAP expression in macrophages leads to mice resistant to viral attack supporting the role of IFN response and an interrelationship between lipid metabolism and type I IFN [64, 65]. Additionally, type I IFNs upregulate microRNAs that control cholesterol biosynthesis. Upregulated expression of miR-342-5p in BMM is shown to be associated with IFN-β stimulation. miR-342-5p targets SREBP2, DHCR7, IDI1, and SC4MOL cholesterol biosynthetic genes [80] (**Table 1**). SC4MOL gene catalyzes demethylation of C4-methylsterols and meiosis-activating sterols (MASs) and encodes methyl sterol oxidase (**Table 1**). Accumulation of C4-methylsterols leads to increased proliferation of cancer cells [63]. Oxysterol, a cholesterol derivative participating in cholesterol metabolic regulation, signaling pathways such as Hedgehog, MAPK, and Wnt, and enzymatic activity playing a major role in cancer metabolism, is upregulated by type I IFNs [81]. Of the most important, 25-hydroxycholesterol (25-HC) and 27-HC (CYP27A1) play a key role in sterol biosynthesis regulation, minimizing accumulation of cholesterol and inhibition of viral spread and replication [68]. Cholesterol-25-hydroxylase (CH25H) encodes 25-HC, which is a soluble oxysterol [39]. Type I and II IFN production in response to Toll-like receptor (TLR) activation leads to the expression of CH25H in dendritic cells and macrophages. 25-HC does this by repressing the activation of SREBP2 or by increasing the expression of miR-185 regulating hepatic homeostasis of lipid [7]. Alternatively, 27-HC has been demonstrated to decrease the cholesterol accumulation in lysosomes and decrease inflammation [7] (**Table 1**). However, oxysterol induced by IFN may also have a damaging role in cancer and other inflammatory diseases. 25-HC amplifies proinflammatory mediator production following infection [82]. 22-HC and 27-HC in cancer are detected in high levels in a majority of tumor cells [83, 84]. They mediate the activation of liver X receptors (LXRs) in tumor, upregulating the efflux of cholesterol while promoting an anti-inflammatory state [85]. Additionally, 22-HC and 27-HC have been reported to enhance the estrogen receptor transcription in breast cancer model, supporting the evidence that it may lead to resistance to hormonal therapy [83] (**Table 1**).

2.4 IFN response and amino acid metabolism

Amino acids serve as a building block for protein synthesis, branched chain fatty acid synthesis, and energy metabolism [27]. Their utilization is associated with metabolic signaling pathway such as nucleotide synthesis and mTOR pathway in tumor cells during immune response. Amino acid metabolism is reprogrammed to meet the biosynthetic and bioenergetic requirements of the cells [27]. However, several other studies have shown the role of amino acid as an important signaling molecule to alter cellular survival and function [27]. For the purpose of the importance of interferons in cancer-associated metabolism, we will focus on arginine and tryptophan metabolism in regulating type I IFN responses.

In response to type I IFN, metabolism of amino acid is tightly regulated against virus causing cancer [86]. A major example of this regulation includes tryptophan metabolism. Tryptophan is one of the nine essential amino acids and is very important in playing a key role in various metabolic pathways. The catabolites of tryptophan play an important role in cancer immunosuppression. Indoleamine-2,3-dioxygenase (IDO), catabolic enzyme converting tryptophan to kynurenine, is the essential rate limiting enzyme expressed in antigen-presenting cells or tumor cells. This metabolic pathway creates an immunosuppressive milieu in tumor-draining lymph nodes and in tumors by inducing apoptosis and T-cell anergy through tryptophan depletion and accumulation of immunosuppressive tryptophan catabolites. Specifically, the synthesis of tryptophan derivatives in kynurenine accounts for more than 80% of tryptophan catabolism. The synthesis of kynurenine is done by the catalytic activity of tryptophan-2,3-dioxygenase (TDO2) and indoleamine-2,3-dioxygenase (IDO1) (**Figure 2**). The expression of ISG (interferon-stimulated gene), IDO1, which is highly effective at controlling and resisting pathogens, is very high across different cell types, whereas TDO2 has a lower affinity for tryptophan and is majorly expressed in hepatocytes [7]. Several studies have shown the development of an immunotolerant state associated with enhanced regulatory response of T cells and suppressed T cell activation and proliferation due to increased tryptophan catabolism [87, 88]. Additionally, metabolites of kynurenine including 3-hydroxyanthranilic acid and quinolinic acid have cytotoxic as well as inflammatory effects [52]. These studies suggest the role of tryptophan catabolism in response to type I IFN in a protective or detrimental manner in cancer. Supporting their protective role, studies have demonstrated that induction of IDO can be important in autoimmune disease prevention and cancer [89]. Consistent with such findings, IDO protein is expressed in varieties of solid tumor and in human malignancies [90] (**Table 1**). These findings and observations highlight the importance of type I IFN in the development of anticancer therapeutics by modulating tryptophan catabolism pathway. In addition to the role of type I IFN in modulating tryptophan metabolism, arginine plays an important role in adaptive and innate immune response [90]. Arginine is catabolized by four different classes of enzyme in various cell types: arginase, arginine: glycine amidinotransferase (AGAT), nitric oxide synthase (NOS), and arginine decarboxylase (ADC) [91]. This catalytic process produces several metabolites, which are biologically important with various functions such as urea, citrulline, glutamate, creatinine, polyamines, and nitric oxide (NO). Arginine is metabolized by arginase and/or NOS pathway [12]. The specific role of arginase or iNOS leads to the functional polarization of these cells into anti-inflammatory M2 phenotypes or M1 inflammatory phenotypes [12]. The expression of iNOS is increased by type I IFN and is linked to enhanced

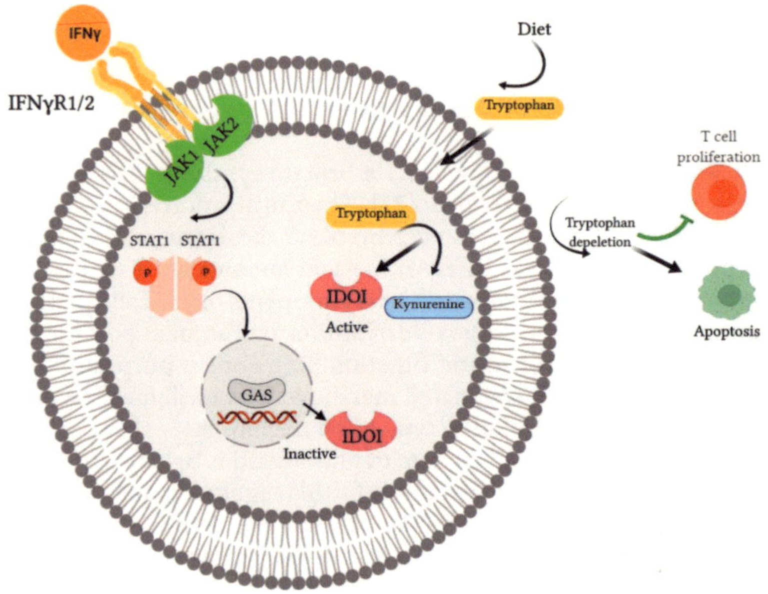

Figure 2.
Tryptophan catabolism in response to type I IFN. The synthesis of kynurenine is done by the catalytic activity TDO2 and IDO1. Increased tryptophan catabolism suppresses T cell activation, and proliferation IDO is an important mediator in metabolism, autoimmune disease prevention, and cancer. TDO2, tryptophan-2,3-dioxygenase (TDO2); IDO1, indoleamine-2,3-dioxygenase.

levels of NO, L-citrulline, and reactive nitrogen species. Furthermore, enhanced glycolysis in tumor cells, TAMs, and other stromal cells, such as CAFs, leads to lactic acid accumulation in the tumor microenvironment. Lactic acid polarizes TAMs to a tumor-promoting phenotype characterized by the expression of arginase1 (ARG1), VEGFA, and several M2 markers via the activation of HIF1α [12]. This metabolic reprogramming results in accumulation of bioactive metabolites and plays a major role in cytotoxic or cytostatic activities against tumor cells. This suggests that type I IFN signaling may play an important role in tumor immune escape, immunosuppression, and immunopathology [12].

3. IFN-γ and cancer metabolism

In cancer, metabolic reprogramming of macrophages has been widely studied, but its relevance in function of inflammatory cell is a current research interest. Considering the role of Warburg phenomenon (aerobic glycolysis) in M1 macrophages, researchers have been dependent on 2-DG, a competitive inhibitor of glucose in the first reaction step. It was found that induction of 2-DG downregulated both aerobic glycolysis and mitochondrial OXPHOS and had a significant effect in a dose-dependent manner on cell viability and ATP levels. Alternatively, they exploited galactose, which is metabolized to glucose-6-phosphate at a very slow rate, thereby significantly downregulating the glycolytic throughput. Additionally, it was observed that there was downregulation in extracellular acidification rate (ECAR) levels with little effect oxygen consumption rate (OCR), thereby facilitating more exclusive evaluation of the importance of glycolysis in M1 macrophages. Certainly, even under those conditions, macrophages were differentiated by IFN-γ into M1 type phenotype depending on the surface marker expression and cytokines such as IL-6 and TNF-α. However, levels of IL-1β and HIF-1α were profoundly downregulated by galactose, similar to the expression and production of NO. Consistent with these findings, it suggests that aerobic glycolysis in cancer is very particular and plays a significant role for two gene transcription pathways in IFN-γ-stimulated macrophages: HIF-1α and STAT-1. In a similar manner, IFN-γ activated JAK/STAT-1 pathway in cancer increased phosphorylation of STAT-1 in M1 macrophages, and this response was inhibited by using 2-DG as a competitive inhibitor. Also, TAMs showed an increased glycolysis, and glycolysis inhibition using a competitive inhibitor 2DG revoked the functional phenotype of cancer cells. Galactose showed a significant inhibitory effect on the phosphorylation of STAT-1, supporting the importance of aerobic glycolysis in JAK/STAT-1 pathway. In the absence of IFN-γ, glucose itself could not stimulate JAK/STAT-1 pathway. These findings highlight the importance of IFN-γ triggering signaling pathway in M1 macrophages altering the metabolism in cancer [12].

4. Conclusions

The interrelationship between immune function and cellular metabolism is increasingly recognized. Apart from providing substrates to meet the biosynthetic and bioenergetic demand, metabolites from metabolic pathway and enzymes regulate transcription and translation, epigenetic processes, signaling pathways to control cellular function. Increasing evidence suggests the importance of interferons in modulating cell metabolism in cancer and contributing to effector functions. However, it is unclear if these processes can be harnessed to elicit specific immune functions and/or prevent the development of pathological side effects. In order to target metabolic processes with some level of specificity, we require an in-depth understanding of how these processes are regulated across cell types and tissues. Therefore, it is important for the in depth understanding to develop novel interventions to treat cancer, chronic inflammatory, and infectious diseases.

Acknowledgements

We sincerely thank all authors for their valuable inputs and carefully reading the chapter.

Conflict of interest

The authors declare no conflict of interest.

Author details

Vaishali Chandel and Dhruv Kumar*
Amity Institute of Molecular Medicine and Stem Cell Research (AIMMSCR), Amity University, Noida, Uttar Pradesh, India

*Address all correspondence to: dhruvbhu@gmail.com; dkumar13@amity.edu

References

[1] Parker BS, Rautela J, Hertzog PJ. Antitumour actions of interferons: Implications for cancer therapy. Nature Reviews. Cancer. 2016;**16**(3):131-144. DOI: 10.1038/nrc.2016.14

[2] Ivashkiv LB, Donlin LT. Regulation of type I interferon responses. Nature Reviews. Immunology. 2014;**14**(1):36-49. DOI: 10.1038/nri3581

[3] Kotenko SV, Gallagher G, Baurin VV, Lewis-Antes A, Shen M, Shah NK, et al. IFN-λs mediate antiviral protection through a distinct class II cytokine receptor complex. Nature Immunology. 2003;**4**(1):69-77. DOI: 10.1038/ni875

[4] Seder RA, Gazzinelli R, Sher A, Paul WE. Interleukin 12 acts directly on CD4+ T cells to enhance priming for interferon gamma production and diminishes interleukin 4 inhibition of such priming. Proceedings of the National Academy of Sciences of the United States of America. 1993;**90**(21):10188-10192. DOI: 10.1073/pnas.90.21.10188

[5] Freund-Brown J, Chirino L, Kambayashi T. Strategies to enhance NK cell function for the treatment of tumors and infections. Critical Reviews in Immunology. 2018;**38**(2):105-130. DOI: 10.1615/CritRevImmunol.2018025248

[6] Weichhart T, Hengstschläger M, Linke M. Regulation of innate immune cell function by mTOR. Nature Reviews. Immunology. 2015;**5**(10):599-614. DOI: 10.1038/nri3901

[7] Fritsch SD, Weichhart T. Effects of Interferons and viruses on metabolism. Frontiers in Immunology. 2016;**7**(630):1-13. DOI: 10.3389/fimmu.2016.00630

[8] Samuel CE. Antiviral actions of Interferons. Clinical Microbiology Reviews. 2001;**14**(4):778-809. DOI: 10.1128/CMR.14.4.778

[9] Levy DE, Garcı A. The virus battles: IFN induction of the antiviral state and mechanisms of viral evasion. Cytokine & Growth Factor Reviews. 2001;**12**(2-3):143-156. DOI: 10.1016/s1359-6101(00)00027-7

[10] Sadler AJ, Williams BR. Interferon-inducible antiviral effectors. Nature Reviews. Immunology. 2008;**8**(7):559-568. DOI: 10.1038/nri2314

[11] Haller O, Staeheli P, Schwemmle M, Kochs G. Mx GTPases: Dynamin-like antiviral machines of innate immunity. Trends in Microbiology. 2015;**23**(3):154-163. DOI: 10.1016/j.tim.2014.12.003

[12] Ahmed D, Cassol E. Role of cellular metabolism in regulating type I interferon responses: Implications for tumour immunology and treatment. Cancer Letters. 2017;**409**:20-29. DOI: 10.1016/j.canlet.2017.08.037

[13] DeBerardinis RJ, Thompson CB. Cellular metabolism and disease: What do metabolic outliers teach us? Cell. 2012;**148**(6):1132-1144. DOI: 10.1016/j.cell.2012.02.032

[14] Hanahan D, Weinberg RA. The hallmarks of cancer. Cell. 2000;**100**(1):57-70. DOI: 10.1016/s0092-8674(00)81683-9

[15] Zheng J. Energy metabolism of cancer: Glycolysis versus oxidative phosphorylation (review). Oncology Letters. 2012;**4**(6):1151-1157. DOI: 10.3892/ol.2012.928

[16] Warburg O, Wind F, Negelein E. The metabolism of tumors in the body. The Journal of General Physiology. 1927;**8**(6):519-530. DOI: 10.1085/jgp.8.6.519

[17] Eales KL, Hollinshead KE, Tennant DA. Hypoxia and metabolic adaptation of cancer cells. Oncogene. 2016;**5**:e190. DOI: 10.1038/oncsis.2015.50

[18] Sandulache VC, Myers JN. Altered metabolism in head and neck squamous cell carcinoma: An opportunity for identification of novel biomarkers and drug targets. Head & Neck. 2012;**34**(2):282-290. DOI: 10.1002/hed.21664

[19] Mathupala SP, Rempel A, Pedersen PL. Glucose catabolism in cancer cells: Identification and characterization of a marked activation response of the type II hexokinase gene to hypoxic conditions. The Journal of Biological Chemistry. 2001;**276**(46):43407-43412. DOI: 10.1074/jbc.M108181200

[20] Maxwell PH, Dachs GU, Gleadle JM, Nicholls LG, Harris AL, Stratford IJ, et al. Hypoxia-inducible factor-1 modulates gene expression in solid tumors and influences both angiogenesis and tumor growth. Proceedings of the National Academy of Sciences of the United States of America. 1997;**94**(15):8104-8109. DOI: 10.1073/pnas.94.15.8104

[21] Ebert BL, Firth JD, Ratcliffe PJ. Hypoxia and mitochondrial inhibitors regulate expression of glucose transporter-1 via distinct cis-acting sequences. The Journal of Biological Chemistry. 1995;**270**(49):29083-29089. DOI: 10.1074/jbc.270.49.29083

[22] Firth JD, Ebert BL, Ratcliffe PJ. Hypoxic regulation of lactate dehydrogenase A. interaction between hypoxia-inducible factor 1 and cAMP response elements. The Journal of Biological Chemistry. 1995;**270**(36):21021-21027. DOI: 10.1074/jbc.270.36.21021

[23] Kim JW, Gao P, Liu YC, Semenza GL, Dang CV. Hypoxia-inducible factor 1 and dysregulated c-Myc cooperatively induce vascular endothelial growth factor and metabolic switches hexokinase 2 and pyruvate dehydrogenase kinase 1. Molecular and Cellular Biology. 2007;**27**(21):7381-7393. DOI: 10.1128/MCB.00440-07

[24] Kumar D. Regulation of glycolysis in head and neck squamous cell carcinoma. Postdoc Journal. 2017;**5**(1):14-28. DOI: 10.14304/surya.jpr.v5n1.4

[25] Meijer TW, Kaanders JH, Span PN, Bussink J. Targeting hypoxia, HIF-1, and tumor glucose metabolism to improve radiotherapy efficacy. Clinical Cancer Research. 2012;**18**(20):5585-5594. DOI: 10.1158/1078-0432

[26] Ward PS, Thompson CB. Metabolic reprogramming: A cancer Hallmark even Warburg did not anticipate. Cancer Cell. 2012;**21**(3):297-308. DOI: 10.1016/j.ccr.2012.02.014

[27] O'Neill LA, Kishton RJ, Rathmell J. A guide to immunometabolism for immunologists. Nature Reviews. Immunology. 2016;**16**(9):553-565. DOI: 10.1038/nri.2016.70.A

[28] Platanias LC. Mechanisms of type-I- and type-II-interferon-mediated signalling. Nature Reviews. Immunology. 2005;**5**(5):375-386. DOI: 10.1038/nri1604

[29] Hedl M, Yan J, Abraham C. IRF5 and IRF5 disease-risk variants increase glycolysis and human M1 macrophage polarization by regulating proximal signaling and Akt2 activation. Cell Reports. 2016;**16**(9):2442-2455. DOI: 10.1016/j.celrep.2016.07.060

[30] Robey RB, Weisz J, Kuemmerle NB, Salzberg AC, Berg A, Brown DG, et al. Metabolic reprogramming and dysregulated metabolism: Cause, consequence and/or enabler of environmental carcinogenesis ?

Carcinogenesis. 2015;(Suppl 1): S203-S231. DOI: 10.1093/carcin/bgv037

[31] Zhao GN, Jiang DS, Li H. Interferon regulatory factors: At the crossroads of immunity, metabolism, and disease. Biochimica et Biophysica Acta. 2015;**1852**(2):365-378. DOI: 10.1016/j.bbadis.2014.04.030

[32] Dinasarapu AR, Gupta S, Ram Maurya M, Fahy E, Min J, Sud M, et al. A combined omics study on activated macrophages—Enhanced role of STATs in apoptosis, immunity and lipid metabolism. Bioinformatics. 2013;**29**(21):2735-2743. DOI: 10.1093/bioinformatics/btt469

[33] Stark GR, Darnell JE Jr. The JAK-STAT pathway at twenty. Immunity. 2012;**36**(4):503-514. DOI: 10.1016/j.immuni.2012.03.013

[34] Camporeale A, Demaria M, Monteleone E, Giorgi C, Wieckowski MR, Pinton P, et al. STAT3 activities and energy metabolism: Dangerous liaisons. Cancers (Basel). 2014;**6**(3):1579-1596. DOI: 10.3390/cancers6031579

[35] Pitroda SP, Khodarev NN, Beckett MA, Kufe DW, Weichselbaum RR. MUC1-induced alterations in a lipid metabolic gene network predict response of human breast cancers to tamoxifen treatment. Proceedings of the National Academy of Sciences of the United States of America. 2009;**106**(14):5837-5841. DOI: 10.1073/pnas.0812029106

[36] Sisler JD, Morgan M, Raje V, Grande RC, Derecka M, Meier J, et al. The signal transducer and activator of transcription 1 (STAT1) inhibits mitochondrial biogenesis in liver and fatty acid oxidation in adipocytes. PLoS One. 2015;**10**(12):e0144444. DOI: 10.1371/journal.pone.0144444

[37] Wegrzyn J, Potla R, Chwae YJ, Sepuri NB, Zhang Q, Koeck T, et al. Function of mitochondrial Stat3 in cellular respiration. Science. 2009;**323**(5915):793-797. DOI: 10.1126/science

[38] Pitroda SP, Wakim BT, Sood RF, Beveridge MG, Beckett MA, MacDermed DM, et al. STAT1-dependent expression of energy metabolic pathways links tumour growth and radioresistance to the Warburg effect. BMC Medicine. 2009;**7**:68. DOI: 10.1186/1741-7015-7-68

[39] Poli V, Camporeale A. STAT3-mediated metabolic reprograming in cellular transformation and implications. Frontiers in Oncology. 2015;**5**:121. DOI: 10.3389/fonc.2015.00121

[40] Oh WJ, Jacinto E. mTOR complex 2 signaling and functions. Cell Cycle. 2011;**10**(14):2305-2316. DOI: 10.4161/cc.10.14.16586

[41] Sarbassov DD, Guertin DA, Ali SM, Sabatini DM. Phosphorylation and regulation of Akt/PKB by the rictor-mTOR complex. Science. 2005;**307**(5712):1098-1101. DOI: 10.1126/science.1106148

[42] Kaur S, Lal L, Sassano A, Majchrzak-Kita B, Srikanth M, Baker DP, et al. Regulatory effects of mammalian target of rapamycin-activated pathways in type I and II interferon signaling. The Journal of Biological Chemistry. 2007;**282**(3):1757-1768. DOI: 10.1074/jbc.M607365200

[43] Kaur S, Sassano A, Majchrzak-Kita B, Baker DP, Su B, Fish EN, et al. Regulatory effects of mTORC2 complexes in type I IFN signaling and in the generation of IFN responses. Proceedings of the National Academy of Sciences of the United States of America. 2012;**109**(20):7723-7728. DOI: 10.1073/pnas.1118122109

[44] Saxton RA, Sabatini DM. mTOR signaling in growth, metabolism, and

disease. Cell. 2017;**168**(6):960-976. DOI: 10.1016/j.cell.2017.02.004

[45] Laplante M, Sabatini DM. mTORC1 activates SREBP-1c and uncouples lipogenesis from gluconeogenesis. Proceedings of the National Academy of Sciences of the United States of America. 2010;**107**(8):3281-3282. DOI: 10.1073/pnas.1000323107

[46] Buel GR, Kim SG, Blenis J. mTORC1 signaling aids in CADalyzing pyrimidine biosynthesis. Cell Metabolism. 2013;**17**(5):633-635. DOI: 10.1016/j.cmet.2013.04.018

[47] Iadevaia V, Liu R, Proud CG. mTORC1 signaling controls multiple steps in ribosome biogenesis. Seminars in Cell & Developmental Biology. 2014;**36**:113-120. DOI: 10.1016/j.semcdb.2014.08.004

[48] Zhang L, Tschumi BO, Corgnac S, Rüegg MA, Hall MN, Mach JP, et al. Mammalian target of rapamycin complex 1 orchestrates invariant NKT cell differentiation and effector function. Journal of Immunology. 2014;**193**(4):1759-1765. DOI: 10.4049/jimmunol.1400769

[49] Pantel A, Teixeira A, Haddad E, Wood EG, Steinman RM, Longhi MP. Direct type I IFN but not MDA5/TLR3 activation of dendritic cells is required for maturation and metabolic shift to glycolysis after poly IC stimulation. PLoS Biology. 2014;**12**(1):e1001759. DOI: 10.1371/journal.pbio.1001759

[50] Man K, Miasari M, Shi W, Xin A, Henstridge DC, Preston S, et al. The transcription factor IRF4 is essential for TCR affinity-mediated metabolic programming and clonal expansion of T cells. Nature Immunology. 2013;**14**(11):1155-1165. DOI: 10.1038/ni.2710

[51] Zhao B, Takami M, Yamada A, Wang X, Koga T, Hu X, et al. Interferon regulatory factor-8 regulates bone metabolism by suppressing osteoclastogenesis. Nature Medicine. 2009;**15**(9):1066-1071. DOI: 10.1038/nm.2007\

[52] Routy JP, Routy B, Graziani GM, Mehraj V. The kynurenine pathway is a double-edged sword in immune-privileged sites and in cancer: Implications for immunotherapy. International Journal of Tryptophan Research. 2016;**9**:67-77. DOI: 10.4137/IJTR.S38355

[53] Bi X, Hameed M, Mirani N, Pimenta EM, Anari J, Barnes BJ. Loss of interferon regulatory factor 5 (IRF5) expression in human ductal carcinoma correlates with disease stage and contributes to metastasis. Breast Cancer Research. 2011;**13**(6):R111. DOI: 10.1186/bcr3053

[54] Mei Z, Wang G, Liang Z, Cui A, Xu A, Liu Y. Prognostic value of IRF-2 expression in colorectal cancer. Oncotarget. 2017;**8**(24):38969-38977. DOI: 10.18632/oncotarget.17163

[55] Alsamman K, El-Masry OS. Interferon regulatory factor 1 inactivation in human cancer. Bioscience Reports. 2018;**38**(3): pii: BSR20171672. DOI: 10.1042/BSR20171672

[56] Sugiura A, Rathmell JC. Metabolic barriers to T cell function in tumors. Journal of Immunology. 2018;**200**(2):400-407. DOI: 10.4049/jimmunol.1701041

[57] Shirai T, Nazarewicz RR, Wallis BB, Yanes RE, Watanabe R, Hilhorst M, et al. The glycolytic enzyme PKM2 bridges metabolic and inflammatory dysfunction in coronary artery disease. The Journal of Experimental Medicine. 2016;**213**(3):337-354. DOI: 10.1084/jem.20150900

[58] Tannahill GM, Curtis AM, Adamik J, Palsson-McDermott EM,

McGettrick AF, Goel G, et al. Succinate is an inflammatory signal that induces IL-1β through HIF-1α. Nature. 2013;**496**(7444):238-242. DOI: 10.1038/nature11986

[59] Boussiotis VA. Targeting T cell metabolism for improvement of cancer immunotherapy. Frontiers in Oncology. 2018;**8**:237. DOI: 10.3389/fonc.2018.00237

[60] Fruman DA, Chiu H, Hopkins BD, Bagrodia S, Cantley LC, Abraham RT. The PI3K pathway in human disease. Cell. 2017;**170**(4):605-635. DOI: 10.1016/j.cell.2017.07.029

[61] Avalle L, Pensa S, Regis G, Novelli F, Poli V. STAT1 and STAT3 in tumorigenesis: A matter of balance. JAKSTAT. 2012;**1**(2):65-72. DOI: 10.4161/jkst.20045

[62] Yi M, Ban Y, Tan Y, Xiong W, Li G, Xiang B. 6-Biphosphatase 3 and 4: A pair of valves for fi ne-tuning of glucose metabolism in human cancer. Molecular Metabolism. 2019;**20**:1-13. DOI: 10.1016/j.molmet.2018.11.013

[63] Sukhanova A, Gorin A, Serebriiskii IG, Gabitova L, Zheng H, Restifo D, et al. Targeting C4-demethylating genes in the cholesterol pathway sensitizes cancer cells to EGF receptor inhibitors via increased EGF receptor degradation. Cancer Discovery. 2013;**3**(1):96-111. DOI: 10.1158/2159-8290

[64] York AG, Williams KJ, Argus JP, Zhou QD, Brar G, Vergnes L, et al. Limiting cholesterol biosynthetic flux spontaneously engages type I IFN signaling. Cell. 2015;**163**(7):1716-1729. DOI: 10.1016/j.cell.2015.11.045

[65] Blanc M, Hsieh WY, Robertson KA, Watterson S, Shui G, Lacaze P, et al. Host defense against viral infection involves interferon mediated down-regulation of sterol biosynthesis. PLoS Biology. 2011;**9**(3):e1000598. DOI: 10.1371/journal.pbio.1000598

[66] Cheng X, Li J, Guo D. SCAP/SREBPs are central players in lipid metabolism and novel metabolic targets in cancer therapy. Current Topics in Medicinal Chemistry. 2018;**18**(6):484-493. DOI: 10.2174/1568026618666180523104541

[67] Eibinger G, Fauler G, Bernhart E, Frank S, Hammer A, Wintersperger A, et al. On the role of 25-hydroxycholesterol synthesis by glioblastoma cell lines. Implications for chemotactic monocyte recruitment. Experimental Cell Research. 2013;**319**(12):1828-1838. DOI: 10.1016/j.yexcr.2013.03.025

[68] Civra A, Francese R, Gamba P, Testa G, Cagno V, Poli G, et al. 25-Hydroxycholesterol and 27-hydroxycholesterol inhibit human rotavirus infection by sequestering viral particles into late endosomes. Redox Biology. 2018;**19**:318-330. DOI: 10.1016/j.redox.2018.09.003

[69] DeBerardinis RJ, Cheng T. Q's next: The diverse functions of glutamine in metabolism, cell biology and cancer. Oncogene. 2010;**29**(3):313-324. DOI: 10.1038/onc.2009.358

[70] Chang CF, Diers AR, Hogg N. Cancer cell metabolism and the modulating effects of nitric oxide. Free Radical Biology & Medicine. 2015;**79**:324-336. DOI: 10.1016/j.freeradbiomed.2014.11.012

[71] Lu J, Tan M, Cai Q. The Warburg effect in tumor progression: Mitochondrial oxidative metabolism as an anti-metastasis mechanism. Cancer Letters. 2015;**356**(2 Pt A):156-164. DOI: 10.1016/j.canlet.2014.04.001

[72] Liberti MV, Locasale JW. The Warburg effect: How does it benefit cancer cells? Trends in Biochemical Sciences. 2016;**41**(3):211-218. DOI: 10.1016/j.tibs.2015.12.001

[73] Lewis JA, Huq A, Najarro P. Inhibition of mitochondrial function by interferon. The Journal of Biological Chemistry. 1996;**271**(22):13184-13190. DOI: 10.1074/jbc.271.22.13184

[74] Haghikia A, Faissner S, Pappas D, Pula B, Akkad DA, Arning L, et al. Interferon-beta affects mitochondrial activity in CD4+ lymphocytes: Implications for mechanism of action in multiple sclerosis. Multiple Sclerosis. 2015;**21**(10):1262-1270. DOI: 10.1074/jbc.271.22.13184

[75] Porporato PE, Filigheddu N, Pedro JMB, Kroemer G, Galluzzi L. Mitochondrial metabolism and cancer. Cell Research;**28**(3):265-280. DOI: 10.1074/jbc.271.22.13184

[76] Bajwa G, DeBerardinis RJ, Shao B, Hall B, Farrar JD, Gill MA. Cutting edge: Critical role of glycolysis in human plasmacytoid dendritic cell antiviral responses. Journal of Immunology. 2016;**196**(5):2004-2009. DOI: 10.4049/jimmunol.1501557

[77] Ganeshan K, Chawla A. Metabolic regulation of immune responses. Annual Review of Immunology. 2014;**32**:609-634. DOI: 10.1146/annurev-immunol-032713-120236

[78] Hubler MJ, Kennedy AJ. Role of lipids in the metabolism and activation of immune cells. The Journal of Nutritional Biochemistry. 2016;**34**:1-7. DOI: 10.1016/j.jnutbio.2015.11.002

[79] Thaker SK, Ch'ng J, Christofk HR. Viral hijacking of cellular metabolism. BMC Biology. 2019;**17**(1):59. DOI: 10.1186/s12915-019-0678-9

[80] Robertson KA, Hsieh WY, Forster T, Blanc M, Lu H, Crick PJ, et al. An interferon regulated microRNA provides broad cell-intrinsic antiviral immunity through multihit host-directed targeting of the sterol pathway. PLoS Biology. 2016;**14**(3):e1002364. DOI: 10.1371/journal.pbio.1002364

[81] Gorin A, Gabitova L, Astsaturov I. Regulation of cholesterol biosynthesis and cancer signaling. Current Opinion in Pharmacology. 2012;**12**(6):710-716. DOI: 10.1016/j.coph.2012.06.011

[82] Gold ES, Diercks AH, Podolsky I, Podyminogin RL, Askovich PS, Treuting PM, et al. 25-Hydroxycholesterol acts as an amplifier of inflammatory signaling. Proceedings of the National Academy of Sciences of the United States of America. 2014;**111**(29):10666-10671. DOI: 10.1073/pnas.1404271111

[83] Kloudova A, Guengerich FP, Soucek P. The role of oxysterols in human cancer. Trends in Endocrinology and Metabolism. 2017;**28**(7):485-496. DOI: 10.1016/j.tem.2017.03.002

[84] Baek AE, Yu YA, He S, Wardell SE, Chang CY, Kwon S, et al. The cholesterol metabolite 27 hydroxycholesterol facilitates breast cancer metastasis through its actions on immune cells. Nature Communications. 2017;**8**(1):864. DOI: 10.1038/s41467-017-00910-z

[85] Raccosta L, Fontana R, Maggioni D, Lanterna C, Villablanca EJ, Paniccia A, et al. The oxysterol-CXCR2 axis plays a key role in the recruitment of tumor-promoting neutrophils. The Journal of Experimental Medicine. 2013;**210**(9):1711-1728. DOI: 10.1084/jem.20130440

[86] Musella M, Manic G, De Maria R, Vitale I, Sistigu A. Type-I-interferons in infection and cancer: Unanticipated dynamics with therapeutic implications. Oncoimmunology. 2017;**6**(5):e1314424. DOI: 10.1080/2162402X

[87] Mellor AL, Munn DH. IDO expression by dendritic cells: Tolerance and tryptophan catabolism. Nature

Reviews. Immunology. 2004;**4**(10):762-774. DOI: 10.1038/nri1457

[88] Taylor MW, Feng GS. Relationship between interferon-gamma, indoleamine 2,3-dioxygenase, and tryptophan catabolism. The FASEB Journal. 1991;**5**(11):2516-2522

[89] Brown RR, Ozaki Y, Datta SP, Borden EC, Sondel PM, Malone DG. Implications of interferon-induced tryptophan catabolism in cancer, auto-immune diseases and AIDS. Advances in Experimental Medicine and Biology. 1991;**294**:425-435. DOI: 10.1007/978-1-4684-5952-4_39

[90] Hornyák L, Dobos N, Koncz G, Karányi Z, Páll D, Szabó Z, et al. The role of indoleamine-2,3-dioxygenase in cancer development, diagnostics, and therapy. Frontiers in Immunology. 2018;**9**:151. DOI: 10.3389/fimmu.2018.00151

[91] Rath M, Müller I, Kropf P, Closs EI, Munder M. Metabolism via arginase or nitric oxide synthase: Two competing arginine pathways in macrophages. Frontiers in Immunology. 2014;**5**:532. DOI: 10.3389/fimmu